Chemistry, Biological Activities and Therapeutic Applications of Medicinal Plants in Ayurveda

Chemistry, Biological Activities and Therapeutic Applications of Medicinal Plants in Ayurveda

Edited by

Augustine Amalraj
Aurea Biolabs Private Limited, India
Email: augustin14amal@gmail.com

Sasikumar Kuttappan
Aurea Biolabs Private Limited, India
Email: sasikumartk@gmail.com

and

Karthik Varma A. C.
Aurea Biolabs Private Limited, India
Email: ackvarmaapm@gmail.com

ROYAL SOCIETY OF CHEMISTRY

Print ISBN: 978-1-83916-403-3
PDF ISBN: 978-1-83916-621-1
EPUB ISBN: 978-1-83916-622-8

A catalogue record for this book is available from the British Library

The Royal Society of Chemistry is a charity, registered in England and Wales, Number 207890, and a company incorporated in England by Royal Charter (Registered No. RC000524), registered office: Burlington House, Piccadilly, London W1J 0BA, UK, Telephone: +44 (0) 20 7437 8656.

Visit our website at www.rsc.org/books

Printed in the United Kingdom by CPI Group (UK) Ltd, Croydon, CR0 4YY, UK

Preface

From ancient days onwards, people have depended greatly on the traditional systems of medicine, which were concentrated in specific regions. These systems of medicines concentrated on the use of the most popular herbs and spices, which are common to the specific geographic locations and were used effectively to treat many serious disorders. Ayurveda, commonly referred to as the Elixir of life in the Indian system of medicine, has gained many accolades from the time it was used to cure medical ailments. Moreover, Ayurveda is one of the oldest traditional health care systems in the world, with more than 2000 medicinal plants in its *Materia Medica*, which highlights health promotion, illness prevention, delivers actions to address sickness, and also deals with pharmaceutical science. As per Ayurvedic systems, life is sustained by a tripod of mental, physical and spiritual factors constituted by body (Sharir), senses (Indriyas), mind (Satwa) and spirit (Atma), and the healing principles of Ayurveda focus on curing the three doshas, vata, pitta and kapha. Ayurvedic preparations are available in numerous versatile delivery platforms like choornas, kashayams, Ghrita, aasava, tablets, decoctions, *etc.* Ayurveda also relies on the concept of polyherbal formulations to make the final formulations.

Ancient literatures have cited the use of herbs/spices, which are of importance in common kitchen use and also for treating many medical disorders. These herbs and their bioactive components act by multiple mechanisms of action to heal disorders. Various plants and their parts are used in different Ayurvedic and Siddha formulations, in particular, turmeric, ginger, boswellia, ashwagandha, pepper, bacopa, amla, galangal, cissus, *etc.* are the most important species, with lengthy therapeutic practice dating back 5000 years. These spices and herbs have enormous amounts of bioactive components, which have shown various biological activities. Several drugs have captured the global market through the exploration of

Chemistry, Biological Activities and Therapeutic Applications of Medicinal Plants in Ayurveda
Edited by Augustine Amalraj, Sasikumar Kuttappan and Karthik Varma A. C.

Published by the Royal Society of Chemistry, www.rsc.org

traditional medicine and ethnopharmacology. Some of these herbs and their important formulations have been clinically validated to heal multiple disorders.

This book provides insights into the most popular herbs and spices, and their bioactive components responsible for the medicinal properties of different herbal formulations and their clinical evaluation. The book highlights chemistry, biological activities and therapeutic applications of significantly used herbs and spices in various Ayurvedic formulations and their utilization and applications in different dietary supplements, and nutraceutical, pharmaceutical and functional foods, and also discusses the advantages, concerns, and challenges of integrating Ayurvedic principles. Furthermore, the book also highlights the interdisciplinary research approach involving traditional knowledge and evidence, history of ethnopharmacology, molecular biology, pharmacoepidemiology and other advanced research areas for the scientific basis to integrate Ayurveda into modern drug discovery.

The book comprises 15 chapters, each written by experts in their respective field and each chapter having a comprehensive bibliography. The editors have attempted to collect and organize as much information as possible about the chemistry, biological activities and therapeutic applications of selected medicinal plants in Ayurveda and they would like to place on record their gratitude to the various chapter authors for their wonderful efforts. This book is directed at researchers, scientists, clinicians and industrialists working in the areas of Ayurveda and herbal medicine, experimental and molecular medicine and general biomedical sciences. We hope that this book will inspire all those who are involved in the research, production, processing and marketing of medicinal plants in Ayurveda for the growth and upgrading of Siddha and Ayurveda.

Augustine Amalraj
Sasikumar Kuttappan
Karthik Varma A. C.

Contents

Chemistry, Biological Activities and Therapeutic Applications of Medicinal Plants in Ayurveda
Edited by Augustine Amalraj, Sasikumar Kuttappan and Karthik Varma A. C.

Published by the Royal Society of Chemistry, www.rsc.org

CHAPTER 1

Introduction to Ayurvedic Formulations: Exploring the Classical Concepts with Modern Science

J. A. A. S. JAYAWEERA

Department of Microbiology, Faculty of Medicine and Allied Sciences, Rajarata University of Sri Lanka, Saliyapura, Sri Lanka,
Emails: jaas071@gmail.com; jaasjayaweera@med.rjt.ac.lk

1.1 Ayurvedic Formulations: Exploring the Classical Concepts

The system of Ayurveda has a long history and is approximately as old as the *Vedic* age. At present, most people prefer Ayurveda as an alternative to allopathic medicines because it is significantly cheaper and has fewer side effects.[1] The products obtained from natural resources have been essential for maintaining and healing life for millennia. They are used following the processing or formation of raw materials from which robust chemical structures with promising biological activity are isolated. Nowadays, products from natural resources are becoming increasingly essential as complementary and alternative medicines and sources of pharmacotherapeutics.[2] Almost half of the population of the United States has tried natural medicines to prevent and/or treat diseases.[3] Chemicals posing medicinal values are considered as active ingredients of natural therapies. For new drug development, natural products have been used as a primary

Chemistry, Biological Activities and Therapeutic Applications of Medicinal Plants in Ayurveda
Edited by Augustine Amalraj, Sasikumar Kuttappan and Karthik Varma A. C.

Published by the Royal Society of Chemistry, www.rsc.org

resource. Nearly half of the Food and Drug Administration (FDA)-approved drugs from the 1940s to the end of 2020 were developed from natural products.[4–6] In natural medicine, the active ingredients are relatively low in quantity, and cumbersome extraction and isolation processes have been the limiting factor in applying natural drug development products. Therefore, the development of efficient methods for extracting and isolating natural bioactive products is appealing.[3]

The minerals and metallic preparations with healing properties occupy a significant area in the *Ayurvedic* pharmacopeia, which has been practiced in different parts of India for many centuries. Interestingly, those preparations are safe and, even in minute doses, are effective, provided that they are manufactured and used following the specified classical guidelines. However, over the past decade, several concerns have been expressed by the scientific community in the West regarding the safety of *Ayurvedic* preparations, including various minerals and metallic medical products.[7] *Ayurvedic* medicines are based on plants, extracts of animals, and by-products of minerals as single drugs and/or compound formulations. Moreover, in Ayurveda, all substances are regarded as a potential source of medicine.[8] With the aid of different manufacturing processes, *Ayurvedic* herbal dosage forms are formulated. *Ayurvedic* compound formulations mainly originate from plants (*Kasthausadhi*) and metals and minerals (*Rasausadhi*). Plant formulations such as *Asavaristra*, *Avleha*, *Grafa Churena*, and *Taila*, and metal and mineral formulations such as *Bhasma*, *Lauha Kapibadkva*, and *Rasayana* are described in Ayurveda.[9]

1.1.1 Arista and Asava

Arista and *Asava* are made by drenching the herbs in the form of powder or decoction in a solution of sugar or jaggery for a specific period. Over a period of time, it undergoes a process of fermentation that leads to the generation of alcohol, which aids in extracting the active constituents in the herbs.[10] To manufacture essential liquid, medicines such as juices or decoctions, as indicated in the classics, are kept for fermentation. Ethyl alcohol is produced as an outcome of the manufacturing procedure and is not included from an outside source. In addition to ethyl alcohol, many other organic compounds, such as alcohol and acetic acid, are generated; in turn, herbal drugs are extracted. These formulations have a longer half-life, more rapid absorption and activity, and higher therapeutic efficacy than other *Ayurvedic* herbal medicines. The *Ayurvedic* system relies on specific dosage forms, *Sadhana Kalpana* (*Asava*, *Arishta*, and *kanji*), to treat diseases in day-to-day practice.[11] Similarly, in European countries, different stimulants such as wine and beer are formulated and used for pleasure. Interestingly, the preparation principles of fermented products were similar in ancient times to modern days, and few differences were observed in machinery, resource, and sterilization techniques. However, a marked improvement is observed in parameters which are tested to assess the quality of the end products.[11]

1.1.2 Rasa Rasayana

In Ayurveda, use of mineral drugs is called *Rasa Rasayana* or *Ras-yoga*. Rasibhavana refers to the ability of *Bhasma* to act on the body, which mimics fluid containing nutrients that are circulating inside the body.[12] The impurified states of mercury, gold, silver, copper, iron, and sulfur are used to convert to *Bhasma* form; subsequently, other herbs are added in aliquots and grounded to form a fine powder. Drugs such as *Gandhaka* and *Manaḥaśila* are used in purified form. *Rasa* and *Gandhakuare*, the components of a formulation of *Kajjalī*, are prepared, and fine powders of other specified drugs are added in specified quantities. The color and smell of the formula depend on the type of metal.[13]

The *Laura Kalpa* is formed following mixing powders of active drugs with *Loha Bhasma*. The powder mixture (*churnas*) is prepared by sieving the dry drugs through a fine cloth. *Laura Kalpa* in *Churna* form is prepared following the addition of finely powdered herbal medicines to the *Lauha Bhasma*. Until a homogeneous mixture is formulated it is triturated.[14] Physicians mastered the art of *Bhasma* preparation relatively safely and effectively. The processing of *Bhasma* was remarkable, with the whole idea being to remove the toxicity of the minerals and metals in the therapeutic doses. To assess the safety of these formulations, physical and chemical tests have been developed, and the ancient texts discuss such tests. Interestingly, they considered the size of the particle, its density, and its chemical and physical stability under high temperature.[8] The characteristics of properly processed *Bhasma* should include: once rubbed between the thumb and the index finger *Bhasma* particles are powdered to settle in the ridges of fingers; it should float on the surface of water; once mixed with the jaggery, fruits of *Abrus precatorious* Linn., ghee, honey, and borax and subjected to intense heating it should never re-appear as the original mineral or metal used to prepare the *Bhasma*. Special attention is required to keep the temperature below the temperature used to test the quality of *Bhasma*.[8]

Moreover, the *Bhasma* is placed in a crucible with an equally weighted piece of silver metal, and the crucible is heated for 3 hours, at the exact temperature used for the preparation of *Bhasma*. Thus, heating should not change the weight of the silver metal pieces, with any change indicating that the *Bhasma* has not been correctly prepared. *Bhasma* that fails to retain the described physical and chemical parameters is considered to be toxic. It is worth noting the harmful and poisonous effects of mercury and other minerals and metals for the therapeutics following improper or inadequate processing.[7,8]

1.1.2.1 Method of Preparation of Bhasma

Shodhana (purification) and *Maarana* are regarded as the two main steps in *Bhasma* preparation. Some minerals and metals, such as *Biotite mica* and copper, may undergo additional steps such as *Amrutikarna* and *Lohitikarna*.[8]

1.1.2.1.1 Shodhana. *Shodhana* is regarded as the process of purification. The aspects of Shodhana are broad and are not restricted to physical or chemical purification and are mainly focused on the therapeutic utility. Sometimes the processed material may contain foreign molecules. A summary of the objectives of *shodhana* is as follows: macroscopically remove foreign matter like dust and gravel; biologically eliminate the harmful activities of the drug; the physical and chemical modifications lead to improvement of its therapeutic action; and finally, preparation for further processing.[8] At the end of *shodhana*, sulfur, hematite, and alum are suitable for use as a mono-drug or material for a herbal mixture. Other minerals and metals require further processing to be suitable as a drug, and this further processing is termed *Maarana*.[8]

1.1.2.1.2 Maarana. Mica, pyrite ores, and metals are hard in consistency and following *shodhana* their consistency is reduced to a reasonable extent; however to make them edible further processing is required. In addition to making the materials soft and palatable, this augments the intended therapeutic efficacy.[8] The objective of *maarana* is to obtain the material in powder form and as *Bhasma* that is as fine and as soft as *collyrium*. *Collyrium* is applied in the eye, is fine and soft in consistency, and does not cause chemical or physical injury to the cornea. To minimize toxic effects, the material is required to be further processed. The prescribed procedure is *maarana*, which is an outcome following the processes of incineration and calcination. It is carried out following wet grinding (*Bhaavana*) and incineration (*Putapaaka*).[8]

The end product of the *shodhana* process is termed *shuddha dravya* (cleaned products), which is used as the raw material for the process of *maarana*. Those raw materials are subjected to wet grinding, and initially, using the prescribed liquid media, the material is wholly soaked until the liquid is evaporated. When the grinder contents became doughy, the process of wet grinding is stopped, and pellets of the required size are cut from the dough. These pellets are dried in air, and further processing, called *Putapaaka*, is carried out. The incineration must be carried out appropriately using pellets. A specific device called a *puta* is used for incineration, and as fuel, cow dung cakes are used.[8] Once dried pellets are obtained, they are sealed in two shallow earthenware plates using an appropriate sealing material. Then, the surrounding cow dung cakes are ignited, left to burn to ashes, and left to cool on their own. The sealed pot is later opened, and the contents are ground using a mortar to prepare fine powder.[15] One *puta* describes the process from wet grinding to incineration. The process must be repeated until *Bhasma* in the required quality is obtained. However, following the prescribed number of *puta*, if the *Bhasma* fails in the prescribed tests, the *puta* process must be continued until *Bhasma* has been reached.[15]

In regards to the therapeutic activity of metal *Bhasma*, metal with mercury contains the highest activity. In contrast, medium activity is ascertained following metal with plant material, and the lowest following metal with sulfur. Metal *Bhasma* prepared by blending metals with opposing characters are thought to be toxic. In contrast, modern medicine strongly prohibits the internal use of mercury and its derivatives. Also, to avoid physical contact with mercury and its compounds, extra caution is recommended.[15] Mercuric sulfide, an inorganic form, is the form that is primarily used in Ayurveda. Occasionally, mercuric chloride is used, and such use is limited and always advised with caution due to its toxicity. A negligible amount of mercuric sulfide is absorbed through the gastric mucosa and hence is nontoxic. In contrast, mercury in organic form has been often highly toxic and known to damage the brain and liver.[16]

1.1.2.2 The Metals Used to Develop Various Herbal Formulas Under Rasa Rasayana

1.1.2.2.1 Mercury (*Parada*). The first reference to mercury and its utility in the classics is controversial, and its use is rarely mentioned in *Charaka*. Only a few scholars interpret *Rasa* in the verse *chikitsasthana* as mercury. In *Dwivraniya Chikitsa*, *Rasa* is interpreted as mercury by *Chakrapani*, the commentator. Interestingly, both these formulations are recommended only for external application.[17] *Rasausadhies*, which were named after mercury, were classified into two separate classes: mercurials and non-mercurials. The use of mercury has changed the direction of the management of disease. Mercury was regarded as a bio-enhancer and not as a drug. When any medication is combined with mercury, it is supposed to increase its efficacy and reduce its dose manyfold.[15] Initially, mercury is purified and converted into an effective, stable, and nontoxic compound. The primary material used to prepare most mercury-based drugs is a combination of mercury and sulfur, which are ground together. This fine powder is black and termed *kajjali* which mimics collyrium due to its color and softness. Depending on the preparation methods, mercury-based herbals are categorized as *Khalvi Rasa*, *Parpati Rasa*, *Kupipakwa Rasa*, and *Pottali Rasa*.[15,16]

1.1.2.2.1.1 Khalvi Rasa. A finely powdered formula is formed using a mortar by grinding mercury and sulfur with other herbal, mineral, and metal ingredients. After being ground in a mortar, termed *Khalva*, in Ayurveda the formulation is classified as *Khalvi Rasa*.[15]

1.1.2.2.1.2 Parpati Rasa. A homogeneous black-colored fine powder termed *Kajjali* is formed using a ground mixture of mercury and sulfur mixed with or without a specified metal *Bhasma*. Then a molten mass is obtained using an iron spatula following the application of heat in a controlled manner. The molten mass then is poured immediately onto a flat soft surface, with a banana leaf placed on a smooth-surfaced platform made using fresh wet cow dung or mud and spread evenly. Then, the poured mass is covered immediately with a banana leaf and hand-pressed to create an even and thin flat sheet of the material termed *parpati*.[15]

1.1.2.2.1.3 Kupipakva Rasa. *Kupipakva Rasa* is formed using mercurial drugs prepared by *kajjali* and is heated using a long narrow-necked glass flask (*kupi*). The drug is designed in the following stages: the *kajjali*, mercury, sulfur, and other required raw materials are subjected to the purification procedure; using a suitable mortar the purified metal is grounded with pure mercury to obtain a homogeneous mass.[15]

Kajjali appropriately mixed with other metals or herbs can alleviate most symptoms and cure diseases. Following administration, the herbs spread rapidly through the body, which in turn clears the obstructed vessels in diseased organs and enhances the properties of other metallic or herbal medications.[18] In Ayurveda the toxicity of mercury is immensely reduced following its combination with sulfur therefore reducing its bioavailability. The preparation of *Kajjali* by mixing purified mercury and sulfur is the initial step in preparing any herb mineral formulations. The gradual and homogeneous application of heat for the preparation of *Kajjali* has been described. A layer of clayed fabric is wrapped around a flask or bottle, assuring that there is no air gap between the clayed fabric and the outer surface of the flask or bottle, and then it is left to dry. Soon after the first layer dries, the second layer of clayed fabric is wrapped around the flask. Successively, the flask is covered with seven layers of clayed fabrics, and once all the wrappings are entirely dried it is ready to use. Heating starts with a low flame at between 100–200 °C, and the apparatus is left to cool on its own. Therefore, the heating must be continued for a number of hours.[15] Safe drug storage is mentioned in an ancient text as being in the form of a fine powder in an airtight container.[15,16]

1.1.2.2.1.4 Pottali Rasa. *Kajjali* is wrapped in a silk cloth to prepare a tight bundle (*pottali*). Sulfur powder in the prescribed quantity is placed in a stainless steel or wide-mouthed earthenware pan, kept on a fire, and allowed to melt. Then the bundle of *kajjali* is immersed in the melted sulfur while an adequate temperature is maintained to keep the sulfur in a molten state for a reasonable period of time. The wrapped silk cloth is burnt and wrapped around the hardened ball. In a circular motion, the hardball is rubbed on a moist stone slab, and the accumulated paste is prescribed for licking. The drug's prescribed dose is decided following several rounds of rubbing.[15]

The *Pottali Rasa* is prescribed in medical emergencies and is regarded as an instantly acting drug formulation. Caution has also been issued for *Rasa rasayana* regarding the toxic effects of mercury and mercury toxicity treatment and its usability in Ayurveda texts has been intensely argued about. In

Rasa therapy, mercury is regarded as the most unstable and chemically reactive substance. Mercury absorbs atmospheric impurities passively and tends to react with other metals and form amalgams quickly. Due to reactions, purified mercury is challenging to obtain.[15–17]

1.1.2.2.2 Gold (*Swarna*). Since antiquity, gold has been described as a noble metal. It can be referred to as *Charaka* and *Sushruta Samhita*. The *Bhasma* form of gold is a mixture of metallic gold (96.76%), silica (1.14%), ferric oxide (0.14%), phosphates (0.78%), potash (0.16%), salt (0.078%), and traces of copper and magnesium. Various formulations of gold are helpful as disease alleviators, particularly in chronic debilitating diseases. The *Swarna Bhasma* dose is between 15–30 mg.[17]

1.1.2.2.3 Silver (*Rajata*). According to the classics, the therapeutic use of silver is limited. Classical alchemy describes silver as clear and heavy with a metallic sheen. Following heating or cutting, it becomes bright white.[16] The *Bhasma* form of silver is a combination of metallic silver (52–59%), free sulfur (0.675%), ferric oxide (14.33%), calcium (10.769%), silver chloride (0.479%), and traces of Na, K, and Al. The silver *Bhasma* dose is from 30 mg to 120 mg.[16,17]

1.1.2.2.4 Copper (*Tamra*). Since pre-*Vedic* times, copper has been routinely used, and it is the best known metal for preparing the more robust brass and bronze alloy metals. A characteristic metallic sheen, smooth and bright reddish color with good tensile strength are observed in copper. Moreover, impurities are rarely identified and so it is very suitable for medicinal purposes. Formulations of copper are used to cure a broad spectrum of diseases including worm infestation and hemorrhoids. The use of copper vessels for various pharmaceutical procedures has been described in *Charaka*. The copper *Bhasma* dose is between 15–60 mg.[16,17]

1.1.2.2.5 Iron (*Aayasa* or *Loha*). After gold, silver, and copper, iron is another well-known metal that has been used for millennia. During *Charaka*, iron was used in various topical and oral formulations for many diseases. Compounds of iron were mainly used for anemia. Further classics describe iron as an excellent rejuvenator that stimulates all organ functions and promotes life, helping to regain mental and physical strength.[16,17] Iron has a large number of therapeutic applications in Ayurveda, and during purification and incineration, proper care should be taken. A procedure to convert thin metal into a fine digestible form was described by *Charaka*. In addition, the use of iron vessels for various pharmaceutical procedures has been described. Iron in *Bhasma* form is a combination of ferric oxide (96.5%), ferrous oxide (2.5%), magnesium oxide (0.8%), copper oxide (0.3%), with trace amounts of phosphorus and potassium. Various iron formulations are used in a broad spectrum of diseases including hemorrhoids and different types of pain. The iron *Bhasma* dose is between 30–240 mg.[16]

1.1.2.2.6 Mandura. In Ayurveda the *Mandura*, another form of iron *Bhasma*, is well known for its use in a wide array of therapeutic procedures. *Madhava Upadhyaya* defines it as a collection of debris around a blacksmith's anvil, collected after heating and beating for many years. If anvils are 100 years old, they are regarded as the most useful, while those 60–80 years old are considered average and less efficacious. *Mandura* is smooth to touch, heavy, firm, devoid of fissures, and useful as a therapeutic.[17,18] Following purification *Mandura* is useful for inflammations, edema, jaundice, and anemia. *Mandura* is a combination of ferric oxide (59.14%), ferrous oxide (26.7%), chlorides (4.4%), magnesium (3.9%), sodium (1.7%), and a few other elements in small quantities. Its constituents play a major role in curing anemia. The *Mandura Bhasma* dose is between 30–240 mg.[17,18]

1.1.2.2.7 Lead (*Naga*/*Sisaka*). Since ancient times lead has been known as an important *loha*. Lead *Bhasma* topically, in particular for rashes, has been advocated by *Charaka*. Lead quickly melts and is heavy, externally black, and when cut shines a bright black. Quantitatively, lead *Bhasma* contains lead oxide (75.6%), ferric oxide (7.5%), and small amounts of calcium and magnesium chlorides and carbonates.[16,17] Various formulations are useful for obesity, pre-diabetes and diabetes, various abdominal lumps, and leucorrhea. The lead *Bhasma* dose is between 30–120 mg.[17,18]

1.1.2.2.8 Tin (*Vanga*/*Trap*). The use of tin, a *loha*, has been well described in Ayurveda. The *Rasa Vagbhata*, an influential *Ayurvedic* writer, described two varieties of tin (*Khurana* and *Mishraka*), with the former being the accepted form of therapy. *Khurana* that is bright white, smooth, melts at low temperature, and heavy is identified as *Khurana* that is preferred for therapy. Tin in the form of *Bhasma* is a combination of stannic oxide (91.4%), ferric oxide (2.9%), potassium (2.9%), calcium oxide (2%), aluminum (2%), and magnesium oxides (0.6%).[19] Tin *Bhasma* formation is used as a therapy for diabetes, respiratory tract disease, asthma, microbes, and bone dislocation. Alone, or in combination, it is beneficial for genitourinary tract pathologies. Also, it is the drug of choice for the treatment of obesity. The tin *Bhasma* dose is between 120–240 mg.[20]

1.1.2.2.9 Brass (*Pittala*). Brass is a mixed metal alloy of copper and zinc. Formulations of brass are beneficial in diseases such as various abdominal lumps, rashes, and jaundice. The brass dose is between 60–120 mg.[21]

1.1.2.2.10 Bronze (*Kamiya*). Bronze, the bell metal, is made by mixing copper and tin. According to Ayurveda, there are two bronze varieties called *Pushpa* and *Talika*, and only *Pushpa* is used for therapy. When tapped bronze produces a sharp sound, and is soft to touch, gray in color, devoid of impurities, and following heating, it turns red. Formulations of bronze are used in the treatment of various types of abdominal lumps and rashes. The bronze *Bhasma* dose is from 60 mg to 120 mg.[22] The method of preparation of bronze *Bhasma* is described by *Charaka*. It is initiated following metal powder preparation and as with *Carla*, a 7.4×7.4 cm thin sheet of metal, penetrable with a thorn prick, is heated until it is red hot. The red-hot sheet is sequentially drenched till a delicate powder is formed in the decoction of *Triphala* (a mixture of dried *Emblica officinalis*, *Terminalia bellirica*, and *Terminalia chebula*), cow excreta (urine), *Hordeum vulgare* (Linn.) ash in water, rock salt, *Balanites aegyptiaca* (Linn.) ash in water, and *Butea frondosa* Koen Ex (Roxb.) ash in water.[23]

The soft powder is subsequently mixed with honey and *Phyllanthus emblica* fruit juice and kept for a year. Once a month it is stirred and yields a semi-solid mixture which is the final form. This applies to gold, silver, copper, or iron as well.[24] In *Sushruta Samhita*, the concepts of *Lohaadi Rasayana*, chemistry in metals, is converted to powder form using a slightly different procedure described by *Charaka*. Here, thin metal sheets are softened using different mineral salts and made red hot following intense heating. These sheets are then drenched to make a powder in *Triphala* and decoction *Shorea robusta* Gaertn. f., *Acacia catechu*, *Betula utilis* D. Don, *Gymnema sylvestre* R. Br. 16 times. The powder is sieved using a fabric that is then used as a drug.[25,26]

1.1.3 Vati or Gutika

The tablet or pill form in Ayurveda is known as *Vati* or *Gutika*. It is made using one or more plant drugs or purified materials from animals or purified minerals. When the dosage is in a circular mass it is called *Gutika*. This is comparable to pills in modern pharmaceutics. *Vati* is similar to a tablet that can be swallowed, while *Ghanavati* has to be chewed before swallowing.[27] Generally, the consumption of *Ayurvedic* medicines is a significant challenge for most patients due to their bitter and astringent taste. Most will accept this challenge to cure their diseases, however some are unable to. *Vati* is the most basic and palatable form of *Ayurvedic* medicine and is very easy to consume due to not having an unpleasant taste.[28] *Vati* can be modified into a long oval shape form called *Varti*. It is suitable for local administration in the anal or vaginal canals, penis, or eye. Another form, *modaka*, is circular and has a considerable size, having a weight of around 20–50 g.[29] Two types of *vati* preparation methods are described in Ayurveda, *Agnisadhya vati* and *Anagnisadhya vati*. *Agnisadhya vati* is made using sugar or jaggery or *Guggulu* on a warm fire. Contrary, *Anagnisadhya vati* is made without heat.[30] The *vati* or *Gulika*, made of herbal raw materials kept in airtight containers, can be used for two years. *Vati* or *gulika*, which contains minerals, can be used indefinitely if preserved properly, meaning that they do not have any expiry date. *Vati* should not lose their original color, smell, taste, or form. When sugar, jaggery, salt, or *kshara* are ingredients, the *vati* should be kept away from moisture.[31]

1.1.4 Churna

The fine powder formulation in Ayurveda is termed *churna*. The herbs are cleaned, dried, and powdered mechanically using a fine mesh. If *churna* is preserved in an airtight container, it can keep for one year without losing its potency.[32]

1.1.5 Avaleha

Semi-solid formulations are called *Avaleha*. These are prepared by adding jaggery, sugar, or sugar candy, and boiled with the prescribed decoction. When cold and well-mixed, honey is added. Various *Avaleha* are mentioned in *Ayurvedic* classics and they are the most popular *Ayurvedic* dosage form. They are easy to administer, palatable, and pose a long shelf-life.[33] According to various formulations, the percentage of sugar, sugar candy, and jaggery varied, and additionally sugar acts as a preservative. Once *Avaleha* is incorporated into a lipid medium using *Ghrita* it becomes soft. Moreover, honey, gingelly oil, salts, alkalis, milk and milk products, and *Bhasma* are added according to formulations and contain specific timeframes to develop it into a finished drug.[34]

1.1.6 Ghrita

When *ghee* is boiled with the prescribed decoction, *Ghrita* is developed. *Ghee* contains 100% fat, and routine consumption may increase the likelihood of coronary artery diseases. Many therapeutic forms are described in *Ayurvedic* classics, such as tablets and powder, and it additionally contains self-generated alcohol. There are more than 55–60 medicated *Ghrita* described in the *Ayurvedic* classics. Drugs given in the form of *Ghrita* are digested and absorbed quickly and reach some of the most distant areas of the body, including the central nervous system.[35] The described preparation procedures used to produce *Ghrita* rely upon heating at temperatures from 105–118 °C to remove the water. Studies have been conducted on *Ayurvedic* medicated *Ghrita* to explore its effects in various conditions. A study conducted on *Saraswati ghrita* (a polyherbal formulation containing *Bacopa monnieri* as the main content with *Curcuma longa*, *Phyllanthus emblica*, *Operculina turpethum*, *Terminalia chebula*, and *Piper longum embelia* ribes and cow's *Ghrita*) demonstrated that *Ghrita* had exquisite potential for the management of disorders like Alzheimer's diseases and dementia.[36] Anticonvulsant activities have also been reported in medicated *Unmadnashak Ghrita* (an *Ayurvedic* formulation containing *Ferula narthex* (6 g).[37]

1.1.7 Taila (Oils)

According to the formula, various oils are formed following boiling with the prescribed decoction. The formation of lipophilic fluid ensures extraction of the active ingredients from the raw herbal materials. *Taila* generally has the color, odor, and taste of the herbs used, and has the consistency of an oil. When milk is used in a considerable quantity in the preparation, the oil becomes thick.[38] In Ayurveda, it is claimed that medicated *Taila* has greater power of healing and half-life than crude *Taila*. *Taila* can increase the level of energy and the body's anabolism.[39] For the preparation of *Taila*, precautions should be taken, such as the intensity of the fire should be maintained throughout the process to assure a consistent temperature. The mixture has to be carefully stirred throughout to ensure that the extract does not stick to the vessel. A moist-free container is used to preserve the *Taila* and to develop a favorable odor when *Taila* is lukewarm, also, the materials added should stir well.[40]

1.1.8 Guggulu

Guggulu is prepared from exudates obtained from the plant *Commiphora mukul*. Small pieces of exudates are wrapped with a cloth and boiled in *Triphala* or other herbals till they pass into the solution through the cloth. Further, it is boiled to form a solid mass and left to dry. Once dried, ghee is added to form a waxy paste.[39,40] *Ayurvedic* physicians have used *Guggulu* for centuries to treat different diseases. It contains oils which vaporize rapidly, Gugu lipids, guggul-sterones, guggul-sterols, mukolol, and other steroids. *Guggulu* is used as an antiseptic, expectorant, antispasmodic, and emmenagogue. It is renowned as a fat-burning agent and is used as a slimming aid. It reduces total body cholesterol and tri-glycerides.[40] It is used to treat rheumatoid arthritis, gout, and sciatica and also for steatohepatitis, neuronal disorders, bronchial mucus congestion, and cardiac and circulatory problems. It stimulates digestion and libido and is useful in treating skin ulcers, local abscess, bone fractures, gynecological problems, and multiple skin diseases.[40]

1.2 Modern Science Perspectives

Extraction is the initial step to select active natural products from raw materials.[41] According to the principle of extraction, different extraction methods are used, and solvent extraction, distillation, pressing, and sublimation are known methods; however, solvent extraction is most the commonly employed method. The liquid solvent has to penetrate the solid matrix, and subsequently, the active ingredients dissolve. Anything which enhances the diffusion and its solubility facilitates the process of extraction.[41] Moreover, the efficiency of extraction relies on the characteristics of the solvent used for the extraction, particle diameter of raw material, solvent-to-solid ratio, temperature used for the extraction, and the duration.[41,42] For efficient solvent extraction, the proper solvent selection is essential, and solubility, cost, and safety characters also are required to be considered. For phytochemical extraction, alcohols are regarded as the universal solvents.[41,42] Currently, based on the required extraction in the phytochemical investigation, alcohol in the appropriate concentration and known strength is used; however, Ayurveda follows unique methods based on different scales and measurements. In *Arista* and *Asava*, alcohol extraction was used to extract essential herbs from the herbal formulary. Moreover, alcohol was formed following fermentation within *Arista* and *Asava* and consumers may take herbals in addition to formed alcohol.[41,42] Once the duration of extraction increases, the extraction efficiency increases; however, once equilibrium is reached between the inside of the solid material and outside the solvent, increased time barely affects the extraction. For the formation of different *Arista* and *Asava* specifically applied periods have been described in the ancient scripts.[43]

The utility of different metals for modern medicine was identified in the 17th century, and was introduced following the use of iron salts for anemia. The roles of metals in the functions of metalloenzymes and metalloproteins are well known, and trace amounts of cobalt, copper, iron, selenium, zinc, and magnesium are essential. The use of selenium, copper, and gold in cancer treatment has been practiced for decades.[44] When considering the concepts of *Rasa Rasayan*, mercury or other metals in the form of *Bhasma*, the bioavailability of the drug is increased. According to modern medicine, mercury use is often limited to a few topical applications, dental fillings, and contact lenses. Compared to the organic form and various compounds, its elementary form is less toxic.[45] The various *Ayurvedic* formulations and their Western medicine counterparts are included in Table 1.1.

Over the centuries, gold in its elementary form was used as an antipruritic agent. Clinical trials have been conducted to assess the therapeutic potential of gold on arthritis and systemic lupus erythematosus.[46] Following the use of gold compounds, the concentration of rheumatic factor is decreased, and its influence on immunological response was remarkable. Compared to glucocorticoids, gold strongly suppresses the release of histamine following a type-1 hypersensitivity reaction. Sodium aurothiomalate, in a water-soluble form, was discovered around two decades ago, and is useful for treating arthritis.[47] However, its pharmacokinetics and dynamics properties are not well established, and its antiinflammatory and stimulatory effects on the reticuloendothelial system could be the reason for the above-described actions. Gold-based medications can be use as an analgesic, and an antioxidant effect also has been described.

Copper-based formulas have liver-protective- antioxidant- and gastroprotective effects. Zinc-based formulas arrest myopia and iron-based ones are hepatic- and gastro-protective. Often, *Bhasma* forms can be regarded as different complex compounds and have been shown to have a number of essential activities, especially a cytoprotective effect against different types of ulcers.[47,48]

Silver has the highest thermal and electrical conductivity and is the most reflective metal. Most importantly, silver can be used to prevent the growth

Table 1.1 The classical formulations and modern medicine counterparts.

Ayurvedic formulations	Modern medicine counterparts
Plant formulations	
Asavaristra—made by drenching the herbs in the form of powder or decoction in a solution of sugar or jaggery for a specific period[10]	For phytochemical extraction, alcohols are regarded as the universal solvent and using different techniques and concentrations the active compounds are extracted[34]
Avaleha—semi-solid formulations are prepared by adding jaggery, sugar or sugar candy and boiled with the prescribed decoction. Is useful for the absorption of drugs both systemically and locally[34]	Sublingual formulations such as glycerin trinitrates[22]
Taila—formed following boiling with the prescribed decoction and often acts as a base and soluble media for lipophilic herbs[40]	Different excipients with lipid and water solubility to enhance drug delivery, dissemination and absorption to tissues[34]
Guggulu—prepared by the exudates from the plant *Commiphora mukul*[39,40]	—
Mineral formulations	
Rasayana—the impurified states of mercury, gold, silver, copper, iron, and sulfur are used to convert *Bhasma* form; subsequently, other herbs are added in aliquots and ground to form a fine powder[14–16]	No such use, however, mercury is often limited to a few topical applications, dental fillings and contact lenses. Also, the elementary form is less toxic and gold, silver and copper are used as antipruritic agents, with antibacterial, antioxidant and gastro-protective effects, respectively[44,45]
Kajjali—homogeneous black-colored fine powder formed using a grounded mixture of mercury and sulfur mixed with or without specified metal *Bhasma* and is more effective due to sustained release[8–12]	Is in line with the concepts of sustained release and forms of gels, implants and devices in addition to tablets and capsules[51,54]
Formed using plant or mineral formulations	
Vati—made up using one or more plant drugs or purified material from animals or purified minerals[27]	Compressing single or multiple powdered ingredients to form a hard tablet or smooth-coated capsule[54]
Churna—herbs are cleaned, dried, and powdered mechanically using a fine mesh[32]	Powder forms with high purity and various powder mixers[62]
Ghrita—ghee is boiled with the prescribed decoction, the *Ghrita* is developed and is an excellent and crude vehicle for the administration of drugs[35]	Liposomal preparations which are used to carry medications to certain body sites. However, liposome composed of one or more phospholipid bilayers resembles cell membrane,[56,57] *e.g.* liposomal amphotericin b

of potentially harmful bacteria. At present, silver nanoparticles are used as antimicrobial products; however, the concepts of infection and infectious agents are not described in detail in Ayurveda.[46–48]

The consumption of mineral materials is impossible due to their hard consistency. However, chemical and physical processing change their consistency to make them smooth and palatable. This applies to the preparation of various modern-day metallic formulations by utilizing the knowledge of chemistry and physics combined with the battery of knowledge from the ancient era. The preparation of metal-based formulations involves the conversion of a thin sheet of metal into a powder. Heat and various acidic and alkaline preparations in a particular sequence are used for this conversion.[48,49]

Moreover, Ayurveda includes various herbs and purifying techniques to increase the quality of the desired herbal formulations. Often these are in the semi-solid phase, and those preparations are topically applied to the eyes as a therapy. However, Western medicine remains reluctant to use topical semisolid preparations on the eyes, only using liquids in the form of eye drops.[50] The use of metal-based herbal formulas for therapy is considered to be controversial. Two schools of thought exist, with practitioners of metallic *Bhasma* representing one while the other is represented by the scientific community holding Western beliefs. The practitioners of metallic *Bhasma* are invariably in favor of its use and argue that it has been used as a medication for thousands of years. However, in *Ayurvedic* classics it is mentioned that these drugs should be used with caution.[51]

General observations illustrate that metals are toxic; however, in therapeutic doses, compounds such as metal sulfides are nontoxic. The main concern is whether the metal-based preparations mentioned in the *Ayurvedic* classics are safe and efficacious.[52] Several studies have been carried out over the years, and many recent studies have shown that toxicity is not generally observed at the therapeutic dose level if used properly.[47–51] However, a great deal of data were gathered in the last century that oppose the above findings.[44–52] Often toxicity potential is inherent, if metal-based products are used inappropriately under unacceptable conditions, with fatal outcomes being relatively high. Such toxicity is generally not related to the active pharmaceutical ingredients and instead is often related to the manufacturing process. Since the manufacturing process is cumbersome, lengthy, and requires a high level of skill, only a few experts have gathered and retained the required knowledge. Following the loss of those experts, the retained knowledge has been gradually lost, resulting in a deficiency in the standard operating procedures and standardization has reduced its reproducibility.[50,51] There is a demand to standardize all aspects of herbal preparation.[53] Moreover, observations and facts are sufficient to claim that metal-based preparations used in Ayurveda contain essential biological effects. Therefore, these should not be discarded without an in-depth analysis to decide on the usability and safety.[54]

Complexes of mercury and sulfur are prepared using different methods. During the preparation of *Kajjali* sulfur is provided in excess compared with that required for the stoichiometric arrangement of HgS. The idea behind this is to minimize oxidation. The processing of mercury reduces its toxicity and the adopted methods of purification are unique. *Allium sativum* (Rasona), which contains sulfur, is used for the purification of mercury. Also, for processing, heat is applied and then it is subsequently boiled in a liquid bath. Later grinding, sublimation, and distillation are carried out.[53,54] The solubility of HgS is low. Therefore, the toxic threshold is also low. While passage along the gastrointestinal tract due to the interaction with enzymes in the digestive tract and a change of pH leads to the development of complex biomolecules in food. As a result, the solubility of HgS would increase greatly. By using scanning electron microscopy and X-ray diffraction, the results of mercury and sulfur mixing have been elucidated. Following 60 min of grinding it slowly forms black HgS, which contains mercury globules. Interestingly, following 90 min of grinding, free mercury was not detected; however, after 120 min of grinding, particles were formed containing 2–15 weight % of mercuric oxide. It was also found that sulfur particles surround the HgS particles. In *Kajjali* free sulfur in excess is detected and often trapped in the mesh of HgS.[55]

Biomolecules including methionine, cysteine, taurine, and enzymes act as antioxidants, and those such as glutathione are rich in sulfur.[54,55] Most Ayurveda formulas contain excess amounts of antioxidants, and their influence on its therapeutic effect is remarkable. Antioxidants protect cells following free radical-induced damage. Based on its antioxidant properties, it is possible to explain the rejuvenating effects and antiaging effects of metal-*Bhasma*. Mercury in *Bhasma* may serve as a transient catalyst also.[53–55]

The trituration of the prescribed powder of herb with *Kajjali* develops a unique molecular layer which is capable of acting as a sustained-release formulation. When consumed, it gradually releases an effective dose, that is adsorbed and acts on the target area. Preparations with sustained-release are advancements in modern medicine, which can improve the therapeutic activity. With regards to the concept of slow-release, Ayurveda classics pose similar concepts.[54,55]

During the formation of metal *Bhasma*, quality assurance and quality control are carried out to assure the quality and efficacy of the formulation. This is compatible with the concepts of modern-day quality control; however, they should be reassessed and reassured. Also, modern techniques could be introduced to make it safer. Also, drug formulation knowledge is kept secret and hidden. Nevertheless, it has been kept and protected through various modes, such as in the form of *sloka* (poetic form in *Sanskrit*) and phrases. These forms provide the methods in the correct order and can be considered as ancient standard operational procedures.[53,54]

Compressed single or multiple powdered ingredients form hard tablets or smooth-coated capsules that following ingestion break down in the digestive tract. Most tablets contain additives and active ingredients, which hold the pill together and improve the taste, texture, and appearance. Tablets can be round, oblong, or disc-shaped. Oblong tablets are known as caplets, which can be easier to swallow. Similarly, in Ayurveda, *vati* has been prepared as a single or mixture of medications required for ingestion. The size and the shape of tablets are considered to be important for ingestion and insertion to the gastrointestinal tract through mouth or anus.[51,55] Moreover, Ayurveda has considered the importance of a base to ensure the stability, taste, and texture of the final product. Most are swallowed, while some have to be chewed before swallowing. This is similar to concepts in modern tablets. However, based on pharmacokinetics and pharmacodynamics properties, in-depth studies on such *Ayurvedic* preparations are required to assess their bioavailability and efficacy.[51,52,55] Other formularies, *churna*, are similar to powdered formulations in modern medicine. Depending on their use, the physical, chemical, and biological characteristics can vary.[56]

The oral transmucosal route is useful for the absorption of drugs, both systemically and locally. *Avala* is a formulation intended to be administered through the oral cavity. Ayurveda has developed such a formulation that can bypass the first-pass metabolism of the liver to enhance its bioavailability.[56,57] *Ghrita* is an almost anhydrous milk fat prominent in the Indian diet and that is given great importance. *Ghrita* is used as a food and is the main ingredient for many Ayurveda medicines and is an excellent vehicle for the administration of drugs. Its spreading nature facilitates the dispersion of all medicinal properties to the deepest tissues. In this context, it is also referred to as a catalytic agent. *Ghrita* made from cows' milk is considered superior to all other *Ghrita*. Like many dairy products, *Ghrita* is composed almost entirely of fat, with 62% saturated fats. Plant-based oils are generally rich in unsaturated fatty acids, while the products derived from animals are rich in saturated fatty acids.[57]

Saturated fatty acids (SFAs) are produced as a result of ruminant biochemical hydrogenation, producing mostly the *cis* configuration of SFAs, and it differs quite extensively from the FAs (fatty acids) produced by industrial hydrogenation, which converts some *cis* FAs to *trans* FAs through *cis–trans* isomerization. The *trans* configuration of SFAs is not recognized by the enzymes present in the biological systems of humans and other animals and so they tend to collect in the body, thus being associated with higher coronary vascular disease risk.[58]

Most of the activities described for *Ghrita* in the classics are known to be imparted by antioxidants and docosahexaenoic acid (DHA), an essential fatty acid. Several studies have proved that using traditional methods to produce *Ghrita* results in a higher amount of DHA and omega-3 long-chain polyunsaturated fatty acid (PFA). Dietary DHA reduces the risk of ischemic heart disease, malignancy, type-2 diabetes mellitus, and arthritis.[59] For years, *Ghrita* has been considered to be fattening and unhealthy and it has been associated with many diseases and increased mortality. Several studies have revealed *Ghrita* has saturated fatty acids and that heated products contain a high level of oxidative products of cholesterol which would increase the prevalence of ischemic heart diseases in Asians, particularly Indians.[60] Studies have shown that up to 10% *Ghrita* in the diet positively affects the serum lipid profile. There was a dose-dependent decrease in triacylglycerol, low-density lipoprotein, and VLDL cholesterol and triacylglycerol when *Ghrita* was given at a rate higher than 2.5% in the diet. The level of 3-hydroxy-3-methylglutaryl coenzyme A reductase activity in microsomes of the liver was not affected by dietary *Ghrita*, indicating that cholesterol biosynthesis was not affected by it, but instead, the excretion of bile constituents was increased by 18–30%, lowering the serum cholesterol levels.[61]

Moreover, conventional herbal extraction methods have several drawbacks, such as excess time consumption, solvents with residual effects, stickiness, and handling difficulty of highly inflammable solvents. Hence, it is necessary to develop a novel method for extracting the phytoconstituents utilizing traditional knowledge. In regard to oil preparations, the decoction was mixed with different seed or plant oils. Depending on the source of the oil, the duration and intensity of heating were prescribed. These oils often act as a base and soluble media for lipophilic herbs, which helps in the better absorption for lipid-soluble substances and allowing greater affinity through the cell membrane when used for different mucosa.[62]

Guggulu contains a mixture of steroids, diterpenoids, triterpenes, aliphatic esters, alcohols, carbohydrates, amino acids, cholesterol, *Guggulu*-sterol, flavonoid, and a variety of inorganic compounds. *Guggulu* has antiplatelet activity, cardiac protective activity, antioxidant activity, reducing the pro-inflammatory cytokines and antibacterial activities. L-arabinose is a good source of sugar ferulic acid esters and has antitumor properties. *Guggulu* and *Guggulu*-lipid are used for cardiovascular diseases, including dyslipidemia and atherosclerosis. *Guggulu* is generally used in compound dosage forms such as Ayurveda tablets, oils, and topical preparations. The different choices of therapy are based on soluble alkaloid and the insoluble resinous gum content of *Guggulu*. The formulation concept is also designed for better pharmacokinetics of *Guggulu* in clinical practice in Ayurveda.[63] Gum *Guggulu* acts as a binder in the preparation of pills and tablets, while it acts as a suspending and emulsifying agent in the case of liquid dosage forms. The solid dosage forms thus have a longer disintegration time, which delays the therapeutic action of *Guggulu* formulations. Possibly considering this in view, the seers of ancient times advised dissolving *Guggulu* completely in suitable liquids before its consumption, and *Ayurvedic* physicians in the present scenario advise breaking *Guggulu* solid dosage forms like pills and tablets before oral administration for proper assimilation and absorption in the gastrointestinal tract.[64,65]

The possible modes of action of *Ayurvedic* formulations can be multi-dimensional and include scavenging of free radicals, antioxidation, anti-microbial, pro-enzymatic, and immunomodulation. The *kajjali* complex is more effective due to its sustained release.[66] In Ayurveda, herbal mixtures have been used, and concepts of the molecule, its size, weight, and inter-molecular interactions were not considered. Moreover, thermodynamics and other characteristics were not considered explicitly in Ayurveda drug preparation. However, for the preparation of different herbal formularies, the intensity and duration of heat were specifically considered following changing characteristics of fire by using different types of firewood and other fire sources.[67]

In Ayurveda, raw herbal plant materials were used by volume to volume, weight to weight, or volume to weight ratios to extract herbs efficiently. The Ayurveda disease management system was mainly practiced on a small scale as a service without industrial measures. Herbal preparations were created using abundant herbal plants; concepts of efficient extraction would not have been considered to any great extent.[68–70] In Ayurveda, the quality of the final drug subjective parameters is assessed, while in modern medicine, multiple analytical and advanced biotechniques are employed.[71] The development of herbal formulations and the methods used to assess their therapeutic usability and validity were not described in Ayurveda and it is generally believed that spiritual activities were relied upon. In contrast, laboratory studies, clinical assays, and trials in various phases have been done to assess their validity and safety.[69,72,73] The Ayurveda system poses different scales and parameters and can be compared easily with those in current scientific scales. In-depth scientific studies are required to ascertain the essence of Ayurveda concepts to make them more applicable to modern medicine.

References

1. S. Parasuraman, G. S. Thing and S. A. Dhanaraj, *Pharmacogn. Rev.*, 2014, **8**, 73.
2. C. Veeresham, *J. Adv. Pharm. Technol. Res.*, 2012, **3**, 200.
3. Q. W. Zhang, L. G. Lin and W. C. Ye, *Chin. Med.*, 2018, **13**, 20.
4. D. J. Newman and G. M. Cragg, *J. Nat. Prod.*, 2020, **83**, 770.
5. A. G. Atanasov, B. Waltenberger, E. M. Pferschy-Wenzig, T. Linder, C. Wawrosch, P. Uhrin, V. Temml, L. Wang, S. Schwaiger, E. H. Heiss, J. M. Rollinger, D. Schuster, J. M. Breuss, V. Bochkov, M. D. Mihovilovic, B. Kopp, R. Bauer, V. M. Dirsch and H. Stuppner, *Biotechnol. Adv.*, 2015, **33**, 1582.
6. G. M. Cragg and D. J. Newman, *Biochim. Biophys. Acta, Gen. Subj.*, 2013, **1830**, 3670.
7. S. Ota, A. Singh, N. Srikanth, B. Sreedhar, G. Ruknuddin and K. S. Dhiman, *Ancient Sci. Life*, 2017, **36**, 207.
8. S. S. Savrikar and B. Ravishankar, *Afr. J. Tradit., Complementary Altern. Med.*, 2011, **8**, 66.
9. N. Kumar, *Int. J. Pharm. Biol. Arch.*, 2013, **4**, 411.
10. C. Das, G. Ghosh, A. Bose and D. Das, *Indian J. Pharm. Sci.*, 2019, **81**, 396.
11. A. Chaudhary, N. Singh, M. Dalvi and A. Wele, *Ayu.*, 2011, **32**, 408.
12. S. Dongre and D. Wadodkar, *Ayurline: Int. J. Res. Indian Med.*, 2018, **2**, 1.
13. Government of India, *The Ayurvedic Formulary of India Part-III*, 2011, p. 287.

14. K. L. Gupta, G. Pallavi, B. J. Patgiri, S. M. Galib and P. K. Prajapati, *J. Ayurveda Integr. Med.*, 2012, **3**, 21.
15. D. J. Sexton, K. E. Powell, J. Liddle, A. Smrek, J. C. Smith and T. W. Clarkson, *Arch. Environ. Health*, 1978, **33**, 186.
16. M. Jaishankar, T. Tseten, N. Anbalagan, B. B. Mathew and K. N. Beeregowda, *Interdiscip. Toxicol.*, 2014, **7**, 60.
17. M. Galib, M. Barve, C. Mashru, B. J. Jagtap and P. K. Prajapati, *J. Ayurveda Integr. Med.*, 2011, **2**, 55.
18. N. Joshi, M. K. Dash and P. K. Panda, *Int. J. Green Pharm.*, 2018, **11**, S112.
19. R. B. Saper, S. N. Kales, J. Paquin, M. J. Burns, D. M. Eisenberg and R. B. Davis, *JAMA*, 2004, **292**, 2868.
20. R. Angadi, Vanga-Vijnaniya Taranga, in *Rasa Tarangini of Sri Sadananda Sarma*, Chaukhamba Surbharati Prakashan, New Delhi, 2015, ch. 18, pp. 289–300.
21. Vaagbhata, *'Rasaratnasamuccaya' 5/142*, Chowkhamba Sanskrit Series Office, Varanasi (India), 5th edn, 1976d.
22. Vaagbhata, *Rasaratnasamuccaya, 5/13*, Chowkhamba Sanskrit Series Office, Varanasi (India), 5th edn, 1976e.
23. Vaagbhata, *Rasaratnasamuccaya*, Chowkhamba Sanskrit Series Office, Varanasi (India), 5th edn, 1976a.
24. R. W. Phelps, T. W. Clarkson, T. G. Kershaw and B. Wheatley, *Arch. Environ. Health*, 1980, **35**(3), 161–168.
25. S. Samhitaa and N. Sangraha, in *Choukhamba Orientalia*, ed. A. J. Trikamaji, Varanasi (India), 5th edn, 1992c, p. 349.
26. M. Banarasidas and S. Sadanand, *Rasatarangini*, Varanasi (India), 1989.
27. V. K. Joshi, *The Ayurvedic Pharmacopoeia of India*, Ministry of Health and Family Welfare, Govt. of India, New Delhi, 1978.
28. R. D. Kulkarni and B. B. Gaitonde, *Indian J. Med. Res.*, 1962, **50**, 715.
29. R. Semwal, *J. Convent. Knowl. Holist. Health.*, 2018, **2**, 12.
30. G. S. Lavekar, B. Ravishankar, S. V. Rao, S. Gaidhani, B. K. Ashok and V. J. Shukla, *Toxicol. Int.*, 2009, **16**, 37.
31. S. S. Savrikar and B. Ravishankar, *Afr. J. Tradit., Complementary Altern. Med.*, 2010, **7**, 174.
32. D. Chamundeeswari, P. Kanimozhi and R. Vasanthkumar, *Sri Ramachandra J. Med.*, 2007, **2**, 39.
33. S. Mahesh, K. B. Sheela, S. Aravind and S. P. Neethu, *Int. J. Res. Ayurveda Pharm*, 2017, **5**, 62.
34. S. Duraipandi and V. Selvakumar, *BMC Complementary Altern. Med.*, 2015, **15**, 62.
35. O. A. Ansari, J. S. Tripathi and S. Ansari, *Int. J. Res. Ayurveda Pharm.*, 2013, **12**, 4194.
36. G. S. Achliya, S. G. Wadodkar and A. K. Dorle, *Indian J. Pharmacol.*, 2005, **37**, 33.
37. V. K. Joshi, *The Ayurvedic Formulary of India*, Govt. of India, Ministry of Health and Family Welfare, New Delhi, 1978, pp. 1–120.
38. Government of India, *The Ayurvedic Formulary of India Part-I*, 2021, p. 202.

39. P. Panda, B. Das, D. S. Sahu, S. K. Meher, G. C. Bhuyna, B. K. Das and M. M. Rao, Taila Kalpana (Medicated Oil) in Ayurveda, *Res. J. Pharmacol. Pharmacodyn.*, 2016, **8**(1), 39–41.
40. R. Tomar, G. Kaur, R. Sannd, H. Singh and B. Sarkar, *Ann. Ayurvedic Med.*, 2014, **3**, 96.
41. M. Bhardwaj, *J. Drug Delivery Ther.*, 2019, **9**, 12.
42. Q. W. Zhang, L. G. Lin and W. C. Ye, *Chinese Med.*, 2018, **13**(1), 108–134.
43. N. Bhatt and M. Deshpande, *World J. Pharm. Res.*, 2016, **5**, 1565.
44. M. Maithani, H. Grover, R. Raturi, V. Gupta and P. Bansal, *Am. J. Drug Alcohol Abuse*, 2019, **45**, 208–216.
45. A. Gupta and P. K. Prajapati, *Ayu.*, 2010, **31**, 520–524.
46. M. R. Murthy, P. K. Ranjekar, C. Ramassamy and M. Deshpande, *Cent. Nerv. Syst. Agents Med. Chem.*, 2010, **10**, 238–246.
47. D. Pillai and N. Pandita, *Indian J. Pharm. Sci.*, 2016, **78**, 129–135.
48. R. Singh and B. Sharma, Metal-Based Therapy in Traditional and Modern Medicine Systems, in *Biomedical Applications of Metals*, ed. M. Rai, A. Ingle and S. Medici, Springer, Cham, 2018, pp. 195–211.
49. K. Iqbal and M. Asmat, *J. Ayub. Med. Coll. Abbottabad*, 2012, **24**, 204.
50. J. Fuchs, *Oxidative Injury in dermatopathology*, Springer, Berlin, Heidelberg, 1992.
51. R. H. Persellin, E. V. Hess and M. Ziff, *Arthritis Rheum.*, 1967, **10**, 99.
52. G. C. Nille and A. K. Chaudhary, *J Ayurveda Integr Med*, 2021, **12**, 172–177.
53. X. F. Zhang, Z. G. Liu, W. Shen and S. Gurunathan, *Int. J. Mol. Sci.*, 2016, **17**, 202.
54. M. Dubald, S. Bourgeois, V. Andrieu and H. Fessi, *Pharmaceutics*, 2018, **10**, 2.
55. D. M. Aronoff and E. G. Neilson, *Am. J. Med.*, 2001, **111**, 304.
56. X. Yuan, R. L. Chapman and Z. Wu, *Phytochem. Anal.*, 2011, **22**, 189.
57. P. K. Mukherjee, R. K. Harwansh, S. Bahadur, S. Banerjee, A. Kar and J. Chanda, *J. Ethnopharmacol.*, 2017, **197**, 10.
58. H. Zaigham, A. Tauheed and A. Ali, *Int. J. Pharm. Sci. Res.*, 2019, **10**, 1640.
59. R. Kumar and M. Sharma, *J. Mat. NanoSci.*, 2018, **5**, 35.
60. A. Bhardwaj, K. K. Jha, A. Panka, A. Agarwal, B. Singh and N. Trivedi, *Indian J. Pharm. Biol. Res.*, 2017, **5**, 39.
61. A. H. Shojaei, *J. Pharm. Pharm. Sci.*, 1998, **1**, 15.
62. V. V. Singh and V. S. Rai, *Tradit. Med. Res*, 2019, **4**, 293.
63. F. Menaa, B. Menaa, B. A. Kahn and A. Menaa, Trans Fats and Risks of Cardiovascular Diseases: Facts or Artifacts? in *Handbook of Lipids in Human Function*, AOCS Press, 2016, vol. 1, p. 21.
64. K. S. Joshi, *J. Ayurveda Integr. Med.*, 2014, **5**, 85.
65. Y. S. Karandikar, A. S. Bansude and E. A. Angadi, *J. Clin. Diagnostic Res.*, 2016, **10**, 11.
66. H. Sharma, X. Zhang and C. Dwivedi, *Ayu*, 2010, **31**, 134.
67. N. Singh and A. Chaudhary, *Ayu*, 2011, **32**, 103.
68. M. M. Hedaoo and T. P. Bhole, *Int. J. Compr. Adv. Pharmacol.*, 2020, **5**, 151.

69. A. Chaube, S. K. Dixit and P. V. Sharma, *Ancient Sci. Life*, 1997, **14**, 161.
70. D. R. Laurence and A. L. Bacharach, *Evaluation of Drug Activities*, New York/London, 1964, Academic Press.
71. J. Pant, R. M. Singh and H. Marwah, *Med. Plant Res*, 2021, **9**, 141.
72. K. Sahil, B. Sudeep and M. Akanksha, *Int. J. Ayurveda Res*, 2011, **2**, 1100.
73. S. Sen and R. Chakraborty, *J. Tradit. Complement Med.*, 2016, 7, 234.

CHAPTER 2

Scientific Approach to the Role of Turmeric in Ayurvedic Formulations and Their Phytochemistry

GURMEET SINGH,* NADANA VINAYAGAN AND ABDUL MATEEN

Centre for Ayurveda Biology and Holistic Nutrition, The University of Trans-Disciplinary Health Sciences and Technology (TDU), Bengaluru, Karnataka 560064, India
*Email: gurmeet.singh@tdu.edu.in

2.1 Introduction

Ayurveda, one of the oldest traditional healthcare systems in the world, has more than 2000 medicinal plants in its *Materia Medica*.[1] Turmeric (*Curcuma longa* L.) is one of the renowned herbs mentioned in Ayurveda and is part of the many medicinal formulations for a wide range of disease conditions. Besides Ayurveda, its medicinal applications are also mentioned in Unani, Siddha and Chinese traditional medicine systems. Turmeric has at least 53 Sanskrit and vernacular names in India, indicating its widespread use across the country and pointing to its cultural importance. Many of these names are suggestive of the therapeutic properties exhibited by the herb.[2]

Turmeric has been a part of the daily diet in India for thousands of years and, more recently, has also transcended into global cuisines and diets, and names such as "golden spice" and "Indian saffron" with which the spice is

Chemistry, Biological Activities and Therapeutic Applications of Medicinal Plants in Ayurveda
Edited by Augustine Amalraj, Sasikumar Kuttappan and Karthik Varma A. C.

Published by the Royal Society of Chemistry, www.rsc.org

referenced are suggestive of the hallowed position it occupies as a culinary ingredient with perceived wellness benefits. Its popularity continues to grow globally through products such as turmeric lattes and turmeric teas, besides its culinary uses as a spice because of its perceived health benefits. Besides foods and beverages, it also finds use in various supplements and cosmetics.[3] India is the largest producer of turmeric. In 2020, 15 888 tons of turmeric were cultivated, which accounted for 80% of the global production.[4] Turmeric is also one of the more researched plants in modern science and has been the subject of study in 4600 publications on PubMed and 200 clinical studies conducted over the last three decades[5] (Figure 2.1). The Curcumin Resource Database shows an even larger number with over 9000 publications and 500 patents listed.

Turmeric is a perennial herb belonging to the family of Zingiberaceae. The rhizome and leaves are the parts used for medicinal and culinary purposes. The rhizome is pungent and bitter in taste. Scientific research on turmeric has focussed on its nutritional profile, phytochemical composition and pharmacological uses. Almost 180 nutritional components and 100 additional phytochemicals[6] and more than 15 pharmacological actions have been reported for turmeric. A number of studies have shown that the herb consists of phytochemicals expressing anticancer,[7] antidiabetic,[8] antiinflammatory,[9]

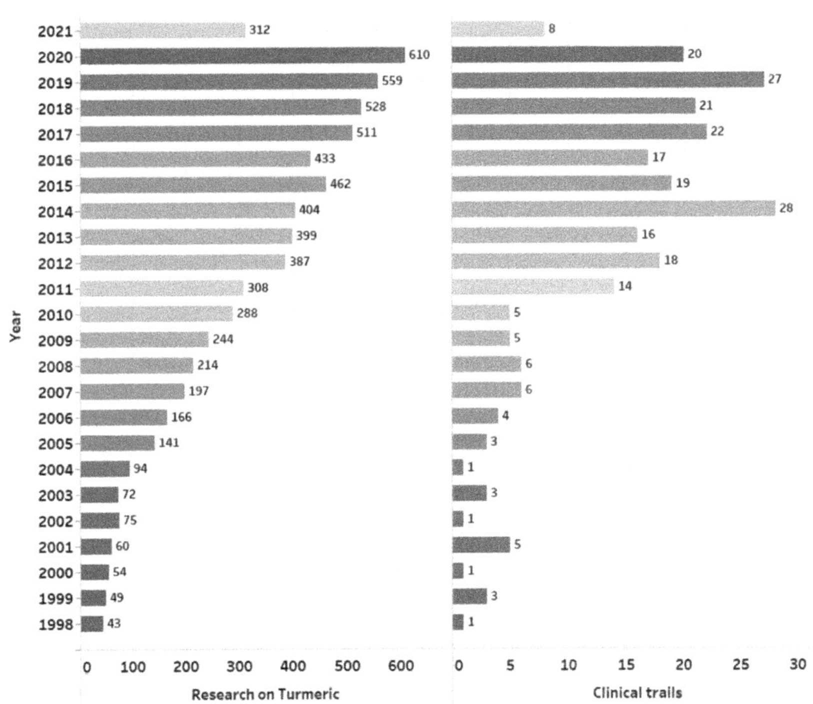

Figure 2.1 The number of references for turmeric in PubMed since 1998: turmeric research (left), clinical trials (right).

antimicrobial,[10] antimutagenic,[11] and cardio- and neuro-protective activities.[12,13] Turmeric and its phytochemical constituents show oxygen-scavenging action in biological assays, inhibit the production of histamine thereby acting as an antimicrobial and anti-allergic herb, and show inhibitory action on lipopolysaccharides in inflammatory conditions. They have also been shown to induce apoptosis, autophagy and mitochondrial action in many cancers and mutagenic conditions.[14] However, this extensive research on its pharmacology has not yet led to convincing evidence of its therapeutic benefits.[15] Much of this research has used reductionist frameworks where purified single molecules (*e.g.* curcumin) obtained from turmeric have been investigated. Turmeric stands at the cross-roads of scientific query, seeking a new holistic approach for understanding its compositions and activity that can help unravel its benefits to human health.

This chapter reviews the use of turmeric in Ayurveda and provides an analysis of its Ayurvedic formulations and pharmacological actions with more recent scientific literature. In addition, this chapter summarizes the major phytochemicals present in the turmeric rhizome based on mining from various online databases, classical literature and research articles, and presents their therapeutic action based on evidence. Furthermore, a molecular mapping is presented based on their polarity and volatility, which may help in deconstructing the formulations and understanding the active molecules in different formulations.

2.2 Data-mining Classical Texts for Insights

The use of turmeric for its health benefits predates Ayurveda. Its reference in Vedic literature (Table 2.1) suggests that it was a symbol of prosperity. It appears to have been highly regarded as a cleansing herb for the entire body and it was used in cosmetic conditions involving depigmentation of the skin and greying of the hair, and therapeutic applications.

The many names of turmeric in the vernacular literature also point to its widespread use and deep cultural embedding. Many of these names and synonyms are based on turmeric's therapeutic action, morphological character and pharmacological properties (Table 2.2).

Table 2.1 Turmeric mentions in Vedic literatures.

S. no.	Vedic literature	Total medicinal plants mentioned
1	Rigveda	67
2	Yajurveda	82
3	Atharva veda	288
4	Brahman grantha	129
5	Upnishidha	31
6	Kalpasutra	319
7	Paniniya Ashtadhyayi	152
8	Vartik	31
9	Patanjal mahabhasaya	109
10	Yaskkrit Nirukta	26

Table 2.2 Some vernacular names and synonyms of turmeric mentioned in Ayurveda classical texts.

Some vernacular names and synonyms used for turmeric	Context	Ref.
Haridra	Plant description: yellow-coloured plant	74–76
Krimighna	Medicinal property: it is antihelminthic in action, which reduces worm infestations	74–76
Nishakhya, Nisha	Plant description: it looks beautiful in the moonlight	76
Mehaghni	Medicinal property: specially denoted that this drug is a choice to treat Prameha (diabetes)	74–76
Vishaghni	Medicinal property: removes toxins and purifies the blood circulation in the body	77–79
Vamavati/vamada/ Vamapradayini, Vamavilasini/ Varyamini	Medicinal property: used as a dyeing agent for cloths and skin and enhancing the complexion	74 and 78–81
Harita	Medicinal property: it is an effective drug for jaundice	77–80
Bhadralata	Plant description: holly climber-like shrub	77, 78 and 80
Vanvilasini	Plant description: grows wildly in forests	74–76 and 78–82
Romashmoolika	Plant description: having hairy roots	74–76 and 78–82
Pindaharidra/ Pindabhadra	Plant description: roots are cylindrical/tubular	80
Pita/pitangi/Gauri/ garangi	Plant description: roots have attractive yellow colour	74–76 and 78–82
Suvama, Kanchani	Plant description: roots are golden yellow colour	76 and 83
Pavitra	Religious significance: holy plant or used in holy events (daily pooja karma)	76 and 83
Mangalya	Religious significance: Laxmi, Bhadra	80
Subhaga	Religious significance: root powder is auspicious and used for religious purposes	80
Hattavilasini	It attracts itself and enhances beauty of the market	76
Yoshitpriya/ Shreevallabha/ Shobhana	Cosmetic use: it has been used by women as a cosmetic (udvartana) and enhances the complexion and colour of skin	77 and 80
Vaishya	It sells daily in markets	77 and 80

Turmeric has been well studied in Ayurveda as indicated by the extent of documentation it has received in the Samhitas (the key classical medical treatises of Ayurveda such as Charaka Samhita, Sushruta Samhita and Ashtanga Hridaya) as well as the Nighantu's (the Materia Medicas of Ayurveda). These texts describe the poly-pharmacological action of drugs on the basis of *rasapanchaka*, the Ayurveda pharmacology. The *rasapanchaka* includes the Rasa (equivalent of in-mouth sensory experience), Guna (equivalent of biophysical properties), Virya (equivalent of potency) and Vipaka (equivalent of post-digestion effect). These properties describe how a medicine works pharmacologically. Turmeric's Rasa is tiktha and katu (bitter and pungent), its Guna is rooksha (dry), its Virya is ushna (hot) and its Vipaka is katu (pungent).[16] In addition, turmeric helps to balance all three doshas: the vata, pitta and kapha. Many of the Nighantus include turmeric under the group of botanicals called "Oshadhi Varga" or medicinal group, suggesting its prominent use in formulations used across a diverse set of diseases. It has been also used as a home remedy for thousands of years, and repeated use has demonstrated its effectiveness in treating a variety of illnesses.

Ten Ayurvedic classical texts, namely Caraka Samhita (CS), Susruta Samhita (SS), Astanga Sangraha (AS), Astanga Hridya (AH), Charakadatta (CD), Sarangadhara Samhita (SAR), Bhaisajyaratnavali (BR), Yogaratnakar (YR), Bhavaprakasa (BP) and Vangasen Samhita (VS), were mined for turmeric-based orally administered formulations, their methods of preparation and their therapeutic uses.

This search resulted in 88 turmeric formulations used for 26 disease conditions (Table 2.3). Among them, vatarakta (inflammatory conditions like arthritis), prameha (diabetes), panduroga (symptoms of iron deficiency) and kustharoga (skin diseases) were more frequently associated with these formulations. Other diseases like visaroga (poisoning), svasa (respiratory diseases) and kamala (jaundice) are also mentioned (Figure 2.2). Further, an analysis of the botanical composition of the 88 formulations was done to understand its pairing. Fifty-eight ingredients were curated from this analysis, where Amla (*Emblica officinalis*), Musta (*Cyperus rotundus*), Pippali (*Piper longum*) and Nimba (*Azadirachta indica*) were found to be the more prevalent with turmeric as shown in Figure 2.3. The improved bioavailability of curcuminoids, a key group of molecules of turmeric and also the most studied in modern science, when used with piperine is well established. Interestingly, turmeric is paired with *Piper longum*, a botanical rich in piperine, in many Ayurveda formulations, indicating that the improved efficacy of turmeric in the presence of *Piper longum* was recognized in Ayurveda. Even more interesting is the fact that this pairing is not used in all the formulations as this suggests specificity of formulations for different treatments. Such analysis can help in deconstructing formulations and arriving at the group of molecules that Ayurveda may be leveraging for different therapeutic conditions.

Deconstruction of various formats used for the formulations can provide additional insights. The different formats mentioned in the classical texts include curna (dry powder), kvatha (decoction), kalka (paste), gutika (tablet),

Table 2.3 Number of turmeric formulations mentioned for various health benefits in a few important Ayurveda classical texts.

S. no.	Diseases	CS	SS	AS	AH	HS	CD	SAR	BR	YR	BP	VS	Total
1	Vatarakta (arthritis)		1	1	2	1		1		5			11
2	Prameha (diabetes)	2	1	4	1	1	1						10
3	Pandu (anaemia)	1	3	2		1		3					10
4	Kusthroga (skin diseases)		1	1	3	1		1	2		1		10
5	Visaroga (poisoning)	1		5		2							8
6	Asthma	2	1	1		1		1					6
7	Kamala (jaundice)	1		1		1		2					5
8	Oedema	1							2				3
9	Kasa (cough)	1		1								1	3
10	Jwara (fever)	2					1						3
11	Dyspnoea									1		2	3
12	Medya (brain health)		1						1				2
13	Balaroga (strength)						1				1		2
14	Trsna (morbid thirst)	1			1								2
15	Slipada (stone-like enlargement of the foot)						1				1		2
16	Purgative			1	1								2
17	Vataroga	1											1
18	Gulma (phantom tumour)	1											1
19	Udara roga (diseases of the abdomen)	1											1
20	Parasites									1			1
21	Atisara (diarrhoea)		1										1
22	Sirsasula							1					1
23	Svarabheda (hoarseness of voice)								1				1
24	Sitalika vikara										1		1
25	Asmari (calculus)										1		1
26	Phiranga (syphilis)										1		1
	Total	15	9	17	8	8	4	9	6	7	6	3	

peya (soup), leha (semi-solid syrup) and ghee macerations (ghrita). Of these kvatha, kalka and peya are the water-based formats, while leha and ghrita are fat-based formats. Overall, curna, kvatha and ghrita formats are more frequently prescribed for treating illnesses (Figure 2.4). This is interesting when it is considered with the fact that many of the turmeric phytochemicals that have been of interest in modern scientific studies are poorly water soluble.

Further, water-based formats are used more often for diseases such as Prameha (diabetes), vatarakta (inflammation conditions like arthritis), inflammation in the abdomen, trsna (morbid thirst) and jvara (fever), whereas fat-based formats are more often used for medhya (brain health), kamala (jaundice), svasa (dyspnoea) and kasa (cough). Both water-based and fat-based formulations seem to be frequently used for diseases such as

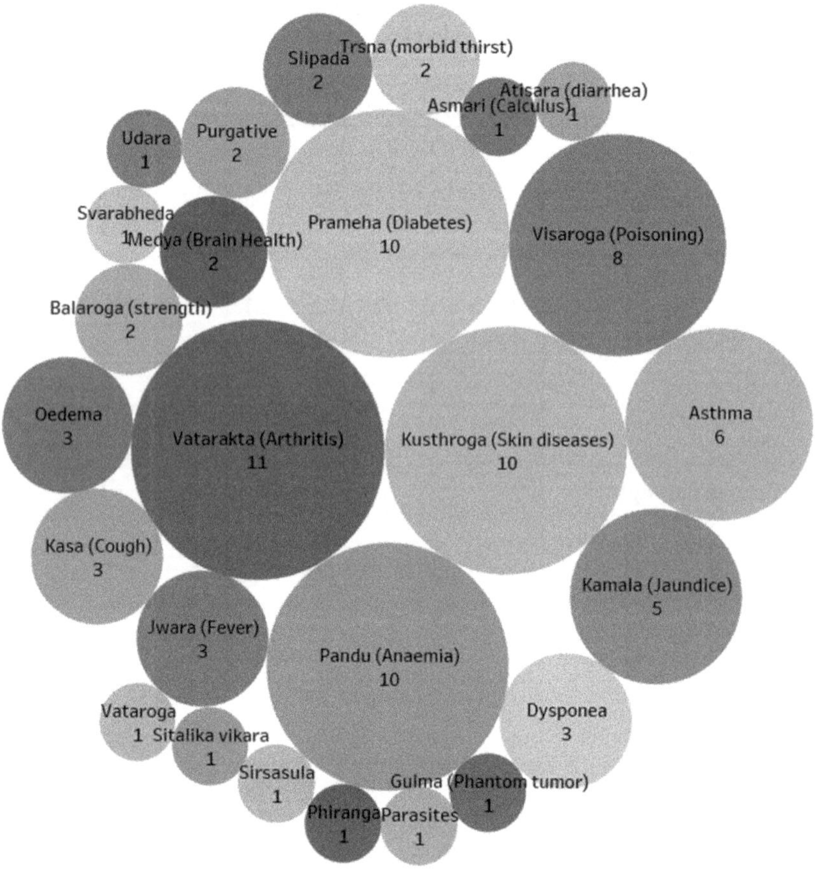

Figure 2.2 Number of turmeric formulations in Ayurveda for various health benefits.

kustharoga (skin diseases), panduroga (anaemia) and balaroga (paediatrics). This type of analysis can help in deconstructing formulations from a molecular partitioning and bioavailability perspective.

2.3 Therapeutic Uses of Turmeric

Turmeric and its phytochemicals have been studied extensively for their health benefits using modern scientific tools. We briefly discuss these in the context of the four lead therapeutics shown in Figure 2.2.

2.3.1 Antidiabetic and Prameha

Turmeric appears in many Ayurveda formulations used for Prameha, a disease with symptoms similar to type-2 diabetes. This has led to a number of enzyme-based, *in vitro*, animal and clinical studies involving extracts of

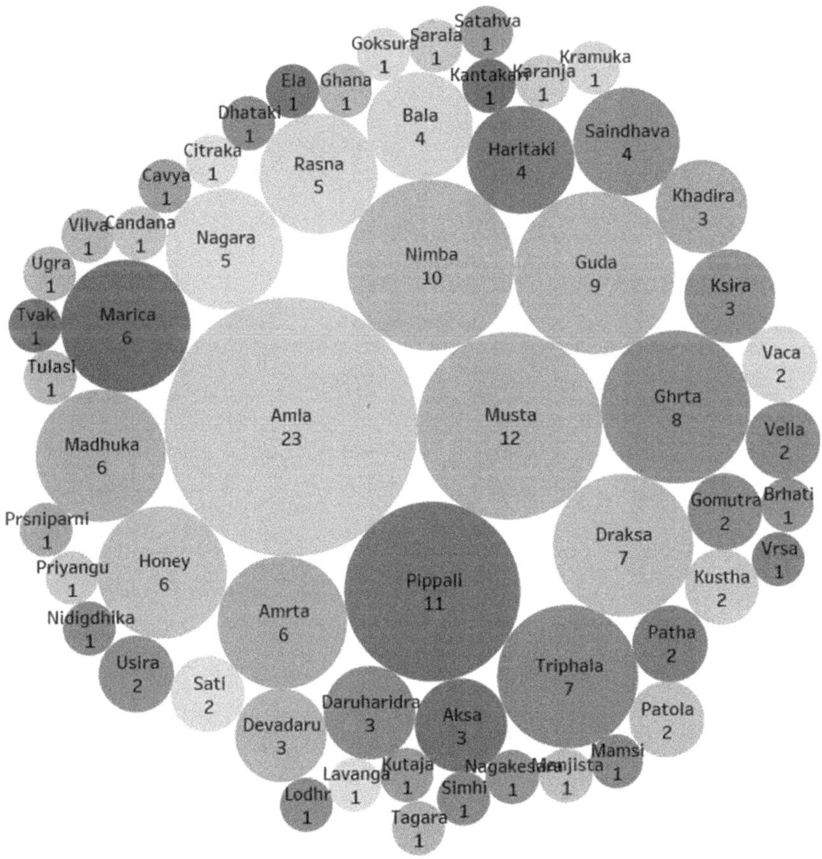

Figure 2.3 Pairing frequency of various ingredients with turmeric in Ayurveda formulations.

turmeric and purified curcumin, one of the key molecules of turmeric, to evaluate their antidiabetic activity. Curcumin-rich extracts of turmeric have been shown to delay diabetes development, enhance β-cell functions, control β-cell death and decrease insulin resistance in animal studies.[17] Curcumin has been shown to reduce the level of blood glucose and HbA1c, and to increase the antioxidant status of pancreatic β-cells. It has been shown to enhance the activation of PPAR-γ, the phosphorylation of AMPK (AMP-activated protein kinase), glycogen synthesis and stimulation of glucose uptake by activating GLUT4, GLUT2 and GLUT3 gene expressions.[18,19] In addition, curcumin appears to aid glucose metabolism in differentiated C_2C_{12} skeletal muscle cells, curcumin-mediated glucose absorption necessitates the activation of the AMPK/ACC pathway and improves insulin sensitivity.[20] Besides, turmeric compounds may also protect against non-insulin-dependent diabetes mellitus through multiple mechanisms such as insulin upregulation, enhancing insulin sensitivity, lipid profile modulation and lowering the level

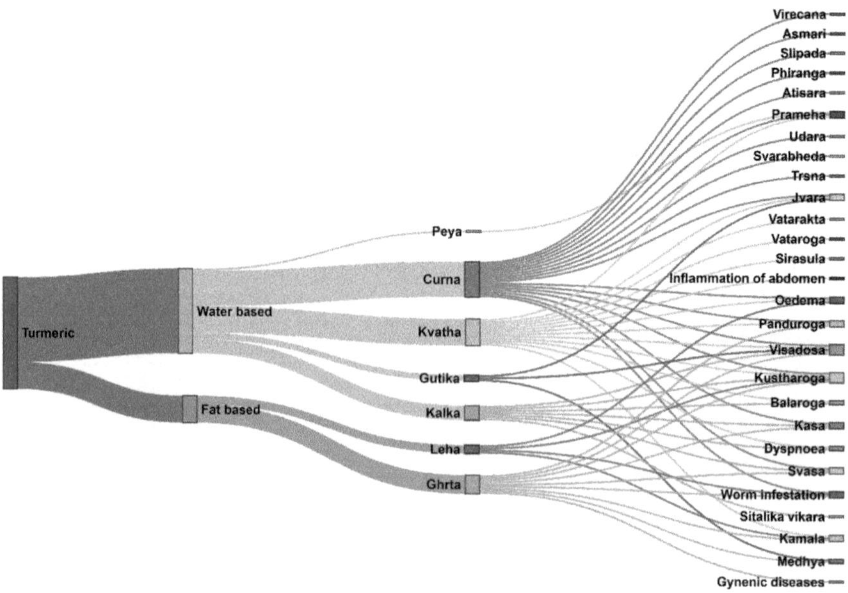

Figure 2.4 Formats of turmeric-based Ayurvedic formulations used in the treatment of diseases.

of glucose at the cellular level.[17,21,22] A study to test the antihyperglycaemic effect of curcumin (80 mg kg^{-1}) in mice subjected to a high-fat diet showed that it rapidly reduced the glucose levels and TNF-α concentrations and significantly inhibited the lowering of glucose tolerance.[23]

Besides curcumin, ar-turmerone, a major volatile compound in the turmeric rhizome, has also been studied for its antidiabetic activity. In one such study it was found to exhibit antidiabetic activity by inhibition of alpha-glucosidase and alpha-amylase.[24]

The ingestion of whole turmeric (6 g) was also found to enhance post-prandial serum insulin levels and alleviate the glycaemic response in healthy adults without affecting plasma glucose levels, and these findings suggest that turmeric may influence insulin secretion.[25,26]

Such whole turmeric studies have also led to interest in understanding the effects of the non-curcuminoid, non-turmerone fractions of turmeric. These minor components, found widely in the botanical world besides turmeric, also show promising activity and could complement or supplement that activity of curcuminoids and turmerones. One such volatile molecule, eugenol, which is more commonly associated with clove and cumin essential oils where it is present at higher levels, but that is also present in turmeric, has one of the highest alpha-amylase inhibition activities.[27] These studies suggest that not only do phytochemicals of turmeric have therapeutic potential for diabetes but also standardized whole turmeric extracts that include the non-volatile curcuminoids and the volatile turmerones along with all the other minor phytochemicals have the potential to be the basis of functional foods and supplements for post-prandial glycaemic modulation.

2.3.2 Antiarthritic and Vatarakta

Turmeric's use for managing inflammation is well documented in Ayurveda and it is also used widely as a home remedy in medical conditions ranging from arthritis to skin inflammations.[28] Arthritis affects over 250 million people globally, with the three most commonly occurring subtypes being gout, osteoarthritis (OS) and rheumatoid arthritis (RA).[29,30] Turmeric extracts and fractions have therefore been widely studied in modern science for their antiinflammatory properties. An *in vivo* antiarthritic efficacy of turmeric study stated that daily administration of turmeric curcuminoids (23 $mg\,kg^{-1}$ per day) showed significantly inhibited joint inflammation in both the acute inflammatory (64% inhibition) and chronic destructive phases (72% inhibition) of arthritis.[31] A comparison study emphasized that turmeric extracts with ginger showed antiarthritic activity by reducing the production of antiinflammatory/pro-inflammatory cytokines and by inducing antioxidant mechanism.[32] Turmeric essential oils of ar-turmerone, r-turmerone and β-turmerone have also been shown to possess inhibitory activities towards arthritic joint inflammation.[33] Eugenol, another volatile compound found in turmeric, inhibits cyclooxygenase-2 (COX-2) expression and tumour necrosis factor (TNF) signalling, while eugenol oligomers prevent inflammatory cytokine expression and NF-kappa B (nuclear factor-kappa B) activation in macrophages.[34,35]

2.3.3 Anaemia and Panduroga

Turmeric powder is very rich in iron (46.08 ± 1.83 mg/100 g). In fact, on a weight basis, it is an ingredient with one of the highest iron contents amongst our daily foods. Its contribution to iron in our diet is usually small as it is mainly used as a spice.[36] However, at therapeutic levels, it could contribute 10–30% of the daily required intake of iron. Here too, the bioavailable iron may be low as the curcuminoid and other polyphenolic constituents of turmeric chelate iron. This was observed in mice where administering turmeric lead to iron deficiency.[37] Interestingly, formulations that include turmeric are used for treating anaemia in Ayurveda.[38] These formulations also contain other botanicals such as Amla (*Emblica officinalis*), Bibhitaki (*Terminalia bellerica*), honey and ghee,[39,40] which may play a role in speciation of iron and making it more absorbable in these formulations than that present in just turmeric. It has also been suggested that iron from such turmeric-based sources may be safer as they avoid situations of iron excess that cause haemochromatosis or haemolytic anaemias.[41] In the context of the continued widespread prevalence of anaemia and the caution around iron overload through supplements, studying turmeric-based iron formulations may provide a new generation of safe yet effective iron supplements.

2.3.4 Skin Diseases and Kushtaroga

Turmeric-based topical and orally administered formulations are used in Ayurveda for a wide range of skin diseases. Turmeric, its extracts and phytochemicals have been studied in modern science for their use in the treatment of dermatological disorders such as acne, skin rashes and warts.[42] The antiinflammatory and antimicrobial properties of turmeric's phytochemicals have been proposed as the mechanism of action to underpin these treatments. In biological assay systems, curcumin has been found to suppress several inflammatory cytokines, including TNF-α and interleukins, as well as downregulate inflammatory targets including lipoxygenase, cyclooxygenase-2 and inducible nitric oxide synthase.[42] Curcumin has been shown to protect against skin disorders like psoriasis in preclinical experiments with experimental animal models and a few human clinical trials, it appears to protect skin by quenching free radicals and decreasing inflammation, with NF-κB suppression being curcumin's major target.[43] Turmeric essential oil showed antimutagenic and anticarcinogenic activity through inhibiting cytochrome p450 enzymes through an *in vitro* study.[44] In addition, a randomized oral administration of turmeric extracts induced a therapeutic response to psoriasis lesions.[45] These results have fuelled sufficient interest in academics to continue to study curcumin and turmeric extracts for their antioxidant, antiinflammatory and antibacterial properties, and a potential effective role in the treatment of acne, psoriasis, dermal wounds, sunburn, premature ageing and melanoma.

2.4 Phytochemical Composition and Pharmacological Activities of Turmeric

Dried turmeric food composition includes carbohydrate (69.4%), protein (6.3%), fat (5.1%), mineral (3.5%) and moisture (13.1%).[5] The remaining 1–10% contains a huge diversity of phytochemicals. We have curated a list of 119 phytochemicals reported in turmeric through literature and database searches. Of these, PubChem[46] IDs along with their physicochemical properties were found for 59 (Table 2.4).

The diversity of these molecules was studied using a hierarchical chemical classification. Hierarchical chemical classification along with information of kingdom, class and superclass for these chemical molecules was done using the open access application "ClassyFire".[47] The classification data generated from Classyfire were used for structural similarity-based molecule clustering using the open access "ChemMine" tools.[48] Similarity indices were generated along with Tanimoto coefficiency for all the molecules and the cluster maps were presented in a circular Newick tree format (Figure 2.5).

The ten major classes found were (1) prenol lipids, (2) diarylheptanoids, (3) fatty acyls, (4) cinnamic acids and derivatives, (5) phenols, (6) unsaturated hydrocarbons, (7) organooxygen compounds, (8) flavonoids, (9) steroids and steroid derivatives and (10) oxanes. Notably, most compounds belong to five

Table 2.4 Phyto-constituents present in turmeric rhizome.

S. no.	Phyto-chemical	Molecular formula	Pubchem Id	Ref.
1	Quercetin	$C_{15}H_{10}O_7$	5280343	84
2	Caffeic acid	$C_9H_8O_4$	689043	85 and 86
3	Cinnamic acid	$C_9H_8O_2$	444539	85 and 86
4	*P*-Coumaric acid	$C_9H_8O_3$	637542	85 and 86
5	Demethoxycurcumin	$C_{20}H_{18}O_5$	5469424	87
6	Arabinose	$C_5H_{10}O_5$	439195	84
7	Caprylic acid	$C_8H_{16}O_2$	379	84
8	Guaiacol	$C_7H_8O_2$	460	84
9	Camphor	$C_{10}H_{16}O$	2537	85 and 86
10	Isoborneol	$C_{10}H_{18}O$	64685	85 and 86
11	*P*-Methoxy cinnamic acid	$C_{10}H_{10}O_3$	699414	84
12	Linalool	$C_{10}H_{18}O$	6549	85 and 86
13	1,5-Bis-(4-hydroxy-3-methoxyphenyl)-1,4-pentadien-3-one	$C_{19}H_{18}O_5$	131752986	84
14	Alpha-pinene	$C_{10}H_{16}$	6654	88
15	Azulene	$C_{10}H_8$	9231	84
16	Beta-pinene	$C_{10}H_{16}$	14896	89
17	Bisacurone	$C_{15}H_{24}O_3$	14287397	85 and 86
18	Camphene	$C_{10}H_{16}$	6616	85 and 86
19	Curcumene	$C_{15}H_{22}$	92139	89
20	Cineol	$C_{10}H_{18}O$	2758	85, 86 and 90
21	Curcumenol	$C_{15}H_{22}O_2$	167812	84
22	Curzerenone-C	$C_{15}H_{18}O_2$	3081930	88
23	D-Camphor	$C_{10}H_{16}O$	159055	85 and 86
24	Guaiane	$C_{15}H_{28}$	9548703	88
25	Isoprocurcumenol	$C_{15}H_{22}O_2$	14543198	84
26	L-Alpha curcumene	$C_{15}H_{22}$	442360	88
27	L-Beta curcumene	$C_{15}H_{24}$	14014430	88
28	Letestuianin A	$C_{20}H_{20}O_5$	638190	84
29	Letestuianin B	$C_{21}H_{22}O_6$	10429233	84
30	Limonene	$C_{10}H_{16}$	22311	88
31	*P*-Cymene	$C_{10}H_{14}$	7463	85 and 86
32	Procurcumenol	$C_{15}H_{22}O_2$	189061	84
33	Terpinene	$C_{10}H_{16}$	7462	84
34	Terpineol	$C_{10}H_{18}O$	17100	90
35	Zedoarondiol	$C_{15}H_{24}O_3$	24834047	84
36	Zingiberene	$C_{15}H_{24}$	92776	90 and 91
37	Curcumin	$C_{21}H_{20}O_6$	969516	87
38	Eugenol	$C_{10}H_{12}O_2$	3314	84
39	Bisdemethoxycurcumin	$C_{19}H_{16}O_4$	5315472	87
40	Ar-turmerone	$C_{15}H_{20}O$	160512	92 and 93
41	(*E*)-4-(4-Hydroxy-3-methoxyphenyl)but-3-en-2-one	$C_{11}H_{12}O_3$	5354238	84
42	Alpha-linolenic acid	$C_{18}H_{30}O_2$	5280934	89
43	Arachidic acid	$C_{20}H_{40}O_2$	10467	84
44	Beta-caryophyllene	$C_{15}H_{24}$	5281515	90
45	Cuminol	$C_{10}H_{14}O$	325	84
46	Docosanoic acid	$C_{22}H_{44}O_2$	8215	84
47	Ferulic acid	$C_{10}H_{10}O_4$	445858	85 and 86
48	Gitoxigenin	$C_{23}H_{34}O_5$	348482	84

Table 2.4 (*Continued*)

S. no.	Phyto-chemical	Molecular formula	Pubchem Id	Ref.
49	Linoleic acid	$C_{18}H_{32}O_2$	5280450	94
50	Oleic acid	$C_{18}H_{34}O_2$	445639	84
51	Palmitic acid	$C_{16}H_{32}O_2$	985	89
52	Pentadecanoic acid	$C_{15}H_{30}O_2$	13849	84
53	Phytol	$C_{20}H_{40}O$	5280435	89
54	Stearic acid	$C_{18}H_{36}O_2$	5281	89
55	Vanillin	$C_8H_8O_3$	1183	84
56	Lignoceric acid	$C_{24}H_{48}O_2$	11197	84
57	Sodium curcuminate	$C_{21}H_{18}Na_2O_6$	6445487	84
58	Beta-elemene	$C_{15}H_{24}$	6918391	88
59	Germacrone-13-Al	$C_{15}H_{20}O_2$	14633002	94

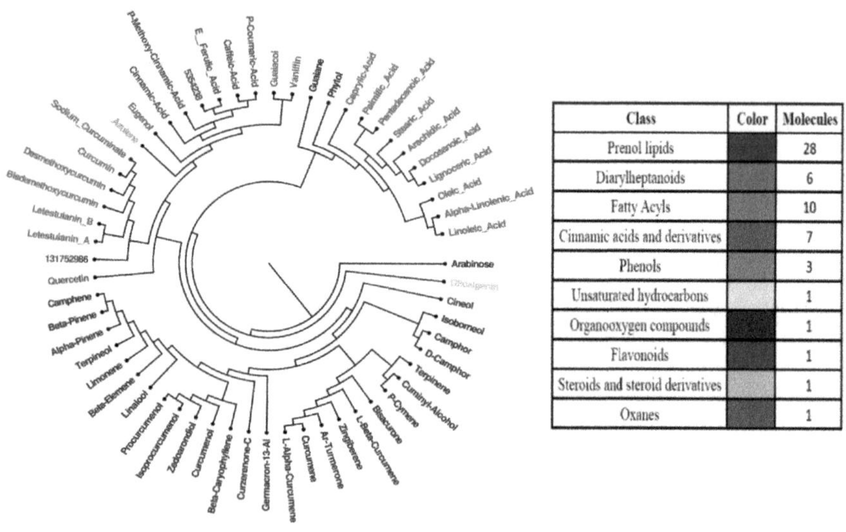

Figure 2.5 Clustering and chemical classification of turmeric compounds. The left panel shows the hierarchical clustering of turmeric molecules (59) into the different classes of molecules and the right panel shows respective count of the molecules.

of these categories: prenol lipids, diarylheptanoids, fatty acyls, cinnamic acids and derivatives, phenols with molecule counts of 28, 6, 10, 7 and 3, respectively. Diarylheptanoids are part of the major class of phenolics with structurally different compounds that include the curcuminoids found in turmeric (curcumin, demethoxycurcumin and bisdemethoxycurcumin). These curcuminoids are the major constituents of turmeric rhizomes and cluster into diarylheptanoids along with three other derivative compounds that have received considerable interest as potential therapeutic agents for the treatment of various diseases and therapeutic activities such as gastroprotective, antiulcerogenic and anticancer.[49] Prenol lipid (fatty acids, triglycerides and phospholipids) compounds are related to lipid metabolism and are responsible for the vital endogenous therapeutic actions of certain pathological conditions.[50] Therefore, molecule clustering results showed that turmeric molecules predominantly group under prenol lipids, diarylheptanoids and fatty acyls that possess wide-ranging therapeutic properties. Of these, the non-volatile diarylheptanoid curcuminoids and the volatile turmerones have received the most attention.

One reason why the *in vitro* biological activity of turmeric has not been translated into substantiated clinical evidence is attributed to the low oral bioavailability of curcuminoids, turmeric's major phytochemical constituents, as described above. Curcuminoids have relatively low absorption in the small intestine. In addition, the absorbed amount undergoes rapid reductive and conjugative metabolism in the liver and elimination through the gall bladder resulting in a very short half-life.[51,52] This has been shown in animal studies where curcuminoids were seen to be rapidly metabolized and conjugated in the liver and excreted in the faeces, as well as in human studies even when administrated at high dosages.[53] An oral dose of 500 $mg\,kg^{-1}$ administered to rats resulted in a plasma concentration of only 1.8 $ng\,kg^{-1}$, though a number of metabolites and conjugated forms were identified such as curcumin glucuronide and curcumin sulfate.[54] Curcumin's low bioavailability is a major question mark in its use as a nutritional supplement. However, animal studies have shown that other compounds such as ar-turmerone, which is a sesquiterpenoid, mainly present in turmeric rhizome, possess good bioavailability after both intravenous and intraperitoneal injection in the mouse, raising the possibility that the multimolecular mix of phytochemicals present in whole extracts of turmeric may collectively be responsible for the *in vivo* activity.[55]

Additionally, several naturally occurring molecules from botanicals typically used along with turmeric in traditional medicinal formulations have been shown to increase the bioavailability of curcuminoids. Piperine, a major active component of black pepper (*Piper nigrum*), the third most-paired botanical with turmeric (Figure 2.3) and known to be an inhibitor of hepatic enzymes, was found to enhance the bioavailability of curcuminoids.[56–58] Several strategies have been developed to enhance curcumin's bioavailability and absorption by using nanoparticles, liposomes, micelles and phospholipid complexes.[57,59] Encapsulation of other lipophilic molecules of turmeric has also been studied to enhance their bioavailability. Besides curcuminoids, work has been done in improving the bioavailability of other phytochemicals of turmeric also. Linalool, one such poorly bioavailable volatile lipophilic molecule found in turmeric and many aromatic botanicals, was found to have improved solubility when encapsulated with beta-cyclodextrin (β-CD).[60] Collectively, these strategies of whole extracts, coextracts through judicious botanical pairings and encapsulation may lead to effective supplements and therapeutics.

However, uses of whole extracts come with their challenges of residual complexity from a phytochemical standardization perspective as well as a pharmacokinetic and pharmacodynamic (PK/PD) perspective. Holistic approaches are required to study these that consider poly-pharmacology and synergy. But before we apply these, we need some simplification strategies that allow for targeting of fractions and botanical combinations to be studied. This is where the formulation network analysis depicted in Figures 2.2 and 2.3 can play a role.

Volatility and polarity mapping of the phytochemicals can additionally help target groups of molecules that may be concentrated in the processing steps involved in the extracts or foods. As an example, a fresh extract of turmeric will be richer in both curcuminoids and turmerones, while a dry powder of turmeric will have curcuminoids but low levels of turmerones. Formulations that combine these with piperine may be attempting to use curcuminoids as the active group by increasing its half-life by slowing the liver's action. Figure 2.6 depicts one such mapping of the 59 phytochemicals of turmeric. The values of $\log p$ and Henry's law constant were retrieved for these molecules from the PubChem and ChemSpider databases, respectively.[46,61] The $\log p$ value is an indicator of water solubility and Henry's law constant is an indicator of volatility. The four quadrants of such a plot I, II, III and IV represent volatile with fat-soluble (non-polar) molecules, volatile with water-soluble type (polar) molecules, non-volatile with water-soluble molecules and non-volatile with fat-soluble molecules, respectively. Quadrant IV (non-volatile with fat soluble molecules) largely includes curcuminoids. Based on such a plot we can deconstruct classical processes and formulations to identify the group of phytochemicals that may be playing an active role in the therapeutic action.

Molecules present in all quadrants showed distinct association with cancer disease that implies turmeric compounds possess promising antitumor properties. Preponderantly, quadrant-IV molecules (curcumin, bisdemethoxycurcumin, sodium curcuminate, desmethoxycurcumin) were associated with cancer and tumour diseases including breast cancer, colon cancer and lung cancer followed by inflammation diseases, Alzheimer's disease, cytotoxicity and diabetes.[62–65] In addition, curcuminoids have been shown to exhibit antiinflammatory and anticancer activities on many cell signalling pathways including TNF, NFKB1, AKT1, CASP3, IL6, TP53, NRF2 and COX-2. Quadrant-IV and quadrant-III molecules also have a similar disease association pattern along with liver damage and cardiovascular disease.[66–68] Quercetin, caffeic acid, ferulic acid and vanillin, which are from quadrant-III, have high bioavailability scores and zero Lipinski rule violation. Molecules present in quadrant-II highly interact with cancer and inflammation pathways in network pharmacology approaches.[69,70] Exclusively, liver cirrhosis and liver diseases were shown in the quadrant-II and eugenol, linalool, caprylic acid, cinnamic acid and isoborneol were the lead responsible molecules that interacted with these diseases. Obesity is placed into quadrant-I, which has antihyperglycaemic compounds such as ar-turmerone, beta-caryophyllene, linoleic acid and palmitic acid. Furthermore, these compounds are strongly associated with fatty acid metabolism,

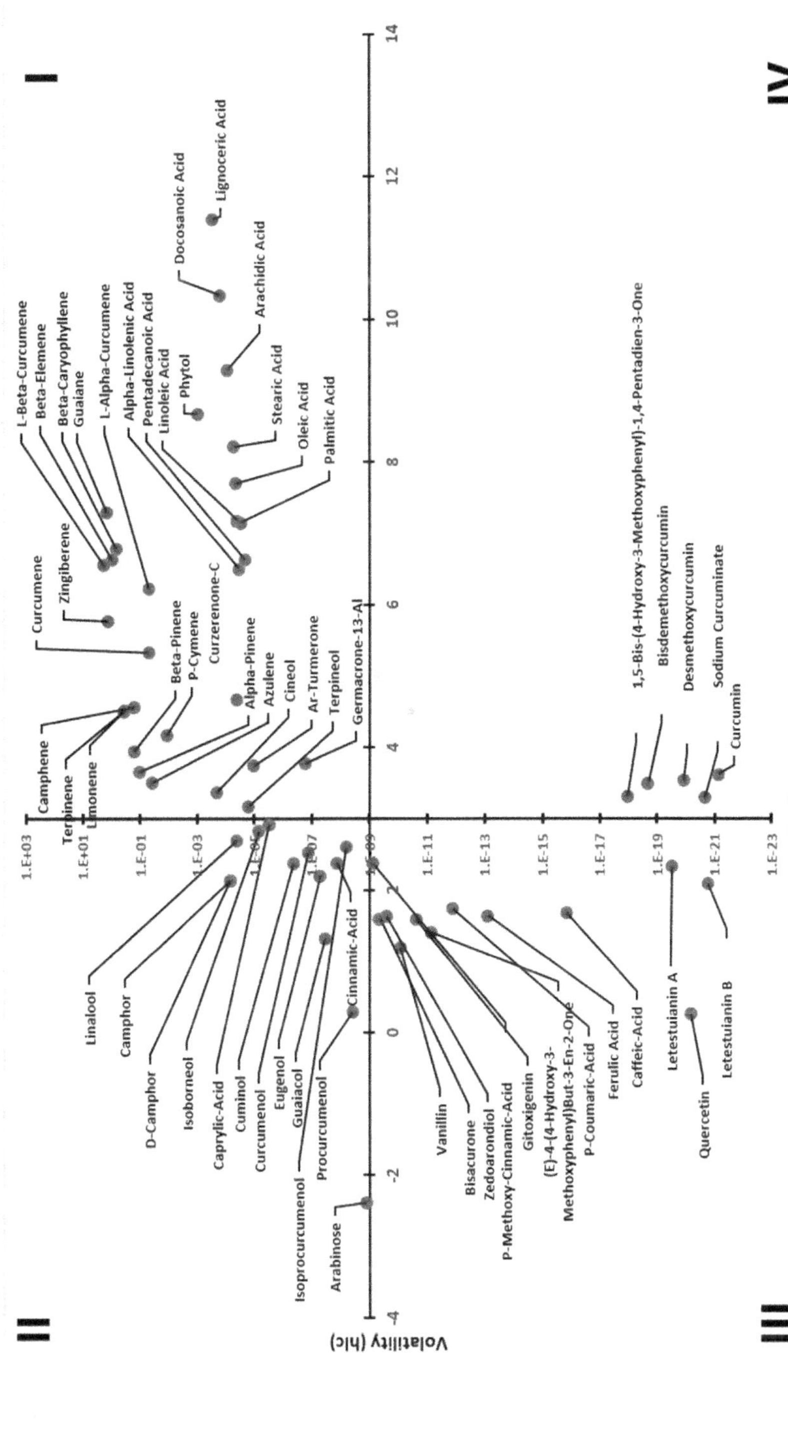

Figure 2.6 Solubility and volatility map of turmeric molecules.

Table 2.5 Therapeutic action and disease association of turmeric molecules.

Quadrant	Molecules	Pharmacological action/disease association	Ref.
I	Alpha-pinene, azulene, beta-pinene, camphene, curcumene, cineol, curzerenone-c, guaiane, L-alpha-curcumene, L-beta-curcumene, limonene, *p*-cymene, terpinene, terpineol, zingiberene, ar-turmerone, alpha-linolenic acid, arachidic acid, beta-caryophyllene, docosanoic acid, linoleic acid, oleic acid, palmitic acid, pentadecanoic acid, phytol, stearic acid, lignoceric acid, beta-elemene, germacrone-13-al	Antitumor activity, obesity	24 and 71–73
II	Cinnamic acid, arabinose, caprylic acid, guaiacol, camphor, isoborneol, linalool, curcumenol, D-camphor, isoprocurcumenol, procurcumenol, eugenol, cuminol	Liver cirrhosis and liver diseases	69 and 70
III	Quercetin, caffeic acid, *p*-coumaric acid, *p*-methoxy cinnamic acid, bisacurone, letestuianin a, letestuianin b, zedoarondiol, (*e*)-4-(4-hydroxy-3-methoxyphenyl) but-3-en-2-one, ferulic acid, gitoxigenin, vanillin, cuminol	Anticancerous properties, liver damage, cardiovascular disease	66–68
IV	Curcumin, bisdemethoxycurcumin, sodium curcuminate desmethoxycurcumin, 1,5-bis-(4-hydroxy-3-methoxyphenyl)-1,4-pentadien-3-one	Breast cancer, colon cancer and lung cancer, inflammation diseases, Alzheimer's disease, cytotoxicity, diabetes	62–65 and 95–98

insulin resistance, glucose metabolism and α-glucosidase that alleviate the blood glucose level in the body.[24,71–73] The disease associations for respective quadrant molecules and their gene associations are given in Table 2.5.

2.5 Conclusions

Turmeric has received the attention of the scientific community stemming from the multiple health benefits attributed to it by traditional medicine systems such as Ayurveda. Much of this pharmacological research has been with purified or synthetic curcumin and has shown promise in enzyme-based model systems, *in vitro* assays and animal studies, but has not led to any conclusive evidence in human clinicals. In fact, curcumin has now been labelled as a PAINS (pan assay interfering compounds)[15] in some scientific literature. The inability to link pure single compounds to human health benefits is not limited to curcumin, and establishing such a link has been a challenge for most other natural molecules. Even the efficacy of vitamins outside of their natural food systems, such as in supplements, is being questioned. While the research based on the single molecule curcumin has not led to any effective clinical evidence, the potential beneficial effects on human health of whole turmeric powders and extracts cannot be ruled out.

The beneficial effects of turmeric in traditional medicine systems are achieved through dietary consumption, which is at low levels over long periods of time or through medicinal formulations in which turmeric extract is paired with other botanicals and undergoes processing steps that select for a group of its phytochemicals based on their separation properties. The complex milieu of phytochemicals in such extracts used in formulations and foods may not seem effective when tested at these concentrations in the standard biological *in vitro* assays as they escape detection and blur or confound our understanding of the mechanism of action.[15] Instead, their therapeutic potential may be supported by the emerging metabolic feedback hypothesis which states that bioactivity, especially of many foodborne phytochemicals, can act *via* weak negative biological feedback mechanisms on multiple targets.

To study turmeric extracts and formulations in a metabolic feedback framework, consideration of holistic approaches needs to be taken. These considerations involve taking into account the complexity of chemical characterization and standardization of the fractions introduced by whole extracts and coextracts as well as their PK/PD complexity. The visualization of phytochemistry of turmeric in terms of its phytochemical properties that govern partitioning such as volatility and water solubility can give new insights, as can network "ingredient pairing" analysis of its formulations. Testing such compound extracts in assays in a poly-pharmacology and synergy framework is a potential direction for future research on turmeric. Multi-omics, network pharmacology, free-living epidemiological studies with tracking devices and machine learning statistical tools are making the holistic scientific framework possible. Collectively, recognition of these factors and holistic methodologies may guide the development of novel approaches that are scientifically rigorous and unravel the complexity and potential health benefits of turmeric and its formulations.

Finally, while much of the research is focussed on creating an IP (intellectual property)-protected block-buster based on phytochemicals of turmeric, its use as a culinary ingredient as an individual spice or in a mix of botanicals as a functional food should not be forgotten. The insights from holistic scientific approaches can give leads to specialized post-harvest processing techniques, extracts and formulations targeting food and non-food moments. They may lead to the development of functional foods and recommendations for their effective home cooking and industrial process and product design that could make "our food our medicine".

Acknowledgements

The authors wish to acknowledge the support of the Rural India Support Trust for funding the research on turmeric.

References

1. P. K. Mukherjee, S. Rai, V. Kumar, K. Mukherjee, P. J. Hylands and R. C. Hider, *Expert Opin. Drug Discovery*, 2007, **2**, 633.
2. S. Prasad and B. B. Aggarwal, *Herbal Medicine: Biomolecular and Clinical Aspects*, 2nd edn, 2011.
3. J. Sharifi-Rad, Y. E. Rayess, A. A. Rizk, C. Sadaka, R. Zgheib, W. Zam, S. Sestito, S. Rapposelli, K. Neffe-Skocińska, D. Zielińska, B. Salehi, W. N. Setzer, N. S. Dosoky, Y. Taheri, M. El Beyrouthy, M. Martorell, E. A. Ostrander, H. A. R. Suleria, W. C. Cho, A. Maroyi and N. Martins, *Front. Pharmacol.*, 2020, **11**, 01021.
4. Agricultural Market Intelligence Centre, PJTSAU, 2021, p. 1.
5. S. Prasad, S. C. Gupta, A. K. Tyagi and B. B. Aggarwal, *Biotechnol. Adv.*, 2014, **32**, 1053.
6. S. Li, W. Yuan, G. Deng, P. Wang, P. Yang and B. Aggarwal, *Pharm. Crop.*, 2011, **2**, 28.
7. K. Mansouri, S. Rasoulpoor, A. Daneshkhah, S. Abolfathi, N. Salari, M. Mohammadi, S. Rasoulpoor and S. Shabani, *BMC Cancer*, 2020, **20**, 791.
8. S. Fazel Nabavi, R. Thiagarajan, L. Rastrelli, M. Daglia, E. Sobarzo-Sanchez, H. Alinezhad and S. Mohammad Nabavi, *Curr. Top. Med. Chem.*, 2015, **15**, 2445.
9. P. Wal, N. Saraswat, R. S. Pal, A. Wal and M. Chaubey, *Open Med. J.*, 2019, **6**, 64.
10. S. Zorofchian Moghadamtousi, H. Abdul Kadir, P. Hassandarvish, H. Tajik, S. Abubakar and K. Zandi, *BioMed Res. Int.*, 2014, **2014**, 186864.
11. U. Kolpe, V. Ramaswamy, B. S. Rao and M. Nagabhushan, *Int. Congr. Ser.*, 2002, **1245**, 327.
12. R. S. Ahmad, M. B. Hussain, M. T. Sultan, M. S. Arshad, M. Waheed, M. A. Shariati, S. Plygun and M. H. Hashempur, *Evidence-Based Integr. Med.*, 2020, **2020**, 7656919.
13. G. P. Eckert, C. Schiborr, S. Hagl, R. Abdel-Kader, W. E. Müller, G. Rimbach and J. Frank, *Neurochem. Int.*, 2013, **62**, 595.

14. M. C. Fadus, C. Lau, J. Bikhchandani and H. T. Lynch, *J. Tradit. Complement. Med.*, 2017, **7**, 339.
15. K. M. Nelson, J. L. Dahlin, J. Bisson, J. Graham, G. F. Pauli and M. A. Walters, *J. Med. Chem.*, 2017, **60**, 1620.
16. D. G. S. Pandey, The Bhavprakash nighantu with elaborated Hindi commentary by Padmashri prof. K.C. Chunekar, 2010.
17. S. Chuengsamarn, S. Rattanamongkolgul, R. Luechapudiporn, C. Phisalaphong and S. Jirawatnotai, *Diabetes Care*, 2012, **35**, 2121.
18. T. Nishiyama, T. Mae, H. Kishida, M. Tsukagawa, Y. Mimaki, M. Kuroda, Y. Sashida, K. Takahashi, T. Kawada and K. Nakagawa, *J. Agric. Food Chem.*, 2005, **53**, 959.
19. Z. Ghorbani, A. Hekmatdoost and P. Mirmiran, *Int. J. Endocrinol. Metab.*, 2014, **12**, e18081.
20. C. Kang and E. Kim, *Food Chem. Toxicol.*, 2010, **48**, 2366.
21. D.-W. Zhang, M. Fu, S.-H. Gao and J.-L. Liu, *Evid. Based Complement. Alternat. Med.*, 2013, **2013**, 636053.
22. Z. Adab, S. Eghtesadi, M. R. Vafa, I. Heydari, A. Shojaii, H. Haqqani, T. Arablou and M. Eghtesadi, *Phytother. Res.*, 2019, **33**, 1173.
23. M. A. El-Moselhy, A. Taye, S. S. Sharkawi, S. F. I. El-Sisi and A. F. Ahmed, *Food Chem. Toxicol.*, 2011, **49**, 1129.
24. P. Lekshmi, R. Arimboor, P. Indulekha and A. Nirmala Menon, *Int. J. Food Sci. Nutr.*, 2012, **63**, 832.
25. J. Wickenberg, S. L. Ingemansson and J. Hlebowicz, *Nutr. J.*, 2010, **9**, 43.
26. R. N. Thota, C. B. Dias, K. A. Abbott, S. H. Acharya and M. L. Garg, *Sci. Rep.*, 2018, **8**, 13679.
27. H. U. Tahir, R. A. Sarfraz, A. Ashraf and S. Adil, *Int. J. Food Prop.*, 2016, **19**, 2156.
28. Z. Zdrojewicz, M. Szyca, E. Popowicz, T. Michalik and B. Śmieszniak, *Polski merkuriusz lekarski: organ Polskiego Towarzystwa Lekarskiego*, 2017, **42**, 227–230.
29. D. J. Hunter, D. Schofield and E. Callander, *Nat. Rev. Rheumatol.*, 2014, **10**, 437.
30. T. Vos, R. M. Barber and B. Bell, *Lancet*, 2015, **386**, 743.
31. J. L. Funk, J. N. Oyarzo, J. B. Frye, G. Chen, R. C. Lantz, S. D. Jolad, A. M. Sólyom and B. N. Timmermann, *J. Nat. Prod.*, 2006, **69**, 351.
32. G. Ramadan, M. A. Al-Kahtani and W. M. El-Sayed, *Inflammation*, 2011, **34**, 291.
33. J. L. Funk, J. B. Frye, J. N. Oyarzo, H. Zhang and B. N. Timmermann, *J. Agric. Food Chem.*, 2010, **58**, 842.
34. C. B. Magalhães, D. R. Riva, L. J. DePaula, A. Brando-Lima, V. L. Koatz, J. H. Leal-Cardoso, W. A. Zin and D. S. Faffe, *J. Appl. Physiol.*, 2010, **108**, 845.
35. T. Koh, Y. Murakami, S. Tanaka, M. Machino and H. Sakagami, *In Vivo*, 2013, **27**, 269.
36. R. A. T. Longvah, K. Bhaskarachary and K. Venkaiah, *Indian Food Composition Tables 2017*, National Institute of Nutrition, Hyderabad, 2017.

37. D. Chin, P. Huebbe, J. Frank, G. Rimbach and K. Pallauf, *Redox Biol.*, 2014, **2**, 563.
38. M. H. Mughal and I. Haq, *Integr. Food Nutr. Metab.*, 2019, **6**, DOI: 10.15761/IFNM.1000254.
39. S. A. Dutta, *Sushruta Samhita*, Chaukhambha Sanskrit Sansthan, 2006, p. 290.
40. K. Govindadas, in *Bhaisajya Ratnavali*, Chaukhambha Prakashan, Varanasi, 2017, pp. 100, 134.
41. T. J. Smith and B. H. Ashar, *Cureus*, 2019, **11**, e3858.
42. N. G. Abraham and A. Kappas, *Pharmacol. Rev.*, 2008, **60**, 79.
43. R. L. Thangapazham, A. Sharma and R. K. Maheshwari, *Adv. Exp. Med. Biol.*, 2007, **595**, 343.
44. V. B. Liju, K. Jeena and R. Kuttan, *Asian Pac. J. Cancer Prev.*, 2014, **15**, 6575.
45. M. Carrion-Gutierrez, A. Ramirez-Bosca, V. Navarro-Lopez, A. Martinez-Andres, M. Asín-Llorca, A. Bernd and J. F. Horga de la Parte, *Eur. J. Dermatol.*, 2015, **25**, 240.
46. S. Kim, P. A. Thiessen, E. E. Bolton, J. Chen, G. Fu, A. Gindulyte, L. Han, J. He, S. He and B. A. Shoemaker, *Nucleic Acids Res.*, 2016, **44**, D1202.
47. Y. Djoumbou Feunang, R. Eisner, C. Knox, L. Chepelev, J. Hastings, G. Owen, E. Fahy, C. Steinbeck, S. Subramanian, E. Bolton, R. Greiner and D. S. Wishart, *J. Cheminf.*, 2016, **8**, 61.
48. T. W. Backman, Y. Cao and T. Girke, *Nucleic Acids Res.*, 2011, **39**, W486.
49. H. Zhou, C. S. Beevers and S. Huang, *Curr. Drug Targets*, 2011, **12**, 332.
50. M. Markovic, S. Ben-Shabat, A. Aponick, E. M. Zimmermann and A. Dahan, *Int. J. Mol. Sci.*, 2020, **21**, 3248.
51. M. Dei Cas and R. Ghidoni, *Nutrients*, 2019, **11**, 2147.
52. S. Prasad, A. K. Tyagi and B. B. Aggarwal, *Cancer Res. Treat.*, 2014, **46**, 2.
53. P. Anand, A. B. Kunnumakkara, R. A. Newman and B. B. Aggarwal, *Mol. Pharm.*, 2007, **4**, 807.
54. C. Ireson, S. Orr, D. J. Jones, R. Verschoyle, C. K. Lim, J. L. Luo, L. Howells, S. Plummer, R. Jukes, M. Williams, W. P. Steward and A. Gescher, *Cancer Res.*, 2001, **61**, 1058.
55. A. M. Orellana-Paucar, T. Afrikanova, J. Thomas, Y. K. Aibuldinov, W. Dehaen, P. A. de Witte and C. V. Esguerra, *PLoS One*, 2013, **8**, e81634.
56. S. J. Hewlings and D. S. Kalman, *Foods*, 2017, **6**, 92.
57. S. Prasad, A. K. Tyagi and B. B. Aggarwal, *Cancer Res. Treat.*, 2014, **46**, 2.
58. V. Sharma, B. Nehru, A. Munshi and A. Jyothy, *Methods Find. Exp. Clin. Pharmacol.*, 2010, **32**, 227.
59. H. Mirzaei, A. Shakeri, B. Rashidi, A. Jalili, Z. Banikazemi and A. Sahebkar, *Biomed. Pharmacother.*, 2017, **85**, 102.
60. P. Bonetti, F. F. de Moraes, G. M. Zanin and R. de Cássia Bergamasco, *Polym. Bull.*, 2016, **73**, 279.
61. H. E. Pence and A. Williams, *J. Chem. Educ.*, 2010, **87**, 1123.
62. N. Aravindan, R. Madhusoodhanan, S. Ahmad, D. Johnson and T. S. Herman, *Cancer Biol. Ther.*, 2008, **7**, 569.

63. T. Ahmed and A.-H. Gilani, *Brain Res.*, 2011, **1400**, 1.
64. J. R. Cashman, S. Ghirmai, K. J. Abel and M. Fiala, *BMC Neurosci.*, 2008, **9**, 1.
65. V. Soetikno, F. R. Sari, P. T. Veeraveedu, R. A. Thandavarayan, M. Harima, V. Sukumaran, A. P. Lakshmanan, K. Suzuki, H. Kawachi and K. Watanabe, *Nutr. Metab.*, 2011, **8**, 1.
66. M. Li, Y. Jiang, W. Jing, B. Sun, C. Miao and L. Ren, *Can. J. Physiol. Pharmacol.*, 2013, **91**, 951.
67. W. J. Oh, M. Endale, S.-C. Park, J. Y. Cho and M. H. Rhee, *Evidence-Based Complementary Altern. Med.*, 2012, **2012**, 485262.
68. Y.-Z. Hou, J. Yang, G.-R. Zhao and Y.-J. Yuan, *Eur. J. Pharmacol.*, 2004, **499**, 85.
69. G. Kaur, M. Athar and M. S. Alam, *Mol. Carcinog.*, 2010, **49**, 290.
70. M. Huo, X. Cui, J. Xue, G. Chi, R. Gao, X. Deng, S. Guan, J. Wei, L. W. Soromou and H. Feng, *J. Surg. Res.*, 2013, **180**, e47.
71. F. Moloney, S. Toomey, E. Noone, A. Nugent, B. Allan, C. E. Loscher and H. M. Roche, *Diabetes*, 2007, **56**, 574.
72. M. P. Hernández-Cáceres, L. Toledo-Valenzuela, F. Díaz-Castro, Y. Ávalos, P. Burgos, C. Narro, D. Peña-Oyarzun, J. Espinoza-Caicedo, F. Cifuentes-Araneda and F. Navarro-Aguad, *Front. Endocrinol.*, 2019, **10**, 176.
73. D. A. Youssef, H. M. El-Fayoumi and M. F. Mahmoud, *Biomed. Pharmacother.*, 2019, **110**, 145.
74. *e-NIGHANTU (Collection of Āyurvedic* Lexicons*)*, 2020, https://niimh.nic.in/ebooks/e-Nighantu/.
75. K. C. Chunekar, in *Bhavaprakash Nighantu*, ChaukhambaBharati Academy, Varanasi, 2013, p. 111.
76. P. T. Aryadasa Kumara Singha, in *Mahaushadhanighantu*, Chaukhambha Bharati Acadamy, 2006, p. 55.
77. Shaligram, in *Shaligrama Nighantu*, Khemaraja Shrikrishnadas Prakashana, Mumbai, 2011, p. 157.
78. P. Sharma, in *Shodhala Nighantu*, Oriental Institute, Baroda, 1978, pp. 12, 103.
79. B. K. Diwedi, in *Dhanvantarinighantu*, Chaukhamba Krishnadas Academy, Varanasi, 2008, p. 25.
80. H. Trpathi, in *Madanpala Nighantu*, Chaukhamba Krishnadas Academy, Varanasi, 2009, p. 47.
81. P. V. Sharma, in *Kaideva Nighantu*, Chaukhamba Orientalia, Varanasi, 2006, p. 205.
82. R. K. Sharam and V. B. Dash, *Caraka Samhita*, Chowkhamba Sanskrit Series Ofice, Varanasi, 2007.
83. P. V. Sharma, in *Hridayadipaka Nighantu*, Chaukhamba Amarbharti Prakashan, Varanasi, 1977, p. 2.
84. J. A. Duke, 1929- and S. M. Beckstrom-Sternberg, 1994.
85. W. Sun, S. Wang, W. Zhao, C. Wu, S. Guo, H. Gao, H. Tao, J. Lu, Y. Wang and X. Chen, *Crit. Rev. Food Sci. Nutr.*, 2017, **57**, 1451.

86. *Herbal Medicine: Biomolecular and Clinical Aspects*, ed. I. F. F. Benzie and S. Wachtel-Galor, CRC Press/Taylor & Francis, Boca Raton (FL), 2nd edn, 2011.
87. R. R. Kotha and D. L. Luthria, *Molecules*, 2019, **24**, 2930.
88. N. Dosoky and W. Setzer, *Nutrients*, 2018, **10**, 1196.
89. X. Ma and D. R. Gang, *J. Agric. Food Chem.*, 2006, **54**, 9573.
90. D. K. Gounder and J. Lingamallu, *Ind. Crops Prod.*, 2012, **38**, 124.
91. D. Yadav, S. K. Yadav, R. K. Khar, M. Mujeeb and M. Akhtar, *Int. J. Green Pharm.*, 2013, 7, 85.
92. M. Jankasem, M. Wuthi-Udomlert and W. Gritsanapan, *Int. Scholarly Res. Not.*, 2013, **2013**, 250597.
93. I.-C. Chao, C.-M. Wang, S.-P. Li, L.-G. Lin, W.-C. Ye and Q.-W. Zhang, *Molecules*, 2018, **23**, 1568.
94. B. B. Aggarwal, W. Yuan, S. Li and S. C. Gupta, *Mol. Nutr. Food Res.*, 2013, **57**, 1529.
95. S. Gunaseelan, A. Balupillai, K. Govindasamy, K. Ramasamy, G. Muthusamy, M. Shanmugam, R. Thangaiyan, B. M. Robert, R. Prasad Nagarajan and V. K. Ponniresan, *PLoS One*, 2017, **12**, e0176699.
96. F. Khan, K. Niaz, F. Maqbool, F. Ismail Hassan, M. Abdollahi, K. C. Nagulapalli Venkata, S. M. Nabavi and A. Bishayee, *Nutrients*, 2016, **8**, 529.
97. G. G. Yue, B. C. Chan, P.-M. Hon, M. Y. Lee, K.-P. Fung, P.-C. Leung and C. B. Lau, *Food Chem. Toxicol.*, 2010, **48**, 2011.
98. A. Sgarbossa, D. Giacomazza and M. Di Carlo, *Nutrients*, 2015, 7, 5764.

CHAPTER 3

Ginger in Ayurvedic Preparations and Its Applications in the Development of Nutraceuticals, Dietary Supplements and Functional Foods and Their Beneficial Effects in Various Biological Activities

AUGUSTINE AMALRAJ,* KARTHIK VARMA A. C., SHINTU JUDE AND SASIKUMAR KUTTAPPAN

R&D Centre, Aurea Biolabs (P) Ltd, Kolenchery, Cochin-682 311, Kerala, India
*Emails: augustin14amal@gmail.com; amalraj.a@plantlipids.com

3.1 Ayurveda

The traditional medicine system is defined as the collected healing knowledge of generations;[1] it is still utilized by nearly 75–80% of the world's population for their principal health maintenance.[2,3] The World Health Organization (WHO) defines traditional medicine as "the summation completeness of the knowledge, practices and skills created on the concepts, philosophies and familiarities indigenous to various philosophies, whether

Chemistry, Biological Activities and Therapeutic Applications of Medicinal Plants in Ayurveda
Edited by Augustine Amalraj, Sasikumar Kuttappan and Karthik Varma A. C.

Published by the Royal Society of Chemistry, www.rsc.org

justifiable or not, utilized in the preservation of healthiness in addition to the inhibition, identification, perfection or management of physical and mental disease".[4] Each culture has its set of ethno-medical beliefs and practices connected with illness and health, which shape diagnosis, treatment and anticipated results.[5] India has numerous classifications of traditional medicine, which have been used for centuries. This comprises Ayurveda, Yoga, Unani, Siddha and Homeopathy (AYUSH). Encouraging these arrangements is the chief aim of the Ministry of AYUSH (Government of India). Ayurveda is an Indian therapeutic arrangement, created around the epistemology of three energies (doshas): vata is the energy of movement, pitta is the energy of digestion or metabolism and kapha is the energy of lubrication and structure. The origin of illness in Ayurveda is noted as a deficiency of appropriate cellular utility owing to an excess or shortage of vata, pitta or kappa. Sickness can also be caused by the existence of toxins. Steadiness in structure is ideal and the natural direction; imbalance causes illness. Health is order, diseases are disorder. Ayurvedic beneficial methodologies comprise meditative practices, phytomedicine, diet, physical management and the environment.[5–7] The ancient Indian system of Ayurveda has been used in India since about 1500 BCE, with the chief intention of this system being the preservation of normal health and avoiding unhealthy conditions. The current elevated problem regarding Ayurveda and alternative medicine is increasing reports of adverse responses associated with herbal medicine; this may be because of a growth in the number of people consuming herbal products both as a nutritional supplement and/or as a medicine.[8]

3.2 Siddha

Siddha is one of the key traditional medical arrangements of India that is mainly experienced in Tamil Nadu and its peripheries;[9] it is also experienced in some other countries, particularly Sri Lanka, which has a large Tamil population.[3] Traditional medicine in Tamil Nadu was earlier known as Thamzhil Maruthuvam (Tamil medicine) and termed as Siddha medicine, which was created after 1923 by the British.[10] Evidence has revealed that from around the 3rd century BCE onwards the Tamils had established their own literature and had oceanic relations with distant of the world.[11] The names of some literature belong to the Sangam period (300 BCE–300 AD), such as Sirupanchamoolam, Thrikadugam and Elathi, which are some therapeutic preparations that are still practiced. Similarly, the 95th section of Thirukural (a Tamil classic) explains the philosophies of medicine and food.[12] Purananuru (a Tamil classic) remarks regarding the antiaging influence of *Phyllanthus emblica*.[13] The literatures used in contemporary Siddha practices started from the 4th to 5th centuries AD and becomes predominant in 16th century, however, the element of healing practices emerged earlier.[14,15] Along with the Siddha principle, the human body is made up of 96 principles (thattuvam). Among these, three humors, namely vata ($\approx$air), pitta ($\approx$fire) and kappa ($\approx$water), are signified as vital principles for health. Any imbalance between these results in illness.[3]

As per a WHO report, up to 80% of the inhabitants in some developing countries practice traditional medicine. Traditional herbal medicine remains a significant section of healthcare in India.[15] Even in the USA, the use of plants and phytomedicines has grown intensely over the last two decades. A National Center for Complementary and Alternative Medicine has been established in the USA. Herbal products have been categorized under "dietary supplements" and are included with minerals, vitamins, amino acids and other products that are added to the diet.

The use of plants as a therapeutic medication is an essential part of South African ethnic life. It is estimated that 27 million South Africans use herbal medicines from in excess of 1020 plant species. Actually, there are numerous curative plants around the world, including India, which are utilized by traditional medicines for the inhibition and cure of several illnesses.[16] Among the different medicinal plants, ginger (*Zingiber officinale* Roscoe) has a long past of being utilized as a herbal medicine since ancient times and it has been utilized in folk medicine from Southeast Asia, China, India, Taiwan, Japan, Australia, Africa, Brazil, Mexico, Greco-Roman traditions, parts of the United States and Middle East, where ginger rhizomes are cultivated for medicinal purposes.

3.3 Ginger and Its Components

Ginger, the rhizome of *Zingiber officinale* Roscoe, has a lengthy history of therapeutic practice dating back 2500 years,[16] and it is utilized as a spice all over the world.[17] It is a member of the Zingiberaceae plant family and is native to Asia, but is nowadays cultivated in Africa, India and other tropical regions.[18] Ginger is a plant with thick roots and vertical, upright stems. This perennial plant has nodal rhizomes with several distributed swells. The rhizome of ginger grows horizontally and its outside skin can be brown or yellow. The inner membrane of the rhizome is yellowish brown and it has several vessel elements and cells, comprising oleoresin. Ginger has an attractive odor and spicy taste, with the dried powder looking a yellowish white to yellowish brown.[19]

Ginger is an extensively utilized food constituent and has been regularly recommended as a therapeutic for various symptoms, such as nausea, common cold, cough, asthma, bleeding and muscle pain in traditional medicine.[20,21] Moreover, it has been applied in several diseases, such as metabolic dysfunction, emesis, cancer, bone disorders and vascular disorders, with clinical evidence.[21–26] More than 400 different components have been recognized in ginger.[27] The amount and quantity of active components of ginger depend on the area of farming and the processing technique utilized and whether it is in fresh or dried form.[19]

3.4 Chemical Constituents of Ginger

The ginger rhizome comprises 2–3% volatile oil, 3–6% fatty oil, 8% ash, 9% protein, 3–8% crude fiber and 60–70% carbohydrates. The distinctive flavor of ginger is due to gingerols, shogaols, zingerone and volatile (essential) oils that include up to 3% of ginger on a fresh weight basis. The volatile fragrant essential oil of ginger comprises chiefly sesquiterpenoids, with α-zingiberene (30–70%) as the core component, and lesser quantities of other sesquiterpenoids, including β-sesquiphellandrene (15–20%), β-bisabolene (10–15%) and α-farnesene, and monoterpenoids (β-phellandrene, camphene, cineol, geraniol, citral, *etc.*). Ginger also comprises diterpenes and ginger glycolipids.[28] The pungency of the fresh rhizome is the reason why the group of phenolic compounds is known as gingerols, which are mixture of an arrangement of structural analogs, particularly 6-gingerol, 8-gingerol and 10-gingerol. Among these, 6-gingerol is the most abundant and principal pungent constituent. The pungency of dried or cooked ginger is a result of nonvolatile phenylpropanoid-derived compounds from gingerols, namely, shogaols, which are mostly 6-shogaol and 8-shogaol. The less pungent zingerone is also formed from gingerols during the drying procedure and has a spicy-sweet aroma. Ginger also comprises acrid resinous substances (5–8%).[29] Other compounds found in ginger include 3-dihydroshogaols, paradols, dihydroparadols, acetylated gingerol derivates, gingerdiols, mono- and diacetyl gingerdiol derivates, 1-dehydrogingerdiols, diaryl heptanoids, epoxide diaryl heptanoids, methyl ether derivates, ferulic acid derivates and terpenes such as zingerones and zingerols.[19,27,28] The chemical structures of important bioactive components of ginger are shown in Figure 3.1.

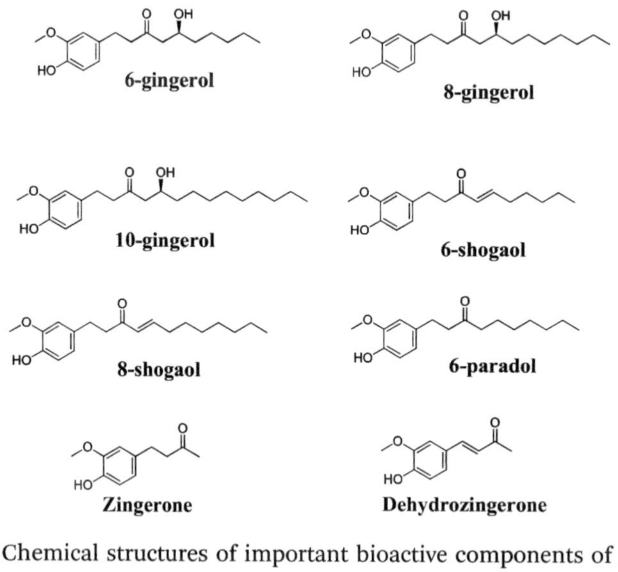

Figure 3.1 Chemical structures of important bioactive components of ginger.

3.5 Classical Ayurvedic Formulations with Ginger and Their Uses

Ayurveda pharmacology and therapeutics offer various sophisticated preparations, which have been utilized for hundreds of years. More than 100 preparations are very prevalent and are utilized even at the community levels as over-the-counter products, with ginger being one of the important constituents. Some of these drugs continue to be utilized for preventive and primary health care by Indian families.[30] There are a few classic mixtures utilized as generic drugs, which are described in Table 3.1. Trikatu is a mixture of three acrid constituents, which are *Piper nigrum* known as black pepper, *Piper longum* known as long pepper and *Zingiber officinale* known as ginger. This mixture is utilized in more than 1500 Ayurveda preparations. Trikatu is utilized as a medication of choice in several diseases of Vata and Kapha. It increases appetite, is a digestive and a fat metabolizer, and is active in gastrointestinal, respiratory and liver diseases, particularly in obstructive jaundice. A good deal of work has been done on Trikatu as a bioavailability enhancer, because it facilitates the absorption of drugs from the gastrointestinal tract, and protects them from being metabolized in the liver.[31] *Ferula asafoetida* L., *Zingiber officinale* and *Glycyrrhiza glabra* are utilized together in traditional treatment as "Chitrakadivati" for the management of flatulence, gut microflora and indigestion.[32]

Divya Pidantaka Kvatha consists of *Cyperus rotundus*, *Nyctanthes arbor-tristis*, *Piper chaba*, *Piper longum*, *Pluchea lanceolata*, *Ricinus communis*, *Trachyspermum ammi*, *Vitex negundo*, *Withania somnifera* and *Zingiber officinale*. It is beneficial in joint pain, sciatica, rheumatoid arthritis, osteoarthritis, gout, muscular and skeletal pains and edema. Divya Churna includes *Foeniculum vulgare*, *Ipomoea nil*, *Rosa centifolia*, *Terminalia chebula* and *Zingiber officinale*. It remedies abdominal pain, flatulence, heaviness and nausea. Divya Peya (Herbal Tea) contains *Adhatoda vasia*, *Bacopa monnieri*, *Boerhaavia diffusa*, *Cinnamomum zeylanicum*, *Convolvulus pluricaulis*, *Cymbopogon martini*, *Cyperus rotundus*, *Elettaria cardamomum*, *Ephedra gerardiana*, *Foeniculum vulgare*, *Myristica fragrans*, *Nelumbo nucifera*, *Ocimum sanctum*, *Phyllanthus niruri*, *Piper chaba*, *Piper longum*, *Piper nigrum*, *Plumbago zeylanica*, *Pterocarpus santalinus*, *Rosa centifolia*, *Santalum album*, *Syzygium aromaticum*, *Terminalia arjuna*, *Tinospora cordifolia*, *Viola odorata*, *Withania somnifera* and *Zingiber officinale*. It controls cholesterol and guards against heart diseases, and also stimulates immunity and promotes digestion.

Divya Pidantaka Taila includes *Aconitum ferox*, *Acorus calamus*, *Aegle marmelos*, *Allium sativum*, *Anethum sowa*, *Asparagus racemosus*, *Berberis aristata*, *Butea mono-sperma*, *Calotropis procera*, *Celastrus paniculatus*, *Cinnamomum zeylanicum*, *Clerodendrum phlomidis*, *Curcuma longa*, *Datura metel*, *Eclipta alba*, *Foeniculum vulgare*, *Glycyrrhiza glabra*, *Gmelina arborea*, *Habenaria intermedia*, *Inula racemosa*, *Lilium polyphyllum*, *Malaxis acuminate*, *Mesua ferrea*, *Nardostachys jatamansi*, *Oroxylum indicum*, *Paderia foetida*, *Piper chaba*, *Piper longum*, *Plantago lanceolata*, *Plumbago zeylanica*, *Polygonatum*

Table 3.1 Various Ayurvedic formulations with ginger (*Zingiber officinale*) and their uses.[31–35]

Formulation	Plants used	Uses
Trikatu	*Piper nigrum*, *Piper longum* and *Zingiber officinale*	Various diseases of Vata and Kapha
Chitrakadivati	*Ferula asafoetida* L., *Zingiber officinale* and *Glycyrrhiza glabra*	Flatulence, gut microflora and indigestion
Divya Pidantaka Kvatha	*Cyperus rotundus*, *Nyctanthes arbortristis*, *Piper chaba*, *Piper longum*, *Pluchea lanceolata*, *Ricinus communis*, *Trachyspermum ammi*, *Vitex negundo*, *Withania somnifera* and *Zingiber officinale*	Joint pain, sciatica, rheumatoid arthritis, osteoarthritis, gout, muscular and skeletal pain and edema
Divya Churna	*Foeniculum vulgare*, *Ipomoea nil*, *Rosa centifolia*, *Terminalia chebula* and *Zingiber officinale*	Abdominal pain, flatulence, heaviness and nausea
Divya Peya (herbal tea)	*Adhatoda vasia*, *Bacopa monnieri*, *Boerhaavia diffusa*, *Cinnamomum zeylanicum*, *Convolvulus pluricaulis*, *Cymbopogon martini*, *Cyperus rotundus*, *Elettaria cardamomum*, *Ephedra gerardiana*, *Foeniculum vulgare*, *Myristica fragrans*, *Nelumbo nucifera*, *Ocimum sanctum*, *Phyllanthus niruri*, *Piper chaba*, *Piper longum*, *Piper nigrum*, *Plumbago zeylanica*, *Pterocarpus santalinus*, *Rosa centifolia*, *Santalum album*, *Syzygium aromaticum*, *Terminalia arjuna*, *Tinospora cordifolia*, *Viola odorata*, *Withania somnifera* and *Zingiber officinale*	Controls cholesterol and guards against heart diseases; stimulates immunity and promotes digestion
Divya Pidantaka Taila	*Aconitum ferox*, *Acorus calamus*, *Aegle marmelos*, *Allium sativum*, *Anethum sowa*, *Asparagus racemosus*, *Berberis aristata*, *Butea monosperma*, *Calotropis procera*, *Celastrus paniculatus*, *Cinnamomum zeylanicum*, *Clerodendrum phlomidis*, *Curcuma longa*, *Datura metel*, *Eclipta alba*, *Foeniculum vulgare*, *Glycyrrhiza glabra*, *Gmelina arborea*, *Habenaria intermedia*, *Inula racemosa*, *Lilium polyphyllum*, *Malaxis acuminate*, *Mesua ferrea*, *Nardostachys jatamansi*, *Oroxylum indicum*, *Paderia foetida*, *Piper chaba*, *Piper longum*,	Relieves pain in the lumbar region, knee joints, cervical spondylitis, edema and inflammation

	Plantago lanceolata, *Plumbago zeylanica*, *Polygonatum verticillatum*, *Pseudarthria viscida*, *Ricinus communis*, *Roscoea alpina*, *Rubia cordifolia*, *Sesamum indicum*, *Sesamum indicum*, *Stereospermum suaveolen*, *Strychnos nuxvomica*, *Trachyspermum ammi*, *Tribulus terrestris*, *Uraria lagopoides*, *Valeriana wallichii*, *Vitex negundo* and *Zingiber officinale*	
Divya Svasari Rasa	*Anacyclus pyrethrum*, *Capparis moonii*, *Cinnamomum zeylanicum*, *Glycyrrhiza glabra*, *Piper longum*, *Piper nigrum*, *Pistacia integerrima*, *Syzygium aromaticum* and *Zingiber officinale*	Bronchitis, cough, coryza, cold, asthma and sinusitis
Divya Vatari Churna	*Moringa oleifera*, *Picrorhiza kurroa*, *Trigonella foenum-graecum*, *Withania somnifera* and *Zingiber officinale*	Rheumatoid arthritis, sciatica, pain in the back and lumbar region
Himalaya Abana	*Acorus calamus*, *Asparagus racemosus*, *Boerhaavia diffusa*, *Carum copticum*, *Celastrus paniculatus*, *Centella asiatica*, *Cinnamomum cassia*, *Commiphora wightii*, *Convolvulus pluricaulis*, *Crocus sativus*, *Cyperus rotundus*, *Eclipta alba*, *Elettaria cardamomum*, *Embelia ribes*, *Emblica officinalis*, *Foeniculum vulgare*, *Glycyrrhiza glabra*, *Nardostachys jatamansi*, *Nepeta hindostana*, *Ocimum sanctum*, *Piper longum*, *Rosa damascena*, *Rosa centifolia*, *Santalum album*, *Syzygium aromaticum*, *Terminalia arjuna*, *Terminalia chebula*, *Tinospora cordifolia*, *Withania somnifera* and *Zingiber officinale*	Hyperlipidemia and hypertension, adjuvant in angina therapy
Himalaya Mentat	*Acorus calamus*, *Bacopa monnieri*, *Celastrus paniculatus*, *Centella asiatica*, *Embelia ribes*, *Elettaria cardamomum*, *Emblica officinalis*, *Evolvulus alsinoides*, *Foeniculum vulgare*, *Ipomoea digitata*, *Mucuna pruriens*, *Myristica fragrans*, *Nardostachys jatamansi*, *Orchis mascula*, *Oroxylum indicum*, *Prunus amygdalus*, *Syzygium aromaticum*, *Terminalia arjuna*, *Terminalia bellirica*, *Terminalia chebula*, *Tinospora cordifolia*, *Valeriana officinalis*, *Withania somnifera* and *Zingiber officinale*	Parkinson's and Alzheimer's diseases

Table 3.1 (*Continued*)

Formulation	Plants used	Uses
Rumalaya gel	*Boswellia serrata*, *Cedrus deodara*, *Cinnamomum zeylanicum*, *Gaultheria fragrantissima*, *Mentha arvensis*, *Pinus roxburghii*, *Vitex negundo* and *Zingiber officinale*	Pain and joint mobility
Aswagandharishtam	*Withania somnifera*, *Chlorophytum tuberosum*, *Rubia cordifolia*, *Terminalia chebula*, *Curcuma longa*, *Berberis aristata*, *Glycyrrhiza glabra*, *Pluchea lanceolate*, *Pueraria tuberosa*, *Terminalia arjuna*, *Cyperus rotundus*, *Ipomoea turpethum*, *Hemidesmus indicus*, *Cryptolepis buchanani*, *Santalum album*, *Pterocarpus santalinus*, *Acorus calamus*, *Plumbago zeylanica*, *Woodfordia fruticose*, *Zingiber officinale*, *Piper nigrum*, *Piper longum*, *Cinnamomum zeylanicum*, *Elettaria cardamomum*, *Cinnamomum tamala*, *Callicarpa macrophylla* and *Mesua ferrea*	Nervous system and digestive system
Sarasvata ghrita	*Terminalia chebula*, *Zingiber officinale*, *Piper nigrum*, *Cissampelos pareira*, *Acorus calamus*, *Piper longum* and *Moringa pterygosperma*	Memory booster

verticillatum, *Pseudarthria viscida*, *Ricinus communis*, *Roscoea alpina*, *Rubia cordifolia*, *Sesamum indicum*, *Sesamum indicum*, *Stereospermum suaveolen*, *Strychnos nuxvomica*, *Trachyspermum ammi*, *Tribulus terrestris*, *Uraria lagopoides*, *Valeriana wallichii*, *Vitex negundo* and *Zingiber officinale*. It relieves pain in the lumbar region, knee joints, cervical spondylitis, edema and inflammation.

Divya Svasari Rasa contains *Anacyclus pyrethrum*, *Capparis moonii*, *Cinnamomum zeylanicum*, *Glycyrrhiza glabra*, *Piper longum*, *Piper nigrum*, *Pistacia integerrima*, *Syzygium aromaticum* and *Zingiber officinale*. It is used in the management of bronchitis, cough, coryza, cold, asthma and sinusitis. Divya Vatari Churna includes *Moringa oleifera*, *Picrorhiza kurroa*, *Trigonella foenum-graecum*, *Withania somnifera* and *Zingiber officinale*. It remedies rheumatoid arthritis, sciatica, and pain in the back and lumbar region.

Himalaya Abana contains *Acorus calamus*, *Asparagus racemosus*, *Boerhaavia diffusa*, *Carum copticum*, *Celastrus paniculatus*, *Centella asiatica*, *Cinnamomum cassia*, *Commiphora wightii*, *Convolvulus pluricaulis*, *Crocus sativus*, *Cyperus rotundus*, *Eclipta alba*, *Elettaria cardamomum*, *Embelia ribes*, *Emblica officinalis*, *Foeniculum vulgare*, *Glycyrrhiza glabra*, *Nardostachys jatamansi*, *Nepeta hindostana*, *Ocimum sanctum*, *Piper longum*, *Rosa damascena*, *Rosa centifolia*, *Santalum album*, *Syzygium aromaticum*, *Terminalia arjuna*, *Terminalia chebula*, *Tinospora cordifolia*, *Withania somnifera* and *Zingiber officinale*. It is used to manage hyperlipidemia and hypertension, as it is an adjuvant in angina therapy. Himalaya Mentat includes *Acorus calamus*, *Bacopa monnieri*, *Celastrus paniculatus*, *Centella asiatica*, *Embelia ribes*, *Elettaria cardamomum*, *Emblica officinalis*, *Evolvulus alsinoides*, *Foeniculum vulgare*, *Ipomoea digitata*, *Mucuna pruriens*, *Myristica fragrans*, *Nardostachys jatamansi*, *Orchis mascula*, *Oroxylum indicum*, *Prunus amygdalus*, *Syzygium aromaticum*, *Terminalia arjuna*, *Terminalia bellirica*, *Terminalia chebula*, *Tinospora cordifolia*, *Valeriana officinalis*, *Withania somnifera* and *Zingiber officinale*. Its adjuvants are used in Parkinson's and Alzheimer's diseases. Rumalaya gel contains *Boswellia serrata*, *Cedrus deodara*, *Cinnamomum zeylanicum*, *Gaultheria fragrantissima*, *Mentha arvensis*, *Pinus roxburghii*, *Vitex negundo* and *Zingiber officinale*. It relieves pain and enhances joint mobility.[33]

Aswagandharishtam is another traditional medicine for illnesses affecting the nervous and digestive systems that is organized around a number of plants and plant parts, and it also contains *Zingiber officinale* as one of the main ingredients. The Ayurvedic preparation contains *Withania somnifera*, *Chlorophytum tuberosum*, *Rubia cordifolia*, *Terminalia chebula*, *Curcuma longa*, *Berberis aristata*, *Glycyrrhiza glabra*, *Pluchea lanceolate*, *Pueraria tuberosa*, *Terminalia arjuna*, *Cyperus rotundus*, *Ipomoea turpethum*, *Hemidesmus indicus*, *Cryptolepis buchanani*, *Santalum album*, *Pterocarpus santalinus*, *Acorus calamus*, *Plumbago zeylanica*, *Woodfordia fruticose*, *Zingiber officinale*, *Piper nigrum*, *Piper longum*, *Cinnamomum zeylanicum*, *Elettaria cardamomum*, *Cinnamomum tamala*, *Callicarpa macrophylla* and *Mesua ferrea*.[34] Sarasvata ghrita is a polyherbal preparation described in Ayurveda and utilized as a memory booster.[35] The preparation includes *Terminalia chebula*, *Zingiber officinale*, *Piper nigrum*, *Cissampelos pareira*, *Acorus calamus*, *Piper longum* and *Moringa pterygosperma*, which all have a memory-boosting activity,[36–41] and the main ingredients of which are chebulinic acid,[42] 6-shogaol,[43] piperine,[44] bebeerine,[45] β-asarone,[46] piperine[47] and carotene.[48]

An Ayurvedic decoction comprising a combination of 26 bioactive components from four selected plants, *viz.* *Zingiber officinale*, *Glycyrrhiza glabra*, *Piper nigrum* and *Ocimum basilicum*, has been shown to be effective in the management of respiratory ailments.[49] Recently, AYUSH has recommended utilizing Kadha (herbal tea/decoction) constituting *Ocimum basilicum*, *Cinnamomum zeylanicum*, *Piper nigrum*, *Zingiber officinale* and *Vitis vinifera* and golden milk to enhance immunity as a prophylactic measure against COVID-19. Because of the multipart configuration of several phytoconstituents from these medicinal plants, the blend of this mixture could modify numerous proteins and enhance immunity, which can be described through a network pharmacology and gene-set enhancement investigation. The network pharmacology and gene set enrichment investigation are well-accepted tactics to recognize a disease target, leading to successive molecules and modulated pathways *via* "multiple component–protein interactions".[50–52] Based on the Ayurvedic and scientific literature, the Ministry of AYUSH, India, delivered an advisory where it recommended the utilization of Kadha for self-care, to improve immunity against severe infection caused by COVID-19. Further, it also suggested the consumption of golden milk, with a half teaspoon of *Curcuma longa* powder in 150 mL hot milk to be taken once or twice a day.[53,54]

3.6 Ayurvedic Herbal Formulations with Ginger for Different Health Issues

Ginger is recommended in traditional medicine for high blood pressure, cardiopathy, palpitations and as a vasodilator to improve the circulation. The decoction of ginger rhizome is extensively utilized in Ayurvedic medicine. As per the explanation of the Ayurvedic Pharmacopoeia of India,[55] the ginger decoction was prepared using 30 g of powdered rhizomes mixed with 1000 mL of boiling water, permitting the volume of water to reach 200 mL. The decoction was cooled at room temperature, then filtered and lyophilized. The effects of the decoction of ginger rhizome were considered on cardiovascular parameters and vascular and intestinal smooth muscle. The ginger decoction displayed weak negative inotropic and chronotropic intrinsic activities with a substantial intrinsic action on smooth muscle with the effectiveness on the ileum greater than that on the aorta.[56] Ginger is a wonderful common spice utilized in Ayurvedic medicine to improve digestion and inhibit nausea. These properties for assisting bowel movements and relaxing the muscles regulate the digestive system. Ginger is considered an adjuvant in various Ayurvedic formulations in which it boosts absorption and inhibits gastrointestinal side effects.[57,58] Some significant Ayurvedic herbal formulations include ginger as one of the essential components and are developed by the percentage dry weight of the various plants for several health problems.

To rectify asthma bronchitis: mixed powder (4 g) of *Solanum xanthocarpum* (whole plant) 25%, *Piper longum* (fruits) 10%, *Adhatoda vasica* (leaves) 25%, *Zingiber officinale* (roots) 10%, *Curcuma zedoaria* (roots) 10%, *Ocimum sanctum* (leaves) 10% and *Phyllanthus emblica* (fruits) should be given to the patient, twice a day (morning and night before sleep) with water.[58,59] To manage blood circulation: powdered mixer (4 g) of *Zingiber officinale* (roots) 20%, *Piper longum* (roots) 10%, *Withania somnifera* (roots) 10%, *Phyllanthus emblica* (fruits) 10%, *Curcuma longa* (roots) 10%, *Terminalia bellirica* (fruits) 10%, *Terminalia chebula* (fruits) 10%, *Ocimum sanctum* (leaves) 10% and *Tephrosia purpurea* (leaves) 10% should be given to the patient, twice daily with water.[58,59]

To cure chronic fever: mixed powder (4 g) of *Tinospora cordifolia* (stems) 15%, *Ocimum sanctum* (leaves) 15%, *Adhatoda vasica* (leaves) 15%, *Azadirachta indica* (leaves) 15%, *Holarrhena antidysenterica* (bark) 10%, *Piper longum* (fruits) 10%, *Zingiber officinale* (roots) 10% and *Terminalia bellirica* (fruits) 10% is given to the patient, twice daily before meals with water.[58,59] To cure cough: mixed powder (3 g) of *Phyllanthus emblica* (fruits) 25%, *Adhatoda vasica* (leaves) 20%, *Ocimum sanctum* (leaves) 10%, *Piper longum* (fruits) 10%, *Zingiber officinale* (roots) 10%, *Glycyrrhiza glabra* (roots) 15% and *Solanum xanthocarpum* (whole plant) 10% should be given to the patient twice daily (morning and at night before going to bed) mixed with lukewarm honey.[58,59]

To manage dysentery and diarrhea: mixed powder (3 g) of *Holarrhena antidysenterica* (bark) 25%, *Aegle marmelos* (fruits) 25%, *Zingiber officinale* (roots) 10%, *Terminalia chebula* (fruits) 10%, *Cyperus rotundus* (roots) 10%, *Syzygium cumini* (seeds) 10% and *Phyllanthus emblica* (fruits) 10% should be given to the patient, three times a day, with curd.[58,59] To cure gastritis: mixed powder (4 g) of *Zingiber officinale* (roots) 10%, *Piper longum* (fruits) 10%, *Mentha piperita* (leaves) 10%, *Terminalia chebula* (fruits) 15%, *Terminalia bellirica* (fruits) 15%, *Phyllanthus emblica* (fruits) 15%, *Plumbago zeylanica* (roots) 10% and *Tinospora cordifolia* (stems) 15% should be given to the patient twice daily, half an hour before meals with water.[58,59]

To rectify high blood pressure: mixed powder (4 g) of *Terminalia arjuna* (bark) 35%, *Terminalia chebula* (fruits) 15%, *Asparagus racemosus* (roots) 15%, *Zingiber officinale* (roots) 10% and *Withania somnifera* (roots) 25% should be given to the patient, twice a day (morning and night) with honey.[58,59] To manage indigestion (lack of appetite): mixed powder (4 g) of *Zingiber officinale* (roots) 10%, *Piper longum* (fruits) 10%, *Phyllanthus emblica* (fruits) 30%, *Terminalia chebula* (fruits) 15%, *Tinospora cordifolia* (stems) 15%, *Cassia angustifolia* (leaves) 10% and *Mentha piperita* (leaves) 10% should be given to the patient, twice a day after meals with water.[58,59]

To rectify male sterility: mixed powder (3 g) of *Withania somnifera* (roots) 15%, *Mucuna pruriens* (seeds) 25%, *Tribulus terrestris* (fruits) 20%, *Glycyrrhiza glabra* (roots) 10%, *Terminalia arjuna* (bark) 10%, *Phyllanthus emblica* (fruits) 10%, *Zingiber officinale* (roots) 5% and *Piper longum* (fruits) 5% should be given to the patient, twice a day with honey.[58,59] To manage obesity: mixed powder (4 g) of *Terminalia chebula* (fruits) 15%, *Terminalia bellirica* (fruits) 15%, *Phyllanthus emblica* (fruits) 10%, *Crateva nurvala* (bark) 25%, *Tribulus terrestris* (fruits) 25% and *Zingiber officinale* (roots) 10% should be given to the patient, twice a day with warm water.[58,59]

To manage paralysis: mixed powder (3 g) of *Curcuma zedoaria* (roots) 20%, *Withania somnifera* (roots) 20%, *Tribulus terrestris* (fruits) 20%, *Zingiber officinale* (roots) 20%, *Piper longum* (fruits) 5%, *Crataeva nurvala* (leaves) 10% and *Plumbago zeylanica* (roots) 5% should be given to the patient, three times a day with honey.[58,59] To rectify prostate enlargement: mixed powder (4 g) of *Tinospora cordifolia* (stems) 15%, *Tribulus terrestris* (fruits) 15%, *Phyllanthus emblica* (fruits) 15%, *Zingiber officinale* (roots) 10%, *Butea monosperma* (seeds) 10%, *Adhatoda vasica* (leaves) 5%, *Terminalia chebula* (fruits) 10%, *Terminalia bellirica* (fruits) 10% and *Glycyrrhiza glabra* (roots) 10% should be given to the patient twice a day, morning and evening, before meals with water.[58,59]

To cure thyroid problems: mixed powder (3 g) of *Crataeva nurvala* (bark) 20%, *Bauhinia variegata* (bark) 20%, *Sida cordifolia* (leaves) 15%, *Terminalia chebula* (fruits) 10%, *Terminalia bellirica* (fruits) 10%, *Glycyrrhiza glabra* (roots) 15% and *Zingiber officinale* (roots) 10% should be given to the patient, twice daily with lukewarm water.[58,59] To manage urinary tract-related issues: mixed powder (4 g) of *Tribulus terrestris* (fruits) 25%, *Zingiber officinale* (roots) 10%, *Solanum xanthocarpum* (whole plant) 10%, *Crataeva nurvala* (bark) 25%, *Tinospora cordifolia* (stems) 10%, *Asparagus racemosus* (roots) 10% and *Tephrosia purpurea* (leaves) 10% should be given to the patient, twice a day with water.[58,59]

Remarkably, various Ayurvedic ginger-based formulations have the ability to treat many chronic diseases such as chronic fever, cough, gastritis, indigestion, dysentery and diarrhea, blood circulation, blood pressure, male sterility, obesity, asthma bronchitis, thyroid problems, paralysis, urinary tract-related issues, *etc.* Unfortunately, owing to the absence of scientific authentication in several conceptions, this treasurable gift from our ancestors is underused. Moreover, the WHO records that "inappropriate utilization of traditional medicines or practices can have negative or dangerous effects" and that additional investigations are required to determine their efficacy and safety. Indeed, there are multidimensional problems associated with the prevalence of traditional medicine, a field contaminated by fraudulent practitioners, adulterated medications, pseudoscience, unsafe supply chains, toxic components and alternative medical approaches vaguely and improperly abstracted from their original use and formulations.[5] Hence, evidence-based research is urgently required for worldwide appreciation and acceptance of Ayurvedic-based formulations, which requires additional improvements in the investigative methods. Therefore, Ayurvedic scientists make should efforts to improve the fundamental capability of Ayurveda without compromising its fundamental principles.[60]

3.7 Evidence-based Formulations in Various Nutraceuticals and Functional Foods

Recently, various studies have reported evidence-based research in different ginger formulations with Ayurvedic applications (Figure 3.2). A polyherbal mixture for prediabetes comprised aqueous extracts of stem of *Tinospora cordifolia*, bark of *Pterocarpus marsupium*, leaves of *Gymnema sylvestre*, rhizome of *Zingiber officinale* and fruit of *Momordica charantia* in equal amounts. The polyherbal formulation was realized to be an effective and safe method for the management of prediabetes and decreased progress to type 2 diabetes mellitus by altering blood glucose levels, serum insulin and lipid profiles without any adverse effects. Ayurveda parameters like Agni, Aahar and Nidra, in short Dincharya, also play a very significant role in the progress from prediabetes to diabetes.[61] The utilization of an ancient, time-tested design for nasal cleansing could perform an essential role in mitigating the contagiousness of SARS-CoV-2 by preventing the multiplication and shedding of the virus in the nasal epithelium. An *in silico* study recommended that the phytochemical components in *Citrus medica* and *Zingiber officinale* may have real potential in decreasing the load of virus and shedding of SARS-CoV-2 in the nasal routes.[62] A study was conducted to improve the nutraceutical properties of ginger extract by using supersaturable self-emulsifying drug-delivery systems. After oral administration of ginger extract in rats, the relative bioavailability of bioactive components such as 6-gingerol and 8-gingerol in the supersaturable self-emulsifying drug-delivery system treated group was threefold higher than the group treated with ginger extract alone. Therefore, the utilization of a supersaturable self-emulsifying drug-delivery system could boost the dissolution and pharmacokinetic properties of active components in ginger extract,

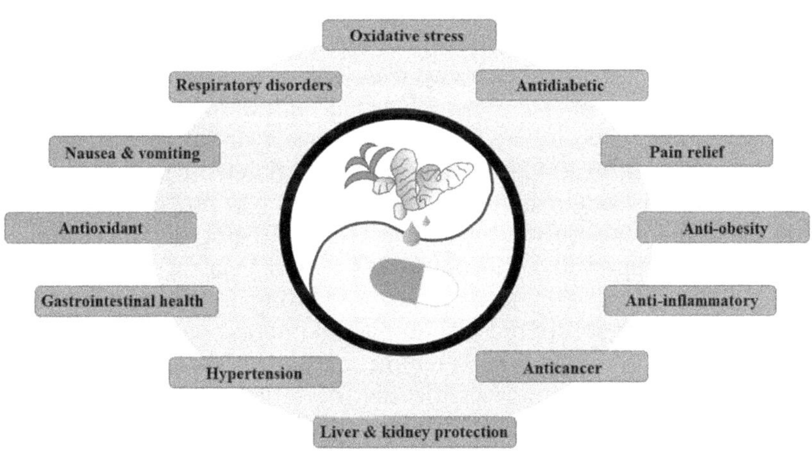

Figure 3.2 Various biological activities of Ayurvedic-based ginger formulations.

contributing to effective hepatoprotective function and other nutraceutical effects.[63]

A superfine ginger food powder was developed using high-energy ball milling utensils, which modifies the internal cohesion, crystal structure and interplane distance of the nanocrystalline food powder. This superfine ginger powder may be appropriate in the pharmaceutical, health and medicine sectors for research and development of innovative composites, or functional materials by reason of its physical behavior and effective antioxidant property.[64] A healthy fruit drink was formulated from a combination of ginger, lemon, orange and honey that was found to be highly nutritive and rich in carbohydrates.[65] Ginger essential oil encapsulated onto polyvinyl alcohol/gum arabic/chitosan composite film and chitosan/gum arabic/polyethylene glycol composite films were developed by the solvent casting method as promising wound-healing and food-packaging materials.[66,67] A powder form of microencapsulated ginger extract was prepared using maltodextrin and/or gum arabic, which could be readily encapsulated into food products, particularly tea and bread.[68]

Dietary supplementation with turmeric and ginger restricted cholinergic and purinergic neurotransmission in the cerebral cortex of hypertensive rats by adaptation of reducing adenosine triphosphate and adenosine monophosphate hydrolysis, and adenosine deaminase and acetylcholinesterase activities in the synaptic cleft of the cerebral cortex. These outcomes recommended that both rhizomes apply a neuroprotective effect under a hypertensive state.[69] Likewise, dietary supplementation with turmeric and ginger rhizome was also related to the renovation of sperm mobility, testosterone level, systolic blood pressure and enhancement of antioxidant status in the epididymides and testes of L-nitro-arginine methyl ester (L-NAME)-induced hypertensive rats. This study recommended that both rhizomes could be connected as functional foods to inhibit hypertension-mediated male reproductive dysfunction.[70] Turmeric and ginger rhizomes modify platelet ectonucleotidase activities in L-NAME hypertensive rats by effecting a rise in adenosine diphosphate and adenosine monophosphate hydrolysis and inhibiting adenosine deaminase activity, thus leading to an enhancement in the level of adenosine, an effective vasodilator and inhibitor of platelet accumulation, which could be utilized as nutraceuticals or functional foods for managing or preventing hypertension related to platelet hyperactivity.[71]

A ginger extract-loaded solid dispersion system was prepared by a freeze-drying method with improved oral absorption of 6-gingerol and 8-gingerol resulting in an improved antihypothermic action in a rat model of hypothermia induced by ethanol treatment.[72] Ginger extracts containing 6-paradol and 6-shogaol suppressed human telomerase reverse transcriptase (hTERT) expression in addition to telomerase activity in A549 cells without any adverse effects. The ginger extract has an anticlastogenic effect against diethylnitrosamine-induced liver micronucleus development in rats, which proposed that the ginger extract could potentially be beneficial in dietary cancer prevention.[73]

Lei *et al.*[74] investigated the effect of dietary gingerol and shogaol-improved ginger extract on plasma total cholesterol and gene expression of intestinal Niemann–Pick C1-like protein 1 (NPC1L1), acyl CoA:cholesterol acyltransferase 1 (ACAT2), microsomal triacylglycerol transport protein (MTP) and ATP-binding cassette transporters 5 and 8 (ABCG5/8) along with hepatic sterol regulatory element-binding protein 2 (SREBP2), liver X receptor (LXRα), 3-hydroxy-3-methylglutaryl-CoA reductase (HMG-CoA-R), low-density lipoprotein receptor and cholesterol-7α-hydroxylase (CYP7A1) in hamsters fed a high-cholesterol diet. The gingerol and shogaol-enriched ginger extract reduced the plasma total cholesterol concentration through a combination of two mechanisms, prevention of cholesterol absorption and improvement of bile acid excretion, leading to considerable decreases in the apparent cholesterol absorption, which was complemented by upregulation of hepatic CYP7A1 and downregulation of the genes of intestinal NPC1L1, ACAT2 and MTP.[74] A novel bioavailable gingerol formulation (Ginactiv) was developed by polar–nonpolar sandwich (PNS) technology with the full spectrum of the ginger matrix, with an improved and sustained release profile of bioactive components and high antioxidant potential. The presence of the synergistic effects of the full spectrum of ginger matrix in Ginactiv specifies its usage as a natural nutritional supplement to maintain gastrointestinal, digestive and gut health and to aid the gut–brain axis.[75]

3.8 Biological Activities of Ginger in Different Uses

3.8.1 Anti-inflammatory Effect

Dysfunction of endothelial cells (ECs) is essential for atherosclerosis progression. The potential of ginger extract containing 6-gingerol and 6-shogaol has been assessed to reverse EC dysfunction and its mechanism of action was examined utilizing cultured human ECs incubated with tumor necrosis factor alpha (TNF-α). Ginger extract reduced monocyte chemoattractant protein-1, vascular cell adhesion molecule-1 and monocyte adhesion, which was connected with the reduction of ninjurin-1 expression; a decrease of TNF-α receptor 1 and the receptor for advanced glycation end-products expression (RAGE), in common with the enhancement of soluble RAGE, reduction of nicotinamide adenine dinucleotide phosphate (NADPH) oxidase subunits expression, stimulation of antioxidant nuclear factor erythroid 2-related factor 2 (Nrf2) and heme oxygenase-1, and inhibition of nuclear factor kappa-light-chain-enhancer of activated B cells (NF-κB). Ginger extract might be a successful substitute to ameliorate sicknesses in which oxidative stress and inflammation are essential.[76] Ho *et al.*[77] demonstrated that ginger and gingerol-associated components, particularly shogaols, are extremely positive nutritional adjuvants for neurodegenerative diseases, because of their effective antineuroinflammatory capability by preventing the creation of nitric oxide, interleukin 1 beta (IL-1β), interleukin 6 (IL-6) and TNF-α in addition to their messenger ribonucleic acid (mRNA) levels in lipopolysaccharide-activated BV2 microglia and blocking NF-κB activation, which was the essential mechanism for preventing proinflammatory gene expression.[77]

The anti-inflammatory activity of ginger is mediated by preventing neutrophil and macrophage activation in addition to reducing leukocyte and monocyte migration. This was demonstrated through the dose-dependent reduction in pro-inflammatory cytokines and chemokines and replacement of the total antioxidant capacity.[78] The 6-gingerol-rich portion of ginger presented a chemo-protective role on chlorpyrifos-induced toxicity, chlorpyrifos-induced oxidative damage, inflammation and apoptosis in the brain, ovary and uterus of rat tissues owing to its antioxidative, anti-inflammatory and antiapoptotic properties.[79] Another bioactive compound of ginger, 6-shogaol, reduced clinical skin severity scores, transepidermal water loss, and epidermal thickness induced by 2,4-dinitrochlorobenzene in mice, and inhibited TNF-α + interferon gamma (IFN-γ)-induced inflammatory mediators in human keratinocytes. Therefore, 6-shogaol could be a suitable candidate for improving skin inflammation associated with allergic dermatitis.[80] The anti-inflammatory effects of ginger and its bioactive compound, zingerone, were evaluated on lipopolysaccharide-induced systemic inflammation through NF-κB bioluminescent imaging. The ginger and zingerone suppressed NF-κB drive luminescence in a number of organs owing to their wide range of anti-inflammatory actions in most organs that reduced the activation of NF-κB, the production of IL-1β and the infiltration of inflammatory cells.[81]

3.8.2 Protective Effect on Liver Health

Piroxicam is a nonsteroidal anti-inflammatory drug extensively utilized in rheumatic diseases. Piroxicam hepatotoxicity has been described as one of its main adverse side effects. Administration of ginger reduced and alleviated serum aspartate aminotransferase (AST), alanine aminotransferase (ALT) and alkaline phosphatase (ALP), and immunoexpression of the proapoptotic protein (Bax), induced by piroxicam. It enhanced immunoexpression of the antiapoptotic protein (Bcl2). It also ameliorated the morphological alterations induced by piroxicam. Accordingly, ginger has defensive effects against piroxicam hepatotoxicity by decreasing serum marker enzymes, liver fibrosis and apoptosis in male mice.[82] The relationship between chronic alcohol consumption and the progress of alcoholic liver disease is an observed occurrence, however the specific core molecular mediators involved in ethanol-induced liver sickness remain elusive. In one study, the rescue effect of ginger extract was measured in improved hepatic fatty acid translocase/cluster of differentiation 36 (FAT/CD36), protein tyrosine phosphatase 1B (PTP1B) and decreased hepatocyte nuclear factor 4 alpha (HNF4A) expression leading to dyslipidemia linked with ethanol-induced liver dysfunction. Chronic ethanol exposure can improve hepatic FAT/CD36, and PTP1B expression, whereas it can reduce HNF4A gene expressions. The alteration of enzyme expression related to lipid metabolism resulted in hepatic lipid metabolism disruption as manifested by increased plasma low-density lipoproteins, cholesterol, triglyceride, as well as AST and ALT in the ethanol group. Consumption of ginger extract along with ethanol, partially ameliorated gene expression modification and histological changes, enhanced the undesirable lipid profile and liver enzyme modifications compared to those in the ethanol group, which revealed that the ginger extract as an antioxidant and anti-inflammatory agent.[83] Aflatoxin B1 (AFB1) is one of the main mycotoxin contaminants in feed and food, causing hepatotoxicity and oxidative stress. Pretreatment with ginger extract protected hepatoma G2 (HepG2) cells from AFB1-induced cytotoxicity through the inhibition of reactive oxygen species generation, deoxyribonucleic acid (DNA) damage and cell death. Likewise, experiments with a rat model further revealed the protective effects of ginger extract against AFB1-induced hepatotoxicity by improving antioxidant enzyme levels and upregulation of the Nrf2/heme oxygenase-1 (HO-1) pathway, which could be because of the synergistic effects of various phenolic compounds present in it. This study suggested that ginger extract might be considered as a potential natural agent for the inhibition of AFB1-induced oxidative stress and hepatotoxicity.[84]

3.8.3 Protective Effect on Kidney Health

Gabr *et al.*[85] assessed the defensive and antioxidant effects of ginger extract against cadmium-induced renal toxicity in animal models and the utilization of ginger as an anti-renal failure natural remedy. Ginger considerably protected renal cell dysfunction, reduced the severity of tubular cells and DNA destruction affected by cadmium intoxication by antioxidant and free radical scavenging activity. Accordingly, ginger extracts could have an effective protective effect against nephrotoxicity induced by several toxicants.[85] Cyclosporine, a common immune-suppressant drug that is utilized in organ transplants, is linked with nephrotoxic effects. Ginger polyphenols decreased some indices of experimental cyclosporine-induced kidney damage in a Wistar rat model by antioxidant and membrane-stabilizing effects in addition to variation of transport proteins. Components of ginger polyphenol extract may have a potential application as future adjuvant usages in renal and other transplant therapies.[86]

3.8.4 Gastrointestinal Health

Necrotizing enterocolitis (NEC) is a most significant gastrointestinal emergency, particularly affecting preterm infants, and causes severe morbidities and mortality. Oxidative stress, inflammation, apoptosis along with prematurity are included in the pathogenesis of this illness. A study evaluated the advantageous effects of ginger on the intestinal injury in an investigational rat model of NEC. The levels of TNF-α, IL-1β and IL-6 in the ginger-treated group were considerably reduced. The glutathione peroxidase and superoxide dismutase levels of the ginger-treated group were considerably conserved in the NEC + ginger group. The tissue xanthine oxidase, malondialdehyde (MDA) and myeloperoxidase levels of the NEC + ginger group were considerably lower than those in the NEC group. Ginger treatment efficiently ameliorated the severity of intestinal damage in NEC and may be a hopeful management choice.[87]

3.8.5 Protective Effect on Ulcerative Colitis

Colitis is the typical pathological lesion of inflammatory bowel diseases. Ginger was more effective as a healing than a prophylactic agent in acetic acid-induced colitis. Ginger ameliorated acetic acid-induced colitis not only through its anti-inflammatory and antioxidant properties but also *via* a decrease in serotonin, which may contribute to the downregulation of NF-κB-dependent TNF-α expression and a decrease in the lipid peroxidation and tissue damage.[88] Pretreatment with ginger essential oil ameliorated the indications of acetic acid-induced colitis in rats because of its antioxidant and antilipoperoxidative actions, which illustrated that ginger may be advantageous in the treatment of colon mucosal injury caused either experimentally or through IBD.[89] Ginger extract and its compound zingerone improved 2,4,6-trinitrobenzene sulfonic acid-induced colitis in mice through modulation of NF-κB activity and the IL-1β signaling pathway.[90]

3.8.6 Pain Relief Effect

Eleven successive days of nutritional supplementation with 2 g of raw and heat-treated ginger reduced muscle pain produced by eccentric exercise, and raw ginger is as effective as heat-treated ginger in attaining this outcome.[91] The effect of ginger powder supplementation on some inflammatory markers was investigated in patients suffering from knee osteoarthritis in a double-blind randomized, placebo-controlled clinical trial with a follow-up period of 3 months that was conducted in 120 outpatients with moderately painful knee osteoarthritis. After 3 months of supplementation, the serum concentration of nitric oxide and hs-C reactive protein decreased in the ginger group. Ginger powder supplementation at a dose of 1 g per day can attenuate inflammatory markers in patients with knee osteoarthritis, and it thus can be suggested as an appropriate supplement for knee osteoarthritis patients.[92]

3.8.7 Anticancer Effect

Cancer-related mortality and morbidity are a major healthcare concern. Developing effective anticancer remedies which are sustainable, affordable and nontoxic is possible with alternative medicine. An aqueous natural extract mixture was prepared from common spices, including turmeric, ginger and garlic, and its apoptotic and anti-proliferative properties examined against both ER-positive and ER-negative breast cancer cell lines (MCF 7, ZR-75 and MDA-MB 231) with or without tamoxifen, which is a selective estrogen receptor modulator utilized in the management of breast cancers. Enhanced effects of tamoxifen when mixed with the natural extracts indicated that the ginger-containing natural extract could be a potential substitute healing tactic in some kinds of cancer.[93]

3.8.8 Management of Chronic Diseases

A large cross-sectional study was conducted to investigate the occurrence of chronic diseases such as hypertension, diabetes, hyperlipidemia, coronary heart disease, fatty liver, cerebrovascular disease, tumor and anemia, along with daily ginger intake. The outcomes indicated that ginger had a potentially protective effect against some chronic diseases, particularly coronary heart disease and hypertension, in addition to its capability to decrease the possibility of sickness by enhancing 1 g per day ginger consumption on the basis of the prior diet. It was advised as a wise dietary decision to suitably boost daily ginger consumption as it was a heart-healthy alternative.[94]

3.8.9 Management of Nausea and Vomiting

One of the most common postoperative complications is nausea and vomiting. Due to numerous morbidities linked with postoperative nausea and vomiting, its inhibition and management have been a matter of urgency for physicians, and so various medications have been suggested for this purpose. In one study, 100 patients with cholelithiasis who has laparoscopic cholecystectomy were the subjects. The patients were separated into two groups: group A included 50 patients who received 500 mg oral ginger 1 h before surgery, and group B comprised 50 patients who received 4 mg intravenous ondansetron before completion of surgery. The nausea severity was considerably lower in the ginger group, which suggested an intake of 500 mg oral ginger 1 h before laparoscopic cholecystectomy, which is tolerable and safe.[95] A randomized controlled trial was conducted to examine the effect of ginger essence on post-nephrectomy nausea and vomiting with 120 nephrectomy patients and they were randomly allotted to either the treatment or control group. Inhaling ginger essence has a positive outcome on postoperative nausea and vomiting. Consequently, ginger essence can be utilized as an active antiemetic agent for treating postoperative nausea and vomiting.[96]

3.8.10 Antidiabetic Effect

The daily intake of 1000 mg ginger decreased the serum fasting glucose, which is a cause of dyslipidemia, peritoneal membrane fibrosis, hyperinsulinemia and cardiovascular disease, in peritoneal dialysis patients.[97]

Daily intake of 3 g of ginger in capsules for 8 weeks by patients with type 2 diabetes leads to depressed fasting blood sugar and hemoglobin A1c (HbA1c), in addition to differences in insulin resistance, fasting insulin, increasing of sensitivity to insulin and quantitative insulin sensitivity check index (QUICKI). Consequently, intake of ginger powder supplements is suitable for diabetic patients.[98]

3.8.11 Neuroprotective and Memory-enhancing Effect

The influence of ginger has been examined on oxidative stress markers in the mitochondrial segments of the cerebellum, cerebral cortex, hypothalamus and hippocampus of diabetic rats. Diabetes exacerbates neuronal damage induced by hyperglycemia-mediated oxidative damage. Decreased activities of antioxidant enzymes in diabetic rats were improved on oral intake of ginger and ginger management exhausted the MDA level, which was improved in the diabetic rats, which recommended that ginger displays a neuroprotective effect by accelerating brain antioxidant defense mechanisms and downregulating the MDA levels to the usual levels in diabetic rats. Accordingly, ginger may be utilized as a healing agent in inhibiting difficulties in diabetic patients.[99] The memory-enhancing effects of dried ginger extract were assessed in a model of scopolamine-induced memory deficits and in normal animals through a novel object recognition test. Ginger extract induced the upregulation of nerve growth factor and enhanced the levels of rabbit monoclonal antiphosphorylated extracellular-signal-regulated kinase (pERK) antibody and rabbit polyclonal antiphosphorylated cyclic adenosine monophosphate response element-binging protein (pCREB) antibody in the hippocampus. Through nerve growth factor-induced activation of extracellular-signal-regulated kinase (ERK)/cyclic adenosine monophosphate response element-binding protein (CREB) signaling, ginger extract may improve synaptogenesis and memory. Taken together, the cognitive improvement of ginger might, in part, be linked with stimulation of nerve growth factor expression and ERK/CREB-mediated synapse development in the hippocampus. Ginger extract could be a promising drug candidate for memory disorders and neurodegenerative diseases.[100]

3.8.12 Protective Effect on Respiratory Disorders

The effect of ginger extract on delayed gastric emptying, developing ventilator-associated pneumonia and clinical outcomes in adult respiratory distress syndrome were evaluated with 32 adult respiratory distress syndrome patients who were in need of mechanical ventilation and fed *via* a nasogastric tube. Gastric feed supplementation with ginger extract reduced and delayed gastric emptying and maintained a reduced incidence of ventilator-associated pneumonia in adult respiratory distress syndrome.[101] The effect of hot water extracts of fresh and dried gingers on human respiratory syncytial virus was verified by plaque reduction assay in both human upper (HEp-2) and lower (A549) respiratory tract cell lines. The capacity of ginger to stimulate antiviral cytokines was assessed by enzyme-linked immunosorbent assay (ELISA). Fresh ginger prevented human respiratory syncytial virus-induced plaque formation in both HEp-2 and A549 cell lines and could motivate mucosal cells to secrete IFN-β that probably participated in countering viral infection. Therefore, fresh ginger is active against human respiratory syncytial virus-induced plaque development on the airway epithelium by hindering viral attachment and internalization.[102] Shariatpanahi *et al.*[103] assessed the impacts of an enteral diet improved with ginger extract on inflammatory factors and respiratory profile, and the results in patients with acute respiratory distress syndrome. Patients fed an enteral diet improved with ginger had considerably lower serum levels of IL-1, IL-6 and TNF-α and higher levels of red blood cell glutathione with considerable enhancement in oxygenation being detected as compared with the control group. The enteral diet supplemented with ginger in patients with acute respiratory distress syndrome may be favorable for gas exchange and could reduce the period of mechanical ventilation and length of stay in the intensive care unit.[103] One study provided an indication that ginger prevents allergen-induced lung inflammation by decreasing airway eosinophilia, T helper type 2 cytokines and allergen-specific antibodies, consequently assisting its anti-inflammatory role during the allergic response in the lung. Ginger is consequently a hopeful novel protective agent to combat allergies.[104]

3.8.13 Effect on Oxidative Stress and Hypertension

Oxidative stress has a significant role in the progression and development of diabetic nephropathy. Ginger extract considerably decreased oxidative stress, apoptosis and inflammation, and improved antioxidant defenses in the diabetic kidney, which highly recommends ginger rhizome extract for use in a defensive role against diabetes-induced renal damage by ameliorating oxidative stress, apoptosis and inflammation.[105] Dietary supplementation of ginger and turmeric rhizomes prevented angiotensin-1 converting enzyme and arginase activities in addition to enhancing nitric oxide production in L-NAME-induced hypertensive rats due to their antihypertensive benefits in traditional medicine.[106] Hydroethanolic extract of ginger has anticonvulsant effects in a model of clonic seizures induced with pentylenetetrazole in male mice, probably *via* an interaction through inhibitory and excitatory systems, oxidative stress, calcium channel inhibition and antioxidant mechanisms.[107]

3.8.14 Antiobesity Effect

Mitochondrial dysfunction and dyslipidemia are related to obesity-associated metabolic disorders. The effect of ginger extract was investigated on muscle mitochondrial biogenesis and high-density lipoprotein-cholesterol (HDL-C) metabolism in rats fed a high-fat diet. Supplementation with ginger extract decreased the final body weight and epididymal adipose tissue mass without disturbing energy ingestion. Ginger extract improved mitochondrial size and mitochondrial DNA content in addition to the expression of important genes linked to mitochondrial biogenesis, comprising peroxisome proliferator-activated receptor γ coactivator-1α, nuclear respiratory factor-1 and transcription factor A in skeletal muscle. Furthermore, ginger extract raised serum HDL-C along with upregulating adenosine triphosphate binding cassette transporter A1, apolipoprotein A-1 and lecithincholesterol acyltransferase mRNA in liver. These outcomes recommend that ginger extract may have an advantageous effect against obesity-related metabolic disorders by elevating muscle mitochondrial biogenesis and serum HDL-C levels.[108] A study has demonstrated that dietary ginger extract reduced diet-induced obesity and enhanced endurance capacity by stimulating fat usage through improved peroxisome proliferator-activated receptor δ (PPARδ) signaling. 6-Shogaol and 6-gingerol in the ginger extract may serve as PPARδ agonists with consequent effects on energy homeostasis, which may be significant for the additional development of PPARδ-based remedies for obesity treatment.[109]

3.9 Conclusion

Traditional Indian medicine (Siddha and Ayurveda) is a holistic methodology for health and personalized medicine based on a knowledge of life. Plant species described in the ancient scripts of Siddha and Ayurveda and other Indian medicine structures could be explored with the current scientific methods for improved healthcare roles. Several drugs have arrived in the global market through the exploration of traditional medicine and ethnopharmacology. Ginger is used globally as a culinary spice, condiment and herbal remedy, and has been as a significant herb in traditional Indian medicine for centuries. More than 400 active ingredients have been identified ginger, among them gingerols, shogaols, paradols and zingerone are the most important bioactive components, which have been shown to have various biological activities in areas such as antioxidant, anti-inflammatory, gastrointestinal health, hepatoprotectivity, anticancer, antiarthritis and pain relief effects, antidiabetic, neuroprotective and memory-enhancing effects, cholesterol and pressure-lowering effects, obesity, *etc.* This chapter has concentrated on the overall application of ginger in Ayurvedic preparations and phytochemical evaluations utilized in the Ayurvedic drug scenario and its upcoming possibilities for advanced scientific exploration. Significant research in chemistry, pharmacology, pharmacognosy and clinical therapeutics has been carried out on ginger-based Ayurvedic formulations, functional foods and various nutraceuticals. Several preclinical and clinical studies have investigated the immunomodulatory, immunoadjuvant and cytoprotective potential of ginger-based Ayurvedic and nutraceutical formulations. The expansion of ginger-based traditional systems of medicines through the perspectives of safety, efficacy and quality will assist not only to preserve this traditional custom but also improve the utilization of natural products in health care. Ayurvedic-based herbal remedies and other nutraceuticals are progressively and comprehensively utilized by a large part of the global population. Collectively, treatment with natural herbal medicines, particularly ginger-based nutraceuticals, functional foods and dietary supplements, is highly recommended.

Acknowledgements

The authors gratefully thank the management of Plant Lipids Private Limited, Cochin, India, for their support and encouragement. They also wish to express their appreciation to their laboratory members for their active help and cooperation.

References

1. V. P. Kamboj, *Curr. Sci.*, 2000, **78**, 35.
2. A. Q. Majaz and I. M. Khurshid, *Int. J. Pharm. Pharmcol.*, 2016, **8**, 1.
3. S. Esakkimuthu, S. Mutheeswaran, P. Elankani, P. Pandikumar and S. Ignacimuthu, *J. Ayurveda Integr. Med.*, 2021, **12**, 58.
4. World Health Organization, *WHO Traditional Medicine Strategy 2014–2023*, 2013.
5. C. Jansen, J. D. Baker, E. Kodaira, L. Ang, A. J. Bacani, J. T. Aldan, L. M. N. Shimoda, M. Salameh, A. L. Small-Howard, A. J. Stokes, H. Turner and C. N. Adra, *J. Ethnopharmacol.*, 2021, **267**, 113477.
6. P. K. Mukherjee, R. K. Harwansh, S. Bahadur, S. Banerjee, A. Kar, J. Chanda, S. Biswas, S. M. Ahmmed and C. K. Katiyar, *J. Ethnopharmacol.*, 2017, **197**, 10.
7. P. K. Mukherjee, S. Banerjee and A. Kar, *Synergy*, 2018, 7, 30.
8. M. Ajanal, S. Nayak, B. S. Prasad and A. Kadam, *J. Young Pharm.*, 2013, **5**, 116.
9. S. S. Shukla, S. Saraf and S. Saraf, Fundamental aspects and basic concepts of Siddha medicines, *Syst. Rev. Pharm.*, 2011, **21**, 48–54.
10. B. Sebastia, *J. Anthopol. Knowledge*, 2011, **5**, 71.
11. A. S. Gaur and Sundaresh, Onshore and near shore explorations along the southern Tamilnadu coast: with a view to locating ancient ports and sub-merged sites, in *Mahasenasiri: Riches of Indian Archaeological and Cultural Studies*, ed. P. C. Reddy, Sharda Publishing House, New Delhi, 2006, pp. 122–130.
12. G. U. Pope, W. H. Drew, J. Lazarus and F. W. Elis, *Thirukkural*, The South India Saiva Siddhandha Works Publishing Society, India, 1982.
13. A. Singaravelu Mudaliyar, *Apithana Cintamani: the Encyclopediain Tamil literature*, Asian Education Service, New Delhi, 1983.

14. K. G. Zysk, *Siddha Medicine in Tamil Nadu*, National Museets Tranquebar Initiative, København, 2008.
15. S. Prabhu, S. Vijayakumar, J. E. Yabesh, K. Ravichandran and B. Sakthivel, *J. Ethnopharmacol.*, 2014, **157**, 7.
16. S. Kumar, T. Jawaid and S. D. Dubey, *Pharmacogn. J.*, 2011, **3**, 1.
17. T. Arablou and N. Aryaeian, *J. Herb. Med.*, 2018, **12**, 11.
18. K. Singletary, *Nutr. Today*, 2010, **45**, 171.
19. T. Arablou, N. Aryaeian, M. Valizadeh, F. Sharifi, A. Hosseini and M. Djalali, *Int. J. Food Sci. Nutr.*, 2014, **65**, 6.
20. N. Mascolo, R. Jain, S. C. Jain and F. Capasso, *J. Ethnopharmacol.*, 1989, **27**, 129.
21. J. G. Choi, S. Y. Kim, M. Jeong and M. S. Oh, *Pharmacol. Ther.*, 2018, **182**, 56.
22. P. Azimi, R. Ghiasvand, A. Feizi, J. Hosseinzadeh, M. Bahreynian, M. Hariri and H. Khosravi-Boroujeni, *Blood Pressure*, 2016, **25**, 133.
23. Y. Jiang, D. K. Turgeon, B. D. Wright, E. Sidahmed, M. T. Ruffin, D. E. Brenner, A. Sen and S. M. Zick, *Eur. J. Cancer Prev.*, 2013, **22**, 455.
24. H. C. Lien, W. M. Sun, Y. H. Chen, H. Kim, W. Hasler and C. Owyang, *Gastrointest. Liver Physiol.*, 2003, **284**, 481.
25. F. Shidfar, A. Rajab, T. Rahideh, N. Khandouzi, S. Hosseini and S. Shidfar, *J. Complementary Integr. Med.*, 2015, **12**, 165.
26. M. T. Sohail, M. I. Chaudhry, M. K. Usman, T. Mian and M. N. Ishaq, *Phytother. Res.*, 2005, **19**, 365.
27. A. Singh, S. Duggal, J. Singh and S. Katekhaye, *Pharm. Globale*, 2010, **2**, 4.
28. B. H. Ali, G. Blunden, M. O. Tanira and A. Nemmar, *Food Chem. Toxicol.*, 2008, **46**, 409.
29. K. Srinivasan, *PharmaNutrition*, 2017, **5**, 18.
30. B. Patwardhan, G. Mutalik and G. Tillu, Drug Discovery and Ayurveda, in *Integrative Approaches for Health*, ed. B. Patwardhan, G. Mutalik and G. Tillu, Academic Press, Elsevier, 2015, pp. 229–258.
31. C. K. Atal, U. Zutshi and P. G. Rao, *J. Ethnopharmacol.*, 1981, **4**, 229.
32. M. M. Ch and P. V. Smitha, *J. Herbs, Spices Med. Plants*, 2011, **17**, 339.
33. B. B. Aggarwal, S. Prasad, S. Reuter, R. Kannappan, V. R. Yadev, B. Park, J. H. Kim, S. C. Gupta, K. Phromnoi, C. Sundaram, S. Prasad, M. M. Chaturvedi and B. Sung, *Curr. Drug Targets*, 2011, **12**, 1595.
34. M. Kotteswari, M. R. K. Rao, S. Kumar, K. Prabhu, R. L. Sundaram and S. Dinakar, *Biomed. Pharmacol. J.*, 2018, **11**, 1061.
35. M. Shelar, S. Nanaware, S. Arulmozhi, S. Lohidasan and K. Mahadik, *J. Ethnopharmacol.*, 2018, **219**, 182.
36. S. Nageswara rao, M. N. Palaksha, S. Satish and Ravishankar, *Asian J. Biomed. Pharm. Sci*, 2013, **3**, 59.
37. H. Joshi and M. Parle, *Afr. J. Tradit., Complementary Altern. Med.*, 2006, **3**, 64.
38. L. Hritcu, J. A. Noumedem, O. Cioanca, M. Hancianu, V. Kuete and M. Mihasan, *Cell. Mol. Neurobiol.*, 2014, **34**, 437.

39. P. D. Kulkarni, M. M. Ghaisas, N. D. Chivate and P. S. Sankpal, *Int. J. Pharm. Pharm. Sci.*, 2011, **3**, 206.
40. G. S. Achliya, U. Barabde, S. Wadodkar and A. Dorle, *Indian J. Pharmacol.*, 2004, **36**, 159.
41. D. Bindu Madhavi, K. L. Senthil Kumar and V. Palanivel, *Int. J. Pharm. Chem. Sci.*, 2012, **1**, 1057.
42. P. C. Gupta, *Int. J. Pharm. Pharm. Sci.*, 2012, **4**, 62.
43. A. K. Ghosh, *Int. J. Pharma. Bio. Sci.*, 2011, **2**, 283.
44. A. Nisar, F. Hina, H. Bilal, F. Shahid, A. Mohammad and A. K. Mubarak, *Asian Pac. J. Trop. Biomed.*, 2012, S1945.
45. M. Arora, T. Sharma, A. Devi, N. Bainsal and A. Siddiqui, *Int. Res. J. Pharm.*, 2012, **3**, 38.
46. G. Divya, S. Gajalakshmi, S. Mythili and A. Sathiavelu, *Asian J. Biochem. Pharm. Res.*, 2011, **1**, 57.
47. M. Zaveri, A. Khandhar, S. Patel and A. Patel, *Int. J. Pharm. Sci. Rev. Res.*, 2010, **5**, 67.
48. R. Mohammed, U. Mohammed, S. D. Barhate and M. D. Abullais, *Int. J. Curr. Pharm. Rev. Res.*, 2012, **3**, 15.
49. R. Shukla, S. Singh, A. Singh and K. Misra, *Eur. J. Int. Med.*, 2021, **43**, 101268.
50. P. Khanal, B. M. Patil, J. Chand and Y. Naaz, *Nat. Prod. Bioprospect.*, 2020, **10**, 325.
51. P. Khanal and B. M. Patil, *Asian Pac. J. Trop. Biomed.*, 2019, **9**, 263.
52. P. Khanal and B. M. Patil, *J. Integr. Med.*, 2021, **19**, 66.
53. Ministry of AYUSH, Ayurveda's immunity boosting measures for self-care during COVID 19 crisis, 2020. Available at: https://www.ayush.gov.in/docs/123.pdf (accessed 12 April 2020).
54. P. Khanal, T. Duyu, B. M. Patil, Y. N. Dey, I. Pasha, M. Wanjari, S. S. Gurav and A. Maity, *J. Ayurveda Integr. Med.*, 2022, **13**, 100374.
55. Ayurvedic Pharmacopoeia Committee, *Zingiber officinale* Government of India, Ministry of Health and Family Welfare, in *The Ayurvedic Pharmacopoeia of India*, Department of Ayurveda, Yoga and Naturopathy, Unani, Siddha and Homoeopathy (AYUSH), New Delhi, 1989, p. 107.
56. A. Leoni, R. Budriesi, F. Poli, M. Lianza, A. Graziadio, A. Venturini, M. Broccoli and M. Micucci, *Nat. Prod. Res.*, 2018, **32**, 2139.
57. S. Nair, R. Nagar and R. Gupta, *Indian Heart J.*, 1996, **48**, 545.
58. R. P. Samy, P. N. Pushparaj and P. Gopalakrishnakone, *Bioinformation*, 2008, **3**, 100.
59. A. Narayana and V. Subhose, *Bull. Indian Inst. Hist. Med.*, 2005, **35**, 21.
60. A. Chauhan, D. K. Semwal, S. P. Mishra and R. B. Semwal, *Ayu*, 2015, **36**, 364.
61. A. Nakanekar, K. Kohli and P. Tatke, *J. Ayurveda Integr. Med.*, 2019, **10**, 284.
62. M. Haridas, V. Sasidhar, P. Nath, J. Abhithaj, A. Sabu and P. Rammanohar, *Future J. Pharm. Sci.*, 2021, **7**, 13.

63. M. Ogino, K. Yakushiji, H. Suzuki, K. Shiokawa, H. Kikuchi, Y. Seto, H. Sato and S. Onoue, *J. Funct. Foods*, 2018, **40**, 156.
64. A. K. Archana, R. K. Aman, N. K. Singh and A. Jabeen, *Mater. Today: Proc.*, 2021, **43**, 3397.
65. B. Tiencheu, D. N. Nji, A. U. Achidi, A. C. Egbe, N. Tenyang, E. F. T. Ngongang, F. T. Djikeng and B. T. Fossi, *Heliyon*, 2021, **7**, e07177.
66. A. Amalraj, J. T. Haponiuk, S. Thomas and S. Gopi, *Int. J. Biol. Macromol.*, 2020, **151**, 366.
67. A. Amalraj, K. K. J. Raj, J. T. Haponiuk, S. Thomas and S. Gopi, *Adv. Compos. Hybrid. Mater.*, 2020, **3**, 485.
68. K. Simon-Brown, K. M. Solval, A. Chotiko, L. Alfaro, V. Reyes, C. Liu, B. Dzandu, E. Kyereh, A. G. Barnaby, I. Thompson, Z. Xu and S. Sathivel, *LWT – Food Sci. Technol.*, 2016, **70**, 119.
69. A. J. Akinyemi, G. R. Thome, V. M. Morsch, N. Stefanello, P. D. Costa, A. Cardoso, J. F. Goularte, A. Belló-Klein, A. A. Akindahunsi, G. Oboh and M. R. C. Schetinger, *J. Appl. Biomed.*, 2016, **14**, 59.
70. A. J. Akinyemi, I. A. Adedara, G. R. Thome, V. M. Morsch, M. T. Rovani, L. Mujica, T. Duarte, M. Duarte, G. Oboh and M. Schetinger, *Toxicol. Rep.*, 2015, **2**, 1357.
71. A. J. Akinyemi, G. R. Thomé, V. M. Morsch, N. B. Bottari, J. Baldissarelli, S. L. de Oliveira, J. F. Goularte, A. Belló-Klein, G. Oboh and M. R. Schetinger, *Phytother. Res.*, 2016, **30**, 1156.
72. H. Sato, M. Ogino, K. Yakushiji, H. Suzuki, K. I. Shiokawa, H. Kikuchi, Y. Seto and S. Onoue, *J. Agric. Food Chem.*, 2017, **65**, 1365.
73. N. Kaewtunjai, R. Wongpoomchai, A. Imsumran, W. Pompimon, A. Athipornchai, A. Suksamrarn, T. R. Lee and W. Tuntiwechapikul, *ACS Omega*, 2018, **3**, 18572.
74. L. Lei, Y. Liu, X. Wang, R. Jiao, K. Y. Ma, Y. M. Li, L. Wang, S. W. Man, S. Sang, Y. Huang and Z. Y. Chen, *J. Agric. Food Chem.*, 2014, **62**, 10515.
75. A. Amalraj, K. Varma, S. Jude, R. V. R. Nair and S. Kuttappan, *ACS Food Sci. Technol.*, 2021, **1**, 1059.
76. L. Toma, M. Raileanu, M. Deleanu, C. S. Stancu and A. V. Sima, *J. Funct. Foods*, 2018, **48**, 654.
77. S. C. Ho, K. S. Chang and C. C. Lin, *Food Chem.*, 2013, **141**, 3183.
78. S. M. Ezzat, M. I. Ezzat, M. M. Okba, E. T. Menze and A. B. Abdel-Naim, *J. Ethnopharmacol.*, 2018, **214**, 113.
79. A. O. Abolaji, M. Ojo, T. T. Afolabi, M. D. Arowoogun, D. Nwawolor and E. O. Farombi, *Chem. – Biol. Interact.*, 2017, **270**, 15.
80. G. Park, D. S. Oh, M. G. Lee, C. E. Lee and Y. U. Kim, *Toxicol. Appl. Pharmacol.*, 2016, **310**, 51.
81. C. Y. Hsiang, H. M. Cheng, H. Y. Lo, C. C. Li, P. C. Chou, Y. C. Lee and T. Y. Ho, *J. Agric. Food Chem.*, 2015, **63**, 6051.
82. M. S. Badawi, *J. Chin. Med. Assoc.*, 2019, **82**, 11.
83. A. Shirpoor, E. Heshmati, F. Kheradmand, F. H. Gharalari, L. Chodari, R. Naderi, F. N. Majd and M. Samadi, *Biomed. Pharmacother.*, 2018, **105**, 144.

84. A. V. Vipin, K. Raksha Rao, N. K. Kurrey, K. A. Anu Appaiah and G. Venkateswaran, *Biomed. Pharmacother.*, 2017, **91**, 415.
85. S. A. Gabr, A. H. Alghadir and G. A. Ghoniem, *Saudi J. Biol. Sci.*, 2019, **26**, 382.
86. I. A. Adekunle, C. E. Imafidon, A. A. Oladele and A. O. Ayoka, *Pathophysiology*, 2018, **25**, 101.
87. U. Cakir, C. Tayman, U. Serkant, H. I. Yakut, E. Cakir, U. Ates, I. Koyuncu and E. Karaogul, *J. Ethnopharmacol.*, 2018, **225**, 297.
88. E. S. Abd Allah, R. Makboul and A. O. Mohamed, *Pathophysiology*, 2016, **23**, 35.
89. A. Rashidian, S. Mehrzadi, A. R. Ghannadi, P. Mahzooni, S. Sadr and M. Minaiyan, *J. Integr. Med.*, 2014, **12**, 115.
90. C. Y. Hsiang, H. Y. Lo, H. C. Huang, C. C. Li, S. L. Wu and T. Y. Ho, *Food Chem.*, 2013, **136**, 170.
91. C. D. Black, M. P. Herring, D. J. Hurley and P. J. O'Connor, *J. Pain*, 2010, **11**, 894.
92. Z. Naderi, H. Mozaffari-Khosravi, A. Dehghan, A. Nadjarzadeh and H. F. Huseini, *J. Tradit. Complementary Med.*, 2015, **6**, 199.
93. S. K. Vemuri, R. R. Banala, G. P. V. Subbaiah, S. K. Srivastava, A. V. G. Reddy and T. Malarvili, *Egypt. J. Basic Appl. Sci.*, 2017, **4**, 332.
94. Y. Wang, H. Yu, X. Zhang, Q. Feng, X. Guo, S. Li, R. Li, D. Chu and Y. Ma, *Nutrition*, 2017, **36**, 79.
95. E. Soltani, A. Jangjoo, M. Afzal Aghaei and A. Dalili, *J. Tradit. Complementary Med.*, 2017, **8**, 387.
96. M. Adib-Hajbaghery and F. S. Hosseini, *Complementary Ther. Med.*, 2015, **23**, 827.
97. H. Imani, H. Tabibi, I. Najafi, S. Atabak, M. Hedayati and L. Rahmani, *Nutrition*, 2015, **31**, 703.
98. H. Mozaffari-Khosravi, B. Talaei, B. A. Jalali, A. Najarzadeh and M. R. Mozayan, *Complementary Ther. Med.*, 2014, **22**, 9.
99. K. R. Shanmugam, K. Mallikarjuna, N. Kesireddy and K. Sathyavelu Reddy, *Food Chem. Toxicol.*, 2011, **49**, 893.
100. S. Lim, M. Moon, H. Oh, H. G. Kim, S. Y. Kim and M. S. Oh, *J. Nutr. Biochem.*, 2014, **25**, 1058.
101. Z. V. Shariatpanahi, F. A. Taleban, M. Mokhtari and S. Shahbazi, *J. Crit. Care*, 2010, **25**, 647.
102. J. S. Chang, K. C. Wang, C. F. Yeh, D. E. Shieh and L. C. Chiang, *J. Ethnopharmacol.*, 2013, **145**, 146.
103. Z. V. Shariatpanahi, M. Mokhtari, F. A. Taleban, F. Alavi, M. H. Salehi Surmaghi, Y. Mehrabi and S. Shahbazi, *J. Crit. Care*, 2013, **28**, 217.e1.
104. M. L. Ahui, P. Champy, A. Ramadan, L. Pham Van, L. Araujo, K. Brou André, S. Diem, D. Damotte, S. Kati-Coulibaly, M. A. Offoumou, M. Dy, N. Thieblemont and A. Herbelin, *Int. Immunopharmacol.*, 2008, **8**, 1626.

105. A. M. Al Hroob, M. H. Abukhalil, R. D. Alghonmeen and A. M. Mahmoud, *Biomed. Pharmacother.*, 2018, **106**, 381.
106. A. J. Akinyemi, G. R. Thome, V. M. Morsch, N. Stefanello, J. F. Goularte, A. Belló-Klein, G. Oboh and M. R. C. Schetinger, *J. Funct. Foods*, 2015, **17**, 792.
107. A. Hosseini and N. Mirazi, *Epilepsy Res.*, 2014, **108**, 411.
108. S. Oha, M.-S. Lee, S. Jung, S. Kim, H. Park, S. Park, S.-Y. Kim, C.-T. Kim, Y.-H. Jo, I.-H. Kim and Y. Kim, *J. Funct. Foods*, 2017, **29**, 193.
109. K. Misawa, K. Hashizume, M. Yamamoto, Y. Minegishi, T. Hase and A. Shimotoyodome, *J. Nutr. Biochem.*, 2015, **26**, 1058.

CHAPTER 4

Most Modern Approach to the Phytochemistry and Therapeutic Potential of Boswellia *in Ayurvedic Formulations*

KARTHIK VARMA A. C.,* SASIKUMAR KUTTAPPAN* AND AUGUSTINE AMALRAJ

R&D Centre, Aurea Biolabs (P) Ltd, Kolenchery, Cochin-682 311, Kerala, India
*Emails: ackvarmaapm@gmail.com; sasikumartk@gmail.com

4.1 Introduction

Traditional systems of medicine have always gained many accolades from the ancient days onwards because of their diverse range of pharmacological activities and efficacy. One such system of medicine, which has gained huge attention, includes the traditional Ayurvedic practices of the Indian medical principles.[1] The principles of Ayurveda have a vast literature in almost all of the Indian languages and especially in Sanskrit, which covers the use of the herbs, their dosages, and the methods of preparation.[2] The treatment of Ayurveda is based on three basic principles called doshas (vata, pitta, and kapha), which are obtained from the five elements of Indian philosophy. Ayurveda's doshas are treated as regulatory control factors for fundamental physiologic processes in biological systems that maintain their identity throughout biologic history. Vata and its subdoshas regulate input/output processes and motion; pitta and its

Chemistry, Biological Activities and Therapeutic Applications of Medicinal Plants in Ayurveda
Edited by Augustine Amalraj, Sasikumar Kuttappan and Karthik Varma A. C.

Published by the Royal Society of Chemistry, www.rsc.org

Table 4.1 Different *Boswellia* species and their geographical locations.

Variety	Geographical location
B. *carterii* Bird.	Somalia
B. *sacra* Flueck Nubia	South Arabia
B. *frereana* Birdw.	Somalia
B. *papyrifera*	Ethiopia
B. *neglecta* S Moore	Somalia
B. *odorata* Hutch	Africa
B. *dalzielii* Hutch	Africa
B. *serrata* Roxeb	India
B. *rivae*	Ethiopia

subdoshas regulate throughput, turnover, and hence energy; and kapha and its subdoshas regulate storage, structure, and lubrication.[3] Studies have proved that almost 95% of Ayurvedic formulations are prepared using plants/herbs.[4] From ancient days onwards, plants and plant-derived phytochemicals have contributed to curing many diseases. As time has progressed, plants became the source of income for traditional communities. One such species which has gained much importance is *Boswellia*.

The genus *Boswellia* belongs to the Burseraceae family, which consist of about 17 genera and around 600 species.[5] Some of the most important species of this genus include *Boswellia serrata*, *Boswellia sacra*, *Boswellia carterii*, *Boswellia papyrifera*, *Boswellia neglecta*, *Boswellia rivae* and *Boswellia frereana*.[6] *Boswellia* gum resin is cultivated in many areas including the Asian and African regions (Table 4.1). The phytochemical compounds mostly contribute to the pharmacological activities, and include essential oil terpenes, resinous materials and other biologically active compounds.[7] In the Ayurvedic system of medicine, the gum resin extract and its bioactive components are used to treat a number of inflammatory diseases affecting skin, eye, gastrointestinal tract (GIT) and respiratory inflammatory disorders, such as asthma, bronchitis and laryngitis, *etc.*[8] *Boswellia* is commonly referred to as Ashvamutri, Kundara, and Shallaki in Sanskrit.[9]

This chapter presents an overview on the taxonomy, phytochemical compounds and uses in Ayurvedic formulations and various pharmacological activities of different *Boswellia* species.

4.2 *Boswellia* in Ayurvedic Texts

Boswellia, commonly referred to as Shallaki in Sanskrit, is cited in many ancient Ayurvedic texts. The Ayurvedic sloka depicts the different uses of *Boswellia* and Shallaki as:

Ath shompichuchannam sushyatam pratisaryet/Pri-yanjan yastavyarodhrachurne samantat//Shallaki phal-churne va shomdhyamain va punaha/Tato vranam yatha yogam badhvaacharkamaadhishet//. Various Ayurvedic literatures like Acharya Charak, Acharya Sushrut, Kaydev Nighantu, Dhanvanthari

Nighantu, Bhavprakash Nighantu, Raj Nighantu, Charak Samhita, Sushrut Samhita, *etc.* position the herb as having potential pharmacological activities. The traditional uses of *Boswellia* in Ayurvedic practices include:

//Badararjunjambuaamshallakivetastwacha. Sharkarash-oudrasanyukta-peeta// – The ash of the flax or the fruit powder are applied to cure wounds.

Palasham shyad shonitam ch anjanarthe shallakya va sharkara shoudra yuktam// – The decoction of the barks of badar, arjun, jambu, Aamra, Shallaki and vetas are used to treat diarrhoea.

Turushksallakinam ch Guggalo Padhmakasya ch// – The smoke of the Shallaki powder is inhaled to relieve the symptoms of asthma.

4.3 Ayurvedic Properties of Shallaki

Ayurvedic literatures describe the various properties of *B. serrata* in different ways (Table 4.2).

4.4 Varg of Shallaki in Different Literature Sources

The use of *Boswellia* has been described in many Ayurvedic literatures (Table 4.3).

4.5 Science Behind the Action of Ayurvedic Formulations

The treatments of Ayurvedic medicines are mainly concentrated on the three basic problems of the body, including Vata, kapha and pitta, which are commonly referred to as doshas in the principle.[10] The majority of the Ayurvedic formulations are based on the principles of poly-herbal combinations.[11] Sarangdhar Samhita, one of the important literatures in Ayurveda,

Table 4.2 Various properties of Shallaki in Ayurveda.

Rasa	Madhur, Katu, Tikta
Guna	Guru, snigdha, Tikshna
Virya	Ushna
Vipaka	Madhur
Karma	Balya, kaphahara, Vaathar, Kaphpitthar, Raktsthambhar, Svedhar

Table 4.3 Varg of Shallaki in Ayurvedic scripts.

Acharya Charak	Purishviranjaniya, Shirovirechan, kashayaskand	Charak Samhita, sutrasthan 4/32, 4/27 Vimansthan 8/144
Acharya Sushrut	Rodraadi, Eladi, Kashyaskand	
Kaydev Nighantu	Aushadhi varg	Shlok 861–863
Dhanvantri Nighantu	Chandanaadi varg	Shlok 122, 123

highlights the use of polyherbal formulations and their role in treating diseases.[12] In some cases of treating diseases, if the individual phytochemical compounds are not effective enough to get the desired result, formulations including multiple herbs are used to increase their efficacy.[13] There are two mechanisms behind this rationale, which involve pharmacokinetic synergy and pharmacodynamic synergy.[14] The idea of pharmacokinetic synergy depends on the concept of the ability of the herbs to absorb, proper distribution, metabolism and elimination of other herbs from the body, and the second concept lies in the principle of the synergistic actions of the ingredients.[15]

4.6 Taxonomy and Phytochemistry

Phytochemical compounds are isolated from the trunks of various species of trees of *Boswellia*. The trees are normally found in the dry mountainous regions of India, northern Africa and the Middle East.[16] The yield and percentage of the bioactive components mainly depend on the climatic and geographical conditions. Scientific studies have validated the presence of valuable phytochemicals in this plant. Studies have proven that the plant contains pentacyclic terpenoids, tetracyclic triterpenoids, macrocyclic diterpenoids and essential oils[17] (Figure 4.1). The major pentacyclic terpenoids of the *Boswellia* species include β-boswellic acid (β-BA), 3-acetyl β-boswellic acid, 11-keto-β-boswellic acid, 3-*o*-acetyl-11-keto beta boswellic acid (AKBA), 12-ursene-2-diketone and other major and minor compounds. The *Boswellia* species also contain tetracyclic triterpenoids like α-elemonic acid, β-elemonic acid, *etc.* (Figure 4.1). Studies have also found the presence of water-soluble arabinogalactan proteins, which are mainly composed of D-galactose units with a molecular weight of 60 mol% in the core chains and branched *via* positions 3 and 6 and in the side chains, which are uronic acids, glucuronic acid (9 mol%) and terminal 4-*O*-methyl-glucuronic acid (13–26 mol%), and arabinose (2–14 mol%).[18]

4.7 Modern Medicinal Uses

The roles of *Boswellia* and its phytochemicals are cited in most modern and many traditional systems of medicine. The modern system of medicine focuses on the bioactive components and their bioavailabilities. The use of innovative technologies has led researchers to focus on developing extracts with the enriched form of bioactives. This has led to a reduction in the dosage of the drug, instead of supplementing the raw herb, the innovative technologies focus on specific bioactive components and their pharmacological activities.

4.7.1 Sandhigata Vata/Arthritis

Sandhigata vata is mentioned as "vatavyadhi" in Ayurvedic texts. Charaka was the first to describe separately "Sandhigata anila", but was not included

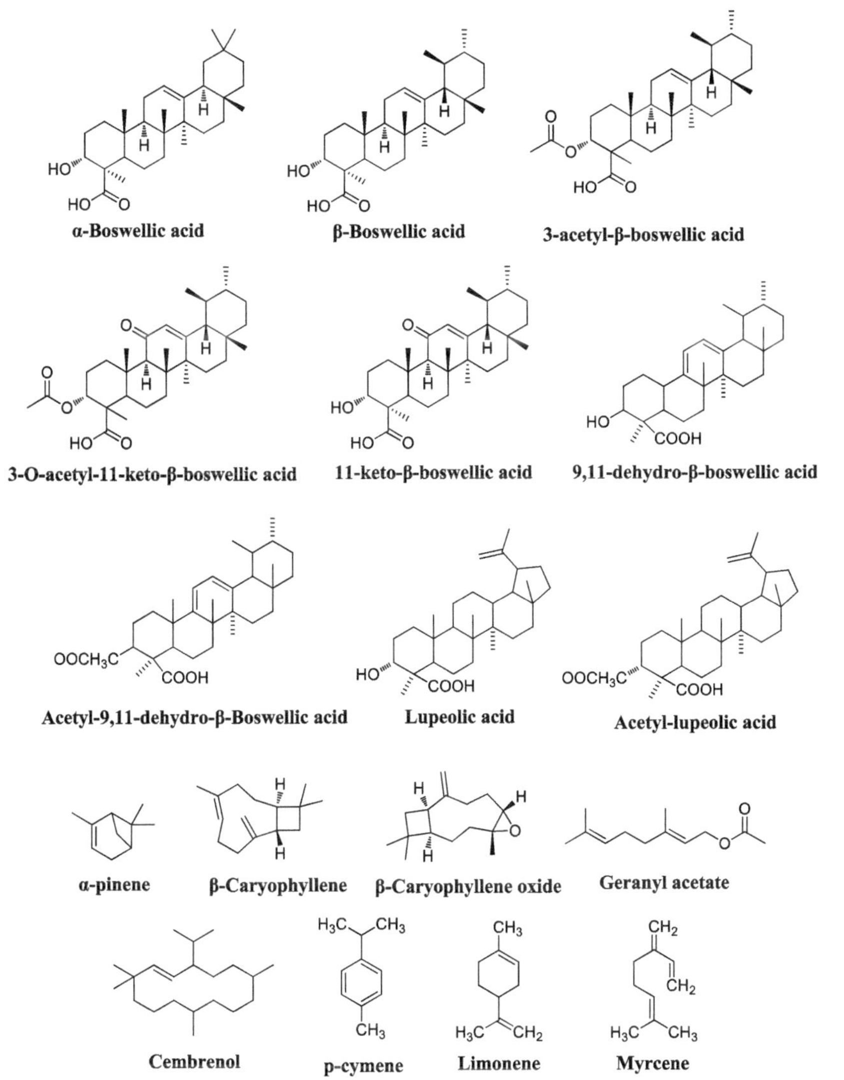

Figure 4.1 Structures of bioactive components in *Boswellia*.

under any of the 80 types of nanatmaja vatavyadhi. In modern terminology, the symptoms associated with Sandhigata vata are more or less equal to osteoarthritis (OA). Most of the studies of the Shallaki/*Boswellia* concentrate on their potential to be used in inflammatory disorders. Gupta *et al.*[19] evaluated the role of Shallaki, in the management of Sandhigata vata/OA. Subjects of 40–70 years of age and having clinical signs and symptoms of Sandhigata, *viz.* shula (joint pain), shotha (joint-swelling), stambha (stiffness), sparsha asahyata (tenderness), sphutana (crepitus) and akunchana prasarana vedana (pain during flexion and extension of the joint) were included in the study. The study included supplementation of Shallaki capsules and Shallaki ointment for topical application. Symptoms like sandhishula (joint pain), Sandhishotha (joint swelling), Akunchana prasaranayoh vedana (pain during movement), Stambha (stiffness), sandhisphutana (crepitations) and Sparsha asahyata (tenderness) all showed significant improvements in both groups with slight variations in the extent of improvement. The effect of therapy on symptoms of vata-vridhi included positive effects reported in gatrasphurana (fasciculation) and asthishula (pain in bones). Among the symptoms of vata-kshaya: alpa chesta, Angasada (looseness of body parts) and apraharsha (unhappiness) all showed positive improvements after the treatment period. The study also evaluated the improvement of symptoms of pittakshaya, which included stambha (stiffness), toda (pricking pain) and gaurava (heaviness), and all of the parameters evaluated showed a positive improvement. The effect of asthivahasrotas, Asthishula (pain in bones) and asthibheda also showed significant improvement after the study duration and paved the way for the use of Shallaki in treating OA and other inflammatory conditions.[19] In another study, a double-blind, placebo-controlled human trial evaluated the efficacy of *Boswellia* extract standardized to AKBA in patients with knee OA. The study, with a duration of 120 days, proved that supplementing with *Boswellia* showed a significant improvement in the physical function of the patients by reducing pain and stiffness compared with placebo. The study evaluated the improved knee joint gap and reduced osteophytes (spur), confirming the efficacy of the extract treatment. The study group also showed significant improvement in the serum levels of high-sensitive C-reactive protein.[20] Sengupta *et al.*[21] also evaluated the role of *Boswellia* in OA. This study showed significant improvements in the pain scores, physical function scores, a reduction in synovial fluid matrix metalloproteinase-3 and proved the safety of the product in OA.[21] Shallaki is also used in combination with other herbs for the management of inflammatory disorders. A randomized double-blind placebo-controlled study was conducted by Kimmatkar *et al.*[22] who evaluated the role of *Boswellia* extract and proved its efficacy in OA. The study proved significant improvement in knee pain, infection, walking distance and frequency of swelling, thus proving its efficacy.[22]

Shallaki is also used as a single herb and in combination with other plant extracts. A polyherbal formulation containing *Boswellia serrata* 240 mg, *Commiphora wightii* (purified) 200 mg, *Alpinia galanga* 70 mg, *Glycyrrhiza glabra* 70 mg, and extracts of *Tribulus terrestris* 60 mg and *Tinospora cordifolia* 60 mg. The study population included 50 enrolments in which one group was supplemented with the formulation and another group was supplemented with a placebo. Subjects were asked for follow-up every 4 weeks for 6 months. At the end of 6 months, subjective (symptomatic) assessment was carried out using the scores below to determine the clinical efficacy of the trial drug. The study showed that there was a 66.28% reduction in the number of joint involvements in the treated group and 51.94% in the placebo group. In a subjective symptomatic evaluation, the study group showed 45.91% improvement in the relief symptoms as compared to 29.1% in the placebo group.[23] A novel formulation containing *Boswellia*, bioavailable form of curcuminoids, *Kaempferia galanga* extract and *Piper nigrum* oil proved its efficacy in controlling OA in patients with knee OA. The study was done in comparison with a combination of glucosamine and chondroitin as the reference product and proved that the test product at a lower dosage of 250 mg per day showed significant results in comparison with the reference product at dosage of 1500 mg + 1200 mg.[24]

Similarly, a formulation containing *Boswellia* extract (300 mg), curcuma extract (100 mg), *Pinus pinaster* (80 mg) and ginger extract (40 mg) was used as an anti-inflammatory drug to reduce joint pain. The pain reduction was observed in a visual analogue scale (VAS) and WOMAC scale as compared to the placebo group. The study showed significant effectiveness in reducing pain and inflammation, with increased joint mobility and patients' quality of life.[25] The clinical efficacy of a herbo-mineral formulation containing roots of *Withania somnifera*, the stem of *Boswellia serrata*, rhizomes of *Curcuma longa* and a zinc complex, was evaluated in a randomized, double-blind, placebo-controlled, cross-over study in subjects with OA. The study included supplementation of 42 subjects with knee OA who were randomly divided into two groups. The clinical efficacy was evaluated every fortnight based on pain symptoms. The study group showed significant improvement in the severity of pain and disability score.[26] LI73014F2, a novel herbal formulation containing fruit extract of *Terminalia chebula*, *Curcuma longa* and *Boswellia serrata* was evaluated for efficacy in relieving joint discomfort in human subjects with knee OA. The study parameters evaluated included VAS, Lequesne's functional index and WOMAC score, which all proved that there was a significant improvement in the pain relief, improved physical function and quality of life in OA subjects.[27] The effectiveness of an Ayurvedic formulation containing *Tinospora cordifolia*, *Zingiber officinale*, *Emblica officinalis* and *Boswellia serrata* was compared with glucosamine sulfate (2 g daily) and celecoxib. In that study, 440 eligible subjects suffering from symptomatic knee OA were enrolled and monitored as per protocol. In the 6-month study, the Ayurvedic formulation proved to have efficacy similar to glucosamine and celecoxib.[28] HFPM-01, a novel formulation containing *Boswellia serrata*, *Zingiber officinale*, *Withania somnifera*, *Curcuma longa* and *Pluchea lanceolata* as key ingredients out of a total of nine ingredients, was evaluated for its efficacy in controlling knee OA. The study involved randomization of 90 subjects into three groups. The study groups received either test product, glucosamine or placebo. The change in the WOMAC score, the increase in Short Form (SF-36) score, the decrease in VAS score, the decrease in C-reactive protein levels and decrease in gastrointestinal symptoms were found to be 33%, 38%, 60.44%, 52% and 40%, respectively, which showed significant improvement in relieving the symptoms associated with OA.[29]

The symptomatology Griva *Sandhi* disease includes shula (pain in cervical region), prasaran akunchana vedana (pain during flexion and extension), vatapurnadriti Sparsha (on palpation of the joints), hanti sandhigatah (loss of function) and aatopa (cracking noise).[30] In modern medical science, this is described as cervical spondylosis. In the case of cervical spondylosis, the degenerative changes in the intervertebral discs lead to secondary changes in the adjacent vertebrae. In one study, which included supplementation of two capsules of Shallaki Niryas in 10 subjects with clinically, pathologically and radiologically diagnosed symptoms were evaluated. The study drug showed significant improvement in the Shula, prasaran akunchanavedana and tingling sensation, thus showing its positive role in treating spondylosis.[31]

4.7.2 Antihyperlipidemic and Antidiabetic Activity

The gum resin of Shallaki was used in many traditional systems of medicine for controlling diabetes mellitus/madhumeha. The anti-inflammatory action of boswellic acids, predominantly the role of proinflammatory cytokines in insulin resistance, is attributed to the antidiabetic activity of Shallaki. Multiple clinical studies have given an insight into the different mechanisms for the antidiabetic activity of the resin extract and its bioactive components. In one study, subjects were supplemented with *Boswellia* gum resin extract and metformin. The study group, in addition to metformin, received 400 mg of the resin orally twice daily for 12 weeks. Compared with the placebo, the study group showed a significant decrease in the fasting blood glucose, haemoglobin A1c (HbA1c) and serum insulin. Moreover, a reduction of serum cholesterol, low-density lipoprotein and triglycerides was reported, and no adverse effects of the study product were recorded.[32] The role of *Boswellia* extract in controlling blood glucose, insulin level and insulin resistance in type 2 diabetic patients was evaluated in 60 subjects with type 2 diabetes. The subjects were divided into two groups in which the subjects were supplemented with either placebo or test products. The 6-week administration of the test product decreased the blood glucose level (173.71 ± 10 *vs.* 147.57 ± 8 mg dl^{-1}, $P < 0.01$). The study also showed that plasma insulin increased while the insulin resistance index did not have a significant effect on the test product.[33]

In addition to the supportive evidence from clinical trials, several preclinical studies have also highlighted the antidiabetic activity of the gum resin extract and its bioactive components, modulated by different pathways. A novel formulation containing *B. serrata* oleo-gum-resin has been reported to produce significant antidiabetic activity by affecting hepatic gluconeogenesis, pyruvate carboxylase and phosphoenol pyruvate carboxykinase.[34] Similarly, a study designed by Shehata *et al.*[35] involved the use of AKBA to prevent the induction of autoimmune reactions, insulitis and hyperglycaemia in a model of multiple low-dose streptozotocin (MLD-STZ) diabetes. In the streptozotocin product-supplemented group there was a significant burst of pro- and anti-inflammatory cytokines in the blood, infiltration of lymphocytes (CD3) into pancreatic islets and the appearance of peri-insular apoptotic cells was recorded. Plasma glucose increased significantly, however the study product group showed a significant decrease in the reduction in pro- and anti-inflammatory cytokines.[35] Not only *B. serrata*, but also *Boswellia ovalifoliolata* has promising antidiabetic activities. The oral administration of 200 and 400 mg kg^{-1} bodyweight for a duration of 7 days significantly reduced the blood glucose level in comparison to a 10 mg dosage for glibenclamide.[36]

4.7.3 Anticancer

The gum resin extract of the Shallaki is extensively used in most modern medicines for anticancer activities. Various activities like cytostatic, anti-apoptotic and antitumor properties have paved the way for the use of *Boswellia* and its positive impact on glioma, leukaemia, melanoma, hepatoma, and prostate and breast cancers.[37–44] Several clinical and preclinical studies have identified their potential in anticancer activities. The anticancer activity of *Boswellia* extract was studied in 44 patients with primary or secondary malignant cerebral tumours. The subjects were randomly allocated to radiotherapy and either *B. serrata* 4200 mg per day or placebo. The primary end of the study was measuring the volume of cerebral oedema in a T2-weighted magnetic resonance imaging sequence. The results of the study indicated that cerebral oedema was reduced by more than 75% in 60% of patients receiving *B. serrata* and 26% of those receiving placebo.[45] Similarly, a clinical study with brain tumour patients has also been conducted, in which *Boswellia* extract was administered to 29 patients with gliomas. The study supplementation involved three different doses of the *Boswellia* extract. After 7 days of treatment, prior to surgical intervention, the reduction in size of perifocal oedema was found to be largest in the case of the group having the highest intake of the extract, to a lesser extent in the group receiving less, and with no effect being seen in the group receiving the smallest dose.[46] Several preclinical studies have also highlighted the potential use of *Boswellia* extract in anticancer activities.[47–49]

4.7.4 Antiulcer

Boswellic acid has a positive impact on the antiulcer activity mediated by the anti-inflammatory pathway. Peptic ulcer in Ayurveda is considered a sort of acid gastritis. This disorder is generally classified under Grahani disorder. Acharya Charaka describes some of the common causes of peptic ulcer such as deranged Agni. Because of the disarranged Agni or pitta vikara, there are many possible causes of gastric disorders with one being ulcer (Annadrava shoola). Several Ayurvedic formulations have proved their superiority in treating peptic ulcers.[50] Singh *et al.*[51] evaluated the effect of boswellic acid-enriched *B. serrata* extract in rats, using models of gastric ulcer induced by pyloric ligation, ethanol–HCl, acetylsalicylic acid, indometacin and cold-restraint stress. This study concentrated on elevating the dosage up to 500 mg kg^{-1}, and found that the extract was found to exhibit dose-dependent antiulcer activity. The proposed mechanism was found to be related to the increase in gastric mucosal resistance, local synthesis of cytoprotective prostaglandins or inhibition of leukotriene synthesis.[51] Similarly, the proulcerogenic effect of water extract of *Boswellia sacra* oleo gum resin in rats was evaluated by Asad *et al.*,[52] and the extract was found to increase the ulcer index and ulcer score. In pylorus-ligated rats, the extract was found to increase the gastric content volume, free acidity, total acidity, ulcer index and pepsin activity. The study also found that the extract was able to exhibit dose-dependent proulcerogenic activity. The GCMS profile of the extract revealed the presence of menthol, 3-cyclohexen-1-ol and octanoic acid, whose presence may have contributed to the proulcerogenic effect.[52] Several other studies have also evaluated the effect of *B. serrata* and its role in the aspirin-induced ulcer model.[53]

4.7.5 Immune Response

The Ayurvedic response is that, if the body is kept in balance and if the immune system is at peak efficiency, the chances of infections and various sorts of inflammatory disorders are reduced. The underlying mechanism of the immunomodulatory action of the herbal extracts includes macrophage stimulation and modulation of the complement system. The modulation of these systems can have a positive impact on both humoral and cellular immune responses. Single or prolonged oral administration of the extract at a dosage of 25–100 mg kg^{-1} per day for a duration of 21 days showed immunosuppressive or immuno-potentiating effects, depending on the dosage and timings of drug administration. The phytochemical constituents enhanced the phagocytic function of the macrophages.[54] The immunomodulatory assessment of *B. carterii* bird wood extract and isolated ingredients at a concentration of 1.00 mg ml^{-1} was evaluated in a lymphocyte blast transformation assay. This study found that the immunostimulatory activity of the combined extract was greater than that of the individual components. The ethanolic extract shows a significant immunostimulatory action on T lymphocytes (90% lymphocyte proliferation).[55] Similarly, a human clinical study in a 50 year-old diabetic was evaluated for prevention/treatment of autoimmune diabetes. The study found that the supplementation of the extract at a dosage of 400 mg for 8.5 weeks did not show any significant improvement in the clinical biomarkers.[56] A study involving a standardized *Boswellia* extract containing 6–44 μg ml^{-1} boswellic acid and lupeolic acid exhibited inhibition of growth of cancer xenografts *in vivo* and released pro-inflammatory cytokines.[57] A standardized extract containing AKBA from *B. carterii* inhibited proliferation, degranulation capacity and secretion of inflammatory mediators of physiologically relevant anti-CD_3 and anti-CD_{28} activated human T lymphocytes in a non-toxic concentration range and exhibited immuno-modulatory activity.[58]

4.7.6 Antithrombotic

The Ayurveda medicine records have highlighted the potential of haemostatic, anti-inflammatory and cardiovascular effects.[59] Kokkiripati *et al.*[60]

studied the antioxidative and antithrombotic properties of stem bark gum resin extracts of *B. serrata*. This study investigated water and hydro-alcoholic extracts of *B. serrata* (7.8 and 9% w/w AKBA) on $FeCl_3$-induced lipid peroxidation (*in vitro*) in rat liver and heart homogenates, H_2O_2-mediated reactive oxygen species (ROS) generation in THP-1 cells, and adenosine diphosphate (ADP)-induced platelet aggregation, plasma coagulation time and human blood-clotting factors Xa and XIa. The study results indicated that both extracts inhibited lipid peroxidation in liver (80%) and heart (50%) tissue homogenates in male Wistar rats.[60]

4.7.7 Skin Disorders

The phytochemical constituents in the essential oil fraction and the non-resinous fraction were cited in different literatures for their effects on skin disorders. A topical administration route is currently used to deliver active compounds in psoriatic and eczematous patients. In a double-blind study, a topical administration route was employed to administer *Boswellia* resin. The subjects were randomly divided into three different groups and subjected to receive test product, *V. myrtillus* seed oil or placebo, and in order to evaluate the changes of scales and erythema from diagnosis to the end of treatment were scored in psoriatic patients, while changes in itch and erythema were analysed for erythematous eczema patients. Psoriasis Area Severity Index and Eczema Area and Severity Index scores were also calculated. The study group showed improvement in scales (70% of cases) and erythema (50% of cases), without any case of worsening.[61] Another study conducted with a topical cream containing *Boswellia* was applied in 15 female volunteers. In this double-blinded study, topical application of creams with or without BAs was done on half of the side of the face once daily for 30 days. Significant improvements of the Dover's global score for photo-aging, tactile roughness and fine lines, as well as with noninvasive diagnostic techniques, an increase in elasticity, a decrease of sebum excretion and a change in echo-graphic parameters were observed in subjects supplemented with the study protocol compared to the placebo, thereby showing the supremacy of the extract in skin disorders (Calzavara-Pinton).[62]

4.7.8 Antidiarrhoeal

Various studies have shown the use of *Boswellia* and its role in antidiarrhoeal activity. In one study, the effect of gum resin extracts on intestinal motility and diarrhoea in rodents was evaluated. The isolated extract depressed electrically, acetylcholine- and barium chloride-induced contractions in the isolated guinea-pig ileum, and showed a considerable effect in inhibiting the contractions induced by acetylcholine and barium chloride.[63]

4.7.9 Hepatoprotective Activity

Diseases like jaundice, cirrhosis and fatty liver are the most common types of liver diseases. There are many factors which contribute to liver diseases, one of the main ones being alcoholism. In ancient scripts like Vedas, the liver is named as "Takima" or "Yakna" and other synonyms are Kalakhanda, Jyotisthana, Yakritkhanda, Yakritpinda, Raktadhara and Raktashaya. The liver can be referred to as Yakrit as per Ayurvedic classics. Most modern medicines also describe different sorts of liver diseases and their correlation with Ayurvedic practices are described in Table 4.4. Many studies have highlighted the potential use of boswellic acids and their role in hepatoprotection. Jyoti *et al.* reported that the hexane extract of *B. serrata* showed good hepatoprotective activity in relatively lower doses (87.5 mg kg^{-1} p.o.) and the effect was comparable to that of Silymarin, the standard drug used for hepatoprotective disorders.[64] Similarly, Odeghe and co-workers reported that the methanolic leaf extract of *Boswellia dalzielii* Hutch exerted hepatoprotective action against carbon tetrachloride-induced hepatotoxicity in rats.[65] The bark extract of *Boswellia ovalifoliolata* was also studied for its potential role in paracetamol (PCM)-induced hepatotoxicity. This study found that it mitigated hepatic damage induced by acute intoxication of PCM marking a significant protective effect against lipid peroxidation, deviated serum enzymatic variables and maintained glutathione status toward control.[66]

4.7.10 Neuroprotective Effects

The anti-inflammatory and antioxidant activities of *Boswellia* account for its usage as a neuroprotective aid. The non-phenolic fraction of ether extract of *Boswellia* was found to exhibit sedative and analgesic effects in albino rats. The phytochemical analysis has established that incensole acetate isolated from the gum resin exerted anxiolytic and antidepressive effects, as well as a sedative effect; these findings suggest the role of *Boswellia* in neuroprotection. The influences of *B. serrata* resin extract on motor dysfunction and oxidative

Table 4.4 Correlation of liver disorders in Ayurveda and modern medicine.

Ayurvedic description	Modern medicine
Vataja Yakrit Vridhi	Steatohepatitis
Kaphaja Yakrit Vridhi	Non-alcoholic fatty liver disease
Pittaj Yakrit Vridhi	Non-alcoholic steatohepatitis
Amaja Yakritvridhi	Metabolic hepatitis
Asatmyaja Yakrit Vridhi	Autoimmune hepatitis
Krimija Yakrit Vridhi	Infective hepatitis
KaphajaYakrit Kshyaya	Cirrhosis liver
Pittaj Yakrit Kshyaya	Hepatic cirrhosis
Madhyaja Yakrit Kshyaya	Alcoholic cirrhosis
Krimija Yakrit Kshyaya	Infective cirrhosis
Visaja Yakrit Kshyaya	Drug-induced cirrhosis

stress markers were investigated in the intrastriatal 6-hydroxydopamine (6-OHDA) model of Parkinson disease. The study used the ethanolic extract at dosages of 125, 250 and 500 mg kg^{-1} for 3 weeks and reference drug neurotoxin 6-OHDA at a dosage of 12.5 μg. The behavioural assessments were done by rotational and elevated narrow beam tests and it was found that the extracts showed a dose-independent activity in the latency and total time.[67] Riazi and coworkers studied the effect of amino acid derivatives of βBA and AKBA on the inhibition of acetylcholinesterase (AChE) and have found that coupling with AKBA and βBA with valine and leucine amino acids can have positive impacts on strengthening and maintaining AChE levels.[68] Multiple preclinical studies have evaluated and proposed different mechanisms for the neuroprotective activities of the extract. Supplementation of *Boswellia* extract at dosages of 68.75 mg and 137.5 mg for a duration of 3 months showed a significant increase in brain ACh, and a significant decrease in brain and serum AChE activity, C-reactive protein, nuclear factor kappa (NF-κB), membrane cofactor protein (MCP-1), leukotriene B4 (LTB4), Bcl-2 and Aβ plaque burden in hippocampus and striatum. The anti-inflammatory action of the *Boswellia* extract illustrated that *Boswellia* acts in the hypothalamus and reduces tumour necrosis factor (TNF-α) levels, lowering PGE2 levels by inhibition of mPGES1, inhibition of cathepsin G and 5-LOX activity, reduction in the brain neurotransmitter levels, and reduction of various inflammatory biomarkers like interleukins IL-1β, IL-6, *etc.*[69]

4.8 Conclusion

The use of Shallaki/*Boswellia* is described in many traditional systems of medicine and also in most modern nutraceutical industries throughout the world. The major bioactive components responsible for these pharmacological activities include boswellic acids, of which the main ones are keto beta boswellic acid and acetyl keto beta boswellic acid. *Boswellia* is used as single herb or in a polyherbal formulation. Most of the pharmacological activities accounted for in *Boswellia* are attributed to the reduction of inflammatory biomarkers.

Acknowledgements

The authors gratefully thank the management of Plant Lipids Private Limited, Cochin, India, for their support and encouragement. they also wish to express their appreciation to their laboratory members for their active help and cooperation.

References

1. R. C. Majumdar, in *Medicine,* ed. D. M. Bose, S. N. Sen and B. V. Subbarayappa, A Concise History of Science in India Indian National Science Academy, New Delhi, 1971, pp. 213–273.
2. S. Sharma, *Realms of Ayurveda*, Arnold-Heinemann, New Delhi, 1979.

3. A. Hankey, *J. Altern. Complementary Med.*, 2001, 7, 567.
4. P. R. Manohar, Clinical Evidence in the Tradition of Avurveda, in *Evidence-Based Practice in Complementary and Alternative Medicine*, ed. S. Rastogi, Springer-Verlag, Berlin, Germany, 2012, pp. 67–78.
5. M. Padhi and S. Mahapatra, *Int. Rev. BioPhys. Chem.*, 2013, **4**, 74.
6. E. J. Blain, A. Y. Ali and V. C. Duance, *Phytother. Res.*, 2010, **24**, 905.
7. A. Sharma, A. S. Mann, V. Gajbhiye and M. D. Kharya, *Pharmacogn. Rev.*, 2007, **1**, 137.
8. M. Z. Siddiqui, *Indian J. Pharm. Sci.*, 2011, **73**, 255.
9. S. P. Ambasta, *The Wealth of India: Raw Material*, Council of Scientific and Industrial Research, New Delhi, 1988, pp. 203–205.
10. A. Chopra and V. V. Doiphode, *Med. Clin. North Am.*, 2002, **86**, 75.
11. S. Parasuraman, G. S. Thing and S. A. Dhanaraj, *Pharmacogn. Rev.*, 2014, **8**, 73.
12. S. Srivastava, V. K. Lal and K. K. Pant, *Phytopharmacology*, 2013, **2**, 1.
13. S. Karole, S. Shrivastava, S. Thomas, B. Soni, S. Khan, J. Dubey, S. P. Dubey, N. Khan and D. K. Jain, *J. Drug Delivery Ther.*, 2019, **9**, 453.
14. M. Spinella, *Altern. Med. Rev.*, 2002, 7, 130.
15. M. S. Chorgade, *Drug Discovery and Development*, Hoboken, NJ, 2007, John Wiley, vol. 2.
16. A. T. Leung and S. Foster, *Encyclopedia of Common Natural Ingredients Used in Food, Drugs, and Cosmetics*, Springer-Verlag, New York, 1996.
17. T. Shen and H. X. Lou, *Chem. Biodiversity*, 2008, **5**, 540.
18. B. Mahajan, S. C. Taneja and V. K. Sethi, *Phytochemistry*, 1995, **39**, 453.
19. P. K. Gupta, S. M. Samarakoon, H. M. Chandola and B. Ravishankar, *Ayu*, 2011, **32**, 478.
20. M. Majeed, S. Majeed, N. K. Narayanan and K. Nagabhushanam, *Phytother. Res.*, 2019, **33**, 1457.
21. K. Sengupta, A. V. Krishnaraju, A. A. Vishal, A. Mishra, G. Trimurtulu, K. V. Sarma, S. K. Raychaudhuri and S. P. Raychaudhuri, *Int. J. Med. Sci.*, 2010, 7, 366.
22. N. Kimmatkar, V. Thawani, L. Hingorani and R. Khiyani, *Phytomedicine*, 2003, **10**, 3.
23. S. Rastogi and S. T. Sivaraman, *Orthop. Today*, 2003, **1**, 63.
24. A. Amalraj, J. Jacob, K. Varma, A. B. Kunnumakkara, C. Divya and S. Gopi, *J. Herb. Med.*, 2019, **17**, 100276.
25. M. Totan, E. Antonescu, L. Duica, C. Roman-Filip and S. C. Silisteanu, *Rev. Chim.*, 2020, **71**, 513.
26. P. D. Kulkarni, N. D. Damle, S. Singh, K. S. Yadav, M. R. Ghante, V. H. Bhaskar, L. Hingorani and V. S. Gota, *Drug Metab. Pers. Ther.*, 2020, **35**, DOI: 10.1515/dmpt-2020-0104.
27. V. Karlapudi, A. V. V. Prasad Mungara, K. Sengupta, B. A. Davis and S. P. Raychaudhuri, *J. Med. Food*, 2018, **21**, 511.
28. A. Chopra, M. Saluja, G. Tillu, S. Sarmukkaddam, A. Venugopalan, G. Narsimulu, R. Handa, V. Sumantran, A. Raut, L. Bichile, K. Joshi and B. Patwardhan, *Rheumatology*, 2013, **52**, 1408.

29. V. Deo, B. Shrikhande, G. Ganu and U. Gajare, *Int. J. Res. Rev.*, 2021, **8**, 2349.
30. Charaka Samhita: Chikitsa Sthana, Adhyaya 28, Shloka-37, Hindi Commentary by Shastri and Chaturvedi, Varanasi, Chaukhambha Bharati Academy, 2006, p. 783.
31. S. Pandey, Y. K. Sharma and A. Kumar, *Int. Ayurvedic Med. J.*, 2013, 1–7.
32. A. Azadmehr, A. Ziaee, L. Ghanei, H. Fallah Huseini, R. Hajiaghaee, B. Tavakoli-Far and G. Kordafshari, *Iran. J. Pharm. Res.*, 2014, **13**, 1003.
33. A. Ahangarpour, H. Heidari, R. A. Fatemeh, M. Pakmehr, H. Shahbazian, I. Ahmadi, Z. Mombeini and B. H. Mehrangiz, *J. Diabetes Metab. Disord.*, 2014, **13**, 29.
34. F. Alawadi, H. Fatania and U. Shamte, *Diabetes Res.*, 1991, **18**, 163.
35. A. M. Shehata, L. Quintanilla-Fend, S. Bettio, J. Jauch, T. Scior, W. A. Scherbaum and H. P. Ammon, *Horm. Metab. Res.*, 2015, **47**, 463.
36. http://hdl.handle.net/10603/144119.
37. S. S. Agrawal, S. Saraswati, R. Mathur and M. Pandey, *Food Chem. Toxicol.*, 2011, **49**, 1924.
38. T. Glaser, S. Winter, P. Groscurth, H. Safayhi, E. R. Sailer, H. P. Ammon, M. Schabet and M. Weller, *Br. J. Cancer*, 1999, **80**, 756.
39. R. F. Hoernlein, T. Orlikowsky, C. Zehrer, D. Niethammer, E. R. Sailer, T. Simmet, G. E. Dannecker and H. P. T. Ammon, *J. Pharmacol. Exp. Ther.*, 1999, **288**, 613.
40. O. Karima, G. Riazi, S. Khodadadi, R. Yousefi, K. Mahnam, F. Mokhtari, T. Cheraghi, E. Hoveizi and A. A. Moosavi-Movahedi, *FEBS Lett.*, 2012, **586**, 4132.
41. M. K. Khan, I. A. Ansari and M. S. Khan, *Pharmacogn. Mag.*, 2013, **9**, 51.
42. O. Olaku and J. D. White, *Eur. J. Cancer*, 2011, **47**, 508.
43. X. Pang, Z. Yi, X. Zhang, B. Sung, W. Qu, X. X. Lian, B. B. Aggarwal and M. Liu, *Cancer Res.*, 2009, **69**, 5893.
44. S. K. Singh, S. Bhusari, R. Singh, A. Saxena, D. Mondhe and G. N. Qazi, *Vasc. Pharmacol.*, 2007, **46**, 333.
45. S. Kirste, M. Treier, S. J. Wehrle, G. Becker, M. Abdel-Tawab, K. Gerbeth, M. J. Hug, B. Lubrich, A. L. Grosu and F. Momm, *Cancer*, 2011, **117**, 3788.
46. J. R. Streffer, M. Bitzer, M. Schabet, J. Dichgans and M. Weller, *Neurology*, 2001, **56**, 1219.
47. X. Pang, Z. Yi, X. Zhang, B. Sung, W. Qu, X. Lian, B. B. Aggarwal and M. Liu, *Cancer Res.*, 2009, **69**, 5893.
48. P. Ravanan, S. K. Singh, G. S. Rao and P. Kondaiah, *J. Biosci.*, 2011, **36**, 297.
49. Y. Jing, S. Nakajo, L. Xia, K. Nakaya, Q. Fang, S. Waxman and R. Han, *Leuk. Res.*, 1999, **23**, 43.
50. G. Vimala and F. Gricilda Shoba, *Int. J. Microbiol.*, 2014, **2014**, 519590.
51. S. Singh, A. Khajuria, S. C. Taneja, R. K. Khajuria, J. Singh, R. K. Johri and G. N. Qazi, *Int. J. Phytother. Phytopharm.*, 2008, **15**, 408.
52. M. Asad and M. Alhomoud, *Pharm. Biol.*, 2016, **54**, 225.

53. K. Zeeyauddin, M. L. Narsu, A. Abid and M. Ibrahim, *J. Med. Allied Sci.*, 2011, **1**, 14.
54. M. L. Sharma, A. Kaul, A. Khajuria, S. Singh and G. B. Singh, *Phytother. Res.*, 1996, **10**, 107.
55. F. A. Badria, B. R. Mikhaeil, G. T. Maatooq and M. M. A. Amer, *Z. Naturforsch., C: J. Biosci.*, 2003, **58**, 505.
56. E. Schrott, S. Laufer, M. Lämmerhofer and H. P. T. Ammon, *Phytomedicine*, 2014, **21**, 786.
57. M. Schmiech, S. J. Lang, J. Ulrich, K. Werner, L. J. Rashan, T. Syrovets and T. Simmet, *Nutrients*, 2019, **11**, 2341.
58. A. M. Zimmermann-Klemd, J. K. Reinhardt, T. Nilsu, A. Morath, C. M. Falanga, W. W. Schamel, R. Huber, M. Hamburger and C. Gründemann, *Fitoterapia*, 2020, **146**, 104694.
59. P. Paranjpe and Sallaki – Boswellia Serrata, in *Indian Medicinal Plants Forgotten Healers: A Guide to Ayurvedic Herbal Medicine*, Chaukhamba Sanskrit Pratishthan Publishers, Delhi, pp. 233–234.
60. P. K. Kokkiripati, L. M. Bhakshu, S. Marri, K. Padmasree, A. T. Row, A. S. Raghavendra and S. D. Tetali, *J. Ethnopharmacol.*, 2011, **137**, 893.
61. S. Togni, G. Maramaldi, F. Di Pierro and M. Biondi, *Cosmet. Invest. Dermatol.*, 2014, 7, 321.
62. A. Petretti, R. Capezzera, C. Zane, E. Facchinetti and P. Calzavara-Pinton, *Planta Med.*, 2010, **76**, 555.
63. F. Borrelli, F. Capasso, R. Capasso, V. Ascione, G. Aviello, R. Longo and A. A. Izzo, *Br. J. Pharmacol.*, 2006, **148**, 553.
64. Y. Jyothi, J. V. Kamath and M. Asad, *Pak. J. Pharm. Sci.*, 2006, **19**, 129.
65. D. A. Onoriose, A. A. Uwakwe, C. C. Monago and O. B. Odeghe, *Indian J. Med. Healthcare*, 2012, **1**, 54.
66. B. U. Mahesh, S. Shrivastava, R. R. Pragada, V. G. Naidu and R. Sistla, *Chin. J. Nat. Med.*, 2014, **12**, 663.
67. P. Doaee, Z. Rajaei, M. Roghani, H. Alaei and M. Kamalinejad, *Avicenna J. Phytomed.*, 2019, **9**, 281.
68. E. Fathi, F. H. Katouli, G. H. Riazi, M. D. Shasaltaneh, E. Parandavar, S. Bayati, A. Afrasiabi and R. Nazari, *NeuroMol. Med.*, 2017, **19**, 136.
69. H. Ahmed, E. Mohamed and S. El-Dsoki, *Int. J. Pharm. Pharm. Sci.*, 2014, **6**, 384.

CHAPTER 5

Phytochemical, Botanical and Therapeutic Potential of Withania somnifera *in Ayurveda*

S. SUBASINI, P. ANANTHI, K. HEMKUMAR AND ANITHA PIUS*

Department of Chemistry, The Gandhigram Rural Institute – Deemed to be University, Gandhigram, Dindigul – 624302, Tamil Nadu, India
*Email: dranithapius@gmail.com

5.1 Introduction

Ayurveda is a 3000 year-old Indian healthcare system. It is a good system of primary healthcare in preventive and curative medicine that has been broadly accepted in many countries in Asia, having emerged in India. Dhanvantari, a physician, attributes Ayurveda to the Gods in Hindu mythology. Its early conceptions were laid down in the Atharvaveda chapter of the Vedas. The Caraka-Samhita and Susruta-Samhita, credited to Caraka, a physician, and Susruta, a surgeon, highlighted the golden era of Indian medicine (800 BCE–1000 CE). A schematic illustration of Ayurvedic components is shown in Figure 5.1. In many regions of the world, wherein Ayurveda is used completely or combined with contemporary medicine, it remains a widespread system of health treatment. Many Ayurveda practitioners engage in rural regions, treating at least 300 million people in India alone; as an analogy, Ayurveda is a significant force in primary healthcare training and deployment.[1]

Winter cherry, *W. somnifera*, popularly called Ashwagandha, Indian ginseng, and poison gooseberry, is an annual evergreen shrub in the family

Chemistry, Biological Activities and Therapeutic Applications of Medicinal Plants in Ayurveda
Edited by Augustine Amalraj, Sasikumar Kuttappan and Karthik Varma A. C.

Published by the Royal Society of Chemistry, www.rsc.org

Figure 5.1 Schematic representation of Ayurveda ingredients.

Solanaceae that grows in the Middle East, India, and some regions of Africa. This small, frail perennial shrub grows to between 14 and 30 inches tall. The leaves are dull green and elliptic, measuring 3.9–4.7 inches long on average. The ripe fruit is orange-red and the flowers are tiny, green, and bell-shaped. "*Somnifera*" is a Latin term meaning "to induce sleep." The word "ashwagandha" was formed as a result of the Sanskrit words "ashwa" (horse) and "gandha" (smell), indicating that the root seems to have a strong horse-like stench. *W. somnifera* is prone to several pests and diseases.[2]

Traditionally, Korean ginseng plays a vital role in Chinese medicine, similar to that of *W. somnifera* which also has a role in Ayurvedic tradition, so it is commonly called Indian ginseng. Ashwagandha has been widely used as an astringent, anticarbuncle, antistress, tonic, narcotic, and diuretic, and has been used in the treatment of worms, piles, goiter, leucoderma, neurological breakdown, sleeplessness, and constipation in Ayurveda, where it is known as the "Sattvic Kapha Rasayana". The root, leaves, and flowers have been used to treat fevers and inflammations, the flowers are often used as antibiotic ointments and diuretics, the fruits have been used to treat skin sores, tumors, and carbuncles, and the seeds have been shown to stimulate sperm count.[3] Ashwagandha is usually consumed as churna, a finely sieved powder blended with water, ghee (clarified butter), or honey. It boosts brain and nervous system function. It stimulates the function of the reproductive system, maintaining a healthy sexual and reproductive balance. As a renowned adaptogen, it improves the body's drought resistance. By boosting cell-mediated immunity, Ashwagandha helps the body resist illness. It also has significant antioxidant capabilities that protect cells from free radical damage. This chapter covers the potential of numerous phytochemicals, botanicals, and therapeutics of *W. somnifera* in Ayurveda.

5.2 Taxonomical classification

Kingdom: Plantae, Plants
Subkingdom: Trachcobionta, Vascular plants
Super division: Spermatophyta, Seeds plants
Division: Angiospermae
Class: Dicotyledons
Order: Tubiflorae
Family: Solanaceae
Genus: Withania
Species: Somnifera dunal

5.3 Distribution

W. somnifera is recognized by a variety of names, including Asgandh in Hindi, winter cherry in English, Gandha in Turangi, Aswagandha in Bengali, Ghodakun and Ghoda in Gujarati, Vajigandha and Pulivendram in Telgu, Vajigandha and Pulivendram in Tamil, Amukkira in Kannada, and Viremaddlinagadde in Goa. This plant flourishes in all of subtropical India's drier regions, such as Madhya Pradesh, Uttar Pradesh, the Punjab plains, and the northwestern states of Gujarat and Rajasthan. The distribution map in Figure 5.2 shows that it may also be found in Congo, South Africa, Egypt, Morocco, Jordan, Pakistan, Afghanistan, and elsewhere. When dried, the fleshy roots are cylindrical, tapering down gradually with a brownish white exterior and a pure white interior when broken. The major parts of the plant employed as a medicinal agent are the roots. The simple, ovate, glabrous

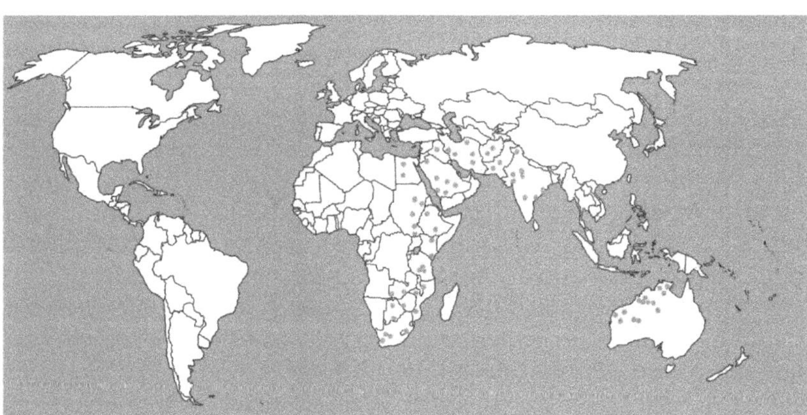

Figure 5.2 Distribution of *W. somnifera* throughout the world.

leaves are 10 cm long, thick underneath and sparse above. There are greenish or lurid-yellow flowers in axillary, umbellate cymes; berries are tiny, globose, orange-red when ripe, encased in the persistent calyx; and it has yellow seeds. The bright red fruit is gathered in the late fall and the seeds are dried for spring sowing.[4]

5.4 Composition

Using separation techniques such as thin-layer chromatography (TLC), column chromatography (CC), and high-performance liquid chromatography (HPLC), several phytochemicals have been extracted from methanolic extracts of various plant components (leaves, twigs, *etc.*) of *W. somnifera*. Phytochemicals such as pubesenolide, jaborosalactone D, 4*b*,27-dihydroxy-L-oxo-22*R*-witha-2,5,24-trienolide, and 2,3-dihydrowithaferin A-3*b*-*O*-sulfate are worthy of mention. Finally, the structures have been confirmed using single-crystal X-ray diffraction (XRD) measurements. Based on anomalous chlorine atom scattering in the crystal, the absolute configuration was determined to be 6*a*-chloro-5*b*,17*a*-dihydroxy withaferin A. A nuclear magnetic resonance (NMR) study has led to the discovery of other chlorinated chemicals. Using a combination of gas chromatography-mass spectrometry (GC–MS) and NMR on *W. somnifera* fruits extracted in *n*-hexane, aqueous methanol, and water, chemically diverse metabolites such as aliphatic and aromatic amino acids, phenolic acids, withanolides, organic acids, sugars, tocopherols and sterols, fatty acids, and polyols have been identified. According to some studies, more than 12 alkaloids, 40 withanolides, and many sitoindosides have been discovered in various areas of WS using different phytochemical assays. Only a few are important players in predicting pharmacological qualities among this wide variety of pharmacologically active principles. Withanolides are steroidal ergostane compounds with an oxidized C-1 position and a lactone performance between C-22 and C-26 atoms. Withanolides are a marker compound found in plants belonging to the Solanaceae family, particularly those belonging to the *Withania* genus. Withanolide D and withaferin A are the two types of withanolides. The plant extract has significant antioxidant properties and also neuroprotection ability under *in vitro* circumstances. Withanolides A, VI and withanoside IV have excellent anticancer and anti-inflammatory abilities. Withanolide A can stimulate nominal networks also. Withanolides suppress acetylcholinesterase and promote dendritic growth in human neuroblastoma cells.[3] Figures 5.3 and 5.4 show the phytochemicals present in *W. somnifera* and structural representations of some predominant phytochemicals.

Withasteroids, which are strong neurostimulants, were highly appreciated in traditional Indian medicine, according to Maumita *et al.* Furthermore, withaferine A and withanolide A, the two most important withanolides, have C-28 steroidal lactones constructed on an intact or rearranged ergostane framework. They can be found in the aerial and root parts of the plant. Withanolides are typically found in leaves, ranging from 0.001 to 0.5% by

Figure 5.3 Phytochemicals in *W. somnifera.*

weight. The production of withasteroid differs from plant to plant. According to tests, the production of the main withasteroids is naturally controlled, using hybrids of different chemotypes of *W. somnifera*. Asteroids with new ring substitutions that aren't found in the parent plants are produced by these hybrids.[5]

Apart from chemically active compounds like flavonoids, alkaloids (tropine and anferine), terpenoids, steroidal lactones (anolides), and saponins in extracts from different parts of the plant, bioactive compounds like ashwagandhine, cuscohygrine, anaferineisopelletierine, sitoindosides, anahygrine, withanamides, and tropine have been isolated and identified from *W. somnifera*. Pure compounds are less efficient in some circumstances than *W. somnifera* extracts containing a range of components. Antifungal and antibacterial activities of *W. somnifera* extracts have been reported.[6]

5.5 Therapeutical Potential

5.5.1 Neuroprotective Effects

W. somnifera has been shown to have neuroprotective properties. Leaf extract of *W. somnifera* and its components reduced scopolamine-induced toxicity in glial and neuronal cells. In neuronal cells, neurofilament-H (NF-H), microtubule-associated protein-2 (MAP-2), postsynaptic density protein-95 (PSD-95), and growth-associated protein-43 (GAP-43) were all decreased. All of these qualities were reduced after pretreatment with varying dosages of withanone. Withanone may be useful as a neuroprotective agent, according to these data.[7] Glycowithanolides from *W. somnifera* demonstrated significant antioxidant activity in the cortex and striatum of the rat brain.

According to Satyan and Bhattacharya, the root powder of *W. somnifera* preserved dying neurons in the Cornu Ammonis 2 (CA2) and Cornu Ammonis 3 (CA3) subareas of rats in the hippocampus region subjected to immobilization stress and rescued mice from streptozotocin-induced oxidative stress. In human neuroblastoma cell lines, *W. somnifera* root extract greatly boosted neurite outgrowth. Withanolide A improved axonal and dendritic

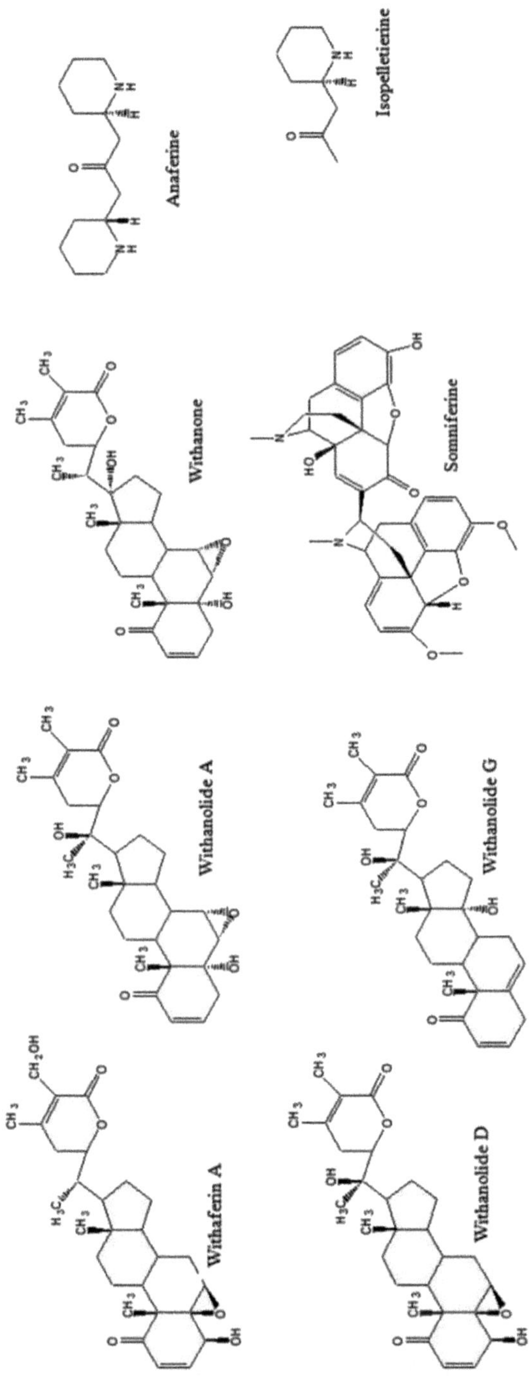

Figure 5.4 Structural representation of predominant phytochemicals.

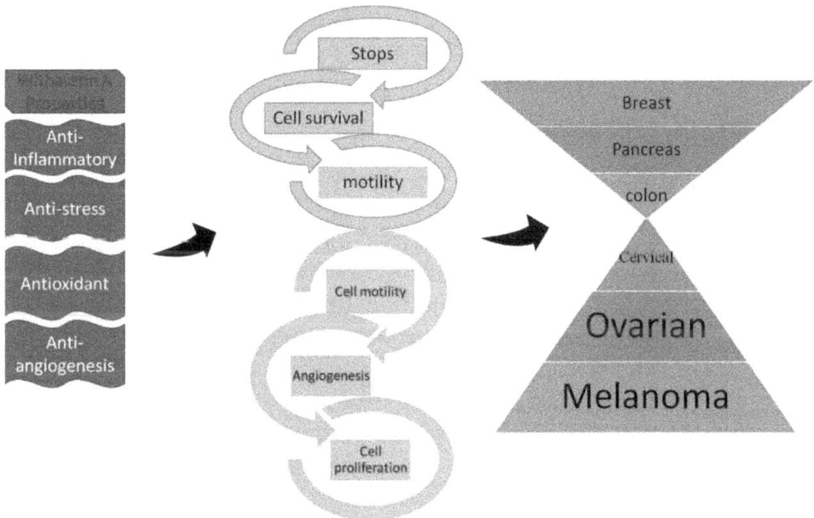

Figure 5.5 Important biological activities of withaferin-A.

modification and synaptic repair in cortical neurons of rats treated with amyloid beta-peptide (Ab).[8] *W. somnifera* extract decreased kainic acid toxicity by decreasing the oxidative damage. Leaf extract of *W. somnifera* inhibited glutamate-induced excitotoxicity in retinoic acid-differentiated C6 (rat glioma cell line) and IMR-32 (human neuroblastoma cell line) cells.[9] Figure 5.5 shows the overall therapeutic properties of phytochemical withaferin-A.

5.5.2 Anti-Parkinson Effects

In rat models of Parkinson's disease, *W. somnifera* extracts were found to protect against functional deficits and pathology caused by 6-hydroxydopamine (6-OHDA). According to one study,[10] the antioxidant mechanisms in *W. somnifera* restored striatal dopamine and its metabolite levels.

W. somnifera root extract restored antioxidant balance, decreased oxidative stress, stabilized catecholamine levels, and reversed functional deficits in the midbrain in rats treated with 1-methyl-4-phenyl-1,2,3,6-tetrahydropyridine. *W. somnifera* extract treated oxidative stress, mitochondrial respiratory chain enzymes, cholinergic dysfunction, and dopamine levels. There was an increase in lethality as well as locomotory function. Due to its antioxidant and antiinflammatory properties and its capacity to restore mitochondrial function, *W. somnifera* root powder reduced rotenone-induced toxicity in the cerebellum and striatum of the rat brain striatal dopamine levels.

In a rat model of Parkinson's disease, an ethanolic root extract of *W. somnifera* reduced toxicity caused by Maneb and Paraquat. Tyrosine hydroxylase expression was preserved, nigral dopamine levels were normalized, and locomotor activity was increased thanks to the antiinflammatory, antiapoptotic, and antioxidant properties. Before glial fibrillary acidic protein (GFAP) levels declined, *W. somnifera* suppressed inducible nitric oxide (NO) synthase expression, downregulated pro-apoptotic BCL2-associated X protein (Bax) and increased antiapoptotic B cell lymphoma (BCL-2) protein expression.[11]

5.5.3 Breast Cancer

According to one study, breast cancer is one of the most common malignancies connected to Withaferin-A (WA) activity. Estrogen has a role in the evolution and growth of breast tumors [through the nuclear estrogen receptor (ER)]. Steroid sex hormones influence breast cancer's onset, course, and prognosis. The significance of ER in breast cancer is widely understood and targeted treatments are now being explored. ER-positive breast cancers account for 75–80% of all breast cancers. Antiestrogens, such as tamoxifen, have been demonstrated to reduce or eliminate tumors by lowering endogenous estrogen levels (*e.g.*, aromatase inhibitors) or interfering with ER activation (antiestrogens, such as tamoxifen).[12]

5.5.4 Cervical Cancer

Cervical cancer is the second most common cancer globally, and it is regarded as a significant worry in women's health. Cervical cancer is caused by chronic infection with particular types of human papillomavirus (HPV). The primary cause of cervical cancer, WA, suppresses human papillomavirus oncogenes and upregulates tumor suppressor proteins, resulting in p53-dependent death of cancer cells in the human cervical region. Various cervical cell lines, such as C33a, SiHa, CaSki, and HeLa, have displayed dose-dependent anticancer activity when exposed to WA. Withaferin therapy resulted in the downregulation of HPV E6 and E7 oncoproteins and increased p53, resulting in the activation of different apoptotic markers. The elevated level of p21 cip1/waf1 and its interaction with proliferating cell nuclear antigen (PCNA) causes G2/M cell cycle arrest (which is connected with the alteration of p34 cdc2, cyclin B1, and PCNA levels).[13]

5.5.5 Ovarian Cancer

Ovarian cancer is the sixth most common malignancy in women and the seventh leading cause of death worldwide.[14] WA has shown exceptional success in treating ovarian cancer, mainly when combined with other cancers. It has been revealed that the anticancer medication doxorubicin (DOX) has improved therapeutic activity in ovarian cancer when coupled with WA *via* reactive oxygen species (ROS)-induced necroptosis. The combination treatment of doxorubicin and withaferin increases cell death and inhibits the proliferation of various epithelial cells such as CaOV3, A2780, and A2780/CP70 in a concentration- and time-dependent manner. Doxorubicin in combination with *W. somifera* increases ROS production, necroptosis induction, and LC3B levels. In *in vivo* xenograft models of ovarian cancer and *in vitro* 3D tumors, tumor cell proliferation was reduced by 70–80%. As a result, the study shows that combining WA with DOX can minimize treatment side effects and dosage.[15] In the case of ovarian cancer, Kakar *et al.* demonstrated that WA inhibits tumor cell formation and spread by inhibiting potential cancer stem cells. The researchers found that in athymic mice, orthotopic ovarian tumors injected with cisplatin and WA in a human ovarian epithelial cancer cell line (A2780) showed a 70–80% reduction in tumor formation and complete suppression of metastasis to other organs as compared to untreated controls. By injecting human ovarian epithelial cancer cell lines (A2780) with WA and cisplatin, the authors demonstrated that in nude mice with orthotopic ovarian tumors, WA alone or combined with cisplatin resulted in a 70–80% reduction in tumor development, as well as complete inhibition of metastases to other organs, as compared to untreated mice.[16]

5.5.6 Role of *W. somnifera* in Anxiety

In the laboratory, many recent investigations have looked into the anti-anxiety qualities of *W. somnifera*. In rats and humans, the plant has been shown to reduce anxiety. The aqueous concentration of roots of *W. somnifera* has demonstrated anxiety effects after extraction with chloroform and spray drying in a dose-dependent manner. The antianxiety effect of *W. somnifera* was comparable to that of lorazepam, a benzodiazepine medicine used to treat anxiety disorders. The gamma-aminobutyric acid (GABA) mimic effect of glycowithanolides of the plant, known as withaferin by HPLC, has been attributed to enabling the antianxiety effect of this extract.[17]

In a recent study, *W. somnifera* root extract, that is commercially available containing 2.8% (w/w) withanolides, was found to be sufficient in decreasing stress generated by electric foot shock in rats. Other phytochemicals found in *W. somnifera*, such as withalongolide-A, an analog of WA and withanamides, are thought to be responsible for this effect. Figure 5.6 shows root and leaf extracts of *W. somnifera* possess anxiolytic properties, which has been validated under different paradigms. In another chronic stress study, rats were subjected to a mild and unexpected foot shock, monitored once a day for 21 days. The roots of *W. somnifera* were extracted with aqueous : ethanol (1 : 1) aqueous : chloroform (1 : 1) to study, the dry powder was combined with 0.3% carboxymethyl cellulose. *W. somnifera* was given once a day for 21 days, 1 hour before footshock. Porsolt's swim stress-induced behavioral despair test revealed that rats given *W. somnifera* had lower levels of behavioral depression. In the stress-induced behavioral despair test, chronic stress resulted in a significant increase in the immobility period and several escape failures, as well as a decrease in the number of avoidance responses, but *W. somnifera*'s adaptogenic properties significantly altered these effects in a dose-dependent manner.

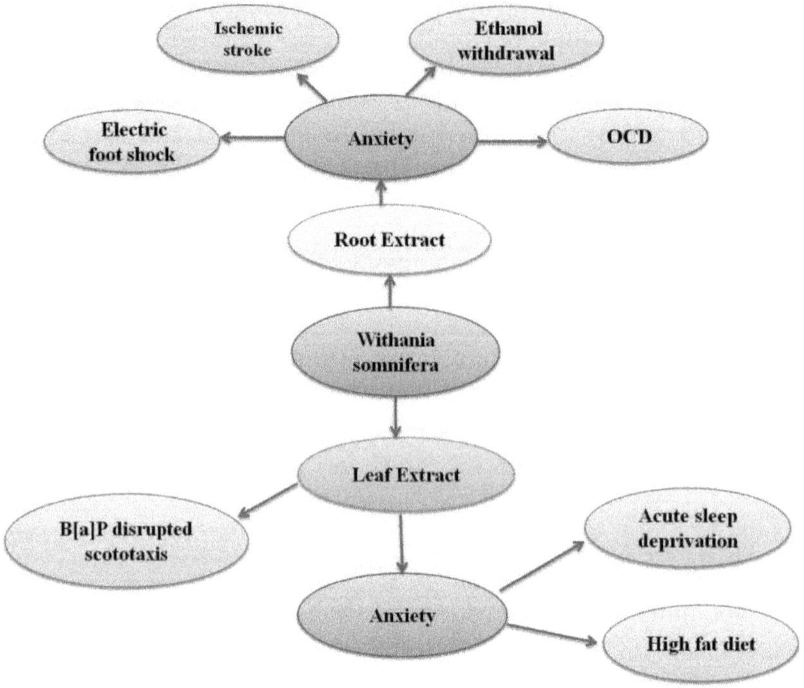

Figure 5.6 Root and leaf extracts of *W. somnifera* possess anxiolytic properties, which have been validated under different paradigms.

Alcoholism is a significant health concern that has a significant negative impact on one's health and general quality of life. An addict's abstinence from alcohol usually results in anxiety and despair. The effect of *W. somnifera* root extract on ethanol suspended in 0.5% (w/v) carboxymethylcellulose was investigated in a rat model of ethanol withdrawal in which the rats were given free access to ethanol for 16 days *via* a liquid diet (4.6% v/v on day 1, 7.5% v/v on day 2, and 9% v/v from the third day onwards for 16 days). Animals were provided an ethanol-free liquid diet starting on day 16. The treatment of *W. somnifera* ethanolic root 3 days after withdrawal lowered anxiety in the mice. The extract's antianxiety impact has been shown to be related to the plant's gamma-aminobutyric acid (GABA) mimic and adaptogenic capabilities, which help to downregulate gamma-aminobutyric acid type ($GABA_A$) receptors. Figure 5.7 shows a graphical abstract depicting pleiotropic roles of Ashwagandha in neuroplasticity mediated by modulation of multiple pathways such as synaptic plasticity, cell survival, and senescence/apoptotic cell death. Obsessive-compulsive disorder (OCD), characterized by irrational thoughts and concerns that lead to compulsive activity, is an anxiety-related mental diseases. Fear of germs or the urge to organize objects precisely are common symptoms in OCD. Antidepressants that block serotonin (5-hydroxytryptamine or 5-HT) reuptake are used to treat OCD. The effect of

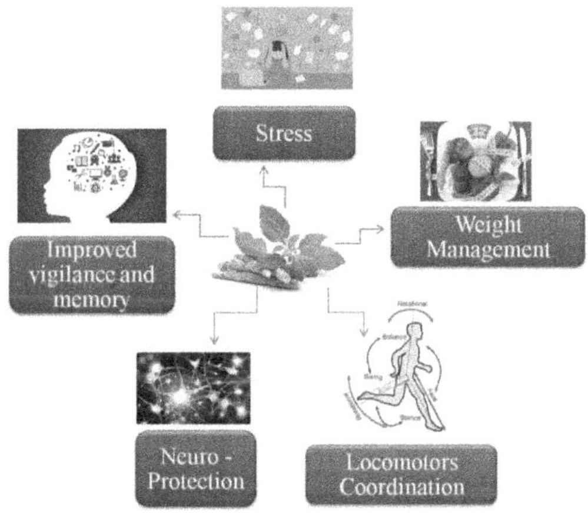

Figure 5.7 A graphical abstract depicting pleiotropic roles of Ashwagandha in neuroplasticity mediated by modulation of multiple pathways such as synaptic plasticity, cell survival, senescence/apoptotic cell death.

W. somnifera methanolic aqueous root extract on mice's marble-burying behavior, which is a familiar model for OCD, has been examined. The results of both parts are comparable to fluoxetine, a common anti-OCD medication that blocks 5-HT uptake. Both aqueous and methanolic extracts are thought to have a similar mechanism of action as fluoxetine. At the level of postsynaptic 5-HT receptors, both sections have been shown to promote serotonergic transmission.[18]

When patients have *Katigraha* (lumbago), they have muscle pain, usually in the lower back. Due to poor postural habits and psychological variables, approximately 60% of the general public and 80% of the working population will experience lower back pain at some point in their lives. One combination to ease *Katigraha* is *Madhur Dravya* (*Ashwagandha Taila*). *Madhur Dravya* is one of the most effective methods for reducing pain and stiffness.[19]

5.5.7 Antivenom

Venom hyaluronidases aid in the rapid spread of toxins in humans by damaging the integrity of the tissue's cellular matrix. From *W. somnifera*, a hyaluronidase resister has been isolated. This glycoprotein inhibits the hyaluronidase activity of viper (*Daboia russelii*) and cobra (*Naja naja*) venoms, as evidenced by a zymogram assay and differential activity staining of skin tissues. At a concentration of 1 : 1 w/w toxin, it entirely blocked the enzyme's training. External application of the plant extract as an antidote to a snakebite has been provided to victims in some rural parts of India, and appears to have a scientific basis. Antitoxin–phospholipase A2 (PLA2) glycoprotein obtained from *W. somnifera* inhibited the PLA2 activity of cobra venom. The relevance of these new PLA2 toxin inhibitor groups in snake biology and the creation of novel therapeutic reagents for snake envenomation has been studied.[20]

5.5.8 Immunity

In normal and cyclophosphamide-treated mice, the roots of *W. somnifera* show cytoprotective and immune-potentiating effects by increasing the levels of interferon (IFN), interleukin (IL)-2, and granulocyte-macrophage colony-stimulating factor. Iron is abundant in the root, which helps to increase the red blood cell count. *W. somnifera* has a more subtle influence on the immune system than merely decreasing the immunological/inflammatory response. The active compound (withanolide A) found in the roots of *W. somnifera* enhances T-helper 1 (Th1) cytokine expression and CD4 and CD8 counts. It also improves natural killer cell activity in immune-suppressed animals, resulting in a faster recovery of CD4+ T cells. Apart from the features listed above, stimulated macrophage function, as seen by increased production of nitrile, IL-2 and TNF-2, reduced IL-4 substantially while having little effect on IL-10, implying that it only influenced the Th1 cytokine profile. The root powder of this plant stimulates cell-mediated immunity, immunoglobulin M (IgM) and immunoglobulin G (IgG) production, and lymphocyte proliferation and differentiation.

Herbal medications are becoming increasingly popular worldwide as they have fewer adverse effects than synthetic drugs. Ayurveda asserts that certain plants have healing properties. A great deal of work continues to be invested into developing this multipurpose drug-producing plant. However, the data are disjointed. As a result, this review has attempted to synthesize all accessible materials. This plant has been the subject of much research for the past three decades. However, further research into this traditional medication's biological activities, efficacies, and mechanisms is still required. *W. somnifera*, *W. ashwagandha*, and *W. coagulans* are the three species of *Withania* discovered in India. The main chemicals discovered in these three species are withanolides, with some withanolides being unique to each of them. *W. somnifera* and *W. ashwagandha* include withaferin A as the main component, but *W. coagulans* has multiple levels of coagulin L. *W. somnifera* has a unique thio-dimer of withanolide called Ashwagandhanolide. This plant has been used for centuries as an antioxidant, adaptogen, aphrodisiac, liver tonic, antiinflammatory agent, anticancer agent, central nervous system depressant, hepatoprotective, and astringent, and more recently as an antibacterial, antihyperglycemic, hypolipidemic, and antitumoral, as well as to treat ulcers, senile dementia, Parkinson's disease, and Alzheimer's disease. In general, it has increased its therapeutic value. The broad activity described for the extracts, fractions, and withanolides obtained from this important medicinal plant bodes well for future studies. Withanolides could become a significant player in the current pharmaceutical industry. The active compounds' pharmacological value will be enhanced through large-scale isolation, chemical transformations, and synthesis. The pharmacophores of withanolides that are pharmacologically active have yet to be discovered. All of these benefits demonstrate *W. somnifera*'s importance in natural product research. Despite their vast therapeutic potential,

Ashwagandha variations require a multifaceted strategy to become more competitive. It is necessary to improve the pharmacological effects of withanolide by selecting and improving chemotypes that produce large amounts of the desired withanolide.[21]

5.5.9 Antiaging

Ashwagandha Rasayana is a well-known Ayurvedic medicine that people of all ages can use. Nerves, bone marrow. and reproductive organs are all revitalized by it. Regular consumption is thought to delay aging, correct sense organ defects and muscular hypotrophy, regenerate reproductive organs, and boost fertility. Stress, weakness, nervous tiredness, sexual debility, geriatric difficulties, memory loss, muscle weakness, sleeplessness, and cough can all benefit from it. It is also said to slow down the aging process and speed up the body's anabolic processes. Ashwagandha Rasayana has been found in preclinical tests to have radioprotective properties, reducing radiation-induced emaciation, decreasing organ weight, and lowering serum alanine amino transaminase levels and lipid peroxidation. Numerous research works have demonstrated that *W. somnifera* promotes physical and mental health and rejuvenates the body in weakened situations, increasing longevity. Ashwagandha also aids in treating epilepsy, stress, and neurodegenerative diseases (such as Parkinson's and Alzheimer's disease, tardive dyskinesia, and cerebral ischemia) and general well-being.[22]

5.5.10 Antidiabetic Effect

The leaf and root extracts of *W. somnifera*, when administered to diabetic rats induced with streptozotocin, decreased the blood glucose level and restored the biochemical parameters. This study showed that the leaf and root extract of *W. somnifera* has significant antidiabetic activity. The methanolic root extract of *W. somnifera* is evident to resist DPP-4 (dipeptidyl peptidase-4), a salient target for antidiabetic activity. Catechin is a crucial compound that is responsible for providing this effect. The root and leaf extracts of *W. somnifera* showed hypoglycemic activity on diabetic rats induced by alloxan, which is similar to the standard glibenclamide. The root extracts of *W. somnifera* made into capsules were given to seven patients with non-insulin-dependent diabetes mellitus. This study revealed this preparation is comparable with the standard oral hypoglycemic drug.[23]

5.6 Ethnomedicinal Practice

In India, *W. somnifera* is widely distributed in the drier parts of the country, especially in Punjab, Gujarat, Uttar Pradesh, Maharashtra, West Bengal, and Rajasthan. The various ethnomedical uses of *W. somnifera* are presented in Table 5.1

Table 5.1 Traditional uses of *W. somnifera*.

S. no.	Location	Part of the plant	Ethnomedicinal practice	Method of preparation	Route of administration	Ref.
1	East Godavari—Andhra Pradesh	Root	Combats anemia; enhances sperm count	Root powder with equal amounts of sugar and ghee	Oral (twice few days); oral (daily once for 40 days)	24
2	Barpeta—Assam	Root	Female infertility	Powder with tablet (Dron, Sidhelota, and Slilax) and Rangajoba juice	Oral	25
3	Buxur—Bihar	Root	Diuretic seed	Seed decoction	Oral	26
4	Chhattisgarh	Root and leaves	Sexual and general weakness, headache, pregnancy care, rheumatism, debility, aging, gonorrhea, ringworm, bedsores, hemorrhoids, abscesses, and smallpox	Root powder, decoction, leaf paste, and ash	Oral, topical	27
5	Raigarh—Chattisgarh	Root	Male sterility	Powdered root with milk	Oral for 7 days	28
6	Jhajjar—Haryana	Root and stem barks	Fever, cough, asthma, migraine	Powder or paste	Oral, twice daily for 7 days	29
7	Keylong—Himachal Pradesh	Root	Tonic, hypnotic, diuretic, narcotic, sedative, abortifacient, rheumatism, cough, delayed aging and graying of hair, provides physical and mental strength	Powder	Oral, 200–500 mg of root powder with one glass of warm milk before sleep for 15 days	30

8	Jharkhand	Latehar Root	Venomous snakebite	Powder, with other plants	Oral, 1 mg of the powder is given to the victim 7 times with 3 min gaps	31
9	Jammu and Kashmir	Leaves	Memory enhancer, obesity	Fresh leaves	Oral, chewing on an empty	32
10	Chikmagalur—Karnataka	Root	Cardiovascular problems	Decoction	Oral, with root decoction of *F. microcarpa*	33
11	Kerala	Seeds	Oligomenorrhea	Pounded and cooked with rice as a form of porridge	Oral, at night time	34
12	Jhabua—Madhya Pradesh	Whole parts of the plant	Fever and weakness	Whole plant used as a tonic, root extract in milk	Oral, daily in the morning	35
13	Pune—Maharashtra	Whole parts of the plant	Menstruation	Powder	5–10 g with warm water twice a day	36
14	Balasore—Orissa	Bark	Piles	Decoction	Oral	37
15	Salem—Tamilnadu	Root, leaves	Anthelmintic	Leaves decoction, root powder mixed with black pepper	Oral	38
16	Adilabad—Telangana	Leaves	Backache	With root bark of *H. suffruticosus* to make small pills	Oral, one pill with goat milk once a day for 3–4 days	39
17	Tripura	Root bark	Leucorrhoea	5 g with *A. berbadensis* juice (5 g), white Shilajit (1 g), *Isapgula* husk (1 g)	Oral, 2 teaspoonsful with water, twice post-meal for 3–5 days	40
18	Aligarh—Uttar Pradesh	Root	Weakness of sexual organs, premature ejaculation	Root powder	Oral, 15 g with 5 ml honey and 200 ml cow milk twice a day	41
19	Dehradun—Uttarakhand	Root and leaves	Urinary disorders, fever, insomnia	Leaf juice, root powder	Oral	42
20	Bankura—West Bengal	Root	Menstrual disorders	Extract with an equal amount of leaf extract of *C. roseus*	Oral mixture for 15 days	43

5.7 Clinical Trials

Ashwagandha's safety and efficacy have clearly been demonstrated in many clinical investigations. According to the findings, Ashwagandha root extract is safe, lowers stress, and improves quality of life. Ashwagandha's traditional use has prompted its clinical use in various illnesses, some of which have been researched in clinical studies. According to a meta-analysis of clinical studies, Ashwagandha supplementation enhances sperm count, semen volume, and sperm motility in oligospermic males. Ashwagandha root extract with a high concentration helps to improve female sexual function. Another clinical trial found that those who use Ashwagandha as a supplement had more muscle growth and strength, which can aid with high-intensity physical exercise. In an open-label exploratory feasibility trial, *W. somnifera* supplemental powder protected the loss of lean muscle mass, decreased tiredness, and delayed weight loss in cancer patients. A purified aqueous extract of *W.* somnifera raised the pain threshold, time, and pain tolerance due to mechanical pain in healthy human volunteers. Patients with schizophrenia benefit from adjunctive therapy with a standardized extract of *W. somnifera*, with significant differences in the Positive and Negative Syndrome Scale (PANSS) total and few adverse effects.

Similarly, in a randomized, placebo-controlled, double-blind research, *W. somnifera* extract demonstrated encouraging benefits in treating sadness and anxiety symptoms in schizophrenia patients. Supplementing a standardized Ashwagandha extract (roots and leaves) enhanced salivary dehydroepiandrosterone sulfate (DHEA-S) and testosterone levels considerably, but not cortisol or estradiol levels, in healthy individuals. In stressed individuals, supplementation with a standardized Ashwagandha root extract resulted in good anxiolytic effects, including a decrease in cortisol and DHEA-S, as well as a positive, albeit non-significant, trend of increased testosterone. In another clinical research, Ashwagandha root extract enhanced sleep quality and anxiety symptoms in individuals with insomnia and anxiety. Supplementing standardized Ashwagandha extract eased post-chemotherapy tiredness conditions in double-blind, randomized research. In clinical individuals, KSM-66 Ashwagandha enhanced oxygen intake and improved quality of life markers. In combination with antituberculosis medications, *W. somnifera* root extract was used as an adjuvant in double-blind placebo-controlled research in HIV patients with pulmonary tuberculosis and with higher CD4 and CD8 counts.[44]

5.8 Market Potential and Marketed Formulations

As the need for dietary supplements has risen, the Ashwagandha market has gone mainstream. Also, with a rise in the natural product sector, customer knowledge and trust in herbal/botanical supplements is expanding. In the United States, the Natural Channel sells most Ashwagandha supplements. Sales in the two primary retail channels have consistently climbed from an estimated \$4.53 million in 2014 to an estimated \$12.24 million in 2017, representing a 39% annual growth. On the Indian market in October 2018, high-quality dried Ashwagandha roots cost between US \$2.46 and \$3.56 (but lower-grade roots, known as tar, were offered for as little as the US \$1.50), with it costing US \$0.34–0.82 for dried Ashwagandha leaves. Ashwagandha is sold as a powder, a liquid extract, and a dried root. Conventional dose forms such as tablets and capsules are common, while contemporary dosage forms are gaining favor. Due to increased health awareness and Internet knowledge, the powder form is employed in health supplements and has maintained a constant growth rate. In Asia Pacific, North America, Japan, Western Europe, and the Middle East, Ashwagandha is widely used. The European Region, particularly Germany, invests the most in botanical research. Recent product launches, such as Gujarat Cooperative Milk Marketing Federation's "Amul Memory Milk," touted as a memory booster, are intriguing (GCMMF, Indian dairy manufacturing company). Ashwagandha with sterilized homogenized flavored toned milk is used in this formulation. KSM-66, AuNutra Industries Inc., Himalaya Worldwide Holdings Ltd, Amax NutraSource, Inc., Carrubba Inc., Sabinsa Corporation, and Banyan Botanicals are just a few of the prominent global firms that share the manufacturing market for *W. somnifera*.[45]

5.9 Conclusion

The multipurpose plant *W. somnifera* has sparked the interest of researchers more than other medicinal plants in the Solanaceae family, thanks to its traditional therapeutic uses, nutraceutical potential, and pharmacological properties. *W. somnifera* is a well-known ethnopharmacological and pharmaceutical therapeutic plant. It has been utilized in folklore medicine in India since antiquity, and has a wide range of therapeutic uses in Indian medical systems. Cancer, arthritis, diabetes, sore eyes, asthma, pyrexia, abscesses, hemorrhoids, ulcers, jaundice, and infections are just a few of the problems that traditional healers can treat with this plant. Multiple preclinical investigations in animal models have found promising antidepressant, antibacterial, antiinflammatory, cardioprotective, antioxidant, antifungal, hepatoprotective, and hypoglycemic properties (in mice, rats, and rabbits). Phytochemists are needed to extract the bioactives accountable for the biological activities so that novel formulations may be created. As a result, it is expected to be a valuable naturopathic remedy or dietary supplement for treating brain disorders, including Parkinson's disease, memory loss, stress-related disorders, cancers, and others. In preclinical investigations, *W. somnifera* is a viable option for modifying plasticity, anxiety, and neuroinflammation in various central nerve system disorders, with overall neuroprotective benefits. *W. somnifera* and its chemical components have been shown in animal experiments to have antiinflammatory, antioxidant, anti-apoptotic, angiogenesis, and ER stress-reducing capabilities. The prospect of using *W. somnifera* as a multipurpose medicinal agent to treat various clinical diseases and disease models is promising.

References

1. https://www.britannica.com/science/Ayurveda (accessed 16 November 2021).
2. Drugs.com [Internet], Ashwagandha Information from Drugs.com, https://www.drugs.com/npp/ashwagandha.html.
3. P. Subhabrata, C. Shreya, A. Uttpal, D. Swarnali, N. Samapika, G. Mimosa, C. S. Suchismita, T. P. Manoj, K. Ramesh, P. Jarosław and D. Abhijit, *Biomed. Pharmacother.*, 2021, **143**, 112175.
4. S. Veena, S. Sadhana, J. Pracheta and P. Ritu, *Int. J. Pharmtech Res.*, 2011, **3**, 187.
5. B. Maumita, J. Sumita and T. David, *Plant Cell Rep.*, 2007, **26**, 599.
6. H. Mervat, Z. M. Mohamed Salem, I. Behiry and M. Ali, *Appl. Surf. Sci.*, 2020, **8**, 113.
7. N. J. Dar, J. A. Bhat, N. K. Satti, P. R. Sharma, A. Hamid and M. Ahmad, *Mol. Neurobiol.*, 2017, **54**, 5061.
8. S. K. Bhattacharya and K. S. Satyan, *Indian J. Exp. Biol.*, 1997, **35**, 565.
9. M. S. Parihar and T. Hemnani, *J. Biosci.*, 2003, **28**, 121.
10. M. K. Ahmad, A. A. Mahdi, K. K. Shukla, N. Islam, S. Rajender, D. Madhukar, S. N. Shankhwar and S. Ahmad, *Fertil. Steril.*, 2010, **94**, 989.
11. J. Prakash, S. Chouhan, S. K. Yadav, S. Westfall, S. N. Rai and S. P. Singh, *Neurochem. Res.*, 2014, **39**, 2527.
12. R. M. O'Regan and V. C. Jordan, *Lancet Oncol.*, 2002, **3**, 207.
13. R. Munagala, H. Kausar, C. Munjal and R. C. Gupta, *Carcinogenesis*, 2011, **32**, 1697.
14. A. Jemal, R. Siegel, J. Xu and E. Ward, *Cancer J. Clin.*, 2011, **61**, 133.
15. S. Vishnu Sankar, M. Reshmi, R. Abbas, B. Simona Bungau, C. Dana, A. Lotfi and M. Delia, *Environ. Sci. Pollut. Res.*, 2020, **27**, 26025.
16. S. S. Kakar, M. Z. Ratajczak, K. S. Powell, M. Moghadamfalahi, D. M. Miller, S. K. Batra and S. K. Singh, *PLoS One*, 2014, **9**, 107596.
17. S. K. Bhattacharya and A. V. Muruganandam, *Pharmacol., Biochem. Behav.*, 2003, **75**, 547.
18. I. Baitharu, V. Jain, S. N. Deep, K. B. Hota, S. K. Hota, D. Prasad and G. Ilavazhagan, *J. Ethnopharmacol.*, 2013, **145**, 431.
19. B. Snehal Vasant and P. Shweta, *Int. J. Ayurvedic Med.*, 2020, **11**, 310.
20. L. Girdhari and A. C. Rana, *Pharmacogn. Rev.*, 2007, **1**, 129.
21. J. Joh, *Int. J. Pharm. Sci. Res.*, 2014, **5**, 2131.
22. M. S. Baliga, S. Meera, A. R. Shivashankara, P. L. Palatty and R. Haniadka, The Health Benefits of Indian Traditional Ayurvedic Rasayana (Anti-aging) Drugs: A Review, in *Foods and Dietary Supplements in the Prevention and Treatment of Disease in Older Adults*, ed. R. R. Watson, Academic Press, 2015, ch. 16, pp. 151–161.
23. R. Udayakumar, S. Kasthurirengan, T. S. Mariashibu, M. Rajesh, V. R. Anbazhagan, S. C. Kim, A. Ganapathi and C. W. Choi, *Int. J. Mol. Sci.*, 2009, **10**, 2367.

24. V. Rampilla and K. S. Mahammad, *Eur. J. Med. Plants*, 2015, **9**, 1–29.
25. V. Singh, D. K. Bhattacharjya and J. Gogoi, *Pleione*, 2019, **13**, 82.
26. A. Singh, M. K. Singh and R. Singh, *J. Pharmacogn. Phytochem.*, 2013, **2**, 2.
27. N. Husain, T. H. Trak and B. Panday, *Indian J. Sci. Res.*, 2018, **9**, 37.
28. S. P. Jain and J. Singh, *Indian J. Nat. Prod. Resour.*, 2010, **1**, 109.
29. J. P. Yadav and S. Kumar, *Environ. Conserv. J.*, 2003, **4**, 59.
30. V. Sharma, B. Singh, R. C. Gupta and H. S. Dhaliwal, *World J. Pharm. Res.*, 2014, **3**, 681.
31. R. R. Marandi, S. J. Britto and T. Arulappan, *World J. Pharm. Res.*, 2015, **4**, 1971.
32. H. Bhatia, Y. P. Sharma, R. K. Manhas and K. Kumar, *J. Ethnopharmacol.*, 2014, **151**, 1005.
33. H. M. Prakasha, M. Krishnappa, Y. L. Krishnamurthy and S. V. Poornima, *Indian J. Tradit. Knowl.*, 2010, **9**, 55.
34. N. P. Rajith, D. V. Ambily, V. M. Dan, P. S. Devi, V. George and P. Pushpangadan, *Indian J. Tradit. Knowl.*, 2012, **11**, 453.
35. A. K. Jain, V. V. Wagh and C. Kadel, *Indian J. Tradit. Knowl.*, 2011, **10**, 538.
36. M. K. Desale, P. B. Bhamare, P. S. Sawant, S. R. Patil and S. Y. Kamble, *Indian J. Tradit. Knowl.*, 2013, **12**, 334.
37. S. S. Mohapatra, J. Sarma, R. K. Roy, S. Panigrahi and S. Ganguly, *Int. J. Cur. Microbiol. Appl. Sci.*, 2018, 7, 1959.
38. C. Alagesaboopathi, *Afr. J. Tradit., Complementary Altern. Med.*, 2009, **6**, 222.
39. N. R. Krishna, C. Saidulu and A. Hindumathi, *World J. Pharm. Res.*, 2014, **3**, 545.
40. P. K. Deb, T. Bhakta, T. K. Ghosh and R. Ghosh, *Indo Am. J. Pharm. Res.*, 2014, **4**, 1221.
41. V. A. Khan and A. A. Khan, *J. Tradit. Knowl.*, 2005, **4**, 317.
42. V. P. Bhatt and G. C. S. Negi, *J. Tradit. Knowl.*, 2006, **5**, 331.
43. S. K. Mallick, P. Banerjee and A. Saha, *Int. J. life Sci.*, 2012, **1**, 82.
44. K. Mukherjee, B. Subhadip, B. Sayan, D. Bhaskar, K. Amit and C. K. Katiyar, *J. Ethnopharmacol.*, 2021, **264**, 113157.
45. S. Tyler, E. Veronica and R. Claire Morton, *HerbalGram*, 2021, **131**, 52.

CHAPTER 6

Scientific Evaluation of the Role of Ocimum sanctum *in Ayurvedic Formulations and Its Phytochemistry and Pharmacological Activities*

HAAMID RASHEED,[a] SUHAIL AHMED,[a] HARJEET SINGH[b] AND ALOK SHARMA*[a]

[a] Deaprtment of Pharmacognosy, ISF College of Pharmacy, Moga, Punjab-142001, India; [b] Central Council for Research in Ayurvedic Sciences, Ministry of AYUSH, Govt of India, Jhansi, UP-284003, India
*Email: alokalok22@gmail.com

6.1 Introduction

Ocimum sanctum is a herbaceous perennial that belongs to the family Lamiaecae (Figure 6.1).[1] The *Ocimum* genus consists of round 160 species that are broadly found across the world's hot climates.[2] *Ocimum sanctum*, *Ocimum gratissimum* (Ram tulsi), *Ocimum canum* (dulal tulsi), *Ocimum basilicum* (Ban tulsi), *Ocimum killimandscharicum*, *Ocimum americanum*, *Ocimum camphora*, and *Ocimum miranthum* are all typical *Ocimum* species.[3] *O. sanctum*, the monarch of medicinal herbs, is the most sanctified and treasured of several recuperation and ill-health-correcting herbs used in this excellent way. Basil, or Tulsi, is revered in regular Ayurvedic and Unani systems.[4] It is extensively viewed as a good-sized supply of remedies and drugs, with several

Chemistry, Biological Activities and Therapeutic Applications of Medicinal Plants in Ayurveda
Edited by Augustine Amalraj, Sasikumar Kuttappan and Karthik Varma A. C.

Published by the Royal Society of Chemistry, www.rsc.org

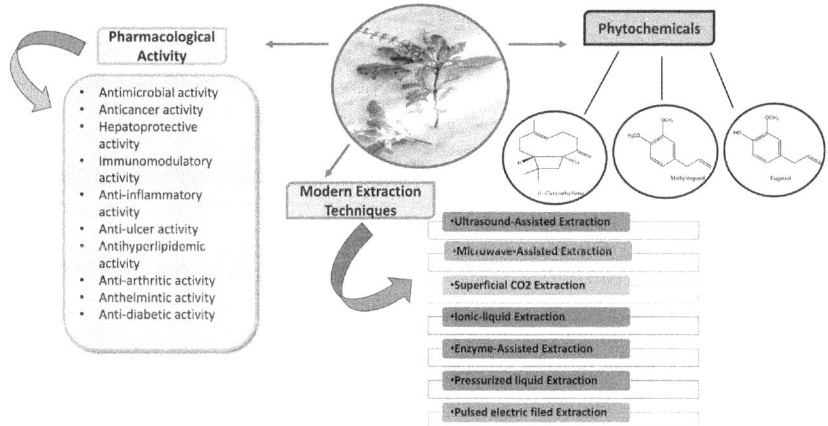

Figure 6.1 *Ocimum sanctum* encompasses pharmacological activity, modern extraction methods, phytochemicals, and their clinical trials.

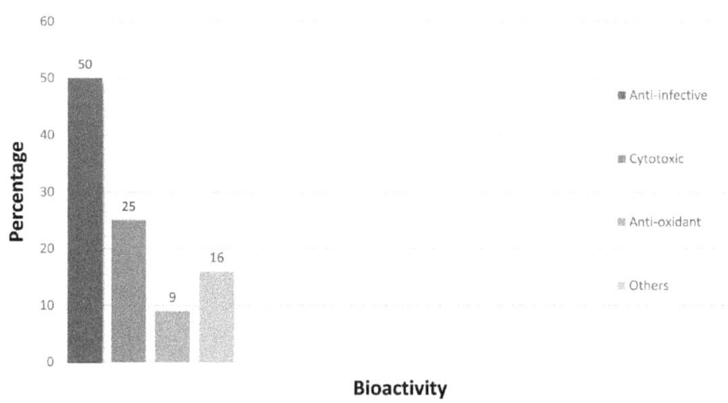

Figure 6.2 Bioactivity profile of compounds of *O. sanctum*.

secondary metabolites and essential oils being used for the remedy of malaria, diarrhea, bronchial asthma, dysentery, bronchitis, pore and skin diseases, arthritis, painful eye diseases, persistent fever, amongst other conditions.[5] Additionally, it demonstrates anticancer, antifungal, antibacterial, antifertility, hepatoprotective, antispasmodic, cardioprotective, antiemetic, antidiabetic, analgesic, adaptogenic, and diaphoretic activities. Tulsi is claimed to be an "Elixir of Life" in Ayurveda due to its healing properties and has been used to deal with a range of common fitness problems.[6] Figure 6.2 represents the bioactivity profile of compounds of *O. sanctum.*

Ayurveda takes a holistic approach to fitness and disease. It does so through promoting ailment resistance rather than cures, through wholesome lifestyle choices. Ayurveda's designs aim at consuming fresh, minimally processed foods. Today, learning about plants has an impact on the world, and evidence has been found that medicinal vegetations have been used in normal systems. As well as their low expense and huge availability, another benefit of medicinal plants in many treatment options is their safety.[7] Among all medicinal plants, Ocimum belonging to the Lamiaceae family is sizable for its therapeutic potential. Tulsi was first protected in standard health claims in the mid-twentieth century. Animal experiments have supplied the majority of scientific proof for Tulsi's therapeutic benefits, with only a few human studies being carried out. Toxicity research highlighted that Tulsi does not elicit toxicity toward people, as evidenced by the absence of damaging results in humans.[8] According to the World Health Organization (WHO), about 65% of the world's populace makes use of plant life as a key supply of therapeutic substances. It is regularly mentioned that 25% of all medicines prescribed nowadays are derived from plants. The aim of this chapter is to furnish an exhaustive examination of Tulsi's common applications, Ayurvedic significance, geographical distribution, Ayurveda formulations, business formulations, phytochemicals, cutting-edge extraction processes, and pharmacological profile.

6.2 Ayurvedic Significance and Ethnopharmacological Relevance

The medicinal properties of *O. sanctum* including its restorative powers used to be well-recorded in Ayurveda and Siddha, as well as in Roman, Greek, and Unani remedies for pore and skin illnesses, frequent chills, headaches, coughs, malaria fever, diarrhea, constipation, and as an antidote to snake bites. Tulsi has been recognized as a Rasayana medicinal drug in historic Ayurvedic writings such as the Charak Samhita, Susrut Samhita, and Rigveda (3500–1600 BC) for the treatment of coughs, respiratory disorders, poisoning, impotence, and arthritis.[9] Tulsi is a regularly occurring manifestation of Ayurveda's holistic strategy toward wellness. It is referenced by using Charaka in the Ayurvedic literature Charaka Samhita. Tulsi has a strong, bitter taste and penetrates deep tissues and dry tissue secretions, thereby balancing kapha and vata.[10] Tulsi supposedly assists in forestalling sickness, increasing health, well-being, and longevity, and helps in coping with day-to-day pressures. Tulsi is additionally viewed as a pore and skin cleanser, a voice sweetener, and a promoter of attractiveness, intelligence, and stamina. The Ayurvedic Pharmacopeia of India recommends 2–3 g leaf powder for the therapy of pratishyaya (common cold), hikka (respiratory disturbances), kasasa (cough), aruci (skin diseases), krimirogas (worm therapy), and parsva sula (chest pain).[11]

Hundreds of scientific studies, consisting of *in vitro*, animal, and human studies, have been carried out on the medicinal advantages of Tulsi. Tulsi leaves have been used in Ayurveda, Greek, Roman, Siddha, and Unani medicinal structures for centuries.[12] Tulsi leaves are used in Ayurvedic remedies to treat a variety of ailments and issues. A massive range of illnesses, including cardiovascular diseases, blood disorders, pore and skin disorders,

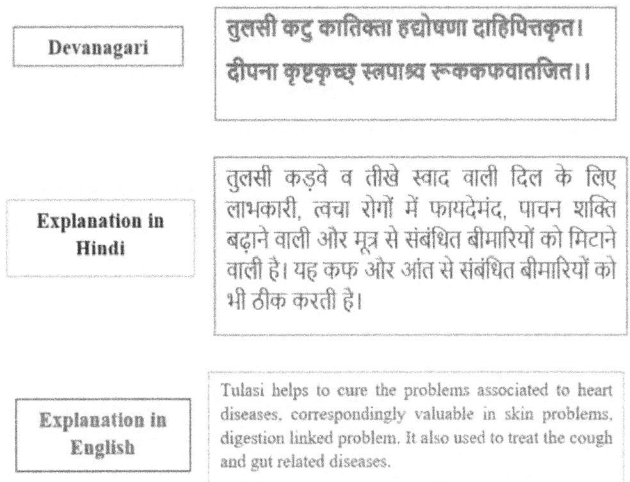

Figure 6.3 As per ancient literature the Rasayana of *O. sanctum*. Reproduced from ref. 85 with permission from Charak Samhita.

asthma, bronchitis, coughing, earache, hiccups, ophthalmia, gastrointestinal problems, genital warts, and scabies, have been treated with it. Tulsi has widely been used to deal with bronchitis, bronchial asthma, malaria, diarrhea, dysentery, pore and skin problems, rheumatoid arthritis, painful eye disorders, powerful fevers, and malicious animal bites. It's additionally utilized to keep gastrointestinal problems at bay. The *O. sanctum* plant and its chemical factors have been proven to have an extensive variety of medicinal effects. The plant has anti-inflammatory, analgesic, antipyretic, antidiabetic, hepatoprotective, hypolipidemic, antistress, and immunomodulatory properties, in addition to a huge variety of organic and pharmacological activities.[13] Taken from historical literature, Figure 6.3 depicts the Rasayana of *O. sanctum.*

6.3 Geographical Distribution

Tulsi's distribution covers India, from elevations of 1800 m in the Himalayas to the Andaman and Nicobar Islands, and down to sea level in the Indian Ocean. This plant is exceedingly adaptable, residing in a variety of distinctive settings.[14] Tulsi has traditionally been an enormously revered and honored plant in India, Southeast Asia, and many tropical and subtropical regions of Asia, and now it is used throughout the world. The plant used to be generally found in Asia, and it has been utilized for at least 3000 years for culinary and medicinal purposes.[15]

6.4 Botanical Description

O. sanctum is an upright, thickly subdivided, and aromatic plant that grows to around 30–60 cm in height. Simple, elliptic, oblong, obtuse, or acute leaves grow to at most 5 cm and have total or sub-serrate or dentate margins. Tulsi blossoms are tiny and crimson to scarlet. On cylindrical spikes, they develop in small compact clusters.[16] Stalk-less heart-shaped bracts are there at the base of each flower cluster. The sepal cup's inside is hairless. The flowers do not often grow more than 5 mm in length, with a bearded calyx tube on the outside and a bearded calyx tube close to the base. The fruits are small, and the seeds are yellow to reddish.[17] The scientific classification of *O. sanctum* is as follows

Kingdom: Plantae
Division: Magnoliophyte
Class: Magnoliopsida
Order: Lamiales
Genus: *Ocimum*
Species: *sanctum*

6.5 Chemical Constituents

The sparkling leaves and stem extract of *O. sanctum* consist of a range of phenolic compounds (antioxidants), which include cirsilineol, circimaritin, isothymusin, apigenin, and rosameric acid, as well as high quantities of eugenol.[18] The leaves of *O. sanctum* contain 0.7 percent unstable oil, which is composed of about 71% eugenol and 20% methyl eugenol. Carvacrol and the sesquiterpine hydrocarbon caryophyllene are additionally common in the oil. Two flavonoids, orientin and vicenin, have been extracted from an aqueous leaf extract of *O. sanctum*.[19] Due to the complexity of *O. sanctum's* chemical composition, uniformity has evaded current science for that reason so far. *O. sanctum* is claimed to include a large range of bioactive compounds, the concentrations of which range considerably between its several species. Numerous investigations have proved the presence of the essential oil, eugenol, in the leaves of *O. sanctum*.[20] The predominant chemical elements of *O. sanctum* leaves are eugenol, eugenol methyl ether, caryophyllene, oleanolic acid, caffeic acid, ursolic acid (triterpenoid), apigenin, luteolin, rosmarinic acid (phenylpropanoid), and chlorophyll. Cirsilineol, cirsimartin, isothymusin, isothymosin, apigenin, rosmarinic acid, and eugenol have all been removed from *O. sanctum* leaves. Additionally, zinc, manganese, and sodium are currently found in small quantities in the leaves. The seeds are claimed to comprise essential oils composed of the following fatty acids: palmitic (11.69%), stearic (3.19%), oleic (13.82%), linoleic (52.23%), and linolenic (52.23%).[21] Vitamins A and C are present, as are minerals such as calcium, iron, and zinc. The plant is believed to comprise alkaloids, glycosides, saponin, tannins, and traces of maleic acid, citric acid, and tartaric acid. Several of these element concentrations are appreciably influenced by means of unknown modifications in growing, harvesting, processing, and storage conditions.[22] Figure 6.4 represents the structure of the exclusive chemical materials of *O. sanctum*.

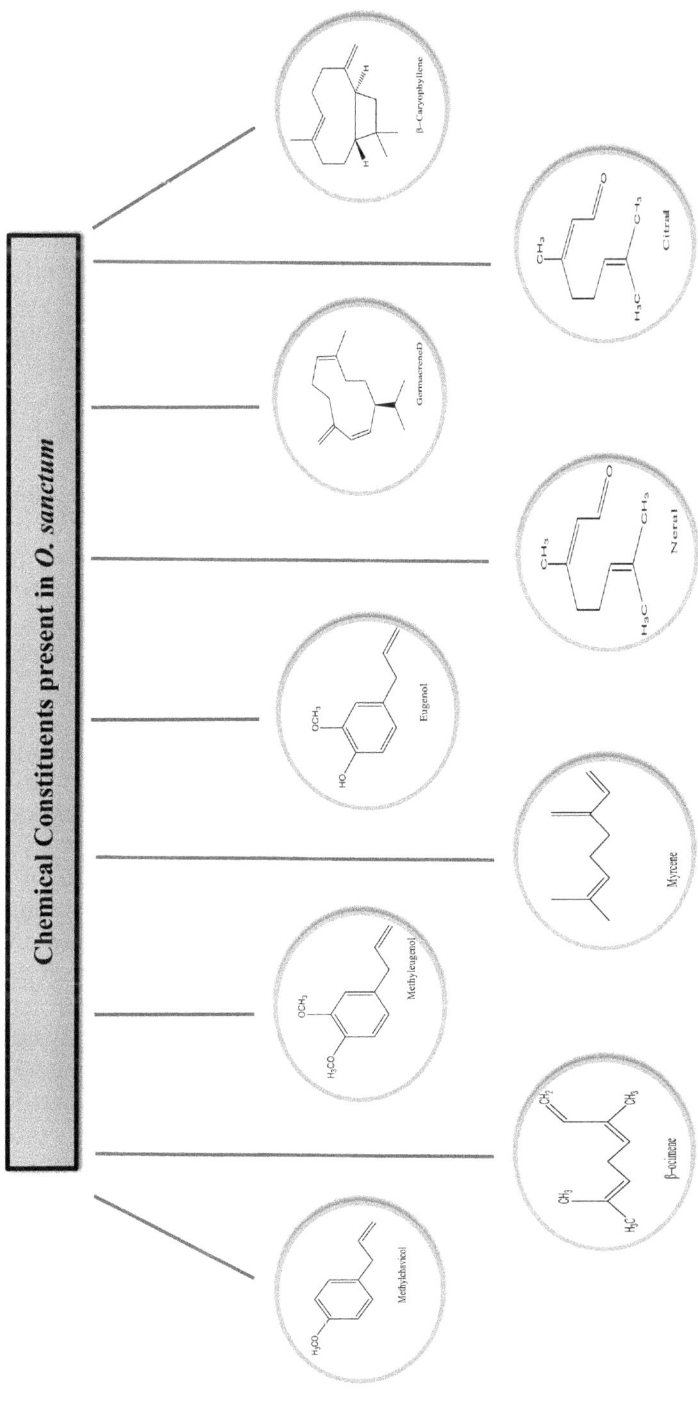

Figure 6.4 Structure of different chemical constituents present in *O. sanctum*.

6.6 Different Ancient Ayurvedic Formulations Containing Tulsi

Tulsi extracts have been utilized in common Indian Ayurvedic remedies for centuries. They have additionally been used in the Unani scientific system. Tulsi is used in Ayurvedic medicinal drugs to treat frequent cold headaches, belly issues, infections, coronary heart disease, poisoning, cataracts, and malaria.[23] Tulsi acts as a good source rejuvenator and helps to reduce stress and improve memory. It is also supports the heart. It serves as an appetizer and aids in digestion and also promotes digestive enzyme secretion and prevents flatulence.[24] Due to its cleaning capabilities, Tulsi cleanses the blood of any pollution that may additionally be present. Tulsi has additionally been shown to be beneficial at decreasing blood sugar levels and controlling diabetes in studies.[25] Additionally, it has been shown to be beneficial in reducing cholesterol levels. Some Ayurvedic formulations and their compositions and therapeutic uses are outlined below.

Vivaldi gulika is an Ayurvedic polyherbal treatment used in the treatment of snake, scorpion, spider, and rodent bites, as well as in the treatment of gastroenteritis, cholera, indigestion, and fever.[26] Poisonous consequences occur in the human physique as a result of an underlying reason or animal-based poisonous chemical. The treatment consists of Bilva, Tulsi, Karanja, Tagara, Devdaru, Haritaki, Bibhitaki, Amla, and some different ingredients. It includes 43 mg of Tulsi in 700 mg of the completed product and the dose per day is 700–1400 mg.

Manasamitra vataka is a mighty Ayurvedic method that incorporates elements that are regarded to assist in the therapy of psychiatric problems, intellectual fatigue, and depression. The very identify of Manasamitra in the time period "Manasamitra Vatakam" is interpreted to be a 'buddy of the mind'. It is an Ayurvedic medication and has long been used as a typical medicinal drug in India to assist with the treatment of low immunity, psychosomatic diseases, nervousness neurosis, and stress.[27] Manasamitra Vatakam is organized as per common texts that use a very massive quantity of elements such as Bala, Nagabala, Bilva, Tulsi, Prishniparni, Pravalapishti, Shankhapushpi, Tamrachuda Padika, Swarna Bhasma, Pushkaramoola Mrigashringa Bhasma, and Vacha Stanya Chandana *etc.* It contains 0.74 mg of Tulsi in a 165 mg of the completed product and the dose per day is a 165–660 mg for adults.

Nasta puspantaka rasa is an Ayurvedic herbomineral complement for increased reproductive health. It is used to treat infertility in men and women. It aids in the stimulation of ovulation in females who are no longer ovulating. It can also be a useful resource to treat menstruation pain, rheumatoid arthritis, and eye problems. It consists of Parada, Gandhaka, Loha bhasma, Vanga bhasma, Tankana, Raupya bhasma, Abhraka bhasma, Tamra bhasma, Guduci, Triphala, Tulsi, and other ingredients.[28] It contains 2.4 mg of Tulsi in 125 mg of the completed product and the dose per day is 250–500 mg.

Tribhuvana kirti ras is an Ayurvedic classical herbo-mineral preparation. It has analgesic and antipyretic properties. It is advantageous in relieving fever of exceptional origins.[29] With Trikatu being among the fundamental ingredients, it is additionally suitable as a digestive medicine. It is also recommended in treating liver and spleen disorders. The key substances of this method consist of Shuddha Hingula, Tankana bhasma, Vatsanabha, Shunthi, Pippali, Maricha, and Tulsi. It contains 3.90 mg of Tulsi in 125 mg of the completed product and the dose per day is 125–250 mg.[30]

6.7 Different Modern Formulations Available Containing Tulsi

Amylcure DS capsule incorporates poly-herbal substances that can defend the liver from a variety of ailments and disorders. Amlycure DS is a combination of amazing herbs such as Kalmegh, Kutaki, Nagarmotha, Raktrohida, Sharpunkha, Brinjasif, Parpatr, Nishoth, Punernava, Kalmegh, Giloe, Tulsi, Sharpunkha, Kutaki, and so forth that have a synergistic impact in correcting liver-associated issues. The selected herbs have been scientifically validated by various *in vivo* and *in vitro* experiments. They are found to restore liver enzymes and have multiple pharmacological effect such as displaying anti-viral, anti-oxidant, anti-inflammatory, immunomodulatory, choleretic, and anti-cholestatic activities.[31] It hastens broken bone repair, protects liver cells from injury due to toxins, and aids in ideal digestion. Each pill incorporates 32 elements out of which Tulsi makes up 20 mg and the dose per day is 1–2 drugs twice or thrice a day.

Decofcyn cough syrup is an Ayurvedic proprietary medicine.[32] It is a cough syrup which can be used with no adverse effects. It enables handy and clean expectoration of phlegm and relieves chest congestion quickly. The elements of the syrup consist of Tulsi, Taleesa patra, Adusha, Ashwagandha, Shatavari, Karchura, Lavanga, Ardraka, Ela, Karpura, *etc.* In each 5 ml of syrup there are sixteen elements and Tulsi makes up 200 mg and the dose per day is 1–2 tablespoons thrice a day.[33]

Feverex syrup is a natural treatment for pyrexia, a condition of increased body temperature. Feverex syrup causes dilation of blood vessels by increasing the blood flow in the superficial vein and in this way it reduces the body temperature through sweating and evaporation. It is a non-toxic and safe herbal remedy. The Tulsi existing in feverex syrup reduces the febrile response due to its anti-pyretic activity.[34] Feverex includes Tulsi which increases sweating due to its Swedjanan (sweat stimulant) properties and it helps to relieve the fever. Each 5 ml of syrup contains extracts of Tulsi, Guduchi, Cincha, Parpita, Sphatika, Saptaparna, Vatsanabha, Rasona, Tankana, *etc.*, incorporating thirteen substances including 100 mg of Tulsi and the dose per day is 2 tablespoons thrice a day.

Cough syrup is an Ayurvedic proprietary medication manufactured through Badariya pharmaceuticals. It is an anti-cough formula. B cough syrup is used in the remedy of coughing with mucus. It loosens and thins the mucus in the nose, windpipe, and lungs. This makes it easier for you to cough it out. It additionally relaxes and widens the muscle tissue in the airways, making it less difficult for you to breathe. B. Cough syrup is organized from Vasa, Tulsi, Vanfasa, Gogihara, Khatmi, Pippali, Jufra, Kantakari, *etc.* It consists of 9 complete substances including 10 g of Tulsi and the dose per day is two tablespoons thrice a day.[35]

Koflet syrup is a herbal syrup that has been tested to be really useful for very prolific and dry coughs. It acts as an antitussive and alleviates bronchial inflammation. It possesses anti-inflammatory, anti-allergic, antibacterial, and immune-suppressive outcomes that furnish relief from coughs precipitated *via* a range of etiologies. Additionally, the anti-allergic, antibacterial, and immunomodulatory characteristics alleviate coughing. Koflet syrup's demulcent influence relaxes respiratory airways. Tulsi is a fine antihistamine that quickly protects from pollen-induced bronchospasms. Koflet cough syrup consists of factors like Madhu, Guggulu, Draksha, Vishnu, Jafa, Vaska, Guduchi, Yashtimadhu, Gojha, Triphala, Trikatu, Taja, Tulsi, Neelapushpa *etc.* It consists of 21 factors out of which Tulsi makes up 25 mg and the dose per day is 1–2 tablespoons 3–4 times a day.[36]

Respicare tablet is formulated to overcome respiratory allergies. It ensures speedy alleviation and prevents reoccurrence of respiratory problems. This medicinal drug is used to manipulate and forestall signs induced by allergies (chronic inflammatory disorder of the lungs characterized by narrowing of the airways). It is used before exercising to prevent respiratory troubles throughout the exercise. It is also used to relieve signs and symptoms of hay fever (a reaction in the nostril in response to allergens in the air) and allergic rhinitis (such as sneezing, stuffy/runny/itchy nose). It consists of herbs like Pushkarmool, Pippali, Karkat shingri, Maricha, Tulsi, Mulethi, Shunthi, Gul banafsa, Talispatra, *etc.* There are 12 substances in the complete range of substances currently in this system and Tulsi makes up 25 mg and the dose per day is 1–2 drugs twice a day taken with luke-warm water.[37]

M-Cof syrup is a proprietary Ayurvedic drug made with the aid of Mukhti prescribed drugs that is used to treat dry coughs. It alleviates hypersensitive reaction signs such as sneezing, runny nose, watery eyes, and infection of the throat. Additionally, it alleviates nasal congestion or stuffiness. It is effective against dry, hacking coughs. Additionally, it alleviates allergic reaction signs and symptoms such as watery eyes, sneezing, runny nose, or throat irritation, permitting you to go about your normal activities with greater comfort. M-Cof syrup consists of components like Soalnum aculea tissimum, Pippali, Yashtimadhu, Vasa, Tulsi, Shallaki, Ela, Twak, Patra, Shunti, Jiraka, Kaempferia galangal, Kapoora, and sugar. It consists of 15 substances and Tulsi makes up 6 g and the dose per day is 1 tablespoon thrice a day.[38]

Kurex light syrup is an Ayurvedic proprietary medicine manufactured by Megha Healthcare Pvt Ltd. It is used for the treatment of respiratory disorders. Kurex light syrup is great for patients suffering from colds, coughs, and bronchitis, with effective expectorant and antitussive action. It provides relief from coughing and liquefies sputum.[39] It has a bronchodilator effect and provides relief in pulmonary congestion. The key ingredients of Kurex light syrup are Vasaka, Tulsi, Yashtimadhu, Pippali, Kantakari, Bharangi, Pippali, Haridra, Navsar, and Sunthi. It consists of 9 ingredients and Tulsi makes up 100 mg and the dose per day is 2 tablespoons thrice a day.[40]

Tussnil syrup is an Ayurvedic herbal mixture that is very natural and safe. This herbal mixture is anti-inflammatory, demulcent, expectorant, mucolytic, bronchodilator, and antispasmodic in action, imparting a speedy and lasting remedy from coughs and respiratory symptoms. The proprietary composition has been clinically proven to provide immediate alleviation from acute coughs, throat irritations, and hypersensitive response symptoms.[41] It promotes convenient expectoration by lowering respiratory tract irritation. Without alcohol or narcotics, it produces no drowsiness or drug dependence. The key factors are Vasa, Kasamarda, Bharngi, Sunthi, Yasthtimadhu, Kapooravalli, Tulsi, Ela, Tvak, Patra, and Lavanga. It consists of eight complete factors out of which Tulsi makes up 20 mg and the dose per day is two tablespoons twice a day.

Pneumonorm syrup is an Ayurvedic proprietary medicinal drug manufactured *via* Dr Pralep's Pharmaceuticals. It is used for the cure of respiratory issues in pediatrics. Pneumonorm syrup is an Ayurvedic formulation. It acts as a bronchodilator.[42] It incorporates Pushkar Mool, Vasa, Yashtimadhu, Shrish, Laghu kantakari, Pippali mool, Shunthi, Tulsi, Pudina satva, Gojivha, Banafsha, Talisa patra, Brihati, Bharangi, and Shatikand. It consists of 15 complete elements out of which Tulsi makes up 1 mg and the dose per day is half a tablespoon thrice a day.

Bresol tablet is an Ayurvedic proprietary medication that alleviates all respiratory disorders. Using haridra, vasa, triphala, trikatu, and tulsi, Bresol is an easy-to-breathe cure for allergic rhinitis and bronchial asthma. Tulsi affords good antihistaminic results that guard the body from pollen-induced bronchospasms.[43] Holy Basil reduces the amount of nitric oxide, resulting in antioxidant properties. This is very effective in the cure of allergic respiratory diseases. The dose of Bresol per day is two drugs twice a day.

Sri Sri Tulasi Arka is distillate of Tulsi manufactured by using Sri Sri Tattva. It has immuno-booster, antioxidant, antiseptic, and antibacterial properties. It is an Ayurvedic medicinal drug that is used to deal with coughs, colds, dyspnea, and hiccups. It additionally strengthens the immune device and provides relief from a sore throat, bronchitis, bronchial allergies, and malaria.[44] The key substances in this system are an aqueous extract of organic Tulsi and potassium sorbate. The dose per day is 10–15 drops, twice a day.

Cough syp Tulasi is a proprietary Ayurvedic cough medication manufactured by using Pankaja Kasturi. It is made of a hundred percent herbs and comes with much of the goodness of Tulsi. It efficaciously treats all kinds of coughs without inflicting any sedation or other characteristic effects. It aims to be a positive decongestant and expectorant which reduces phlegm and is a strong immune modulator especially for children. Also, it is beneficial in relieving frequent blood loss and coughing. The key components of this method are Maricham, Pippali, Viswa, Ela, Twak, Vasa, Kasamardha, Kandakari, Karpoora, and Tulsi. The dose per day is 1–2 teaspoons four times a day.[45]

Dabur Tulsi drops have antioxidant properties which help improve immunity. They are a remedy for coughing and help in relieving mucus and have antimicrobial properties. They incorporate the extracts of 5 uncommon Tulsi which have a lot of benefits. The consumption of Tulsi is a first-rate method of gut cleaning. It additionally helps in stress management. It is suitable for skin, joint, and liver health. It is made from the extracts of 5 uncommon sorts of Tulsi herbs.[46] Ideally, 5–10 drops of this syrup are advocated to be taken 2–3 times a day to improve immunity.

Crux cough syrup is an Ayurvedic proprietary medicine manufactured by BAN Labs Ltd. It is a cough syrup formulated with Tulsi as an ingredient. Crux Syrup is an Ayurvedic product which is a natural and safe remedy for colds and all types of coughs, sore throat, and choked voice. It has a soothing effect in allergic coughs, dry irritating coughs, viral infections, and seasonal coughs. It contains herbal ingredients such as Piparminta Sat, Navsar, Nilgiri Oil, Tulsi, Adusi, and Yashtimadhu. These ingredients have a cooling, pungent effect that soothes throat irritation, is effective for colds, coughs, asthma, and fevers, and helps in expectoration. It also possesses antiseptic and antimicrobial properties.[47] The dose per day is 2 teaspoons, three times a day for adults and $\frac{1}{2}$–1 teaspoon three times a day for children. Table 6.1 includes different marketed formulations, their composition, dosage, and therapeutic uses.

6.8 Modern Techniques of Extraction

O. sanctum, popularly known as Tulsi in India, is a medicinal plant used for a variety of dietary and pharmaceutical purposes. Due to the presence of chemicals such as eugenol, methyl eugenol, -caryophyllene, humulene, and ursolic acid, the leaves of the plant are reported to exhibit substantial nutraceutical characteristics, including antioxidative, antibacterial, and anti-inflammatory qualities. Aromatically, these natural oils are extracted from Tulsi leaves *via* distillation and decoction.[62] However, these extraction procedures are susceptible to biodegradation and the retention of biomolecules in the solvent, as well as thermal degradation and hydrolysis. In light of the extract's intended use in food and medicinal applications, alternative extraction processes such as supercritical carbon dioxide extraction, ultrasonic assisted extraction, microwave assisted extraction, solvent extraction, and hydro distillation are better regarded. Figure 6.5 shows the different extraction methods used to extract bioactive components from *O. sanctum*.

6.8.1 Ultrasound-assisted Extraction

Ultrasonic extraction is the primary approach for separating bioactive components from botanicals. Sonication completes the extraction process,

Table 6.1 Different marketed formulations, their composition, dosage, and therapeutic uses.

Formulation	Composition	Dosage	Uses	Ref.
Amylcure DS capsule	Tulsi, Nishoth, Punernava, Kalmegh, Giloe *etc.*	1–2 capsules twice or thrice a day	Anti-viral, antioxidant, anti-inflammatory, and anti-cholestatic.	48
Decofcyn cough syrup	Tulsi, Taleesa patra, Adusha, Ashwagandha, Shatavari, Lavanga *etc.*	1–2 tablespoon thrice a day	Expectorant	49
Feverex syrup	Tulsi, Guduchi, Cincha, Parpita, Sphatika *etc.*	2 tablespoons thrice a day	Anti-pyretic activity.	50
B. cough syrup	Tulsi, Vanfasa, Gogihara, Khatmi, Pippali *etc.*	2 tablespoons thrice a day	Anti-cough formula.	51
Koflet syrup	Tulsi, Madhu, Guggulu, Draksha, Vishnu *etc.*	1–2 tablets twice a day	Anti-inflammatory property, anti-allergic, antimicrobial properties.	52
Respicare tablet	Tulsi, Mulethi, Shunthi, Gul banafsa, Talispatra *etc.*	1–2 tablets twice a day	Control and prevent symptoms caused by asthma	53
M-Cof syrup	Tulsi, Shallaki, Ela, Twak, Patra *etc.*	1 tablespoon thrice a day	Treatment of dry cough	54
Kurex light syrup	Tulsi, Yashtimadhu, Pippali, Kantakari, Bharangi *etc.*	2 tablespoons thrice a day	Cough and bronchitis, with expectorant and antitussive action.	55
Tussnil syrup	Tulsi, Ela, Tvak, Patra, Lavanga *etc.*	2 tablespoons twice a day	Demulcent, expectorant, mucolytic, bronchodilator.	56
Pneumonorm syrup	Tulsi, Pudina satva, Gojivha, Banafsha, Talisa patra *etc.*	$\frac{1}{2}$ tablespoon thrice a day	Treatment of respiratory disorders in pediatrics.	57
Bresol tablet	Tulsi, haridra, vasa, triphala, trikatu, *etc.*	2 tablets twice a day	Allergic rhinitis, bronchial asthma.	58
Dabur tulsi drops	The extracts of 5 rare Tulsi	5–10 drops 2–3 times a day	Relief from cough and cold.	59
Crux cough syrup	Tulsi, Navsar, Nilgiri Oil, Adusi, and Yashtimadhu.	2 teaspoonful, three times a day	Cough, sore throat, and choked voice.	60
Cough syp Tulasi	Tulsi, Maricham, Pippali, Viswa, Ela *etc.*	1–2 teaspoon 4 times daily	All types of coughs without causing sedation.	61

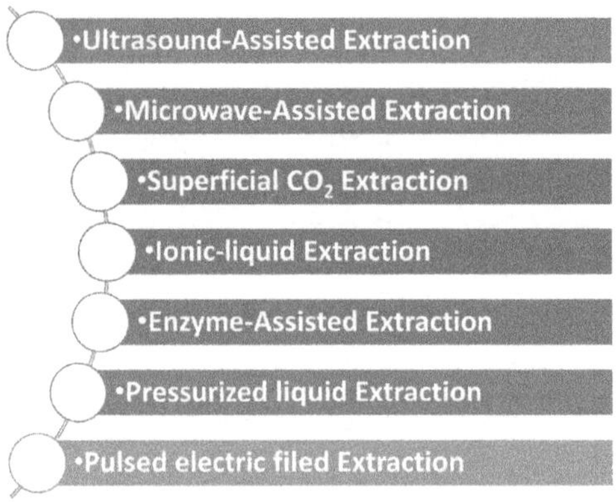

Figure 6.5 Different extraction methods used to extract bioactive components from *O. sanctum.*

resulting in higher extract yields in a very short extraction time. Ultrasound assisted extraction (UAE) is a versatile food processing technology that is used to extract bioactive components from plant materials. Ultrasound frequencies greater than 20 kHz are utilized to disrupt the plant's cell walls, resulting in increased solvent penetration and extraction rate.[63] UAE can process at low operating temperatures while maintaining high-quality chemical extracts. UAE is well-known for its application in traditional laboratory instruments, such as in ultrasonic baths, as one of the most straightforward extraction procedures. A sample is combined and placed in an ultrasonic bath containing the appropriate solvent, while the temperature and time are controlled. Ultrasound extraction is a very effective method of extraction that saves money and time while producing high-quality extracts for food, medicines, and supplements. The ultrasound-assisted extraction technique is based on the acoustic or ultrasonic cavitation principle. While various non-traditional procedures effectively alleviate the disadvantages of traditional methods, their extraction efficacy varies.[64]

Ultrasound-assisted extraction was recently established and showed considerable acceleration of the extraction process, hence lowering the energy consumption. Additionally, the use of ultrasound-assisted extraction has several advantages, including ease of extraction handling, rapid execution, no residue, high output, environmental friendliness, better quality, and prevention of extract deterioration. Alexandru *et al.* evaluated the efficiency of ultrasonic-assisted extraction and traditional maceration for the extraction, and three distinct modes of US extraction were carefully chosen (450, 900, and 1350 ml per min), with the 1350 ml per min mode achieving the highest extraction rate.[65] Similarly, Tekin *et al.* developed a central composite design to extract essential oils utilizing the US extraction technique. The study's independent variables were the temperature (32–52 °C), extraction time (30–60 minutes), and plant concentrations (3–7%), whereas the extract was the dependent variable.[66] The frequency of the ultrasonic bath was set to 53 kHz. The temperature had a substantial effect on extract production, according to the study's findings. The extract produced eugenol, -caryophyllene, and 2-methoxy-4-(2 propenyl) phenol acetate. This research concluded that ultrasonic extraction is an effective instrument for paving the road for the commercialization of this method. Izadiyan and Hemmatee-nejad used central composite design approaches to investigate the extraction of antioxidant-rich compounds from Iranian *O. basilicum* utilizing ultrasound-assisted extraction and the effect of extraction variables (including temperature, methanol to water ratio, and sonication time).[67] They concluded that the most critical factors of the UAE process were the methanol to water ratio and the extraction temperature, while the sonication time was unimportant.

6.8.2 Microwave-assisted Extraction

Hydrolysis, heat degradation, and leaching of specific fragrance components are all problems of old tools for extracting eugenol from various plant sources. To address these concerns, a variety of modern extraction technologies have been developed that provide a high extraction yield while requiring minimal processing time and energy. Microwave-assisted extraction is one of these technologies that is considered a green extraction method since it can produce eugenol and other essential oils with the same sensory qualities and quality as those obtained by traditional methods.[68] Furthermore, compared to normal processes, our technology yields a higher extraction rate at a lower cost. The direct effect of microwaves on molecules *via* ionic conduction and dipole rotation is linked to microwave energy extraction of bioactive compounds from plants. Microwave energy is converted to heat when it passes through a solvent. The electrophoretic movement of charged ions (produced by the microwave's electric field) results in the generation of heat as a result of friction between the ions and the medium is referred to as ionic conduction. Collisions between dipolar molecules cause dipole rotation, which produces heat. During microwave extraction, microwave energy is rapidly absorbed by polar molecules in the cell matrix. As a result, the cells' inner temperature is drastically raised (superheated), enabling liquid vaporization and further cell wall breach. The mass transport of solvent or phytochemicals into the cell is enabled by cell rupture. Despite the fact that microwave radiation has a huge heating potential, microwave extraction technology in analytical laboratories is still relatively new.[69]

In 1975, Abu-Samra *et al.* were the first researchers to use a microwave home oven in the laboratory for trace metal detection in biological materials.[70] It took over ten years for the first edition to be published, in 1986. Using a few milliliters of solvent and irradiating for 30 seconds up to seven times in a domestic oven, Ganzler *et al.* developed extraction procedures for lipids, antinutrients, and pesticides from soils, seeds, foods, and feeds (1140 W).[71] Microwave digestion methods for extracting a number of sample types, comprising botanical, biological, environmental, geological, or metallic matrices, have been developed since then. Furthermore, the extraction time was greatly reduced compared to the traditional extraction method. To summarize, microwave-assisted extraction methods have evolved enough to be used on an industrial scale in place of traditional eugenol extraction methods.[72] MAE has a number of advantages, including a quick extraction time, low energy usage, high extraction efficiency, and minimum target component degradation. While MAE provides the benefits listed above, it also has a few disadvantages. MAE is difficult to perform on fully dry and extremely moist materials when a nonpolar solvent such as hexane is used as the extracting solvent. Furthermore, when the target compounds and/or solvents are non-polar, microwave efficiency can be lowered and rendered useless when the extraction fluid has a high viscosity.[73]

6.8.3 Supercritical Carbon Dioxide Extraction

Supercritical carbon dioxide (SC-CO_2) extraction is another successful method for obtaining eugenol from clove buds. Extensive research has proven that SC-CO_2 extraction is highly effective. An example of the result of the study conducted by Gopalakrishnan *et al.* is that SC-CO_2 extraction extracts more clove oil than liquid extraction does. This study examined two different experiments, in which they discovered that increasing the flow rate of CO_2 during supercritical extraction yields higher eugenol yields. The aforementioned two tests by Wenqiang *et al.* examined the clove bud oil composition and found that it was quite similar whether extracted using SC-CO_2 or other traditional extraction procedures (solvent extraction, steam distillation, and molecular distillation).[74] Though numerous scientific studies have proven the ability of SC-CO_2 extraction to efficiently harvest clove oil, less research has been done to maximize the SC-CO_2 extracted eugenol yield. To prove the efficacy of the SC-CO_2 extraction procedure for the recovery of eugenol from clove bud oil, further research is required. SC-CO_2 is cheap and comes in large quantities, ensuring that no changes are made to bioactive molecules and that their curative or functional characteristics are not compromised. It's also capable of solubilizing lipophilic compounds and is easily removed from products after the process is complete and is a generally accepted environmentally beneficial technique.[75]

A simplified process-scale SFE system consists of two extraction tanks: one equipped with temperature controllers or pressure valves to maintain the appropriate extraction conditions, and the other, which is fitted on both ends with temperature controllers as well as pressure valves to keep the extraction conditions consistent. In order to extract the fluid from the extraction tank, pumps are utilized.[76] Pumps utilized for circulation of fluid are found in the systems, too. Wettable powders can be stored for long periods of time due to their fast hydration process. A valve is placed in the separator's bottom part to collect the product that flows from the separator. Supercritical fluid extraction is now seen as a feasible alternative to traditional extraction methods. Solvents in the supercritical state have physical-chemical properties that are halfway between liquid and gas, increasing the solvent's extractive capacity. Due to their density, these fluids have a high solvation power, while their high diffusion and low viscosity values offer the necessary penetration power into the solid matrix. The solvents used in constructing an SFE process are crucial, and a wide range of compounds can be used as solvents in this approach.[77] Super critical extraction is now widely recognized as a more ecologically friendly alternative to traditional phase separation that avoids the majority of the disadvantages. Separating moderately polar compounds can also be carried out with organic modifiers such as methanol, isopropanol, and acetonitrile. On the other hand, despite the fact that its use is still moderate, it has risen steadily over the last few years. This pattern exemplifies both contemporary technical advances and the general advantages of this separation technique. Polyphenols, biogenic amines, and polyacetylenes were among the natural compounds studied in vegetable-derived diets in addition to lipids. In addition, multiple lipid classes were investigated simultaneously in complex vegetal extracts and bilberry using the supercritical approach.[78] As the food component industry catches up to its materials engineering and pharmaceutical counterparts and grows beyond extraction, SC-CO_2 extraction technology is gaining appeal as a high-value, environmentally safe, and effective extraction method. SC-CO_2 addresses issues, including product purity, process efficiency, wellness, and environmental impact that aren't addressed by traditional methods.[79]

6.8.4 Ionic-liquid Extraction Method

Recently, researchers have concentrated their efforts on ionic liquid applications (ILs). Numerous studies have focused on extracting active components from traditional herbs using ILs as alternative solvents to circumvent the difficulties associated with standard methods of extracting natural products (NPs) using organic chemical reagents. In recent years, the use of ionic liquids (ILs) for analytical applications has evolved, resulting in improvements in the quality and efficacy of extraction. Ionic liquids are a type of liquid organic salt that interacts selectively with specific polar and non-polar compounds *via* ion exchanges, stacking interactions, hydrophobic interactions, or hydrogen bonds, thereby enhancing the selectivity of the extraction process. For example, when paired with hydrophilic amino acids, 1-butyl-3-methylimidazolium hexafluorophosphate ([BMIM][PF_6]) extraction efficiency is enhanced in contrast to standard organic solvent extraction. Because ILs are employed in a number of extractive processes, notably MAE and UAE, they must be stable enough to resist the high temperatures associated with these techniques. Tetraalkylammonium cations and imidazolium ions, such as tetrafluoroborate [BF_4], bis(trifluoromethylsulfonyl)imide [NTf_2], trifluoromethanesulfonate [CF_3SO_3], and hexafluorophosphate [PF_6], exhibit excellent heat stability in this regard; additionally, organic anions have been shown to be more stable.[80]

The IL extraction method is a viable alternative for recovering organic, inorganic, and metal ions in plants and herbs with a high recovery rate. Extensive research has been undertaken over the last two decades on IL-based extraction, isolation, and purification strategies for bioactive compounds from biomass. ILs and their mixes with water or organic solvents can be used directly in solid–liquid extraction (SLE) of biomass-derived value-added chemicals. Because ILs are ionic, they can interact with electromagnetic waves; thus, in addition to even more conventional SLE techniques, IL-based microwave- or ultrasonic-assisted extraction (MAE) as well as ultrasonic-assisted extraction (UAE) may also be preferred due to their shorter extraction times and higher extraction efficiencies. When the true matrix is a liquid, such as a water-rich biomass extract, it is possible to perform IL-based liquid–liquid extractions (LLE). Additionally, ILs may be integrated into a solid matrix to minimize the process's overall IL consumption, to increase the extraction efficiency, and to simplify solvent/material recycling (IL modified materials for solid-phase extractions, SPE). Indeed, this family of ILs has been shown to be effective for extracting and purifying bioactive chemicals from biomass, amino acids, and proteins in water and raw materials, as well as stabilizing proteins, enzymes, and nucleic acids. It was concluded that the predominant solvent for hydrophilic ILs should be IL aqueous solutions. This is an especially interesting approach since it overcomes the high viscosity of some ILs while increasing the solvent's biocompatibility and lowering the overall solvent cost. Additionally, ILs have been shown to be effective at extracting small extractable compounds by increasing their solubility in aqueous solutions through a hydrotropic or micelle-mediated effect, whilst also avoiding the dissolution and the extraction of the lignocellulosic fraction, as well as extracting proteins, that have a low solubility in water, according to the works reviewed here. The development of the more cost-effective and sustainable extraction as well as separation processes is important for the recovery and commercialization of novel and low-cost bioactive products for such nutraceutical, skincare products, and pharmaceutical industries, with the goal of improving modern society's quality in the near future, and in which ILs could play a significant role as a catalyst.[81]

6.8.5 Enzyme-assisted Extraction

A unique and convenient alternative to standard solvent extraction processes is enzyme-assisted extraction (EAE): the method is novel and convenient since the enzymes catalyze specific reactions without subjecting the targeted products to severe conditions that could result in degradation. Additionally, proteins such as cellulases, hemicellulases, and pectinases hydrolyze the cell wall, increasing permeability and enabling metabolite release from plants. By utilizing enzymes such as lipases, proteases, and phospholipases, the amount of solvent utilized in the extraction can be reduced. The most often used enzymes in the extraction of oil from plants are cellulase, α-amylase, and pectinase. These proteins can be obtained from fungi, bacteria, animals, or plants or engineered, and their selective catalysis enables them to be used to recover a specific bioactive chemical in high yields and with minimal energy waste. However, some limitations exist due to the high cost of the enzymatic strategy, the incomplete disruption of the cell wall, and also the difficulty of commercial application owing to the enzymes' variable actions in response to environmental changes such as oxygen concentration, nutrient variety, as well as operating temperature. Natural product extraction is hampered by the structure of the cell membrane and cell wall, the micelles formed by macromolecules like polysaccharides and proteins, and the aggregation and denaturation of proteins at high temperatures during extraction.[82]

EAE improves the extraction efficiency by hydrolyzing the cell membrane and membrane components, and also macromolecules within the cell, allowing the natural product to be released more easily. EAE is mostly composed of cellulose, α-amylase, and pectinase. Polysaccharide is a bioactive component of Traditional Chinese Medicine (TCM) Astragali Radix. Zhang *et al.* reviewed and noted that the EAE of polysaccharide isolated from *Astragalus membranaceus* radix using a variety of enzymes and discovered that glucose oxidase degraded polysaccharide more effectively than the other seven enzymes examined (amyloglucosidase hemicellulase, bacterial amylase, fungal amylase, pectinase, cellulose, and vinozyme). By adjusting the EAE setting with glucose oxidase, polysaccharide synthesis increased by more than 250 percent as compared to the non-enzyme treated approach. When chlorogenic acid was extracted from *Eucommia ulmoides* leaves using cellulase and ionic solutions, the yield of chlorogenic acid was significantly increased. Zhang *et al.* described in their review that using pectinase and cellulase enzymes enhanced the yield of carotenoid and lycopene extraction from tomato waste. In comparison to untreated solvent extraction, cellulase and pectinase-treated samples produced sixfold and tenfold more of the two target chemicals, respectively.[83]

6.8.6 Pressurized Liquid Extraction

Numerous research organizations have referred to pressurized liquid extraction (PLE) as accelerated solvent extraction, improved solvent extraction, pressurized fluid extraction, rapid fluid extraction, and high-pressure solvent extraction. PLE extracts use high pressure. Due to the fact that high pressure maintains solvent liquids above their boiling point, they have a high solubility and diffusion rate of lipid solutes, as well as high solvent penetration into the matrix.[84] When compared to other extraction methods, PLE significantly reduced the extraction time and solvent consumption and showed increased repeatability. Using pressurized liquid extraction, researchers from the University of Macau and other institutions successfully isolated a variety of natural components from the TCM, including saponins, flavonoids, and essential oil. Some researchers believed PLE could not be used to extract thermolabile chemicals due to the high extraction temperature, while others believed it could be employed due to the shorter extraction time in PLE. When PLE was employed at 200 °C to extract antioxidants from grape pomace, Maillard reactions occurred. Gizir *et al.* successfully used PLE to obtain the anthocyanin-rich extract from black carrots because anthocyanin degradation is time-dependent and the high temperature and short duration of the PLE extraction conditions could offset the disadvantage of the high temperature used in the extraction.[85] This approach frequently uses water as the solvent, eliminating the environmental and health hazards associated with organic solvents. By operating at elevated temperatures (up to 200 °C) and pressures (35–200 bar), PLE improves the quality of the extraction, which can be performed dynamically, with the solvent being pushed continuously through the vessel, or statically, with the solvent being refilled more often. At elevated temperatures, the solvent's viscosity decreases, allowing it to penetrate the matrix more effectively and extract the desired analytes; however, this method cannot be used for thermolabile compounds and may result in the co-extraction of other compounds due to the reduced selectivity of extraction at higher temperatures. The increased pressure maintains the solvent in the liquid phase and breaks the cell walls of the plants, putting strain on the matrix. Non-ionic surfactants, antioxidants, including ascorbic acid, CO_2 to lower the aqueous pH, and drying agents are all alternatives to organic solvents. Additionally, the PLE apparatus prevents light- and oxygen-sensitive compounds from degradation and can be combined with other extraction techniques like UAE to maximize its effectiveness.[86]

6.8.7 Pulsed Electric Field Extraction

By altering membrane structures during extraction, pulsed electric field extraction (PEF) significantly increases the extraction yield and minimizes the extraction time. The success of PEF treatment is reliant on a number of parameters, including the field intensity, the specific energy input, the pulse count, and the treatment temperature. PEF extraction is a non-thermal method that reduces the breakdown of thermolabile chemicals. Hou *et al.* obtained the highest yield of ginsenosides (12.69 $mg\,g^{-1}$) utilizing PEF at a 20 $kV\,cm^{-1}$ electric field intensity, a 6000 Hz frequency, a 70% ethanol–water solution, and a 150 $l\,h^{-1}$ velocity. The PEF extraction method yields more ginsenosides than the MAE, heat reflux, UAE, or PLE methods. The entire process of PEF extraction took less than a second and was much faster than the other methods evaluated. Bouras *et al.* discovered that the PEF treatment improved the phenolic content and antioxidant activity of Norway spruce bark by eight-fold and thirty-fold, respectively, as compared to untreated samples.[87]

6.9 Pharmacological Profile

6.9.1 Antimicrobial Properties

In 1952, Joshi CG and Magar examined *O. sanctum's* antibiotic activity against *Escherichia coli*, *Staphylococcus aureus*, *Bacillus anthracis*, *Salmonella* spp., *Vibrio* spp., and *Mycobacterium tuberculosis*.[88] Jabeen *et al.* discovered antimicrobial activity against *Pasturella multocida*, *E. coli*, *S. aureus*, *B. subtilis*, and *Salmonella typhimurium*, as well as *E. coli*, *Klebsiella* spp., *B. subtilis*, and *S. aureus*.[89] Kumar *et al.* conducted *in vitro* trials against specific infections under laboratory conditions and examined the pathogen's development or suppression, as well as observed effects over a set time period.[90] Tulsi essential oil inhibits *E. coli*, *Bacillus subtilis*, *Bacillus anthracis*, *Staphylococcus aureus*, *Pseudomonas vulgaris*, and *Pseudomonas aeruginosa* significantly. These essential oils are high in caryophyllene, eugenol, and methyl eugenol, which are all efficient against Anthrobacterglobiformis, Bacillus megatherium, *E. coli*, and Pseudomonas sp. Eugenol is the most therapeutic component of Tulsi, according to Grover and Rao in 1977.[91] Leaf extracts, both aqueous and alcoholic, have been shown to have antimicrobial properties. The extract is effective against a number of enteric illnesses, including *E. coli*, *Klebsiella aerogens*, *Shigella dysenteriae*, *Pseudomonas aeruginosa*, and *Staphylococcus aureus*. Additionally, the antibacterial activity of seeds was examined, and it was observed that crude, supernatant, residual, and dialyzed seed samples inhibited the growth of *P. multocida*, *E. coli*, *B. subtilis*, and *S. aureus*. Figure 6.6. represents the pharmacological activities of *O. sanctum*. Table 6.2 describes the clinical studies of *O. sanctum* extracts.

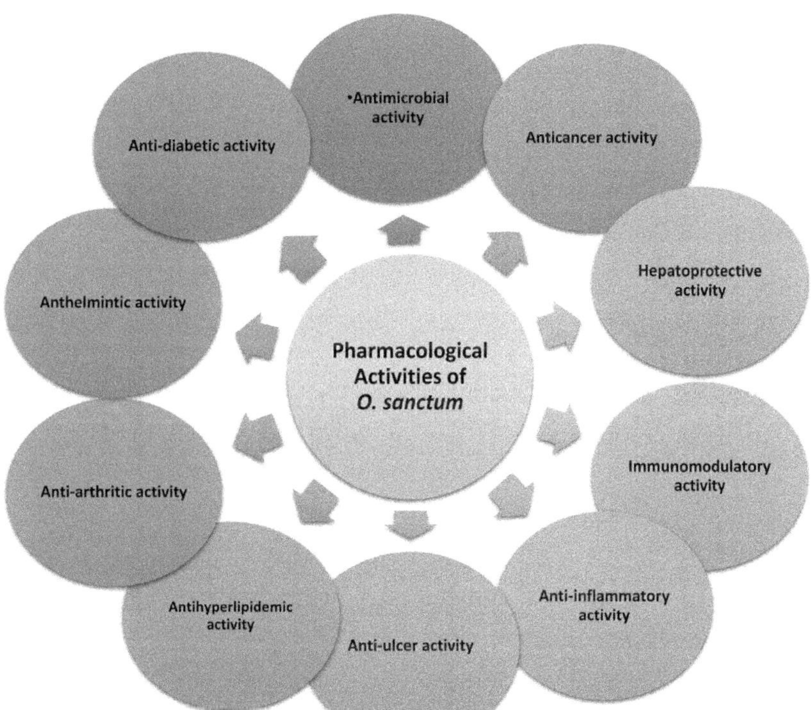

Figure 6.6 Pharmacological activities of *O. sanctum*.

Table 6.2 Clinical studies involving *Ocimum sanctum* extract.[a]

Tulsi extract	Participants	Duration	Dosage	Outcomes	Ref.
Tulsi leaves caps	40 male adults T2DM (45–55 year)	6.5 weeks	3 g per day before meal	Significant↓ postprandial glucose & fasting blood glucose	92
Tulasi leaves caps	30 adults obesity (17–30 year)	8 weeks	250 mg per day 2 × daily before meal	Improved BMI, lipid profile (except TC), TG, and IR	93
Tulsi powder leaves	30 adults T2DM	2 weeks	2 g per day	Significant↓ post-prandial glucose & fasting blood	94
Fresh tulsi leaves	3 adults T2DM	5 weeks	Fresh leaves 3 × daily	Considerable decrease in blood glucose reaching near normal levels	95
Tincture from tulsi	200 adults Gouty Arthritis	12 weeks	10 drops 3 times per day	Significant reduction in serum uric acid	96
Aqueous tulsi leaves	100 adults MetS (≥40 year)	12 weeks	5 ml/ × 2 day before meals	Improved lipid profile, fasting blood glucose, and BP	97
Tulsi leaves + glibenclamide drug	60 adults T2DM (30–65 year)	13 weeks	300 mg per day Tulsi + 5 mg glibenclamide	Significant ↓fasting blood & postprandial glucose reduced HBA1c	98
Aqueous tulsi leaves	40 adults T2DM (45–55 year)	8 weeks	500 g per day	Significant improvement in lipid profile	99
Powder tulsi leaves	90 male adults T2DM/MetS (40–60 year)	12 weeks	2 gper day	Improved T2DM symptoms: ↓polydipsia, ↓polyphagia, & BP	100
Whole plant decoction	10 adults T2DM	12 weeks	14 g per day	Reduced blood glucose in 9 T2DM adults	101

Powder whole plant tulsi	5 adults psychosomatic (60–80 year)	12 weeks	3 g × 2 per day	Significant improvement in lipid profile	102
Fresh juice tulsi leaves	50 female adults, hypotensive (18–30 year)	4 weeks	Juice × 2 per day	Significant changes in blood pressure	103
Ethanolic tulsi	22 healthy adults (22–37 year)	4 weeks	300 mg per day before food	Reduction in lipid profile, in 6 participants	104
Fresh juice 75% tulsi	20 adults, hypertension (45–64 year)	10 days	30 ml per day	Significant ↓blood pressure	105
Tulsi powder leaves	27 adults T2DM/ MeS (45–65 year)	4 weeks	1 g per day in morning before meal	Improved lipid profile, blood glucose, glycated proteins (HbA1c) & UA.	105

[a]BMI = body mass index measured by weight (kg)/height (m^2); BP = blood pressure; HbA1C = glycosylated hemoglobin; IR = insulin resistance; MetS = metabolic syndrome; T2DM = type 2 diabetes mellitus; TC = total cholesterol, TG = triglycerides; UA = uric acid.

Tulsi extract was found to be highly effective against an antibiotic-resistant strain of *Staphylococcus aureus*. Its leaf extract showed considerable antibacterial activity against drug-resistant *Salmonella enterica* serovar Typhi when coupled with chloramphenicol (C) and trimethoprim (Tm) (*S. typhi*). When coupled with C and Tm and *O. sanctum* leaf extract exhibited synergistic activity against S. typhi isolates. Eugenol (1-hydroxy-2-methoxy-4-allylbenzene), the active ingredient in *O. sanctum*, has been demonstrated to contribute significantly to Tulsi's antibacterial therapeutic properties. Tulsi solvents and water extracts exhibited antibacterial activity against multidrug-resistant bacteria *S. aureus*, with MIC values ranging from 1.56 to 6.25 mg ml^{-1}, while multidrug resistant *Klebsiella pneumoniae* and *Escherichia coli* exhibited higher values (6.25–25 mg ml^{-1}). The antimicrobial activity of Tulsi extract against *Streptococcus mutans* was discovered. Eugenol, methyl eugenol, linalool, 1, 8-cineole, and Tulsi oils have all been shown to be extremely cytotoxic toward Candida species.[106]

6.9.2 Anticancer

An ethanolic extract of *O. sanctum* has been shown to inhibit the formation of Sarcoma-180 solid tumors in mice and to increase their life expectancy. The anticancer activity of *O. sanctum* in the Lewis lung carcinoma animal model validated this conclusion. Ursolic acid is naturally anti-cancerous. *O. sanctum* shields DNA from radiation damage. *O. sanctum* is quite useful for treating a variety of tumorigenesis stages. The treatment of mice with Solid Sarcoma-180 tumors with *O. sanctum* aqueous and alcoholic extracts leads to a considerable reduction in tumor size.[107] Additionally, Tulsi extract has anti-inflammatory and radical scavenging characteristics, which are crucial for limiting the growth of some types of malignant cells. Ursolic acid and curcumin found in Tulsi are anti-inflammatory, cyclooxygenase inhibitory, and antioxidant. *O. sanctum* increases glutathione synthesis and storage in mice and glutathione-*S*-transferase activity by approximately 78%. The oil of *O. sanctum* is an effective chemopreventive agent against the development of fibro sarcoma in Swiss albino mice following subcutaneous injections of 20-methylcholanthrene. When given to mice, this resulted in an increase in survival and a delay in the onset of tumors. When mice were given oil, the end product lipid peroxidation, liver enzymatic activity, and malondialdehyde levels were drastically altered in contrast to untreated mice that received 20-methylcholanthrene injections.[108]

On LNCaP prostate cancer cells, Tulsi extract induces apoptosis. After 24 and 48 hours of treatment with various doses of 70% ethanolic extract of Tulsi, LNCaP prostate cancer cells exhibit cytotoxicity. When combined with docetaxel (DTL), the flavonoid vicenin-2 (VCN-2) inhibits prostate cancer (CaP). VCN-2 efficiently inhibits proliferation, angiogenesis, and apoptosis in CaP cells (PC-3, DU-145 and LNCaP). *In vitro*, VCN-2 inhibits the EGFR/Akt/mTOR/p70S6K pathway, as well as c-Myc, cyclin D1, cyclin B1, CDK4, PCNA, and hTERT. Tulsi extracts also have anti-ulcerogenic activity in rats treated with pyloric acid and aspirin. Upon acute and chronic administration, the OSL extract decreased the ligated and pyloric ligated ulcer indexes, as well as free and total acidity. Pre-treatment with the medication for seven days enhanced mucous secretion and decreased acid discharge. OG inhibits the growth and progression of breast cancer and functions as a natural inhibitor of metalloproteases found in matrices.[109]

6.9.3 Hepatoprotective Activity

By acting as a component of the detoxifying system, it stimulates the removal of dangerous compounds and works as a hepatoprotective agent. The study showed that when an alcoholic extract of Tulsi was administered orally, it protected albino rats' livers from paracetamol, carbon tetrachloride, and antituberculosis drug-induced liver impairment. The level of enzymes decreased when *O. sanctum* extract was used to reduce the number of enzymes in male albino rats weighing 100–150 g of the Wistar strain (5–6 weeks). The Biometry Research Unit of the Indian Statistical Institute discovered that when albino rats were given Tulsi extract orally for six days, it had a hepatotonic effect on paracetamol and carbon tetrachloride. When coupled with silymarin, Tulsi extract exhibits a substantial hepatoprotective effect.[110]

Tulsi extracts were found to be efficient in preventing oxidative stress in the liver by boosting glutathione peroxidase and catalase levels. They were also found to be effective in preventing hepatic steatosis. Eugenol, carvacrol, ursolic acid (UA), -caryophyllene, and rosmarinic acid were found to have anti-inflammatory, gastrointestinal, and hepatoprotective effects. The oil has anti-inflammatory properties as a result of its combined suppression of arachidonate metabolism and antihistaminic activity. Antipyretic efficacy is a result of prostaglandin inhibition, and the oil also acts as a peripheral analgesic. The oil was found to be helpful against formaldehyde or adjuvant-induced arthritis in rats, as well as turpentine oil-induced joint edema.[111]

6.9.4 Immunomodulatory Activity

When rats were treated with distilled extracts of fresh leaves, changes in the humoral immune response were observed. These changes were ascribed to processes such as antibody production, tissue responses, and the release of hypersensitive mediators of specific organs. It has been demonstrated that seed oil can affect both cellular and humoral immune responses. GABA-mediated immunomodulatory effects have been reported. Tulsi has been demonstrated to enhance cellular as well as humoral immunity. *O. sanctum* aqueous extract revealed immunotherapeutic potential in bovine subcutaneous aqueous extract infusions. Additionally, the aqueous extract of *O. sanctum* reduces the total number of bacteria, increases the number of neutrophils and lymphocytes, and has a high phagocytic capacity. Mediratta *et al.* examined the immunomodulatory effects of *O. sanctum* seed oil on a variety of immunological markers in healthy and stressed mice.[112] As a result, it was shown that Tulsi seed exerts control over both humoral and cellular immunological responses *via* the GABAergic system.

Godhwani *et al.* investigated the immunoregulatory effect of a methanolic extract and an aqueous suspension of Tulsi leaves in the treatment of antigenic challenge induced by *Salmonella typhosa* and sheep erythrocytes, as well as the antibody agglutination in the Widal and sheep erythrocyte agglutination tests in albino rats.[113] The results indicated that an immunological stimulation of the humoral immunogenic response occurred as a result of an increased antibody titer in the Widal and sheep erythrocyte agglutination tests. Immunomodulation restores and enhances the body's reaction to antigens such as bacteria, germs, viruses, and allergies. Tulsi is a powerful immunomodulatory herb. The distilled extract of fresh leaves was found to alter the humoral immune response. Aqueous extracts of leaves had both proliferative and inhibiting effects on splenocytes *in vitro*. When splenocytes were cultured with 31.25, 62.5, or 125 $g\,ml^{-1}$, a 42.17, 55.42, and 47.38 percent increase in spleen cell proliferation was seen in contrast to the negative control. The methanolic extract of *O. sanctum* leaves in combination with an aqueous suspension induced a clinically detectable immunostimulation of the humoral immune response.

6.9.5 Antiinflammatory Activity

O. sanctum reduces acute and chronic inflammation as the oils extracted from fresh leaves and seeds of *O. sanctum* were discovered to exhibit anti-inflammatory properties in experimental rats suffering from hind paw edema produced by carrageenan, serotonin, histamine, and prostaglandin-E-2.[114] *O. sanctum's* anti-inflammatory properties may be achieved through inhibition of cyclooxygenase and lipoxygenase. The main goal of this study was to evaluate the anti-inflammatory properties of *Ocimum* species extracts, with differing quantities of unsaturated fatty acids and different susceptibility to inflammation triggered by phlogistic agent administration. According to recent research, the most efficient plant to reduce paw edema is *O. basilicum*, which contains the highest concentration of linolenic acid. Fixed oil present in specific *Ocimum* species inhibits improved vascular permeability and leukocyte migration is illustrated by carrageenan-induced inflammation. The seeds of these three plants were investigated for their anti-inflammatory activities in rats with paw edema caused by carrageenan, leukotrine, or arachidonic acid. At a maximum proportion, *O. sanctum* seed oil prevented leukotrine-induced paw edema.[115]

Tulsi extracts, as well as petroleum ether and chloroform fractions, were tested for their anti-inflammatory activity in a Carrageenan-induced model of acute inflammation. Additionally, the effect of methanolic extract on lipopolysaccharide-induced nitric oxide (NO) generation in macrophages was investigated.[116] The results indicated that a methanolic extract of Tulsi (100, 200, and 400 $mg\,kg^{-1}$ body weight) considerably reduced Carrageenan-induced paw edema; the chloroform fraction was the most effective (66%, *p* 0.001). The extract was non-toxic up to 125 mg ml^{-1} in mouse peritoneal macrophages and inhibited lipopolysaccharide-induced NO generation at concentrations of 0–100 μm ml^{-1}. The methanolic extract of Tulsi was found to have substantial anti-inflammatory properties. The chloroform fraction of the Tulsi methanolic extract demonstrated the greatest anti-inflammatory effect.[117]

6.9.6 Antiulcer Activity

O. sanctum is an anti-inflammatory and ulcer-fighting plant. The fixed oil included in it contained an ulcer-fighting component. And it did so in rats by obstructing the stomach's output with a pylorus ligation. It's possible that its anti-ulcer properties stem from its effects on lipoxygenase, antihistaminic, and antisecretory enzymes.[118] Leukotriene antagonists and 5-lipoxygenase inhibitors have both been proven to be effective in reducing ulcers caused by alcohol and nonsteroidal anti-inflammatory drugs (NSAID) in investigations. Alcohol and NSAID-induced ulcers, as well as the 5-lipoxygenase pathway and leukotriene antagonism, are all treated with *O. sanctum* fixed oil. Histamine is known to cause stomach ulcers due to increased stomach acid secretion and histamine's vasospastic effect. This explains why *O. sanctum* fixed oil inhibited histamine-induced vasospasm and histamine-mediated gastric output. Mast cell degranulation and histamine release have been linked to gastric ulcers caused by reserpine exposure, which is aided by a cholinergic mechanism. The antiulcer properties of *O. sanctum* fixed oil are most likely due to histamine and anticholinergic components.[119]

In a model of indomethacin-induced ulceration, the gastroprotective effect of *O. sanctum* was investigated and compared to that of omeprazole. This two-year study was undertaken at Navodaya Medical College and Research Centre. The gastroprotective efficacy of an aqueous extract of *O. sanctum* was investigated and compared to that of omeprazole utilizing an indomethacin-induced gastric injury model. In different groups of rats, *O. sanctum* (200 mg kg^{-1} or 400 mg kg^{-1}) or omeprazole (10 mg kg^{-1}) was delivered alone. When compared to a control, *O. sanctum* had a substantial (*p* 0.05) preventive effect against indomethacin-induced stomach ulcers. *O. sanctum* had a gastroprotective effect equivalent to that of the conventional medication omeprazole. *O. sanctum* had a strong gastroprotective effect against indomethacin-induced stomach injury, which may be related to its inhibitory effect on 5-lipoxygenase, mucoprotective activity, and antisecretory function.[120]

6.9.7 Effect on the Central Nervous System

Various stem, leaf, and stem callus extracts (made on a slightly modified Murashige and Skoog's medium supplemented with 2,4-dichlorophenonyacetic acid and kinetin) were assessed for anticonvulsant efficacy using the maximal electroshock paradigm with Phenytoin as the standard. Extracts of stem, leaf, and stem calli were found to be effective at suppressing tonic convulsions induced by transcorneal electroshock.[121] In 'open field' studies, an ethanolic extract of *O. sanctum* leaves prolonged the period of loss of reflex in mice given pentobarbital, decreased the incidence and intensity of electroshock and pentylenetetrazole-induced convulsions, and reduced the fighting time and ambulation induced by apomorphine. In the forced swimming behavioral despair model, the extract reduced immobility in a manner similar to Imipramine. This activity was suppressed by haloperidol and sulpiride, indicating that it may involve dopaminergic neurons. When the extract was combined with bromocriptine, a potent D2-receptor agonist, a synergistic effect was reported in comparable trials. Nootropic drugs are a relatively new class of pharmaceuticals used to treat organic impairments of learning capacity. Joshi and Parle conducted a study in mice to determine the nootropic and anti-amnesic effects of *O. sanctum* extract.[122] An aqueous extract of the complete *O. sanctum* plant ameliorated the amnesic effects of scopolamine (0.04 $mg\,kg^{-1}$), diazepam (1 $mg\,kg^{-1}$), and aging-induced memory deficits in rats. The raised plus maze and passive avoidance paradigms were used to study exteroceptive behavioral models. *O. sanctum* extract significantly decreased transfer latency and increased step-down latency as compared to the control (piracetam-treated), scopolamine-treated, and aged groups of mice. Thus, *O. sanctum* formulations may be beneficial in the treatment of cognitive disorders like dementia and Alzheimer's disease. A 400 $mg\,kg^{-1}$ (ip) methanolic extract of *O. sanctum* root extract considerably increases mice's swimming time in a despair swim test model, showing that *O. sanctum* possesses stimulant and/or anti-stress properties.[123]

6.9.8 Antihyperlipidemic Activity

The fixed oils are used to lower the lipid levels. *O. sanctum* has been shown to reduce lipid levels in rats fed a high-fat diet in research. It restores lipids to their normal condition and inhibits the lipid peroxidation process, which causes lipids to degrade due to oxidation.[124] This influence was primarily caused by linoleic and linolenic acids. Another study found that seed oil and fresh leaves of *O. sanctum* reduce cholesterol, VLDL, and triacylglycerol levels in rabbits and albino rats, while raising HDL cholesterol and net sterol content in feces. An aqueous extract of the plant was given to Streptozotocin-induced diabetic rats for eight weeks, and the lipid profile was reported to be improved. In rats treated with triton, Amrani *et al.* (2006) discovered that the aqueous extract of *O. sanctum* had a significant hypolipidemic effect.[125] The presence of hypolipidemic and antioxidant chemicals in *O. sanctum*'s extract was linked to its anti-hyperlipidemic activity. These findings demonstrate that the active components in *O. sanctum* extract can be used therapeutically in hyperlipidemic people, hinting that they may be advantageous and paving the way for future study in the fields of hyperlipidemia, atherosclerosis, and associated illnesses.[126]

In rats fed a high fat diet, the antihyperlipidemic and cardioprotective effects of a fixed oil produced from the leaves of *O. sanctum* were examined. The findings revealed that the fixed oil in *O. sanctum* included five distinct fatty acids, with alpha-linolenic acid being the most abundant. The antihyperlipidemic effect of fixed oil was mostly due to a decrease in hepatic lipid synthesis. Linolenic and linoleic acids present in *O. sanctum* fixed oil may be responsible for the oil's lipid-lowering and cardioprotective effects.[127]

6.9.9 Antiarthritic Activity

In rats, the antiarthritic activity of the fixed oil obtained from *O. sanctum* seeds was evaluated by Freund's adjuvant arthritis, formaldehyde-induced arthritis, or turpentine oil-induced joint edema. When the fixed oil was utilized to inhibit the generation of inflammatory mediators, a significant antiarthritic effect was reported in both arthritis models and turpentine oil-induced joint edema (serotonin, histamine, bradykinin and PGE2). This experiment established *O. sanctum*'s medicinal potential in rheumatoid arthritis. As per a study performed on rats suffering from formaldehyde-induced arthritis, fixed oils of OS acted as an anti-inflammatory on the damaged paw. After 10 days of daily intraperitoneal administration of fixed oils to rats with arthritis, significant improvement was observed.[128]

6.9.10 Anthelmintic Activity

In vitro testing of the essential oil of *O. sanctum* and eugenol revealed significant anthelmintic action in a *Caenorhabditis elegans* model. Euenol's ED50 value was 62.1 $g\,ml^{-1}$. Eugenol, the essential oil's primary constituent, is proposed to be the probable anthelmintic principle. When compared to albendazole, ursolic acid contained in Tulsi has a high ability to paralyze and kill worms. The purpose of one study was to extract ursolic acid (UA) from *O. sanctum* and to test its anthelmintic and antibacterial efficacy against *E. coli*. Ursolic acid, a significant bioactive molecule, was isolated from an ethanolic extract of the *O. tenuiflorum*, or *O. sanctum*, plant. Ursolic acid was extracted from an ethanolic extract of the plant *O. sanctum* in this study. The disk diffusion and well diffusion methods were used to determine its anthelmintic and antibacterial activities against *E. coli* and compare ursolic acid's antihelmintic and antibacterial activity with that of the conventional drugs albendazole and tetracycline, respectively.[129]

The results indicated that the compound's efficacy increased dosage-dependently at concentrations of 15, 20, and 25 $mg\,ml^{-1}$. The results were comparable to those obtained with the conventional antibiotic albendazole at the same concentration. The disk diffusion and well diffusion methods were used to screen ursolic acid for antibacterial activity. Antimicrobial *in vitro* investigations indicated that ursolic acid may be active against a variety of bacteria species and has a larger biocidal effect than ligand fragments. The work demonstrated the significant anthelmintic and antibacterial activity of ursolic acid, which was effective against worms. Ursolic acid was found to have the capacity to paralyze and kill parasitic worms.[130]

6.9.11 Antidiabetic Activity

O. sanctum has anti-diabetic effects, and an aqueous extract of Tulsi was found to significantly lower blood glucose levels in rats in one study. Similarly, multiple studies show that Tulsi's hypoglycemic effect aids in the reduction of fasting blood sugar and HBA1c. Tulsi can also be used to treat metabolic issues and can be used in combination with other therapies. It aids in the biochemical breakdown of contaminants and is useful in the treatment of liver disease.[131] Despite the fact that anti-diabetic efficacy has been studied in a variety of animal models, clinical trials have been sparse. The usage of *O. sanctum* in the treatment of diabetes mellitus is attracting a lot of interest. Plant fibers, particularly soluble fibers, have been shown to help with hyperglycemia and hyperlipidemia treatment. The antidiabetic activity of *O. sanctum* is thought to be due to the plant's extract stimulating adenylate cyclase or phosphatidylinositol, or having a direct effect on pancreatic cells by increasing calcium entry and therefore triggering insulin release. One of the earliest randomized controlled clinical trials looked at antidiabetic properties in 40 people with non-insulin dependent diabetes mellitus (NIDDM). It was discovered that taking 2.5 g dried Tulsi leaf powder orally twice daily on an empty stomach reduced fasting blood sugar levels by up to 21 $mg\,dl^{-1}$ and postprandial blood glucose levels by 15.8 $mg\,dl^{-1}$. Another study used an aqueous extract of *O. canum* to help isolated pancreatic β-cells release insulin in the treatment of type 2 diabetes mellitus.[132]

It was found that rats given fructose and treated with *O. sanctum* extract for 30 days had considerably lower serum glucose levels than rats not treated with the extract.[133] On the other hand, *O. sanctum* extract had no noticeable effect on hyperinsulinemia. Vats *et al.* examined the effect of *O. sanctum* on 3 important carbohydrate metabolism enzymes glucokinase (gk), hexokinase (hk), and phosphofructokinase (PFK), and also the glycogen content of insulin-dependent (liver and skeletal muscle) and insulin-independent (kidneys and brain) tissues in rats for 30 days using a streptozotocin (STZ). On the 15th and 30th days after 30 days of treatment of *O. sanctum* extracts at a dose of 200 $mg\,kg^{-1}$, plasma glucose levels reduced by roughly 9.06 and 24.4 percent, respectively. *O. sanctum* considerably decreased renal glycogen concentration in any tissue, but not liver glycogen concentration; also, *O. sanctum* effectively corrected the action of glucokinase (gk), hexokinase (hk), and phosphofructokinase (PFK) distributed throughout the diabetic control.[134]

6.9.12 Miscellaneous Activities

O. sanctum is a cardiac tonic that helps to avoid heart attacks, relieve tension, and restore normal blood pressure. It has anti-arthritic and anabolic effects, boosting muscle mass strength by accelerating protein synthesis. It relieves sore eyes and helps to alleviate night blindness. Tulsi juice and honey are used as an eyewash and anti-conjunctivitis medication to cure conjunctivitis. It regulates and balances neurotransmitter levels in the brain, as well as the neurochemistry of the brain. Tulsi extract aids in memory enhancement. The essential oil of *O. sanctum* is responsible for the anthelmintic activity seen in a Caenorhabditis elegans model *in vitro*.[135] *O. sanctum* is said to be a mosquito repellent with considerable larvicidal activity. Anopheles stephensi, Aedes aegypti, and Culex quinquefasciatus are among the mosquito species that are resistant to it. The time it takes for blood to coagulate is prolonged by *O. sanctum* fixed oil, and the percentage increase is comparable to that of aspirin, which could be due to platelet aggregation inhibition. Furthermore, the oil prolonged pentobarbitone-induced sleep-in rats, showing that the oil may have an inhibitory effect on the cytochromic enzyme responsible for pentobarbitone metabolism in the liver. In 9 day-old rat pups (25 $\mu M\ kg^{-1}$ body weight), Gupta evaluated the anticataract effects of *O. sanctum* extract on selenite-induced cataract. *O. sanctum* (5 and 10 $mg\ kg^{-1}$ body weight) administered intraperitoneally four hours before a selenite challenge lowers the incidence of selenite cataract by 20% and 60%, respectively, and inhibits protein insolubilization.[136] Table 6.3 shows the biological activity of various chemical compounds present in Tulsi plants.

6.10 Conclusion

While thousands of herbal plants exist globally, Tulsi is considered the queen of herbs due to its exceptional medicinal capabilities. Taking the entire study into account, *O. sanctum* or Tulsi is a traditionally and clinically proved medicinal herb in terms of both its uses and efficacy. *O. sanctum* possesses a broad spectrum of pharmacological activity, and it can be considered a medical gift from nature. The use of medicinal plants to cure a variety of ailments dates all the way back to ancient India and other emerging countries. Recent trends indicate a dramatic increase in the use of herbal medications to treat a variety of ailments among individuals in developed countries, as they give superior therapeutic efficacy with fewer adverse effects. Numerous medicinal plants have been evaluated for their potential efficacy in the treatment of major sick situations, and even the WHO has proposed herbal ingredients for further study in circumstances where current medications are not effective. It is highly regarded among herbs with different biological potentials and offers tremendous potential for further research in new areas. Historically, crude extracts of numerous plant components have been utilized for their anti-diabetic, antioxidant, stress-relieving, antihyperlipidemic, and antibacterial activities. In order to produce novel compounds with novel pharmacological potential, more research on the chemical characteristics of *O. sanctum* is required. *O. sanctum*'s long history of traditional use, broad spectrum of pharmacological properties, and toxicity testing all indicate that it is a safe and valuable herb for clinical uses. The present compilation of chemical components and their

Table 6.3 Chemical constituents found in Tulsi plant with their biological activity.

Chemical constituent	Characteristics	Activity	Ref.
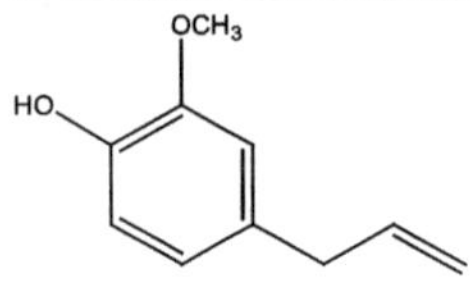 Eugenol	Phenylpropene, an allyl chain-substituted guaiacol	Germs, fungi and infection with antibacterial and anti-inflammatory activities	137
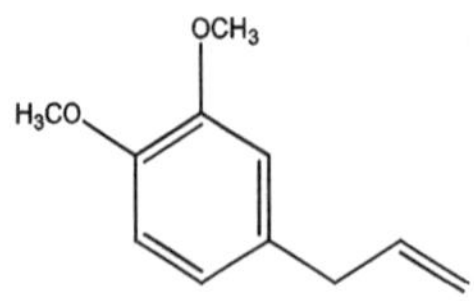 Methyleugenol	Phenylpropene	Used in insect traps and lure products to attract certain fruit flies	138
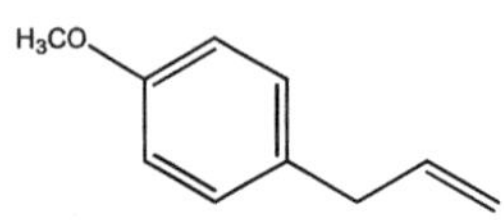 Methylchavicol	Phenylpropene	Excellent tonic for the mind and nerves. Also good for tiredness, headaches, digestive problems, and muscular aches	139
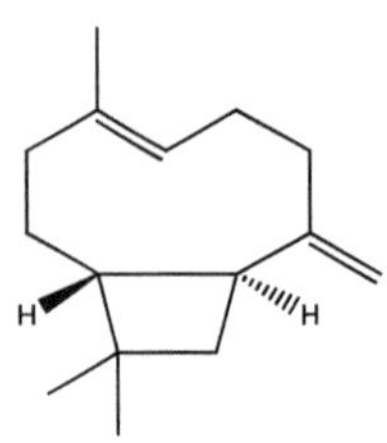 β–Caryophyllene	Multiple components	Multiple biological activity, highly therapeutic	140
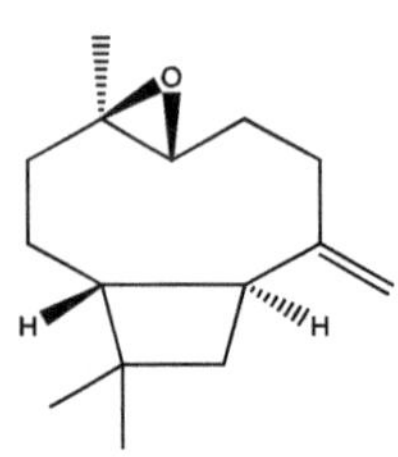 Caryophyllene oxide	Natural bicyclic sesquiterpene	An oxygenated terpenoid, well known as a preservative in food, drugs and cosmetics, has been tested *in vitro* as an antifungal against dermatophytes	141
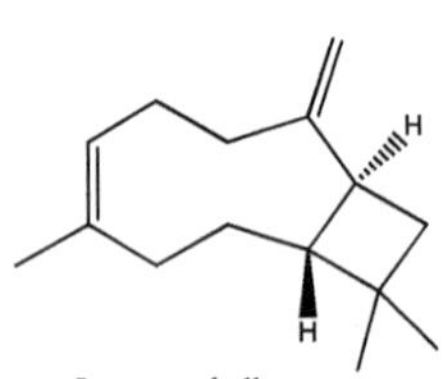 Iso caryophyllene	Natural bicyclic sesquiterpene	Cannabinoid-like properties	142

Table 6.3 (*Continued*)

Chemical constituent	Characteristics	Activity	Ref.
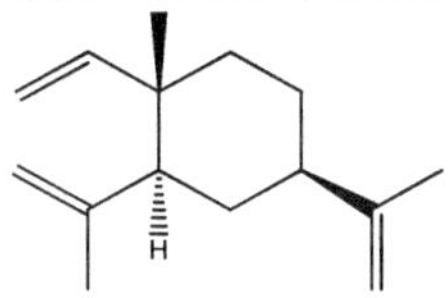 β–Elemene	Sesquiterpenes	Anti-proliferative effect	143
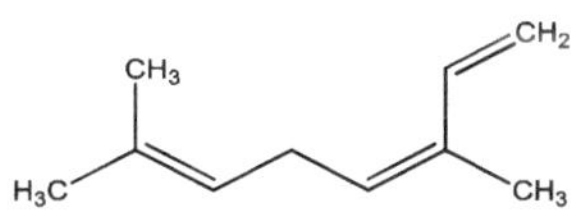 β–ocimene	Monoterpenes	Herbal scent and acts as a plant defense and has anti-fungal properties	144
 1 8-Cineole	Cyclic ether and a monoterpenoid	Used in flavorings, fragrances, and cosmetics	145
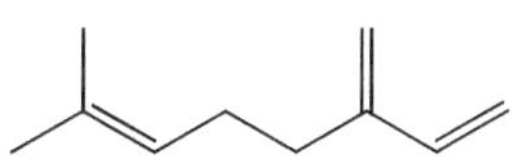 Myrcene	An olefinic natural organic hydrocarbon	Analgesic activity	146
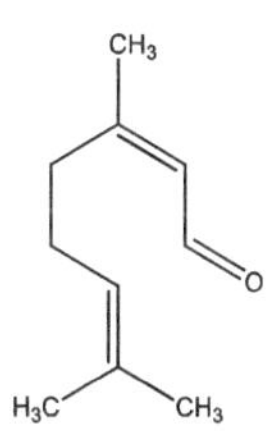 Neral	Geranial and citral B	Cure of skin and hair problems	147
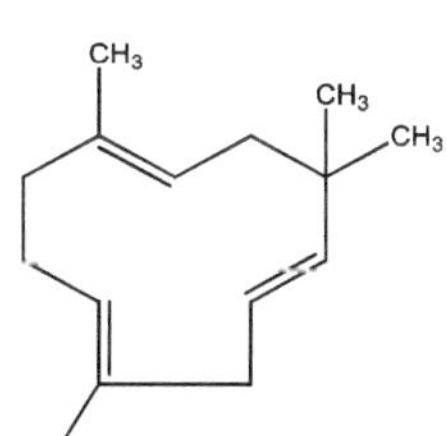 α–humulene	Sesquiterpenes	Pronounced anti-inflammatory	148

Table 6.3 (*Continued*)

Chemical constituent	Characteristics	Activity	Ref.
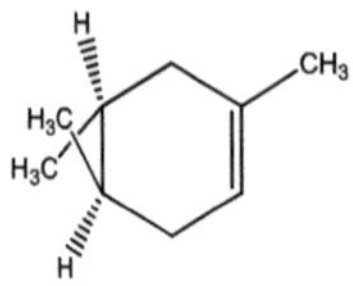 3-carene	Bicyclic monoterpene	Effective anti-inflammatory	149
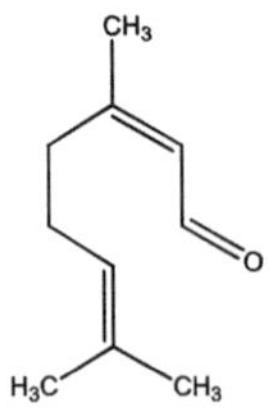 Citral	3,7-Dimethyl-2,6-octadienal or lemonal	Holistic and alternative medicine, potent antiseptic and may prove useful in treating gastro-intestinal infections, including *H. pylori*	150
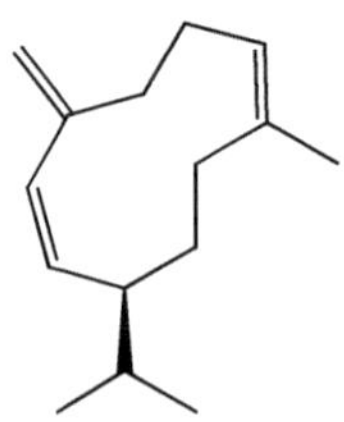 GermacreneD	Sesquiterpene	Hyperglycemia, obesity	151
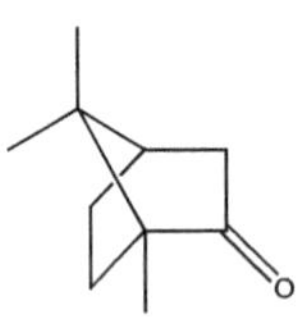 Camphor	Essential oil mainly contains multiple components	Counters elevated blood sugar levels and is highly beneficial in diabetes, cancer, and chronic bronchitis	152

pharmacological properties will benefit future study on the *O. sanctum* plant and the identification of novel therapeutic targets.

Funding

No funding has been provided.

Conflict of Interest

The authors declare that there is no conflict of interest regarding this publication.

Acknowledgements

The authors are thankful to the Department of Pharmacognosy, ISF College of Pharmacy, for providing the necessary facilities.

References

1. R. Borah and S. P. Biswas, *Int. J. Agric. Environ. Biotechnol.*, 2018, **3**, 1732.
2. G. Nahak, R. C. Mishra and R. K. Sahu, *Drug Invent. Today*, 2011, **3**, 95.
3. R. K. Joshi, *J. Pharmacogn. Phtoch.*, 2017, **6**, 261.
4. S. K. Gupta, J. Prakash and Srivastava, *Indian. J. Exp. Biol.*, 2002, **40**, 765.
5. S. Poddar, T. Sarkar, S. Choudury, S. Chatterjee and P. Ghosh, *Int. J. Botany Stud*, 2020, **5**, 174.
6. G. MusTafa, R. Arif, A. Atta, S. Sharif and A. Jamil, *Matrix Sci. Pharma*, 2017, **1**, 17.
7. T. Y. Low, K. O. Wong, A. L. L. Yap, L. H. J. De Haan and I. M. C. M. Rietjens, *Compr. Rev. Food Sci. Food Saf.*, 2017, **16**, 821.
8. B. Koul, Role of Ayurveda in Cancer Treatment, in *Herbs for Cancer Treatment*, Springer, 2019, pp. 151–191.
9. H. Bellebuono, *An Herbalist's Guide to Formulary: The Art & Science of Creating Effective Herbal Remedies*, Llewellyn Worldwide, 2017.
10. R. Kulkarni, R. Pushpan, C. Srilakshmi, J. Nayan and N. S. Sarin, *Int. J. AYUSH*, 2020, **9**, 1.
11. R. Rupani and A. Chavez, *Clin. Dermatol.*, 2018, **36**, 306.
12. R. K. Upadhyay, *Int. Green Pharm.*, 2017, **11**, S1.
13. P. Nath, *Climate of Conquest: War, Environment, and Empire in Mughal North India*, Oxford University Press, 2019.
14. I. Giannenas, E. Sidiropoulou, E. Bonos, E. Christaki and P. Florou-Paneri, The history of herbs, medicinal and aromatic plants, and their extracts: Past, current situation and future perspectives, in *Feed Additives*, Elsevier, 2020, pp. 1–18.
15. R. Sinha, *Our trees*, Publications Division Ministry of Information & Broadcasting, 2017.
16. N. S. Guhane, C. D. Ghode, A. G. Jadhao and P. A. Patil, *J. Pharmacogn. Phytochem.*, 2021, **10**, 1427.
17. A. N. Bhooshitha, A. R. Ghosh, H. M. Chandan, H. S. Nandhini, B. R. Pramod and K. L. Krishna, *J. Pharm. Sci. Res.*, 2020, **12**, 420.
18. S. A. Almatroodi, A. A. Mohammed, A. Ahmad and R. A. Husain, *Pharmacogn. J.*, 2020, **12**, 1198.
19. A. A. Khalil, U. Rahman, M. R. Khan, A. Sahar, T. Mehmoodac and M. Khana, *RSC Adv.*, 2017, 7, 32669.
20. E. Ivanišová and M. J. B. C. Kačániová, Plant adaptogens: an important source of bioactive compounds, in *Bioactive Compounds: Sources, Properties and Applications*, ed. R. Porter and N. Parker, Nova Science Publishers, 2017, p. 91.

21. L. Bell, H. N. Yahya, O. O. Oloyede, L. Methven and C. Wagstaff, *Food Chem.*, 2017, **15**, 521.
22. A. Thakur, V. Sharma and A. Thakur, *Int. J. Chem. Stud.*, 2019, 7, 2472.
23. L. Sethi and P. Bhadra, *Indian J. Nat. Sci.*, 2020, **10**, 20854.
24. K. P. S. Kumar, D. Bhowmik, Biswajit, Chiranjib, Pankaj, K. K. Tripathi and M. Chandira, *Res. J. Pharmacogn. Phytochem.*, 2010, **2**, 93.
25. V. Adiani and P. S. Variyar, Pepper Berries (*Piper nigrum* L.)–Drupes with Therapeutic and Nutraceutical Potential, in *Asian Berries*, ed. G. Xiao, Y. Xu and Y. Yu, CRC Press, 2020, pp. 381–397.
26. M. Sellami, O. Slimeni, A. Pokrywka, G. Kuvačić, L. D. Hayes, M. Milic and J. Padulo, *J. Int. Soc. Sports Nutr.*, 2018, **15**, 14.
27. I. J. Ancheril, *An Interpretation of The Vishachikitsa Tradition in Kerala*, 2020.
28. P. K. Gupta, K. Sonewane, M. Rajan, N. J. Patil, T. Agrawal, E. R. Banerjee, N. S. Chauhan and A. Kumar, *Adv. Tradit. Med.*, 2021, **29**, 1.
29. R. Siddique, N. M. Sureshbabu, J. Somasundaram, B. Jacob and D. Selvam, *J. Conserv. Dent.*, 2019, **22**, 40.
30. P. Gupta, D. K. Yadav, K. B. Siripurapu, G. Palit and R. Maurya, *J. Nat. Prod.*, 2007, **70**, 1410.
31. K. V. Kulkarni and B. V. Adavirao, *J. Med. Plants Stud.*, 2018, **6**, 106.
32. V. Aruna, G. V. Amruthavalli, R. Soundharya and R. Gayathri, *Int. J. Med. Res.*, 2018, **5**, 1.
33. S. Patil, *J. Ayurveda. Integr. Med.*, 2021, **20**, 100420.
34. S. Ahmad, S. Zahiruddin, B. Parveen, P. Basist, A. Parveen, G. Gautam, R. Parveen and M. Ahmad, *Front. Pharmacol.*, 2021, **11**, 578970.
35. P. Pramanick, A. Roy, S. Roy and A. Mitra, *Parana J. Sci. Educ.*, 2017, **3**, 9.
36. C. D. Lakshman, *Adv. Plants Agric. Res.*, 2016, **5**, 561.
37. Y. B. Tripathi, N. Joshi, S. K. Dubey, P. S. Byadgi, S. Bhat, R. Prasad, B. K. Dwivedi, K. K. Pandey, J. S. Tripathi, C. S. Pandey, K. H. H. V. S. S. N. Murthy, S. Suman, A. K. Pandey, V. Srivastava, V. Jaisawal and K. N. Dwivedi, *Indian J. Tradit. Know.*, 2021, **19**, S-69.
38. P. K. Kumar, M. R. Kumar, K. Kavitha, J. Singh and R. Khan, *Int. J. Adv. Pharm. Biol. Chem.*, 2012, **1**, 406.
39. S. Thangapandiyan and M. Selvarani, An in-vitro Bactericidal Efficacy of *Ocimum tenuiflorum* (Tulsi), a Known Indian Folk Medicinal Plant, in *Technological Innovation in Pharmaceutical Research*, ed. P. K. Sharma, B P International, India, 2021, pp. 42–53.
40. S. Banik, R. Mukherjee, P. Ghosh, S. Karmakar and S. Chatterjee, *J. Pharmacogn. Phtoch.*, 2018, 7, 2681.
41. P. Bellavite, M. Marzotto and B. Andreoli, *BMC Complementary Altern. Med.*, 2019, **5**, 073.
42. F. N. Sani, M. K. Ns, F. N. Sanim and M. K. Ns, *Laboratory Surgical Medical Nursing Practicum Module I*, 2020.
43. G. Dash, K. T. Kumar, G. R. Mohanty, D. Sahoo, G. Mahalik and S. Parida, *Int. J. Herb. Med.*, 2018, **6**, 57.

44. M. Vinaya, B. L. Kudagi, M. A. Kamdod and M. Swamy, *Int. J. Basic Clin. Pharmacol.*, 2017, **6**, 511.
45. J. Witonsky, R. Abraham, J. Toh, T. Desai, M. Shum, D. Rosenstreich and S. P. Jariwala, *J. Asthma*, 2019, **56**, 927.
46. A. A. Meshram, S. Srihari and S. Parwe, *Int. J. Res. Pharm. Sci.*, 2020, **11**, 1219.
47. M. Tyagi, N. Tyagi and J. Tyagi, *Int. Res. J.*, 2021, **3**, 370.
48. Amylcure, https://ayurcentralonline.com/en/liver-spleen-health/1710-amlycure-ds-capsule-10caps-aimil.html.
49. Decofcyn, https://www.ayurmedinfo.com/2012/08/09/dekofcyn-cough-syrup-benefits-dosage-ingredients-side-effects/.
50. Feverex syrup, https://www.ayurmedinfo.com/2018/02/19/feverex-syrup/, https://medimartus.com/product/feverex-syrup-100ml/.
51. B. cough, https://www.ayurmedinfo.com/2016/09/22/b-cough-syrup/, https://www.1mg.com/drugs/indikof-b-syrup-306901.
52. Koflet syrup, https://ayurcentralonline.com/en/cough/1107-koflet-syrup-100ml-himalaya.html.
53. Respicare tab, https://www.bacfo.com/Products/product/RESPICARE, https://dawaai.pk/medicine/respi-care-10mg-1682.html.
54. M-cof, https://www.ayurmedinfo.com/2012/08/08/m-cof-syrup-benefits-dosage-ingredients-side-effects/.
55. Kurex syrup, https://www.1mg.com/otc/kurex-light-syrup-otc250050, https://www.ayurmedinfo.com/2019/03/23/kurex-light-syrup/.
56. Tussnil syrup, https://ayurcentralonline.com/en/asthma/250-tussnil-syrup.html.
57. Pneumonorm Syrup, https://www.1mg.com/otc/pneumonorm-syrup-otc319796.
58. Bresol Tablet, https://www.1mg.com/otc/himalaya-bresol-tablet-otc131450.
59. Dabur Tulsi Drops, https://www.dabur.com/in/en-us/ayurvedic-herbal-products/dabur-tulsi-drops.
60. Crux Cough Syrup, https://www.1mg.com/otc/ban-labs-crux-cough-syrup-otc321485.
61. Cough Syp Tulasi, https://www.pankajakasthuri.in/pankajakasthuri-cough-syrup-with-tulsi.html.
62. Q. W. Zhang, L. G. Lin and W. C. Ye, *Chin. Med.*, 2018, **13**, 20.
63. C. Wen, J. Zhang, H. Zhang, C. S. Dzah, M. Zandile, Y. Duan, H. Ma and X. Luo, *Ultrason. Sonochem.*, 2018, **48**, 538.
64. B. Albero, J. L. Tadeo and R. A. Pérez, *Trend. Anal. Chem.*, 2019, **118**, 739.
65. L. Alexandru, G. Cravotto, L. Giordana, A. Binello and F. Chemat, *Innov. Food Sci. Emerg. Technol.*, 2013, **20**, 167.
66. K. Tekin, M. K. Akalın and M. G. Şeker, *Ind. Crop. Prod.*, 2015, 77, 954.
67. P. Izadiyan and B. Hemmateenejad, *Food Chem.*, 2016, **190**, 864.
68. A. Carreira-Casais, C. Lourenço-Lopes, P. Otero, M. Carpena, A. G. Pereira, J. Echave, A. Soria-Lopez, F. Chamorro, M. A. Prieto and J. Simal-Gandara, Application of Green Extraction Techniques for Natural

Additives Production, in *Food Additives,* IntechOpen, London, United Kingdom, 2021, Available: https://www.intechopen.com/online-first/78721, DOI: 10.5772/intechopen.100320.
69. A. Gupta, M. Naraniwal and V. Kothari, *Int. J. Appl. Nat. Sci.*, 2012, **1**, 8.
70. A. Abu-Samra, J. S. Morris and S. R. Koirtyohann, *Anal. Chem.*, 1975, **47**, 1475.
71. K. Ganzler, A. Salgo and K. J. Valko, *J. Chromatogr.*, 1986, **371**, 299.
72. M. Vinatoru, T. J. Mason and I. Calinescu, *TrAC-Trend. Anal. Chem.*, 2017, **97**, 159.
73. D. Skarpalezos and A. Detsi, *Appl. Sci.*, 2019, **9**, 4169.
74. G. Wenqiang, L. Shufen, Y. Ruixiang, T. Shaokun and Q. Can, *Food Chem.*, 2007, **101**, 1558.
75. M. de Oliveira, B. E. F. Frihling, J. Velasques, F. J. C. M. Filho, P. S. Cavalheri and L. Migliolo, *Sci. Total Environ.*, 2020, **705**, 135568.
76. J. F. Osorio-Tobón, P. I. N. Carvalho, M. A. Rostagno, A. J. Petenate and M. A. A. Meireles, *J. Supercrit. Fluid.*, 2014, **95**, 167.
77. N. N. Azwanida, *Med. Aromat. Plants*, 2015, **4**, 196.
78. I. Ignat, I. Volf and V. I. Popa, *Food Chem.*, 2011, **126**, 1821.
79. N. Cheikhyoussef, The Application of Supercritical Carbon Dioxide in the Extraction of Biomolecules, in *Advanced Nanotechnology and Application of Supercritical Fluids*, ed. A. A. Inamuddin, Nanotechnology in the Life Sciences, Springer, Cham., 2020, pp. 141–164.
80. M. R. Thakker, J. K. Parikh and M. A. Desai, *Chem. Eng. Res. Des.*, 2018, **134**, 162.
81. P. Otero, P. Garcia-Oliveira, M. Carpena, M. Barral-Martinez, F. Chamorro, J. Echave, P. Garcia-Perez, H. Cao, J. Xiao, J. Simal-Gandara and M. A. Prieto, *Trends Food Sci. Technol.*, 2021, **116**, 1084.
82. D. C. Panadare and V. K. Rathod, *Process Biochem.*, 2017, **61**, 195.
83. Q. W. Zhang, L. G. Lin and W. C. Ye, *Chin. Med.*, 2018, **13**, 20.
84. A. Mustafa and C. Turner, *Anal. Chim. Acta*, 2011, **703**, 8.
85. A. M. Gizir, N. Turker and E. Artuvan, *Eur. Food Res. Technol.*, 2008, **226**, 363.
86. T. Belwal, S. M. Ezzat, L. Rastrelli, I. D. Bhatt, M. Daglia, A. Baldi, H. P. Devkota, I. E. Orhan, J. K. Patra, G. Das, C. Anandharamakrishnan, L. Gomez-Gomez, S. F. Nabavi, S. M. Nabavi and A. G. Atanasov, *TrAC-Trend Anal. Chem.*, 2018, **100**, 82.
87. M. Bouras, N. Grimi, O. Bals and E. Vorobiev, *Ind. Crops Prod.*, 2016, **80**, 50.
88. C. G. Joshi and N. G. Magar, *J. Sci. Ind. Res.*, 1952, **11b**, 261.
89. R. Jabeen, M. Shahid, A. Jamil and M. Ashraf, *Pak. J. Bot.*, 2008, **40**, 1349.
90. M. Kumar, A. Kumar, R. Nagpal, D. Mohania, P. Behare, V. Verma, P. Kumar, D. Poddar, P. K. Aggarwal, C. J. Henry, S. Jain and H. Yadav, *Int. J. Food Sci. Nutr.*, 2010, **61**, 473.
91. G. S. Grover and J. T. Rao, *Perfum. Kosmet.*, 1977, **58**, 326.
92. R. Gandhi, B. Chauhan and G. Jadeja, *Indian J. Appl. Res.*, 2016, **6**, 62.

93. S. Satapathy, N. Das, D. Bandyopadhyay, S. C. Mahapatra, D. S. Sahu and M. Meda, *Ind. J. Clin. Biochem.*, 2017, **32**, 357.
94. P. Venkatesan and R. Sengupta, *Int. J. Pure Appl. Biosci.*, 2015, **3**, 331.
95. Ch. S. Prasadacharyulu, *Euro. J. Exp. Bio.*, 2015, **5**, 1–5.
96. M. Ahmad, A. A. Faraazi and M. N. Aamir, *Bull. Chem. Soc. Ethiop.*, 2013, **27**, 469.
97. D. K. Devra, K. C. Mathur, R. P. Agrawal, I. Bhadu, S. Goyal and V. Agarwal, *J. Nat. Remedies*, 2012, **12**, 63.
98. G. Somasundaram, K. Manimekalai, K. J. Salwe and J. Pandiamunian, *Int. J. Life Sci. Pharm. Res.*, 2012, **5**, 75.
99. B. Dineshkumar, M. Analava and M. Manjunatha, *Int. J. Diabetes Metab.*, 2010, **18**, 59.
100. A. Kochhar, N. Sharma and R. Sachdeva, *Stud. EthnoMed.*, 2009, **3**, 5.
101. N. Luthy, *Ohio J. Sci.*, 1964, **64**, 223.
102. A. K. Verma, G. P. Dubey and A. Agrawal, *Res. J. Pharm. Technol.*, 2012, **5**, 791.
103. A. Bhargava, L. Gangwar and H. S. Grewal, *Int. Conference Food Agric. Sci.*, 2013, **55**, 83.
104. S. Mondal, B. Mirdha, M. Padhi and S. Mahapatra, *J. Preventive Cardiology*, 2012, **1**, 177.
105. S. Sarvaiya, Studies on hypertension with special reference to hypotensive effect of Ocimum sanctum, PhD thesis, Sardar Patel University, Anand, India, 1986.
106. A. Khan, A. Ahmad, N. Manzoor and L. A. Khan, *Nat. Prod. Commun.*, 2010, **5**, 345.
107. N. Bano, A. Ahmed, M. Tanveer, G. M. Khan and M. T. Ansari, *J. Bioequiv. Availab.*, 2017, **9**, 387.
108. S. Darakhshan, A. Bidmeshki Pour, A. Hosseinzadeh Colagar and S. Sisakhtnezhad, *Pharmacol. Res.*, 2015, **95–96**, 138.
109. P. Nangia-Makker, T. Raz, L. Tait, M. P. Shekhar, H. Li, V. Balan, H. Makker, R. Fridman, K. Maddipati and A. Raz, *Cancer Biol. Ther.*, 2013, **14**, 417.
110. M. Z. Shareef, P. Malathy and G. S. Babu, *J. Global. Trends Pharm. Sci.*, 2018, **9**, 5983.
111. S. Singh and D. K. Majumdar, *Int. J. Pharmacogn.*, 1996, **34**, 218.
112. P. K. Medirata, K. K. Sharma and S. Singh, *J. Ethanopharmacol.*, 2002, **80**, 15.
113. S. Godhwani, J. L. Godhwani and D. S. Vyas, *J. Ethnopharmacol.*, 1988, **24**, 193.
114. D. A. dos Santos, J. Fukui Mde, N. P. Dhammika Nanayakkara, S. I. Khan, J. P. Sousa, J. K. Bastos, S. F. Andrade, A. A. da Silva Filho and N. L. Quintão, *J. Ethnopharmacol.*, 2010, **127**, 543.
115. P. Sharma, G. Singh and G. Goyal, *World J. Pharm. Sci.*, 2017, **6**, 369.
116. O. Sanctum, *Drug Discov.*, 2013, **5**, 23.
117. A. Kumar, A. Rahal, S. Chakraborty, R. Tiwari, S. K. Latheef and K. Dhama, *Int. J. Agron. Plant Prod.*, 2013, **4**, 1580.

118. A. Mazumder, B. Yadav and H. Sharma, *Plant Sci. Today*, 2021, **8**, 1006.
119. S. Singh, M. Taneja and D. K. Majumdar, *Indian J. Exp. Biol.*, 2007, **45**, 403.
120. C. Venkateswararao and K. Venkataramana, *J. Global Trends Pharmaceut. Sci.*, 2013, **4**, 1118.
121. R. K. Jaggi, R. Madaan and B. Singh, *Indian J. Exp. Biol.*, 2003, **41**, 1329.
122. H. Joshi and M. Parle, *Indian J. Exp. Biol.*, 2006, **44**, 133.
123. P. Khurana, *J. Med. Pharm. Allied Sci.*, 2020, **9**, 2509.
124. I. F. Benzie, *Int. J. Food Sci. Nutr.*, 1996, **47**, 233.
125. S. Amrani, H. Harnafi, D. Gadi, H. Mekhfi, A. Legssyer, M. Aziz, F. Martin-Nizard and L. Bosca, *J. Ethnopharmacol.*, 2009, **125**, 157.
126. *Natural Medicines: Clinical Efficacy, Safety and Quality*, ed. D. Ghosh and P. K. Mukherjee, CRC Press, 2019.
127. T. Suanarunsawat, T. Boonnak, W. D. Na Ayutthaya and S. Thirawarapan, *J. Basic Clin. Physiol. Pharmacol.*, 2010, **21**, 387.
128. S. Singh and D. K. Majumdar, *Int. J. Pharmacogn.*, 1996, **34**, 218.
129. G. Hrckova and S. Velebny, *Pharmacological potential of selected natural compounds in the control of parasitic diseases*, Springer Science & Business Media, 2012.
130. P. Pandey, A. Garg, V. Singh and A. Shukla, *Asian J. Pharm. Pharmacol.*, 2016, **2**, 67.
131. R. Tiwari, S. K. Latheef, I. Ahmed, H. M. N. Iqbal, M. H. Bule, K. Dhama, H. A. Samad, K. Karthik, M. Alagawany, M. E. A. El-Hack, M. I. Yatoo and M. R. Farag, *Curr. Drug Metab.*, 2018, **19**, 264.
132. A. K. Nyarko, H. Asare-Anane, M. Ofosuhene and M. E. Addy, *Phytomedicine*, 2002, **9**, 346.
133. J. K. Grover, V. Vats and S. S. Yadav, *Diabetes, Obes. Metab.*, 2005, **7**, 414.
134. V. Vats, S. P. Yadav and J. K. Grover, *J. Ethnopharmacol.*, 2004, **90**, 155.
135. L. M. Pessoa, S. M. Morais, C. M. Bevilaqua and J. H. Luciano, *Vet. Parasitol.*, 2002, **109**, 59.
136. S. K. Gupta, S. Srivastava, D. Trivedi, S. Joshi and N. Halder, *Curr. Eye Res.*, 2005, **30**, 583.
137. A. Marchese, R. Barbieri, E. Coppo, I. E. Orhan, M. Daglia, S. F. Nabavi, M. Izadi, M. Abdollahi, S. M. Nabavi and M. Ajami, *Crit. Rev. Microbiol.*, 2017, **43**, 668.
138. R. I. Vargas, J. D. Stark, M. H. Kido, H. M. Ketter and L. C. Whitehand, *J. Econ. Entomol.*, 2000, **93**, 81.
139. R. K. Upadhyay, *Int. J. Green Pharm.*, 2017, **11**, S1.
140. C. Sharma, J. M. Al Kaabi, S. M. Nurulain, S. N. Goyal, M. A. Kamal and S. Ojha, *Curr. Pharm. Des.*, 2016, **22**, 3237.
141. D. Yang, L. Michel, J. P. Chaumont and J. Millet-Clerc, *Mycopathologia*, 1999, **148**, 79.
142. J. Gertsch, M. Leonti, S. Raduner, I. Racz, J. Z. Chen, X. Q. Xie, K. H. Altmann, M. Karsak and A. Zimmer, *Proc. Natl. Acad. Sci. U. S. A.*, 2008, **105**, 9099.

143. L. Xu, S. Tao, X. Wang, Z. Yu, M. Wang, D. Chen, Y. Jing and J. Dong, *Bioorg. Med. Chem.*, 2006, **14**, 5351.
144. A. Balea, M. Pojar-Fenesan and I. Ciotlaus, *Rev. Chim.*, 2020, **71**, 59.
145. M. R. Gomes-Carneiro, I. Felzenszwalb and F. J. Paumgartten, *Mutat. Res.*, 1998, **416**, 129.
146. L. I. Paula-Freire, G. R. Molska, M. L. Andersen and E. L. Carlini, *Planta Med.*, 2016, **82**, 211.
147. H. Moradkhani, E. Sargsyan, H. Bibak, B. Naseri, M. Sadat-Hosseini, A. Fayazi-Barjin and H. Meftahizade, *J. Med. Plant. Res.*, 2010, **4**, 2753.
148. A. P. Rogerio, E. L. Andrade, D. F. Leite, C. P. Figueiredo and J. B. Calixto, *Br. J. Pharmacol.*, 2009, **158**, 1074.
149. N. Srinivas, K. Sali and A. A. Bajoria, *J. Indian Acad. Oral Med. Radiol.*, 2016, **28**, 17.
150. C. M. Uritu, C. T. Mihai, G. D. Stanciu, G. Dodi, T. Alexa-Stratulat, A. Luca, M. M. Leon-Constantin, R. Stefanescu, V. Bild, S. Melnic and B. I. Tamba, *Pain Res. Manag.*, 2018, **2018**, 7801543.
151. K. Okaiyeto, R. I. Adeoye and O. O. Oguntibeju, *J. King Saud Univ. Sci.*, 2021, **33**, 101548.
152. N. A. Jaradat, *J. Al-Aqsa Univ.*, 2005, **9**, 1.

CHAPTER 7

Most Modern Approach to the Phytochemical Evaluation and Use of Pepper Species in Ayurvedic Formulations

RASHMI SINGH,[a] HAAMID RASHEED,[a] SUHAIL AHMED,[a] HARJEET SINGH[b] AND ALOK SHARMA*[a]

[a] ISF College of Pharmacy, Moga Ghal Kalan Ferozpur GT Road, Moga, Punjab 142001, India; [b] Central Council for Research in Ayurvedic Sciences, Ministry of AYUSH, Jhansi 263645, India
*Email: alokalok22@gmail.com

7.1 Background: Black Pepper

Piper nigrum (*P. nigrum*) belongs to the family Piperaceae, and is commonly known as the king of spices. Black pepper is one of the most important and widely used spices in the world, and is cultivated in over 26 countries, which together produce about 315 000–320 000 tons of pepper (black and white).[1] It is indigenous to India and broadly cultivated in Assam, Kerala, Karnataka, and Maharashtra. It is recognized as a well-known seasoning ingredient, which provides a vigorous, spicy flavour to dishes (Figure 7.1). Among the phytoconstituents in black pepper are alkaloids, volatile oils, carbohydrates, proteins, and starches. Numerous studies have found that piperine has anticarcinogenic, antioxidant, antiulcer, antithyroid, antiinflammatory, and antibacterial properties.[2] Black pepper also protects against diabetes-induced oxidative stress, decreases mitochondrial lipid peroxidation, inhibits aryl

Chemistry, Biological Activities and Therapeutic Applications of Medicinal Plants in Ayurveda
Edited by Augustine Amalraj, Sasikumar Kuttappan and Karthik Varma A. C.

Published by the Royal Society of Chemistry, www.rsc.org

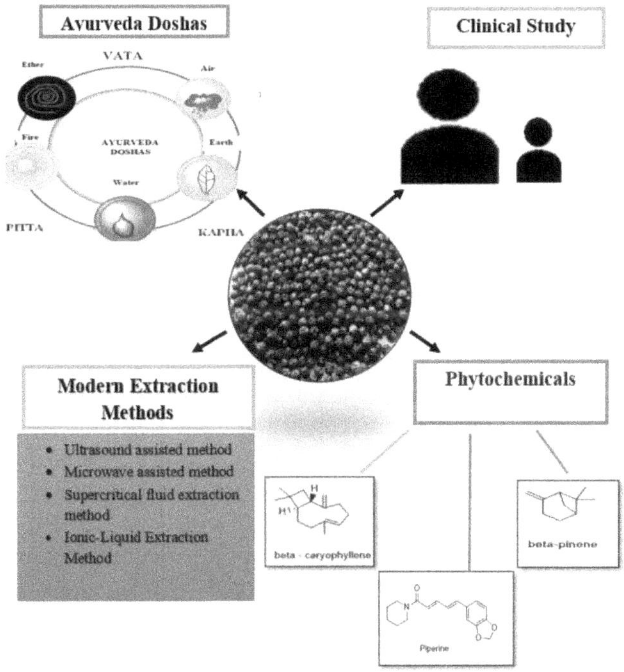

Figure 7.1 Black pepper encompasses Ayurveda, modern extraction methods, phytochemicals, and their clinical trials.

hydroxylation, piperine influenced and activates the biomembrane to absorb a variety of active agents, increased serum concentration, reducing mutational events, and tumour inhibitory activity. Additionally, piperine inhibits mitochondrial oxidative phosphorylation, stimulates growth, and has a chemopreventive effect.[2]

7.2 Ayurvedic Importance and Its Ethnopharmacological Relevance

Biologically, *P. nigrum* is a very important species. The biological role of this species has been explained in different experiments in which peppercorn and secondary metabolites of *P. nigrum* can be used as antiapoptotic, antibacterial, antidepressant, antidiarrhoeal, antiinflammatory, antimutagenic, antioxidative, antispasmodic, antitumor, gastric ailments, hepatoprotective, insecticidal activity, fever, and larvicidal activity.[3] In addition to respiratory conditions, microorganism-related conditions have also been treated with it. Traditional uses of pepper species were reported by Badmaev *et al.*[4] including to treat gastrointestinal issues for a long time. In the 1920s, Bose, an acclaimed author of "Pharmacographia Indica", discovered that an

Ayurvedic mixture containing vasaka (*Adhatoda vasica*) and long pepper (*Piper longum*) had a stronger antiasthmatic effect.[5] An Ayurvedic formulation, Trikatu, is made up of three ingredients: black pepper (*P. nigrum*), long pepper (*P. longum*), and ginger (*Zingiber officinale*). This combination is used in over 1500 Ayurvedic medications. Trikatu is a remedy for a variety of Vata and Kapha imbalances. In addition to stimulating appetite and digestion, it can improve fat metabolism, and it also can be used to treat respiratory problems, gastrointestinal problems, and liver problems, including obstructive jaundice.[6]

In Ayurveda, Trikatu comes from "tri" in Sanskrit meaning three, and "katu" meaning acrids.[1] Maricha (black pepper), Peepli (long Pepper), and Sunthi (ginger) are used in a combination of these three herbs to form an Ayurvedic formulation. In Bhaisajyaratnawali, the Trikatu formulation is formed by the three acrid herbs, together with Maricha (black pepper), Sunthi (ginger), and Peepli (long Pepper), for the management of a lack of taste (Arocaka), for disturbed digestion (Agnimandya and Amadosa), diseases of the nose (Pinasa) and upper respiratory tract (Gala and swasa roga, Kasa), excessive and frequent urination (Meha), oedema (Gulma), obesity (Sthaulya), filariasis (Slipada), and skin diseases (Tvakroga) as described in the Ayurvedic system.[7] As per ancient literature, the rasayana of *P. nigrum* is shown in Figure 7.2.[8]

7.2.1 Balancing the Doshas with Black Pepper

The three essential principles, known as doshas (Vata, Pitta and Kapha), are drawn from five Indian philosophical concepts. As per Ayurveda, Vata and its subdoshas regulate input/output processes and motion, Pitta and its subdoshas regulate throughput, turnover, and consequently energy, and Kapha and its subdoshas regulate structure, lubrication, and storage.[9] Black pepper is a spicy condiment that adds to the spiciness of dishes. It helps calm Kapha, balance Vata, and boost Pitta. Because of the plant's ability to increase heat, pepper is used to enhance Pitta dosha and diminish Vata and Kapha doshas. Pitta dosha regulates digestion, metabolism, and energy production, which is associated with inflammation. Pitta doshas produce digestive fire, metabolism, and energy, and Pitta qualities have a fiery nature. An unbalanced Pitta can trigger skin rashes, burning sensations, peptic ulcers, excessive body heat, and indigestion. Thus, too much pepper can cause heat and lead to unbalanced results.[10]

7.3 Geographical Distribution

The genus *Piper* is the largest of the Piperaceae, with about 3000 species identified, and is widely distributed in tropical and subtropical regions. It is used as a traditional medicine in several countries, mainly in Asia, Africa, and South America. India, Brazil, Indonesia, Malaysia, Vietnam, China, and Sri Lanka are the largest producers of black pepper. Since ancient times, its

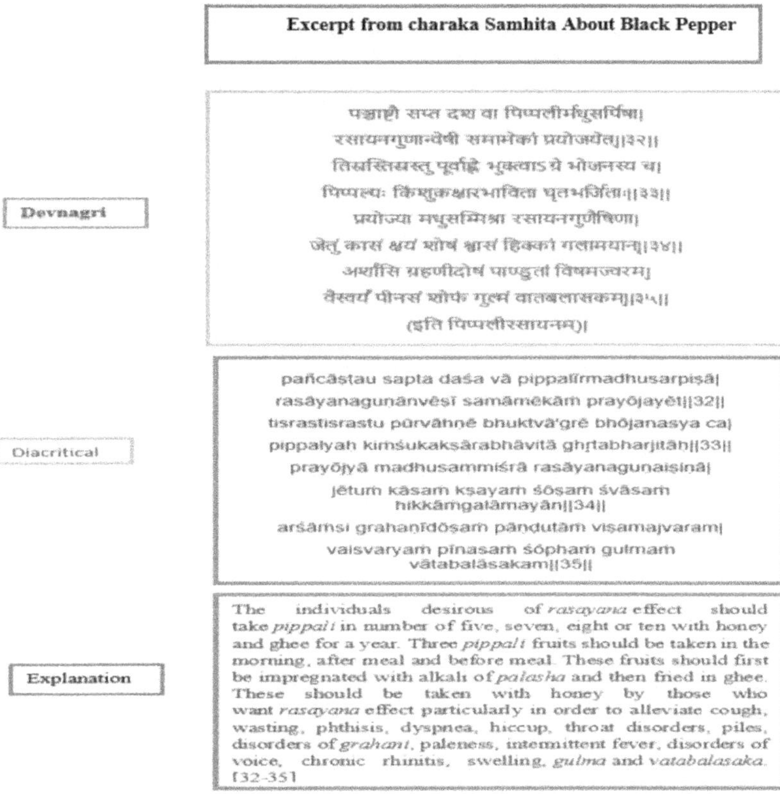

Figure 7.2 As per ancient literature the rasayana of *Piper nigrum*. Reproduced from ref. 85 with permission from Caraka Samhita.

fruits have been used in India as flavouring agents. Geographically, it is native to India and cultivated widely in Assam, Kerala, and Karnataka. Furthermore, the Western Ghats of South India are the primary location for the cultivation of black pepper.[11]

7.4 Botanical Description

Peppercorns (*P. nigrum*) are the dried fruit of a flowering vine, and are widely used as a culinary ingredient. Spices are produced from the fruits, which yield black, white, and green peppercorns. The flowers are oval and produce clusters, or spikes, plants bear alternate leaves and 50 to 150 flowers. The fruits are spherical, green and ripen to red, and the fruits mature on the flower spike. Black pepper can grow to 10 m in height, but under cultivation it is typically limited to 4 m. The perennial plant can live for over 30 years, but the commercial lifespan is about 12–20 years. A native of South and Southeast Asia, Madagascar pepper is also known as black pepper.[12] Dried unripe fruit of the black pepper is shown in Figure 7.3.

Figure 7.3 Dried unripe black pepper fruit.

7.5 Different Ancient Ayurvedic Formulations Prepared from Black Pepper

The antiasthmatic activity is reported to be enhanced by the combination of *Adhatoda vasica* with Pippli (*P. longum*). As a part of Ayurveda, Bose first introduced a concept known as "yogavahi" in 1929. This concept increases the efficacy, tissue distribution, and bioavailability of drugs. Piperine is extensively used to improve the bioavailability of xenobiotic agents in the treatment of the most serious of diseases.[4] Table 7.1 shows various Ayurvedic formulations based on *P. nigrum* and their therapeutic uses. Some Ayurvedic formulations are mentioned, along with their composition. The traditional use of Trikatu Churna, a mixture of black pepper, long black pepper and ginger, is to treat Kapha, Vata and Pitta as well as a number of other disorders. The bioavailability is enhanced with the use of vasicine and spartine.[13] Pippalyëdyësava is another Ayurvedic formulation containing *piper chaba*, long pepper and black pepper and used to treat piles, urticaria, anaemia, and malabsorption syndrome.[14] Jawa rishai and thurushi contain the ingredients black pepper, ginger, vidanga, peppermint, and lemon juice and are used as stimulants, carminative expectorants, and for the treatment of genitourinary diseases and gonorrhoea, and as a diuretic.[15] Furthermore, Kanakasava contains the composition Pippali (Indian long pepper; *P. longum*), dried ginger powder (ginger rhizome; *Z. officinale*), Mulethi (liquorice), Vasaka (Malabar nut), thornapples (purified datura metal; all parts) and is used as a bronchodilator, expectorant, analgesic, antispasmodic, sedative, cardiac stimulant, soporific (sleep-inducing) and digestive aid.[16] The Panchakola Churna and Vardhamana Pippali Rasayana contain ingredients such as Indian long pepper (*P. longum*), Chavak (*Piper retrofractum*), Ceylon (*Plumbago zeylanica*), dried ginger (*Z. officinale*), honey and ghee (clarified butter) and are used in the treatment of bronchitis, pulmonary tuberculosis, asthma, piles, anaemia, chronic rhinitis, intermittent fever, splenic disorders, obstinate abdominal diseases, and rheumatoid arthritis.[17] Avipattikara Churna contains a composition of *P. nigrum*, *Z. officinale*, Indian gooseberry (*Emblica officinalis*), Bhenda (*Terminalia belerica*),

Table 7.1 Ayurvedic formulation, composition and its therapeutic uses.

Ayurvedic formulation	Composition	Therapeutic uses	Ref.
Trikatu Churna (major decoction)	Black pepper (*P. nigrum*), long pepper (*P. longum*), Ginger (*Zingiber officinale* Rosc.)	Traditionally used for the restoration of Kapha, Vata and Pitta. In various other types of fevers, disorders related to abdomen and gastrointestinal tract, pyrexia, dysentery, bronchitis, asthma and sleep disorders. As a bioavailability enhancer for Vasicine, Sparteine	13
Pippalyëdyësava	Long pepper (*P. longum*), black pepper (*P. nigrum*), *Piper chaba*	Malabsorption syndrome, abdominal lumps, emaciation, phthisis, piles, urticaria, anaemia	14
Jawa rishai Thurushi	Black pepper (*P. nigrum*), ginger (*Zingiber officinale*), *Embilia ribes*, mint (*Menthe sativa*), and lemon juice	Stimulant, carminative, expectorant, genitourinary diseases, diuretic, gonorrhoea	15
Kanakasava	Pippali (long pepper) – *P. longum*, Sonth (ginger rhizome) – *Zingiber officinale*, Mulethi – *Glycyrrhiza glabra*, Vasaka – Adhatoda Vasica, Shuddha Datura Panchang – Purified Datura Metal (All Parts)	Bronchodilator, expectorant, analgesic, antispasmodic, sedative, cardiac stimulant, soporific (sleep inducing), digestive	16
Panchakola Churna Vardhamana Pippali Rasayana	Pippali (*P. longum*), Pippalimoola (*P. longum*), Chavya (*Piper retrofractum*), Chitraka (*Plumbago zeylanica*), Shunti (*Zingiber officinale*) Pippali (*P. longum*), Honey, ghee (clarified butter)	Bronchitis, asthma, piles, anemia, chronic rhinitis, intermittent fever, splenic disorders, obstinate abdominal diseases, rheumatoid arthritis	15
Avipattikara Churna	Black pepper (*P. nigrum*), Ginger (*Zingiber officinale*), Amla (*Emblica officinalis*), Behada (*Terminalia belerica*), cardamom (*Elettaria cardamomum*), *Syzygium aromaticum*	Diarrhoea, gastritis, heartburn, indigestion, ulcers, peptic ulcers	16 and 17
Decoction (seeds)	*Piper guineense* Linn.	Fungal infections, asthma, antipsychotic, respiratory infections, infertility (female), bronchitis, syphilis	20–24
Cold maceration (fruits and leaves)	*Piper guineense*	Skin diseases, mouth infections	25
Talisapatradi churnam	Talisa (*Abies webbiana*), Maricha (*P. nigrum*), Shunti (*Zingiber officinale*), Pippali (*P. longum*), Vamshalochana (*Bambusa bambos*)	Respiratory diseases, cough, cold, burning sensation in palms and feet, low digestion strength, loss of sensation in tongue, pain in abdomen and flanks, anorexia, bleeding from the nose	23

Badi Elaichi (*Elettaria cardamomum*) and Eugenia aromatica (*Syzygium aromaticum*), which are used therapeutically in diarrhoea, gastritis, heartburn, indigestion, ulcers and peptic ulcers.[18,19] A decoction (seeds) contains the composition *Piper guineense* Linn. that is used therapeutically for fungal infections, asthma, neuroleptic, breathing infections, sterility (female), bronchitis and syphilis.[20–24] A cold maceration of the fruits and leaves possesses the ingredients *Piper guineense* and is used therapeutically in skin diseases and mouth infections.[25] Talisapatradi churnam (an Ayurvedic powder form) Talisa (Himorinda), Maricha (*P. nigrum*), Shunti (*Z. officinale*), Pippali (*P. longum*) and Vamshalochana (*Bambusa arundinacea*) are therapeutically used in flanks, anorexia, and bleeding from the nose.[23]

The Unani system of medicine regards *P. nigrum*, also known as filfil siyah, as a vital component for treating various ailments. In addition, according to the Siddha system of medicine, *P. nigrum* is the oldest spice and a vital component of many Siddha medicines. Some Unani formulations are mentioned, along with their composition and their therapeutic uses, that contain black pepper, such as Azaaraaqi includes Kuchla Biryan, Mirch Siyan (*P. nigrum*) Nervine tonic and stimulant, used explicitly in paralysis.[26] Jawarish Kamoni Sada contains the ingredients Berg Sadab (*Ruta graveolence* leaf), Boora amreniac (*Arminium bole*), Filfil Siya (black pepper fruit), ginger (*Z. officinalis* rhizome) and Zeera Siyah Mudabbar (Caraway detoxified), and is used therapeutically in cases of constipation, stomach ache and colic pain, and also improves digestion.[27] Habbe Kabid Naushadri contains Naushader (ammonium chloride), Namak Tuam (NaCl), Namak Siyahm (Na_2SO_4 with NaCl), Namak lahori (sodium chlorate), Haleela Siyah (Haritaki unripe fruit), Post Haleela Kabuli (Haritaki ripe fruit), Baobarang (mugwort seed), Filfil Siyah (*P. nigrum* fruit), Zanjabil (*Z. officinalis* rhizome), Suhaga (Borax) and Zaranbad (turmeric) and is used to treat hepatitis, dyspepsia, and constipation.[28] Jawarish Zarooni Sada contains Maghz-Tukhm-Kharpazah, Maghz-Tukhm-Kheyar, Tukhm-e-Karaf Post-Beekh-e-Karafs, Post-Beekh-e-Karafs, Nankhah, Badiyan, Qaranfal and Filfi and is used therapeutically as a diuretic and neuroprotective, in treating kidney dysfunction, weakness of the digestive system, liver dysfunction, indigestion, excessive urination and hydrospermia.[27] Habbe Gule Aakh contains Zanjabeel (*Z. officinalis*), FilfilSiyah (*P. nigrum*), GuleMadar (*Calotropis gigantea* Linn.) and Barge Bans (*Bambusa arundinacea* Wild) and is used therapeutically in cases of backache and cervical spondylosis, and has antiinflammatory and analgesic activities.[27] Table 7.2 shows details of the Unani formulations of *P. nigrum* along with their compositions and therapeutic uses.

Siddha formulations contain black pepper, including Trikadagu Choornam possesses Chukku (*Rhizome zingiberis*), Milagu (long pepper) and Thippili (long pepper) and are used therapeutically for their analgesic and antiinflammatory properties and also in the treatment of indigestion, gas trouble, constipation and bloating, and as a carminative.[28] Kesari Lehyam contains lemon fruit, milagu (long pepper), daniya (coriander), koshtam (spiral flag), chukku (dried ginger), thaleesapatri (yew), tippili (*P. longum*),

Table 7.2 Unani formulations containing piperine.

Formulations	Compositions	Therapeutic uses	Ref.
Habb-e-Azaraqi	Kuchla Biryan, Mirch Siyan (*P. nigrum*)	Nervine tonic and stimulant. Specifically used in paralysis	26
Jawarish Kamoni Sada	Berg Suddab (Ruta graveolence leaf), Boora armani (Arminium bole), Filfil Siyah (*P. nigrum* fruit), Zanjabil (*Zingiber officinalis* rhizome), Zeera Siyah Mudabbar (Carum carvi seed detoxified)	Stomachache, colic pain. Improves digestion and resolves constipation	27
Habbe Kabid Naushadri	Naushader (ammonium chloride) Namak Tuam (sodium chloride) Namak Siyah (sodium sulphate with sodium chloride) Namak lahori (sodium chlorate) Haleela Siyah (*Terminelia chebula* unripe fruit) Post Haleela Kabuli (*Terminelia chebula* ripe fruit) Baobarang (*Artemisia vulgaris* seed) Filfil Siyah (*P. nigrum* fruit) Zanjabil (*Zingiber officinalis* rhizome) Suhaga (Borax) Zaranbad (*Curcuma zedoria* root)	Hepatitis, dyspepsia, constipation	28
Jawarish Zarooni Sada	Maghz-Tukhm-Kharpazah, Maghz-Tukhm-Kheyar, Tukhm-e-Karaf,Post-Beekh-e-Karafs, Post-Beekh-e-Karafs, Nankhah, Badiyan, Qaranfal, Filfi	Diuretic and neuroprotective, kidney dysfunction, weakness of the digestive system, liver dysfunction, indigestion, excessive urination and hypospermia	27
Habbe Gule Aakh	Zanjabeel; *Zingiber officinalis*, FilfilSiyah; *P. nigrum*, GuleMadar; *Calotropis gigantea* Linn., Barge Bans; *Bambusa arundinacea* Willd.	Backache, cervical spondylosis, antiinflammatory and analgesic activity	27

jathikkai (Buah Pala), jathipathri (Mace), ghee and sugar, and is used therapeutically for vomiting, dizziness, indigestion, anorexia, pitha diseases, a loss of taste and flatulence.[29] Mezhugu contains Chukku (dried ginger), Omam (ajwain), Manjal (turmeric), Vaividangam (false black pepper), Vasambu (sweet-flag), Lavangam (cinnamon), Parangi pattai (acorn squash), Seran kottai (bhallatak), Kadukkai thole (myrobalan Retz.), Karum seeragam (black cumin), Kaattu seeragam (Kuntze.), Siruthekku (f. muell.), Thaaleesapathiri (yew), Thiraatchai (grape vine), Thippili (*P. longum* L.), Sitraraththai (shell ginger) and kuth and is used therapeutically for painful conditions, fissures, leprosy, scrofula, rashes, alopecia, itches, hot flush, glandular enlargements, cancer infections, piles, severe pain in the legs, slipped discs, back injury, skin diseases and psoriasis.[30] Amukkara choornam includes Indian ginseng (*Withania somnifera*), ginger (*Z. officinale*), Thippili (long pepper) (*P. longum*), Milagu (black pepper) (*P. nigrum*), Elam (Badi Elaichi), Sirunagapoo, Dalchini (*Cinnamomum verum*) and Kirambu Laung (clove) (*Syzygium aromaticum*).[29] Table 7.3 shows details of the Siddha formulations of *P. nigrum* along with their compositions and therapeutic uses.

7.6 Different Modern Formulations Available Containing Black Pepper

Cadila Pharmaceutics Ltd. in Ahmadabad manufactured the marketed formulation risorine developed by Regional Research Laboratory, Jammu, and marketed in India that contains about 200 mg rifampicin, 300 mg isoniazid, and 10 mg piperine. This medication has been accepted for the treatment of pulmonary tuberculosis. A study of the combination of isoniazid with piperine showed the dose of rifampicin could be reduced to 200 mg which was nearly half the original amount. A pilot study involved using risorine to treat pulmonary tuberculosis in human volunteers. Risorine kit was used to treat 27 patients who had pulmonary tuberculosis. The medication contains risorine and ethambutol, at approximately 800 mg, as well as pyrazinamide, approximately 750 mg, each per day for 2 months. The patients received one risorine tablet per month for the following 4 months. Pure curcumin from Simply Herbal is a concentrated form of curcumin with a combination of 400 mg of natural curcumin extract, 5 mg of piperine, and 395 mg of turmeric powder. It is prescribed as a powerful antioxidant and antiinflammatory compound. According to Gaurav Marketing, curcumin and piperine capsules promote antiallergic effects, are natural detoxifiers, have neuroprotective properties, cholesterol and hepatoprotective properties, and increase gastric acid secretion. A Pukhraj product is also used therapeutically to improve the health of the liver, aid digestion, and ease discomfort. Curcumin 3 is a health supplement manufactured by Health Aid containing biotin (1000 mg), bamboo extract (20 mg) and piperine (5 mg) that is also found in Herbal's premium Biotin capsules.[31] Other different marketed formulations are listed in Table 7.4.

Table 7.3 Siddha formulations containing piperine.

Formulations	Compositions	Therapeutic uses	Ref.
Trikadagu Choornam	Chukku (dried ginger), Milagu (pepper) and Thippili (long pepper)	Analgesic and antiinflammatory properties, digestive, carminative and used in treatment of indigestion, gas problems, constipation and bloating	24
Kesari Lehyam	Lemon fruit, Milagu (*P. nigrum*), Daniya (*Coriandrum sativum*), Koshtam (*Costus speciosus*), Chukku (Dried Ginger), Thaleesapatri (*Taxus baccata*), Tippili (*P. longum*), Jathikkai (*Myristica fragrans*), Jathipathri (Mace), ghee, sugar	Vomiting, dizziness, indigestion, anorexia, pitha diseases, loss of taste and flatulence	29
Mezhugu	Chukku (*Zingiber officinale*), Omam (*Trachyspermum ammi* L.), Manjal (*Curcuma longa* L.), Vaividangam (*Embelia ribes* Burm.f.), Vasambu (*Acorus calamus* L.), Lavangam (*Cinnamomum zeylanicum*), Parangi pattai (*Cucurbita pepo* L.), Seran kottai (*Semecarpus anarcadium* L.), Kadukkai thole (*Terminalia chebula* Retz.), Karum seeragam (*Nigella sativa* L.), Kaattu seeragam (*Centratherum anthelminticum* L.), Siruthekku (*Premna herbacea* Roxb.), Thaaleesapathiri (*Taxus baccata*), Thiraatchai (*Vitis vinifera* L.), Thippili (*P. longum*) Sitraraththai (*Alpinia speciosa* L.), Kottam	Painful conditions, fissures, leprosy, scrofula, rashes, alopecia, itches, hot flush, glandular enlargements, cancer, infections, piles, excessive pain in the legs, slipped disc, back injury, skin diseases and psoriasis	30
Amukkara choornam	Ashwagandha (*Withania somnifera*), Chukku (*Zingiber officinale*), Thippili (long pepper) (*P. Longum*), Milagu (black pepper) (*P. Nigrum*), Elam (cardamom) (*Elettaria Cardamomum*), Sirunagapoo; Dalchini (*Cinnamomum verum*), Kirambu Laung (Clove) (*Syzygium aromaticum*)	Nervine tonic, antiinflammatory, antitussive, Rasayana (rejuvenating), aphrodisiac, antidepressant, immunomodulatory, antioxidant, anodyne, cardioprotective, mild antiasthmatic, mild anabolic effect	29

Table 7.4 Herbal formulations containing piperine in the Indian pharmaceutical market.

Formulation	Manufacturer	Ingredients and quantity	Therapeutic uses	Ref.
Pure curcumin(curcumin and piperine extract)	Simply Herbal	Turmeric powder (395 mg), natural curcumin extract (400 mg) and piperine (5 mg)	Antiinflammatory, powerful antioxidant, prescribed as a dietary supplement	9
Curcumin and piperine capsule	Gaurav Marketing	Curcumin and piperine	Antiallergic, natural detoxifier, neuroprotective, cholesterol, hepatoprotective, gastric acid secretion	
Premium curcumin	Pukhraj	Not mentioned	Weight loss, cancer, protects liver health, aids digestion, eases discomfort, inflammation and prevents arthritis	
Curcumin 3	Health aid	Curcumin, standardized piperine (95%)	Health supplement	
Premium biotin capsules	Simply herbal	Biotin (1000 mg), bamboo extract (20 mg), piperine (5 mg)	Health supplement	

7.7 Phytochemicals

Bioactive components are significant phytochemicals owing to their potential for healing various ailments. Functional foods usually provide additional physiological benefits or prevent chronic illness in addition to basic nutrition. The antioxidant, antimicrobial, and gastro-protective properties of black pepper (*P. nigrum* L.) make it an important food source. Piperine is the active component in black pepper, which has rich phytochemistry, including volatile oil, oleoresins, alkaloids, *etc.* Perakis *et al.* evaluated the *P. nigrum* ground seed, peppercorn, or dried berry as having 11 wt% oleoresins, 6 wt% piperine and 2% (v/w) essential oil.[32] Meanwhile, Singh *et al.*[33] evaluated and found the foremost compounds to be piperine (34%), piperolein B (14%), anonymous alkaloid constituents (6% and 7%) and other recognized piperamides (<4%) in *P. nigrum* essential oil acetone extract.[33] Perakis *et al.*[32] demonstrated the supercritical fluid extraction (SFE) of essential oil of *P. nigrum* seed and *P. tuberculatum* leaf extract comprises five piperamides, together with piperine, commonly existing and used as standards. The different chemical structures of *P. nigrum* are illustrated in Figure 7.4.

7.7.1 Whole Plant

The entire plant parts were individually extracted with petroleum ether, chloroform, and butanol, sequentially, by Liu *et al.*[34] for the isolation of compounds and their structured ethanol extracts. Additionally, chromatographic analysis of chloroform fractions revealed various compounds such as MS, ^{1}H-NMR and ^{13}C-NMR. Liu *et al.* carried out reported some of the compounds such as (4*E*)-7-(4-hydroxy-3-methoxyphenyl)-1-phenylhept-4-en-3-one, Aristolactam FI, isopiperine, cerulenin, 2-hepten-methylenealpha-(1-methylethyl)-1*H*-indene-1-methanolmonodemethoxycurcumin.[34]

7.7.2 Fruits

Different phytoconstituents including (2*E*,8*E*)-9-(2*H*-1,3-benzodioxol-5-yl)-*N*-(2-methylpropyl)nona-2,8-dienamide, retrofractamide B, (*E*,*E*)-*N*-(2-methylpropyl)-2,4-decadienamide2,4-decadienamide, pipyahyine, dehydropipernoline, piperloein B, (2*E*,8*E*)-9-(1,3-benzodioxol-5-yl)-1-piperidin-1-ylnona-2,8-dien-1-one and 1-piperoylpiperidine have been identified from the methanol extract by mass spectroscopy. Additionally, dehydropipernoline, piperloein B and (2*E*,8*E*)-9-(1,3-benzodioxol-5-yl)-1-piperidin-1-ylnona-2,8-dien-1-one had an inhibitory effect on Stat3-dependent luciferase activities induced by IL-6.[35] Sarmentine is a chemical compound derived from the ethanolic extract of *P. longum*, which produces phytotoxic effects and functions as a herbicide.[36]

On HL-60 cell lines, the moieties had cytotoxic activity.[37] Piperlongimin B (2*E*-octadecenoylpiperidine), 2*E*,4*E*-*N*-isobutyl-dodecenamide and 2*E*,4*E*,12*E*,13-(3,4-methylenedioxyphenyl)-trideca-trienoic acid isobutylamide were isolated

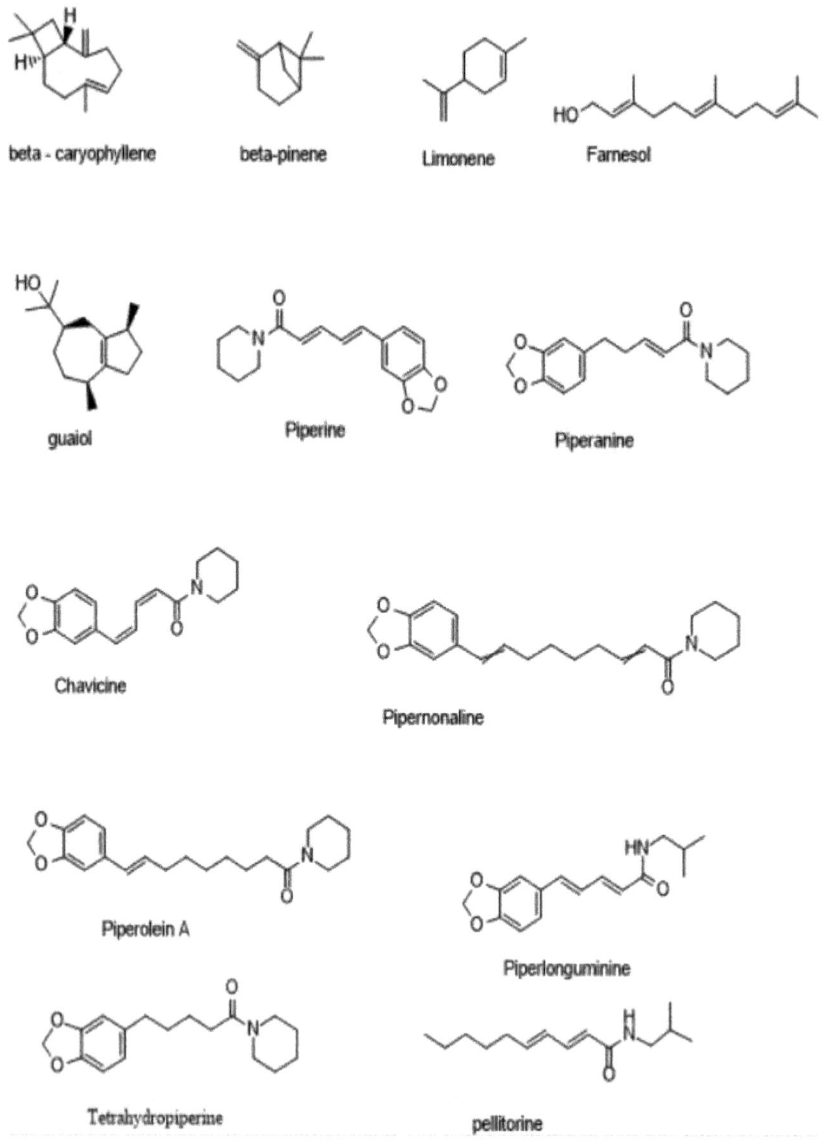

Figure 7.4 The different chemical structures of *Piper nigrum*.

moieties described by Sahi *et al.* as compounds that seemed to have activity against *Leishmania donovani*.[38] A new alkaloid amide derivative called *P. longumide* was identified by Ghosal *et al.* Furthermore, they developed leishmanicidal promastigotes and axenic amastigotes of *Leishmania donovani*.[39] Jiang *et al.*[40] evaluated the isolated amide alkaloids, *P. longumamide* A, *P. longumamide* B and *Piper longumamide* C. The already identified compounds which were isolated included piperine, (2*Z*,4*E*)-5-(1,3-benzodioxol-5-yl)-1-piperidin-1-ylpenta-2,4-dien-1-one, *cis*-piperine, piperlonguminine,

(2*E*,4*E*)-*N*-(2-methylpropyl)deca-2,4-dienamide,(2*E*,4*E*)-15-(1,3-benzodioxol-5-yl)-*N*-(2 methylpropyl)pentadeca-2,4 dienamide, pipyahyine, piperchabamide B, pipernonaline and piperolein B.[40]

7.7.3 Roots

The syntheses of piperlongumine (Piplartine) and piperlonguminine, a rich alkaloid of *Piper longum* L., were explored along with their anticancer properties.[41] The ethanol extract of roots of *P. longum* yielded cefradine, norcepharadione B and 2-hydroxy-1-methoxy-4*H*-dibenzo quinoline-4,5(6*H*)-dione. Also, three new alkaloids were discovered: piperolactam A, piperolactam B and piperadione.[42]

7.7.4 Different Habitats and Diversity in Phytochemicals

An established and widely distributed pantropical taxon of aromatic plants, the *Piper* genus has historically been used both as a food and medicine. Essential oils (EOs) are found in a wide range of tissues and organs of the pepper plant, including the fruits, seeds, leaves, branches, roots and stems.[43,44] Although the *Piper* species have been studied extensively since the 1960s, Dyer[45] claims that only 10% of the EOs have been phytochemically studied. As a result of the variety of chemical components and the variety of tissues from which *Piper* EOs are obtained, it is difficult to summarize the nature of *Piper* EOs. Mgbeahuruioke *et al.*[46] identified more than 270 compounds in *Piper* EOs. Xiang *et al.*[47] stated in their review that over 80 compounds were found in Chinese *Piper* spp., and that these compounds were mostly hydrocarbons, followed by aldehydes, alcohols, acids, ketones, esters and phenols. The secondary metabolites in some species might appear simple, while those in others, such as *P. nigrum*, *P. betle* and *P. auritum*, may contain large amounts of different compounds.

According to Mgbeahuruioke *et al.*[46] the African *Piper* spp. EOs show great chemical diversity and presence of chemotypes. These authors cite seven chemotypes of *Piper guineense* from Nigeria (β-caryophyllene/germacrene D; asaricin; α-pinene/β-pinene/germacrene B; β-pinene/α-pinene/β-caryophyllene; β-pinene/α-pinene/1,8-cineole; linalool) and three chemotypes of *P. guineense* from Cameroon (α-pinene/β-pinene; β-caryophyllene/limonene// pinenes; β-caryophyllene).[25]

In addition to chemical differences, many authors have claimed that chemotypes are responsible for these differences. As suggested by da Silva *et al.*, at least seven species of neotropical insects exhibit chemotypes. Notable differences between neotropical and African general compositions are the greater quantitative importance of the monoterpene hydrocarbons (β-pinene, α-phellandrene and (*Z*)-β-ocimene) and of the sesquiterpene hydrocarbons (β-elemene, β-sesquiphellandrene, (*Z*)-β-bisabolene, (*Z*,*Z*)-α-farnesene, *ar*-curcumeme, α-zingiberene and δ-cadinene). On the other hand, the African species are less rich in the monoterpene hydrocarbons (myrcene, α-terpinene and *p*-cymene) and in the sesquiterpene hydrocarbon (α-cubebene).[48]

7.8 Modern Techniques Used in the Extraction of *Piper nigrum*

A variety of plants contain compounds that can treat human disorders. A key issue in obtaining high yields of therapeutic products is that it requires thorough recovery using an environmentally safe, reliable and time-saving extraction technique due to the low levels of therapeutic compounds in plants. When choosing an extraction method, consideration must be given to the thermolability of the active chemicals, for which the temperature and other parameters must be adjusted. Modern extraction techniques utilized in the extraction of *P. nigrum* are shown in Figure 7.5, and different extraction methods along with extraction time, yields, advantages and disadvantages are listed in Table 7.5.

7.8.1 Hydrotropic Extraction Method

Hydrotropy refers to the phenomenon of water-insoluble or sparsely water-soluble organic compounds becoming more soluble in aqueous solutions if hydrotropes are present. Hydrotropes are organic salts that are very water soluble. Increases in solubility are strongly influenced by hydrotrope concentration and are governed by the nature and composition of the solute. Hydrotrope concentrations can be manipulated to recover a solute, which means that dilutions with water precipitate the solute out of a hydrotropic solution. In aqueous solutions,[49] solubility is enhanced by hydrotropic

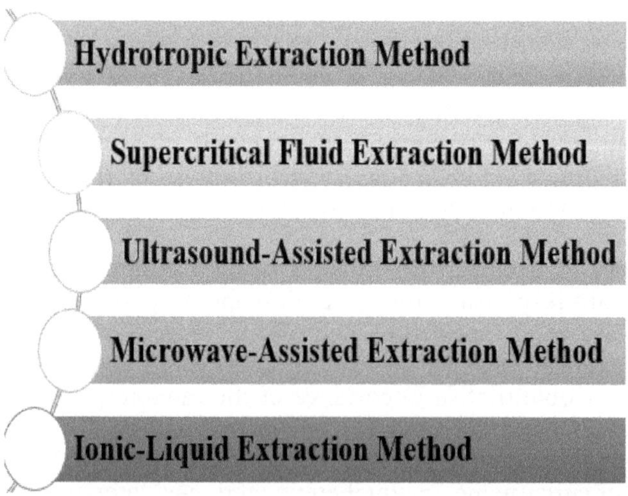

Figure 7.5 Modern extraction techniques used in the extraction of *Piper nigrum*.

Table 7.5 Different extraction methods and their extraction yields.

Extraction methods	Extraction time	Yields	Advantages	Disadvantages	Ref.
Ionic liquid (IL) based ultrasonication-assisted extraction. Detection method: ultra power liquid chromatography (UPLC analysis)	Different ionic liquids time: 30 min	1.96%	Cost effective, sensitive, efficient and eco-friendly technique	Not mentioned	61
Microwave-assisted extraction	2 min 2 min	$92 \pm 1\%$ 94% (extraction efficiency)	Good reproducibility, minimum sample manipulation	An additional filtration assembly to separate the residue	66
Supercritical fluid extraction (SFE)	2–5 h 2–5 h 2.25 h	81–98% 6.7–7.6% $1.23\% \pm 0.05\%$ (batch) $1.36\% \pm 0.02\%$ (continuous)	Short processing time, minimal solid residue	High pressure is required	51
Ultrasound-assisted extraction	5 h 18 min	3.9% 0.58%	Inexpensive	Lacking in the uniformity of the distribution in the sonication energy	61
Hydrotropic solubilization	2 h	90% to 96%	Few purification steps, little or no foaming unlike the surfactant solutions	Not mentioned	50

additives. Amphiphilic hydrotrope molecules can aggregate or accumulate around hydrophobic molecules.[50] The greatest solubilization power is possessed by insoluble organic active compounds due to the high extraction capacity of hydrotropes. Hydrotropic solubilization indicates equilibrium between hydrophobicity and high solubility in water. Hydrotrope molecules have difficulties in diffusion into the cellulosic layer, followed by diffusion into the phospholipid bilayer, because they can enter the dispersed piperine within the cell previously. Osmotic pressure exerts high pressures on the cell wall that enhances the concentration of hydrotrope in the water, resulting in greater penetration of fluid into the cell matrix. This aqueous solution transported piperine through the external hydrotrope medium to the cellular matrix, allowing penetration through rapid stages.[49]

In an experimental study, a new hydrotrope-based extraction process was used to selectively extract curcuminoids from turmeric. Another study found that piperine was preferentially isolated from *P. nigrum* fruits using sodium alkyl benzene sulphonates and sodium butyl monoglycol sulfate. The extraction rate with hydrotrope and water is higher because the molecules penetrate cellular structures and cells. Water-soluble hydrotrope molecules disrupt the cell wall and bilayer membrane structure, which allows piperine to be removed more quickly. A hydrotrope solution was used to extract piperine selectively and quickly from black pepper. When it was analysed, the recovered piperine was 90% pure and free of oleoresins. Extraction is influenced by a variety of elements, including hydrotrope type/nature/temperature/concentration and particle size.[50]

7.8.2 Supercritical Fluid (SCF) Extraction Method

Fluids used in supercritical extraction are heated above their critical point of temperature. Supercritical fluids display a similar viscosity to gases, despite having a density close to that of liquids. By contrast, it is only capable of diffusing gases or liquids. These characteristics allow supercritical fluids to penetrate solid matrices deeper and faster. Separation techniques have been created recently due to their low cost and safety; carbon dioxide is the most commonly used solvent in SCF extraction. Extraction and separation of compounds are carried out using supercritical fluid. In recent years, this technique has been used by industries that sell aromatic crops and within the spice industry as a quick, selective, innovative and clean method.

Pepper oleoresin was extracted using SCF from *P. nigrum* fruit. Using Hildebrandt's solubility equation, the solubility parameters of piperine and carbon dioxide were calculated for maximal extraction. The effects of stimulants such as alcohol and methanol were investigated using various theories. the experimental results were in line with predictions. The maximum relative extraction of oleoresin and piperine at 250 bar/50 °C/60 minutes was determined to be 61.7 and 87.51 percent. Adding 30 percent (vol/wt) ethanol as a cosolvent increased the yield from 95.9 to 184.7 percent under SCF extraction conditions. These oleoresins have demonstrated comparable antioxidant, antiinflammatory and antibacterial properties to their respective control compounds. SCF extraction of pepper uses ethanol as a co-solvent.

As a solvent, pulverized carbon dioxide ($SCCO_2$) was used to extract pepper oil from ground black pepper. In bench-scale investigations, the extraction rate was examined in relation to extraction pressure (90, 100 and 150 bar), temperature (40, 50 °C), and solvent flow rate (1, 2 and 3 kg h^{-1}). Increasing the pressure or reducing the temperature improved the extraction rate, as has been statistically proven. The fluid flow rate had a comparable effect.[51]

7.8.3 Ultrasound-assisted Extraction (UAE) Method

In an ultrasound-assisted extraction method, ultrasound is used to create cavitation when applied to an extraction solvent in contact with a sample. Ultrasound is used to cause cavitation bubbles (in the solvent) that collapse and cause pressure and temperature changes. This increases the amount of mass transfer of analytes into the solvent.[52] There are numerous applications of ultrasound extraction in the disciplines of chemistry, pharmaceuticals, cosmetics and nutrition in the 21st century.[10] Soxhlet extraction and batch extraction techniques, which are conventional methods, include solvent wasting, labour intensity and solvent wasting. Due to the disadvantages associated with conventional extraction methods, a new method has been developed. Microjets are formed and cavitation bubbles collapse to cause higher cell disruption due to the collapse of asymmetrical bubbles to a solid surface.[53] Different parameters can distress the reactor, including its nature, temperature, gas atmosphere, acoustical power, frequency and geometry. As these bubbles collapse, they create high pressure and temperature of about 1000 bar and 5000k, as well as approximately stimulating physical effects.[54] The breakage of cell walls, the circulation of liquid and the reduction in particle size and cell membranes allow mass transfer to take place.[55] Thus, the plant body progress and cell breakage results in the penetration of the solvent and, as a result, augmentation of the extraction yield in UAE is realized. UAE produces higher yields, lower extraction times and lower temperatures as well as better solvent penetration.[53,56]

In order to scale up and improve the extraction method, these factors make UAE appealing. Certain factors interfere with the effectiveness of extractions, such as solvent, time of extraction, solid–solvent ratio, duty cycle and ultrasound power. The polarity index, viscosity, surface tension and vapor pressure of the solvent are measured to determine an appropriate solvent.[57] Lower pressure causes fewer bubbles to form, and higher force implodes the bubbles, thereby extracting the target compound. By increasing the ultrasound power, the bubbles collapse more explosively in a continuous mode that is not possible with ultrasound, and the material is degraded as a result. In response to the decline of viscosity,[58,59] an increase in solubility and development occurs during mass transfer, which is caused by the thermal product. Rathod and Rathod[60] evaluated the three solvents:

acetone, ethanol and hexane. The highest yield of 4.5 mg g^{-1} followed by ethanol (4.3 mg g^{-1}) and hexane (4 mg g^{-1}) was obtained from the solvent acetone. Because of the lower viscosity of acetone, the yield was better diffused into the solid matrix.[60] Determination of optimal conditions such as ethanol as extracting solvent, extraction time of 18 min, solid to solvent ratio of 1 : 10, ultrasound power 125 W, 80% duty cycle, ultrasound frequency 25 kHz and temperature 48.9 °C helped to enhance the yield of piperine by using UAE from fruits of *P. longum* as evaluated by Rathod and Rathod.[60] The maximum yield was found to be 5.79 mg g^{-1}.[60] The optimized method increased the top extraction productivity (from 1.9% to 3.6%) within the shortest extraction time (from 2 h to 30 min) with average recoveries of piperine of between 9.4% and 99% and spiking at levels 0.05, 0.1 and 0.2 mg mL^{-1} with an RSD lower than 5%. In another study, the extraction yield was 4% in 5 h and 0.6% in 18 minutes.[61]

7.8.4 Microwave-assisted Extraction (MAE) Method

Assays that isolate active compounds from medicinal plants use microwave energy to partition the components of a matrix sample into a solvent. As a result of its ability to quickly heat the sample solvent mixture, MAE is used for a wide variety of extraction applications, including the extraction of thermally unstable analytes.[62] This technology can be used for many different purposes, such as extracting bioactive ingredients from matrices more quickly, reducing equipment size, reducing temperature gradients and increasing the extraction yields. A microwave energy approach allows you to extract soluble compounds from various materials into a fluid. By using this method, the volatiles were also condensed.[63]

As a result of the powerful extraction methods such as MAE, polar molecules with a high dielectric constant are targeted during the extraction, which causes rapid soaking and at a power level of 151 W, the solvent loss decreased to 7.9% while the usage of high microwave powers of 289 and 455 W increased the solvent loss up to 20%.[64]

Karpakavalli *et al.*[65] examined how lower volumes of solvent, shorter duration and additional yield can be obtained from positive extraction compared to the reflux method. It showed higher yields of 2.2% of piperine.[65] By adding 40 mL solvent in 150 mL vessels at 250 W extraction experiments were carried out in duplicate. Meanwhile another study showed an extraction yield of 92 ± 1% and 94% in 2 minutes by adding solvent (40 mL) into 150 mL vessels at 250 W extraction experiments were carried out in duplicate.[66]

7.8.5 Ionic-liquid Extraction Method

In a wide range of applications, ionic liquids (ILs) have been suggested as a viable medium for separating and extracting bioactive chemicals. Using an ultrasonic extraction process that uses ionic liquids (ILUAE), white pepper powder can be extracted. A study has been carried out to determine the

Table 7.6 Various methods of Ayurvedic extraction according to Ayurveda.

Method	Part	Species
Decoction	Seeds	*Piper guineense* Linn.
Decoction	Leaves	*Piper guineense*
Cold maceration	Fruits and leaves	*Piper guineense*

extraction efficiency of ionic liquids containing a wide range of alkyl chains and anions. This study found that anions and cations have different extraction rates. The ILUAE approach using ultrasonics was also optimized for extraction time and solid–liquid ratio. The new technique increased the extraction efficiency from 1.950 percent to 3.577 percent in less time than traditional heat-reflux extraction (HRE) and regular UAE (from 2 h to 30 min). The recovery (R^2) and reproducibility (RSD, $n = 3$) of the created technique were also validated. At spike levels of 0.05, 0.10 and 0.2 mg mL^{-1}, between 93.5 and 987 percent of piperine was recovered with an RSD of less than 5%. This method is linear for concentrations within the examined ranges, with R^2 of 0.9994.[61]

P. nigrum contains piperine, a bioactive compound that provides several therapeutic benefits. Piperine has been separated from black pepper using a microwave-ultrasound extraction method in one study. Piperine extraction yields were evaluated according to a number of factors, such as particle size, solvent-to-solids ratio, microwave power and time, and ultrasound temperature and time. A total of 46.6 mg g^{-1} piperine was extracted while using ethanol as the solvent with a particle size of 0.15 mm, a solvent-to-solids ratio of 20 : 1, microwave power of 100 W for 1 minute, and ultrasonic temperature of 50 °C for 30 minutes. The Soxhlet (39.1 mg g^{-1}), microwave (38.8 mg g^{-1}) and ultrasonic (37.0 mg g^{-1}) extraction yields were higher as a result, and 81.4% of the piperine extracted was pure according to an HPLC analysis of the extracted substance. The structural elucidation of piperine was confirmed using FTIR and UV-vis after it was purified. Table 7.6 describes the various methods of Ayurvedic preparations available in Ayurveda that contain an extraction method to increase the bioavailability of poorly absorbed drugs.

7.9 Clinical Studies on Black Pepper

In human clinical studies, piperine was co-administered with many drugs primarily to increase the bioavailability. During these trials, piperine was administered with other diverse groups of medicines in order to be more potent and stay in the plasma longer. Human oral bioavailability of phenytoin may be increased by piperine, which occurs in *P. longum*, *P. nigrum* and *Z. officinalis*. Piperine was tested to see how it affected the steady-state phenytoin pharmacokinetics in patients with uncontrolled epilepsy given a single dose. Ten patients were divided into two groups, each taking 150 mg or 200 mg of phenytoin twice daily. After the night dose of phenytoin was delivered, venous blood samples were collected at 0, 0.5, 1, 2, 4, 6, 9 and 12 hours. On the second study day, piperine 20 mg was given with phenytoin, and samples were collected in the same way. A Student's *t*-test was used to compare mean plasma drug concentrations at different time points and pharmacokinetic features before and after piperine administration. Piperine significantly raised the mean plasma concentration of phenytoin in both dose groups at most time periods. The changes in $AUC_{(0-12\,h)}$ ($p<0.01$), C_{max} ($p<0.001$) and K_a ($p<0.05$) were all statistically significant, but not the changes in K_{el} and T_{max}.[67]

Piperine dramatically increased phenytoin bioavailability, most likely by boosting absorption, according to the researchers. Experiments were carried out to identify the scientific foundation for utilizing the trikatu group of acrids (long pepper, black pepper and ginger) in Ayurvedic therapies. [^{3}H] vasicine and [^{3}H] sparteine were utilized as test substances. *P. longum* raised vasicine levels in the blood by 233 percent. The active ingredient in piper species, piperine, increased sparteine blood levels by more than 100 percent.[68] According to these findings, these acrids have the capacity to improve the bioavailability of a variety of medicines. Trikatu drugs may increase bioavailability by enhancing absorption from the gastrointestinal tract, preventing metabolization/oxidation during the first passage through the liver after absorption, or utilizing a combination of these strategies.

The co-administration of theophylline and piperine improved the C_{max}, extensive elimination half-life and AUC. Shoba *et al.* evaluated a bioavailability of 2000%,[69] and when piperine and curcumin were administered together the resulting bioactive compound was much more concentrated as a result. Dexibuprofen and Lipicur[70] were administered in patients with a diagnosis of lumbar sciatica, lumbar disk herniation, and/or lumber canal steneosis, or with carpal tunnel syndrome. In both, the condition *i.e.*, the treatment with the multi-ingredient formula (Lipicur) decreased the neuropathic pain by more than 66%. Additionally, the treatment with dexibuprofen decreased neuropathic pain by about 40%. According to the dose and time dependence,[71] piperine supplementation may result in enhanced coenzyme Q10 absorption. When co-supplemented with piperine, phytoalexin resveratrol is more bioavailable. Treatment of DIC with PIP suggestively enhanced maximum plasma concentration C_{max}, the area under the curve (AUC), half-life time ($T_{1/2}$), and also decreased the elimination rate constant (K_{el}), and apparent oral clearance (CL/F) compared to that of control phase.[69] Consequently, the combination therapy of DIC laterally with PIP may signify a novel approach to decreased dosage and result in the lessened incidence of gastrointestinal side effects seen with DIC alone at higher doses. The cerebral blood flow increased during task performance compared to placebo and resveratrol alone, but without negatively affecting mood, cognition or blood pressure.[72] C_{last} values of nevirapine increased by approximately 119.5%, 166.5%, 170.1% and 145.9%, respectively,[73] when given along with ethambutol and pyrazinamide.[73] As a component of tuberculosis treatment with Risorine, ethambutol and pyrazinamide, piperine proved to be highly effective.[74] A better safety profile, an increased sputum conversion rate and a more rapid cure rate are all characteristics of chronic therapy as compared to conventional rifampicin therapy, which results in a very powerful interaction between the two compounds.[75] The authors studied the effect of piperine on pharmacokinetics of chlorzoxazone in volunteers and observed significant interaction between two compounds. In another study, authors found the same effect of piperine intake on the bioavailability of fexofenadine and observed that piperine treatment significantly increased C_{max} of fexofenadine and AUC time curve. Combined with piperine, cannabidiol has a 2.3-fold increase in AUC and a 4-fold increase in C_{max}. Diverse clinical trials related to the use of *P. nigrum* are listed in Table 7.7.

7.9.1 Piperine Effect on the Digestive Tract

In 40 dysphagia patients, the treatment of dysphagia was randomized and further divided into two clusters to obtain a piperine dose of about 150 μM through pure piperine. The trial was demonstrated by Rofes *et al.*[76] by supplementing the alimentary bolus with piperine that improved the swallow response in dysphagic patients. An evaluation was made of the effects of systemically administered extracts of curcumin, lycopene, and piperine as an adjunct to scaling and root planing (SRP) in patients with moderate gingivitis.[77] After 21 days of using paired t-tests in both treatments, the group shows a statistically substantial decrease in clinical parameters, but the test group showed a greater reduction as compared to the control group ($P < 0.05$). The results showed that systemically administered antioxidants could cause better resolution of inflammation when used as an adjunct to SRP.

7.9.2 Vitiligo

Application of *P. nigrum* fruit extract and pure piperine liniments for 11–12 weeks on skin areas of patients with vitiligo was studied.[78] In another study, piperine has reported a significant impact on melanocyte proliferation and found side effects less than synthetic drugs. The objective of the study was focused to assess the effect of piperine with narrowband ultraviolet B (NB–UVB) on vitiligo treatment. A total 63 patients with facial vitiligo were randomly divided into 2 groups in this double-blind clinical trial and treated with piperine and placebo. Besides, both groups received NB–UVB phototherapy every other day for 3 months. Based on the clinical results, the combination therapy with NB–UVB/topical piperine was found more influence on vitiligo as compared to NB–UVB alone. To conclude the application of NB–UVB and topical piperine has a notable effect on vitiligo.[79] After 6 months of treatment, 40% of the patients were symptom-free. Pungent spices have long been implicated as a cause of gastric mucosal injury; while their long-term effect on the gastric mucosa remains less known. In a single-dose study, the results of black pepper on the gastric mucosa were assessed using double-blind intragastric administration of the spice (1.5 g) to healthy human volunteers, with aspirin (655 mg) as a positive control.

7.9.3 Other Conditions

Treatment with a combination of piperine and curcuminoids was done for 4 weeks to improve the quality of life (HRQoL). On measuring the oxidative

Table 7.7 Clinical trials on the effect of piperine.

Effect of piperine	No of patients/study design	Trial length	Treatment, dose and formulation of piperine	Ref.
Phenytoin ↑ $AUC_{0-12\,h}$ ↑ C_{max} ↑ K_a K_{el} and T_{max} unchanged	2 groups of 10 patients	A single dose	Phase I: Group 1 received 150 mg of phenytoin twice daily. Group 2 received a 200 mg dose of phenytoin twice daily Phase II: 20 mg of piperine was administered along with phenytoin (150 or 200 mg)	67
Propranolol and theophylline Piperine with propranolol: ↑ C_{max} ↑ AUC Piperine with theophylline: ↑ C_{max} ↑ $T_{1/2}$ el ↑ AUC	6 healthy male volunteers in a randomized crossover study	7 days	Phase I: Group 1 received a single oral dose of propranolol (40 mg). Group 2 received a single oral dose of theophylline (150 mg) Phase II: Piperine 20 mg daily for 7 days following administration of drugs. One week was permitted as a washout period between the two treatments	69
Curcumin: concomitant administration of piperine (20 mg) produced much higher concentrations from 0.25 to 1h post drug and an increase in bioavailability of 2000%	10 healthy male volunteers in a randomized crossover study	Four weeks	Phase I: 2 g of pure curcumin powder (4×500 mg) was given with 150 mL of water Phase II: 2 g of pure curcumin powder combined with 20 mg of pure piperine powder (4×500 mg curcumin $+ 5$ mg piperine each) was given with 150 mL of water	69
Dexibuprofen: multi-ingredient formula Lipicur containing piperine reduced neuropathic pain by more than 66% in both conditions.	141 subjects affected by neuropathic pain in an open randomized control clinical study	Eight weeks	Group 1: received two tablets/day of Seractil containing dexibuprofen (400 mg per tablet) Group 2: received two tablets/day of Seractil plus two tablets/day of Tiobec 400 containing lipoic acid, 400 mg per tablet	70

The treatment reduced dexibuprofen use by about 40%			Group 3: received two tablets/day of Seractil plus two tablets/day of Lipicur containing 400 mg lipoic acid, 400 mg curcumin phytosome and 4 mg piperine	
Coenzyme Q10: ↑ AUC (30%)	12 healthy adult male subjects	21 days	Control group: coenzyme Q10 (30 mg per capsule) + placebo, administered together in a single dose (90 mg per day), 14 day (90 mg per day) and 21 day supplementation (120 mg of Q10 per day) Treatment group: coenzyme Q10 + 5 mg piperine	71
Risorine: 89% of the sputum-positive patients became sputum negative during the first 2 months of treatment	33 patients with pulmonary tuberculosis in a pilot study	6 months	Months 1 and 2: daily oral therapy consisting of one capsule of Risorine, one tablet of ethambutol (800 mg) and two tablets of Pyrazinamide (750 mg each) Months 3–6: daily oral therapy consisting of one capsule of Risorine	26
Risorine: treatment resulted with: higher sputum conversion rate (93%) *vs.* control group (84%) at 4 weeks. Higher cure rate at the end of 24 weeks (92%) *vs.* control group (82%). Decreased side effects (3 patients *vs.* 9 patients in control group)	216 patients with tuberculosis in a randomized, triple-blind, parallel-group, multicenter, comparative clinical phase III study	6 months	Patients were randomized into: control group, receiving a conventional anti-TB therapy ($n = 117$) or Treatment group, receiving a similar regimen + risorine ($n = 99$)	74

stress and inflammation in patients with Mets, the results demonstrated that supplementation with curcuminoid-piperine combination positively enhanced serum SOD activities and reduced malondialdehyde (MDA), CRP plasma C-reactive protein concentration compared with placebo.[80] A total of 37 clinical trials evaluated the significant influence of a single consumption of a CBFI (capsaicinoids, epigallocatechin gallate, piperine and L-carnitine) on satiety, resting energy expenditure (REE), respiratory quotient, glucagon-like peptide-1 (GLP-1), free fatty acids (FFA) and glycerol in overweight adults, as described by Rondanelli *et al.*[81]

Abbreviations

P. nigrum	*Piper nigrum*
P. longum	*Piper longum*
Z. officinale	*Zingiber officinale*
T. chebula	*Terminalia chebula*
P. tuberculatum	*Piper tuberculatum*
NaNBBS	Sodium *n*-butyl benzene sulphonate
NaCS	Sodium cumene sulphonate
NaXS	Sodium xylene sulphonate
NaPTS	Sodium *p*-toluene sulphonate
SFE	Supercritical fluid extraction
UAE	Ultrasound-assisted extraction
ILUAE	Ionic liquid-based ultrasonic-assisted extraction
HRE	Heat-reflux extraction
MAE	Microwave-assisted extraction
ILs	Ionic liquids
AUC	Area under the curve
K_{el}	Elimination rate constant
PLC	Piperine, lycopene and curcumin
SRP	Scaling and root planing
SGRQ	St. George Respiratory Questionnaire
CAT	COPD Assessment Test
GSH	Glutathione
HRQoL	Health-related quality of life
CRP	C-Reactive protein
REE	Resting energy expenditure
FFA	Free fatty acids

Conflict of Interest

The authors declare that they have no conflict of interest.

Acknowledgements

The authors are thankful to the Department of Pharmacognosy, ISF College of Pharmacy, Moga, Punjab, for providing the necessary facilities.

References

1. https://indianculture.gov.in/food-and-culture/spices-herbs/black-pepper-king-spices.
2. K. Ashokkumar, M. Murugan, M. K. Dhanya, A. Pandian and T. D. Warkentin, *Clin. Phytosci.*, 2021, 7, 1.
3. A. Chahardoli, F. Jalilian, Z. Memariani, M. H. Farzaei and Y. Shokoohinia, Analysis of organic acids, in *Recent Advances in Natural Products Analysis*, ed. A. S. Silva, S. F. Nabavi, M. Saeedi and S. M. Nabavi, Elsevier, USA, 2020, ch. 26, p. 767.
4. V. Badmaev, M. Majeed and E. P. Norkus, *Nutr. Res.*, 1999, **19**, 381.
5. K. G. Bose, *Pharmacopoeia India*, Bose Laboratories, Calcutta, India, 1929.
6. Trikatu - Shanti Yoga, https://www.shantiyoga.com.au/trikatu.
7. A. Lakshmipati, *One hundred Useful Drugs*, Arogya Ashram Samithi, Madras, India, 3rd edn, 1946.
8. https://www.carakasamhitaonline.com/index.php?title=Rasayana_Adhyaya#Pippali_Rasayana.
9. A. Hankey, *J. Altern. Complement. Med.*, 2001, 7, 567.
10. https://vikaspedia.in/agriculture/crop-production/package-of-practices/spices/black-pepper.
11. https://plantvillage.psu.edu/topics/black-pepper/infos.
12. B. Salehi, Z. A. Zakaria, R. Gyawali, S. A. Ibrahim, J. Rajkovic, Z. K. Shinwari, T. Khan, J. Sharifi-Rad, A. Ozleyen, E. Turkdonmez, M. Valussi, T. B. Tumer, L. Monzote Fidalgo, M. Martorell and W. N. Setzer, *Molecules*, 2019, **24**, 1364.
13. S. N. Dornala and S. S. Dornala, *J. Ayurveda Integr. Med.*, 2012, **3**, 158.
14. R. K. Johri and U. Zutshi, *J. Ethnopharmacol.*, 1992, **37**, 85.
15. Anonymous, *The Ayurvedic Formulary of India Part-II*, Government of India, Ministry of Health and Family Welfare, Department of Indian Systems of Medicine & Homeopathy, New Delhi, 1st English edn, 2000.
16. K. M. Nadkarni, *Piper nigrum L. Indian Materia Medica*, Popular Prakashan Press, Bombay, 1986, p. 969.
17. K. Usmanghani, A. Saeed and M. T. Alam, *Indusyunic Medicine*, University of Karachi Press, Karachi, 1997.
18. S. Gyawali, G. M. Khan, S. Lamichane, J. Gautam, S. Ghimire, R. Adhikari and R. Lamsal, *J. Clin. Diagn. Res.*, 2013, 7, 1135.
19. A. Soni, K. Patel and S. N. Gupta, *Ayu.*, 2011, **32**, 177.
20. M. Zaveri, A. Khandhar, S. Patel and A. Patel, *Int. J. Pharm. Sci. Rev. Res.*, 2010, **5**, 67.
21. A. N. Ngane, L. Biyiti, P. H. Bouchet, A. Nkengfack and P. H. A. Zollo, *Fitoterapia*, 2003, **74**, 464.
22. J. P. Dzoycm and V. Kuete, Review of the Antifungal Potential of African Medicinal Plants, in *Antifungal Metabolites from Plants*, Springer, 2013, p. 79.
23. J. P. Dzoyem, R. T. Tchuenguem, J. R. Kuiate, G. N. Teke, F. A. Kechia and V. Kuete, *BMC Complementary Altern. Med.*, 2014, **14**, 58.

24. R. U. B. Ebana, U. O. Edet, U. M. Ekanemesang, G. M. Ikon, C. A. Etok and A. P. Edet, *Asian J. Med. Health*, 2016, 1.
25. E. E. Mgbeahuruike, Y. Holm, H. Vuorela, C. Amandikwa and P. Fyhrquist, *J. Ethnopharmacol.*, 2019, **229**, 157.
26. Sharangdhara samhitha Madhyamakhanda 6/130, AFI.
27. K. A. Ansari and M. Akram, *J. Pharmacogn. Phytochem.*, 2014, **2**, 219.
28. B. Ahmad and J. Akhtar, *Pharmacogn. Rev.*, 2007, **1**, 210.
29. N. S. Jeganathan, S. Anbazhagan, P. Aruna, D. Deepa and A. Kalaivani, *Int. J. Adv. Res. Biol. Sci.*, 2016, **3**, 180.
30. A. Riyasdeen, V. S. Periasamy, P. Paul, A. A. Alshatwi and M. A. Akbarsha, *Evidence-Based Complementary Altern. Med.*, 2012, **2012**, 136527.
31. A. Vora, S. Patel and K. Patel, *J. Assoc. Physicians India*, 2016, **64**, 20.
32. C. Perakis, V. Louli and K. Magoulas, *J. Food Eng.*, 2005, **71**, 386.
33. G. Singh, P. Marimuthu, C. Catalan and M. de Lampasona, *J. Sci. Food Agric.*, 2004, **84**, 1878.
34. W. Liu, Z. Jiang, J. Chen, X. Zhang and Y. Ma, *Zhongguo Zhong Yao Za Zhi.*, 2009, **34**, 2891.
35. S. A. Lee, S. S. Hong, X. H. Han, J. Hwang, G. J. Oh, K. S. Lee, M. K. Lee, B. Y. Hwang and J. S. Ro, *Chem. Pharm. Bull.*, 2005, **53**, 832.
36. H. Huang, M. Yu, T. E. Akie, T. B. Moran, A. J. Woo, N. Tu, Z. Waldon, Y. Y. Lin, H. Steen and A. B. Cantor, *Mol. Cell. Biol.*, 2009, **29**, 4103.
37. P. Mishra, S. Sinha, S. K. Guru, S. Bhushan, R. Vishwakarma and S. Ghosal, *J. Asian Nat. Prod. Res.*, 2011, **13**, 143.
38. S. Sahi, P. Tewatia and S. Ghosal, *J. Mol. Model.*, 2012, **18**, 5065.
39. S. Ghosal, A. Deb, P. Mishra and R. Vishwakarma, *Planta Med.*, 2012, **78**, 906.
40. Z. Y. Jiang, W. F. Liu, C. G. Huang and X. Z. Huang, *Fitoterapia*, 2013, **84**, 222.
41. A. Chatterjee and C. P. Dutta, *Tetrahedron*, 1967, **23**, 1769.
42. S. J. Desai, B. R. Prabhu and N. B. Mulchandani, *Phytochemistry*, 1988, **27**, 1511.
43. J. K. Da Silva, R. da Trindade, N. S. Alves, P. L. Figueiredo, J. G. S. Maia and W. N. Setzer, *Int. J. Mol. Sci.*, 2017, **18**, 2571.
44. E. E. Mgbeahuruike, T. Yrjönen, H. Vuorela and Y. Holm, *S. Afr. J. Bot.*, 2017, **112**, 54.
45. L. Dyer and A. Palmer, *Piper: A Model Genus for Studies of Phytochemistry, Ecology, and Evolution*, 2004.
46. E. E. Mgbeahuruike, Y. Holm, H. Vuorela, C. Amandikwa and P. Fyhrquist, *J. Ethnopharmacol.*, 2019, **229**, 157.
47. C. P. Xiang, J. X. Han, X. C. Li, Y. H. Li, Y. Zhang, L. Chen, Y. Qu, C. Y. Hao, H. Z. Li, C. R. Yang, S. Zhao and M. Xu, *J. Agric. Food Chem.*, 2017, **65**, 3702.
48. J. K. Da Silva, R. da Trindade, N. S. Alves, P. L. Figueiredo, J. G. S. Maia and W. N. Setzer, *Int. J. Mol. Sci.*, 2017, **18**, 2571.
49. G. Raman and V. G. Gaikar, *Ind. Eng. Chem. Res.*, 2002, **41**, 2966.
50. M. Bhat and V. G. Gaikar, *Langmuir*, 1999, **15**, 4740.
51. N. Nagavekar and R. S. Singhal, *J. Food Process Eng.*, 2018, **41**, e12670.

52. I. Pacheco-Fernández and V. Pino, Extraction with Ionic Liquids-Organic Compounds, in *Handbooks in Separation Science, Liquid-Phase Extraction*, ed. C. F. Poole, Elsevier, 2020, ch. 17, pp. 499–537.
53. T. Mason, F. Chemat and M. Vinatoru, *Curr. Org. Chem.*, 2011, **15**, 237.
54. P. Cintas and J.-L. Luche, *Green Chem.*, 1999, **1**, 115.
55. Y. Yang and F. Zhang, *Ultrason. Sonochem.*, 2008, **15**, 308.
56. W. H. Lee, C. Y. Loo, M. Bebawy, F. Luk, R. Mason and R. Rohanizadeh, *Curr. Neuropharmacol.*, 2013, **11**, 338.
57. Y. Sun, D. Liu, J. Chen, X. Ye and D. Yu, *Ultrason. Sonochem.*, 2011, **18**, 243.
58. J. Raso, P. Mañas, R. Pagán and F. J. Sala, *Ultrason. Sonochem.*, 1999, **5**, 157.
59. M. Entezari and P. Kruss, *Ultrason. Sonochem.*, 1996, **3**, 19.
60. S. S. Rathod and V. K. Rathod, *Ind. Crop. Prod.*, 2014, **58**, 259.
61. X. Cao, X. Ye, Y. Lu, Y. Yu and W. Mo, *Anal. Chim. Acta*, 2009, **640**, 47.
62. H. Kataoka, *TrAC, Trends Anal. Chem.*, 2003, **22**, 232.
63. C. H. Chan, R. Yusoff, G. C. Ngoh and F. W. Kung, *J. Chromatogr. A*, 2011, **1218**, 6213.
64. N. N. Azwanida, *Med. Aromat. Plants*, 2015, **4**, 1000196.
65. M. Karpakavalli, K. R. Sini and I. Arth, *Pharm. Globale: Int. J. Compr. Pharm.*, 2012, **5**, 1.
66. F. Gogus, M. Z. Özel, H. Keskin, D. K. Yanık and A. C. Lewis, *Int. J. Pept.*, 2015, **18**, 1625.
67. G. B. Dudhatra, S. K. Mody, M. M. Awale, H. B. Patel, C. M. Modi, A. Kumar, D. R. Kamani and B. N. Chauhan, *Sci. World J.*, 2012, **2012**, 637953.
68. C. K. Atal, U. Zutshi and P. G. Rao, *J. Ethnopharmacol.*, 1981, **4**, 229.
69. G. Shoba, D. Joy, T. Joseph, M. Majeed, R. Rajendran and P. S. S. R. Srinivas, *Planta Med.*, 1998, **64**, 353.
70. F. Di Pierro and R. Settembre, *J. Pain Res.*, 2013, **6**, 497.
71. V. Badmaev, M. Majeed and L. Prakash, *J. Nutr. Biochem.*, 2000, **11**, 109.
72. E. L. Wightman, J. L. Reay, C. F. Haskell, G. Williamson, T. P. Dew and D. O. Kennedy, *Br. J. Nutr.*, 2014, **112**, 203.
73. R. Kasibhatta and M. U. Naidu, *Drugs R&D*, 2007, **8**, 383.
74. N. Patel, K. Jagannath, A. Vora, M. Patel and A. Patel, *J. Assoc. Physicians India*, 2017, **65**, 48.
75. S. K. Bedada and P. K. Boga, *Xenobiotica*, 2017, **47**, 1035.
76. L. Rofes, V. Arreola, A. Martin and P. Clavé, *J. Gastroenterol.*, 2014, **49**, 1517.
77. S. Kaur, R. Sharma, V. Sarangal, P. Kaur and P. Prashar, *Ayu*, 2017, **38**, 117.
78. B. Mihăilă, R. M. Dinică, A. L. Tatu and O. D. Buzia, *Exp. Ther. Med.*, 2019, **17**, 1039.
79. A. Shafiee, M. Hoormand, M. Shahidi-Dadras and A. Abadi, *Phytother. Res.*, 2018, **32**, 1812.
80. Y. Panahi, M. Ghanei, A. Hajhashemi and A. Sahebkar, *J. Diet. Suppl.*, 2016, **13**, 93.
81. M. Rondanelli, A. Opizzi, S. Perna, M. Faliva, S. B. Solerte, M. Fioravanti, C. Klersy, E. Cava, M. Paolini, L. Scavone, P. Ceccarelli, E. Castellaneta, C. Savina and L. M. Donini, *Endocrine*, 2013, **44**, 391.

CHAPTER 8

Bacopa monnieri (*Brahmi*): *Phytochemistry and Use in Traditional Ayurvedic Formulations*

RENJU KUNJUMON,[a,b] ANIL JOHN JOHNSON[a] AND SABULAL BABY*[a]

[a] Phytochemistry and Phytopharmacology Division, Jawaharlal Nehru Tropical Botanic Garden and Research Institute, Pacha-Palode 695562, Thiruvananthapuram, Kerala, India; [b] University of Kerala, Thiruvananthapuram 695034, Kerala, India
*Emails: sabulal@jntbgri.res.in; sabulal@gmail.com

8.1 Ayurveda and *Bacopa monnieri*

India has one of the oldest folk traditions using medicinal herbs, and the early records of their use are found in the Sanskrit literature.[1] The main literary sources of the ancient era are the four *Vedas* (Sanskrit word meaning 'knowledge' or 'wisdom') *viz.*, '*Rig Veda*', '*Yajur Veda*', '*Sama Veda*' and '*Atharva Veda*'. '*Rig Veda*' and '*Atharva Veda*', written between 1600 and 400 BCE, describe the uses of plant drugs. '*Ayurveda*' (considered as '*Upa Veda*' or applied knowledge; Sanskrit meaning 'knowledge of life') deals with the understanding of the medicinal properties of plant products and their roles in the treatment of various diseases.[2–4] The initial treatise on '*Ayurveda*', '*Charaka Samhita*' written by Charaka in ∼900 BCE, listing 341 plants and their products, is an authentic reference to illnesses and medical strategies.

Chemistry, Biological Activities and Therapeutic Applications of Medicinal Plants in Ayurveda
Edited by Augustine Amalraj, Sasikumar Kuttappan and Karthik Varma A. C.

Published by the Royal Society of Chemistry, www.rsc.org

'*Susruta Samhita*' written in ~600 BCE by Susruta, an ancient surgeon and teacher of '*Ayurveda*', is a more inclusive and authentic work which describes 1120 diseases, 700 medicinal plants and 121 preparations.[3,5–7] Ayurveda, founded on the collective wisdom of ancient Hindu saints and healers, thrives in the 21st century as a lively healthcare system in India, with a recent estimate of over 6000 crores (1.3 billion USD) worth of drugs produced annually.[8,9]

Ayurveda emphasizes disease prevention and the promotion of good health, by adopting a proper lifestyle and assuming measures that rejuvenate the cells of the body. It is based on the concepts of five elements ('*panchabhuta*') *viz.*, '*Vayu*' (air), '*Jala*' (water), '*Prithvi*' (earth), '*Teja*' (fire) and '*Akasha*' (space), which constitute the physical universe including the human body, and the three '*doshas*' (Sanskrit word meaning 'fault' or 'defect') *viz.*, '*Vata*', '*Pitta*' and '*Kapha*' (*Tridoshas*), which broadly refer to the functions of motion, digestion and cumulation.[10,11] In other words, *Tridoshas* ('*Vata*', '*Pitta*' and '*Kapha*') are the three psychobiological dimensions (energy) or biological rhythms synchronizing the functioning of the human body. The disparity between the three '*doshas*' leads to diseases. The dominance of '*Vata*' dosh causes neurological disorders, manifested by memory loss, poor blood circulation, anxiety and related symptoms. '*Pitta*' imbalance gives rise to gastric troubles and imbalanced '*Kapha*' results in cholesterol-related diseases. Ayurveda provides medicines of herbal origin to balance the '*doshas*', aiming to cure diseases.[4,12]

Bacopa monnieri (L.) Wettst. (*Brahmi*; family Plantaginaceae) is a nootropic herb distributed throughout the wetlands of the tropics and subtropics.[13,14] The name '*Brahmi*' is derived from the word '*Brahma*', the mythical 'creator' in the Hindu religion.[15] *B. monnieri* is one of the herbs used in various formulations administered in Ayurveda against neurological issues.[4,16] Ancient Vedic scholars often used *B. monnieri* to memorize lengthy divine scriptures and hymns.[13,17] In India, in Ayurvedic prescriptions, *B. monnieri* has been consumed as '*Medhya rasayana*' (in Sanskrit, '*medhya*' – intellect or cognition, '*rasayana*' – rejuvenation).[18,19] In *Charaka Samhita*, *B. monnieri* is mentioned as a cure for controlling mental dysfunctions such as poor cognition, lack of concentration and anxiety.[13,15,20–22] *Brahmi* and its polyherbal medications are also applied as treatments for various other diseases in Ayurveda.[21,23]

A search of the keywords '*Bacopa monnieri*' and '*Brahmi*' in Google Scholar in November 2021 retrieved 14 900 and 40 500 results, whereas '*Bacopa monnieri* + Ayurveda' and '*Brahmi* + *Ayurveda*' resulted in 4740 and 5770 articles, respectively. The National Medicinal Plants Board (NMPB) of India estimated the annual market demand of *B. monnieri* in India as 1000 tons in the year 2000.[24–26] Again, in 2004–2005, the NMPB survey quantified the annual demand of *B. monnieri* as 6621.8 tons, with an annual growth rate of 7%.[27] Recent literature reports on the demand of *B. monnieri* in India reflect inconsistencies,[16,22] but its demand remains ever-increasing and it is logical to estimate the annual requirement as above 10 000 tons; and

Ayurvedic drugs contribute significantly to this growing demand.[25] NMPB also listed *B. monnieri* as one of the 'medicinal plant species in high trade' in 2007.[28] Further, the conservation and cultivation of *B. monnieri* are encouraged by NMPB and Technology Information Forecasting and Assessment Council (TIFAC), Department of Science and Technology (DST), Government of India.[25] These data indicate the surging research activities on *B. monnieri*, including significant contributions from the Ayurvedic drug industry. This review deals with the phytochemistry of *B. monnieri* and its uses in Ayurveda. The need for correlation of the phytochemistry and analytical techniques (chemical fingerprinting, quantification) in defining the role of bioactive molecules in *Brahmi*-based and other Ayurvedic formulations is also discussed. The neuroprotection of *B. monnieri* (extracts, bioactive molecules) achieved through modulation of brain stress hormones, suppression of neuronal oxidative stress, attenuation of oxidative damage, pro-cholinergic effect, acetylcholinesterase inhibition, suppression of pro-inflammatory cytokines, calcium channel blocking, β-amyloid reduction, neurotransmitter modulation, and its other biological effects such as anti-inflammatory, cardioprotective, hepatoprotective, antitumor, antiarthritic and antidiabetic activities, have been thoroughly reviewed by various authors,[13,15,21,29–35] and their elaborate assessment is not under the scope of this chapter.

8.2 *B. monnieri*-based Ayurvedic Formulations

In Ayurveda, rejuvenation of the cells is addressed in the *Rasayana shastra*. *Rasayana* (Sanskrit meaning '*path of essence*') is defined as a therapeutic measure which prevents aging, promotes longevity, provides positive health and mental faculties, increases memory, and imparts immunity and resistance against diseases. *Rasayanas* are specialized procedures practiced in the form of rejuvenative ingredients, dietary regimens and definite health-promoting activities and behavior.[5,8,36] They are basically molecular nutrients and nutrition-enhancing agents acting through three fundamental mechanisms *viz.*, *Rasa* (strengthening nutrient effect), *Agni* (promoters of digestion and metabolism) and *Srotas* (purifying agents or stimulants of microcirculation and tissue perfusion). These three events individually or collectively lead to enhanced nutritional status in the body leading to refined cells and tissues which sustain aging and stress.[36] *Rasayanas* promote resistance against infections and other diseases by maintaining the balance of the '*Tridoshas*', *viz.*, '*Vata*', '*Pitta*' and '*Kapha*'. Otherwise, *Rasayana* drugs provide actions such as (i) increase in lifespan, (ii) delay the aging process, (iii) anabolic properties to strengthen the body, (iv) improvement of cognitive ability and (v) gaining immunity and cure from diseases.[37] Doshi and co-workers provided a list of 34 *Rasayana* herbs.[38] Some of these *Rasayanas* are organ- and tissue-specific; *Medhya rasayanas* are those specific to the brain tissues (or neuronutrients). They are a group of drugs widely used in Ayurveda since antiquity with several benefits specifically in improving memory and intellect. *Brahmi* (*B. monnieri*) is one of the *Medhya rasayanas*.[9,36,39]

Other important *Medhya rasayana* plants are *Centella asiatica* (*Mandukaparni*), *Convolvulus pluricaulis* (*Shankhpushpi*), *Acorus calamus* (*Vacha*), *Withania somnifera* (*Ashwagandha*), *Celastrus paniculatus* (*Jyotishmati*) and *Nardostachys jatamansi* (*Jatamansi*). Many Ayurvedic preparations advised for cognitive dysfunctions contain *B. monnieri* as a prime constituent.

Brahmi ghrita, which contains *B. monnieri*, *Acorus calamus*, *Convolvulus pluricaulis*, *Saussurea lappu* and 10 year-old cows' clarified butter, is used to treat memory disorders in Ayurveda. Yadav and co-workers demonstrated the beneficial effects of *Brahmi ghrita* in learning and memory at 400 and 800 mg kg^{-1} body weight in rats. These effects were inferred as due to an increase in cerebral glutamic acid and gamma-amino butyric acid (GABA), achieved through the synergistic action of the herbal ingredients.[40] Achliya and co-workers investigated the effect of *Brahmi ghrita* (30, 50 and 100 mg kg^{-1}, p.o.) on learning and memory processes by various assays in rats and mice. *Brahmi ghrita* administration (50 and 100 mg kg^{-1}, p.o.) showed a significant decrease in transfer latency and escape latency. This study has not elucidated the exact mechanism of action of *Brahmi ghrita*, but inferred that it may act as a memory enhancer formulation and is useful as a supportive adjuvant in the treatment of impaired memory functions.[41] Bhalerao and co-workers[42] evaluated the efficacy of *Brahmi ghrita* against attention deficit hyperactivity disorder (ADHD) in children. In the pilot study *Brahmi ghrita* showed 66% decrease in the total ADHD score. In the therapeutic confirmatory study, 16% improvement was seen with *Brahmi ghrita*, which was comparable to the standard drug (methylphenidate). No side effects were reported by the authors.

The antiamnesic activity of *Brahmi ghrita* (400 and 800 mg kg^{-1}, p.o.) was evaluated against scopolamine-induced amnesia in rats using elevated plus maze, passive avoidance and active avoidance tests, with piracetam (500 mg kg^{-1}, p.o.) as the standard drug. *Brahmi ghrita* and piracetam displayed significant beneficial effects on scopolamine-induced amnesia.[43] The antidepressant and psychoactive effects of *Brahmi ghrita* were studied by randomized, placebo-controlled clinical trials in 42 patients, and it demonstrated moderate improvement in 40% patients and mild improvement in 55% patients.[44] In another recent clinical study, 72 persons of average age 63.3 ± 6.44 years received either *Brahmi ghrita* or integrated yoga or combined intervention for 8 weeks. Both *Brahmi ghrita* and integrated yoga intervention were found to be effective in improving cognitive abilities, but the combined intervention resulted in significantly better improvement in attention, learning and working memory compared to individual interventions among elderly persons with mild cognitive impairment.[19]

Brahmi rasayana, a rejuvenator, is an Ayurvedic medicinal mixture with dried *Brahmi* leaves as its main constituent, along with *Eugenia caryophyllus* (flower buds), *Elettaria cardamomum* (seeds), *Cinnamomum zeylanicum* (inner bark of shoots), *Piper nigrum* (fruits) and *P. longum* (fruits). It is commonly used as a nerve tonic and in the treatment of epilepsy. Joshi and Parle[42] demonstrated that *Brahmi rasayana* improves learning and memory in young mice and it reverses amnesia induced by both scopolamine (0.4 mg kg^{-1}, i.p.) and natural aging. *Brahmi rasayana* significantly decreased whole-brain acetylcholinesterase activity, and it could be a useful memory-restorative agent in the treatment of dementia in the elderly. The memory-improving activity of *Brahmi rasayana* may be attributed to the phytochemicals in various ingredients of the multiherbal preparation which exert antioxidant, anti-inflammatory, neuroprotective, pro-cholinergic and antiacetylcholinesterase properties, and it may be of use in delaying the onset and reducing the severity of Alzheimer's disease.[45] The anti-inflammatory effects of *Brahmi rasayana* were studied by Jain and co-workers in rodents (1 and 10 g kg^{-1}, p.o.); it suppressed various experimentally induced inflammatory reactions and displayed activity comparable to indometacin. *Brahmi rasayana* may partially mediate its anti-inflammatory activity by interfering with the action and/or synthesis of prostaglandins and possibly by stabilization of the lysosomal membranes.[46] Shukia and co-workers studied the effects of *Brahmi rasayana* on the central nervous system in mice and rats (1 and 30 g kg^{-1}, p.o.); it displayed significant antinociceptive effect, coupled with the ability to protect against electroshock seizures and chemoconvulsions. *Brahmi rasayana* also showed the potential to antagonize haloperidol-induced catalepsy. These findings suggested the role of the GABA-ergic system in mediation of the central nervous system effects of *Brahmi rasayana*.[47]

Brahmi vati is another Ayurvedic formulation comprised of *Brahmi* and another 37 herbal and animal ingredients along with nine bhasmas. The ingredients of *Brahmi vati* have been listed by Mishra *et al.*[48] It is one of the most important formulations primarily used to cure mental disorders, epilepsy and to improve memory. Mishra and co-workers demonstrated the memory-enhancing, anticonvulsant and antischizophrenia activities of *Brahmi vati*, when compared with bacoside A-rich fraction of *B. monnieri* and control groups, and it caused notably higher brain glutathione levels. Acetylcholinesterase activity was found to be significantly lower in the *Brahmi vati*-treated group. This study suggested *Brahmi vati* as a suitable polyherbo-mineral formulation for the treatment of schizophrenia and memory impairments associated with seizures.[49] A randomized, double-blind clinical study of *Brahmi vati* in 68 patients meeting the criteria of essential hypertension (EHTN) (either sex, age group 20–60 years) illustrated it as effective and safe. *Brahmi vati* (and similar Ayurvedic formulations) could act on multiple domains leading to a comprehensive management strategy for EHTN.[50]

Saraswata ghrita is a poly-herbal drug consisting of mainly *Brahmi* and Haridra (*Curcuma longa*), Haritaki (*Terminalia chebula*), Amlaki (*Emblica officinalis*), Vidanga (*Embelia ribes*), Pippali (*Piper longum*), Vacha (*Acorus calamus*) and Kushta (*Saussurea lappa*). It is well-known for its nootropic and memory-enhancing properties. An ultra-fast liquid chromatography/mass spectrometric (UFLC/MS) investigation of *Brahmi*, *Brahmi vati*, *Brahmi ghrita* and *Saraswata ghrita* solutions added to differentiated IMR 32 cells displayed enhanced levels of neurotransmitters (serotonin, dopamine, GABA).

Thus, these Ayurvedic medicines when prescribed to children with autism spectrum disorder (ASD) might alleviate the abnormal behavioural symptoms by maintaining neurotransmitter homeostasis. *Brahmi* in polyherbal forms increased neurotransmitter levels considerably more than its pure form.[51]

Saraswatarishta, with 18 plant ingredients including *Medhya rasayanas* (*Brahmi*, Ashwagandha, Shatavari), is traditionally used as a brain tonic to improve memory. It is also used to treat neurosis, psychosis, insomnia, stammering, epilepsy and memory loss, and *in vivo* assays in rats demonstrated its antidepressant effects.[52] Giramkar and co-workers studied the anticonvulsant activity of *Brahmi ghrita* and *Saraswatarishta* against seizures induced by maximal electroshock (MES) in rats with phenytoin (25 $mg\,kg^{-1}$, i.p.) as the standard. The results indicated that both these formulations are effective in promoting restorative and neuroprotective actions in convulsion.[53]

Recently, an increasing number of studies elucidating the function and mechanisms of Ayurvedic formulations have emerged. *Brahmi*-based formulations are subjected to studies illustrating their neuroprotective and memory-enhancing credentials.[4,20]

8.3 Phytochemistry of *B. monnieri*

So far, there are no comprehensive reviews on the chemistry of *B. monnieri*. Over the last eight decades, phytochemistry of *B. monnieri* has been studied extensively by several groups (Tables 8.1 and 8.2, and references therein). An alkaloid [brahmine (**76**)] was first reported in 1931,[54] and thereafter two more alkaloids [herpestine (**77**), nicotine (**78**)] were isolated from *B. monnieri*.[55,56] Chatterji and co-workers deduced the structure of bacoside A in 1965.[57] Kawai and Shibata in 1978 reported ebelin lactone and bacogenin A_1 obtained by acid hydrolysis of bacoside A and the sapogenins (jujubogenin, psuedojujubogenin) by its Smith-de Mayo degradation.[58] They also inferred bacoside A as leavo- and bacoside B as dextro-rotatory based on the configurations of their carbohydrate chains.[26,58]

Dammarane-type triterpenoid saponins with jujubogenin and psuedojujubogenin aglycone moieties (**1–34**) are the specialized compounds found in *B. monnieri* (Table 8.1). Bacoside A is the most pharmacologically active constituent reported from *B. monnieri*. First described in 1965,[57] bacoside A was later identified as a mixture of four saponins [bacopaside II (**2**), bacopasaponin C (**9**), bacopaside X (**20**) and bacoside A_3 (**29**)][66] (Table 8.1). Bacoside B was reported by Basu and co-workers in 1967;[97] and in 2005 Sivaramakrishna and co-workers indicated it as a mixture of four saponins *viz.*, bacopaside V (**4**), bacopaside N2 (**6**), bacopaside IV (**17**) and bacopaside N1 (**22**).[61] However, Deepak and Amit (2013)[79] mentioned that the details given by Sivaramakrishna and co-workers[61] contradicted the original reports[97,98] describing the structure and chromatography of bacoside B. Chakravarty and co-workers in 2001 reported two new psuedojujubogenin glycosides, bacopaside I (**1**) and bacopaside II (**2**), from *B. monnieri* methanol

Table 8.1 Compounds isolated from *B. monnieri*.[a]

No.	Compound	Reference[b]
[A] Dammarane type triterpenoid saponins		
[A1] Psuedojujubogenin derivatives		
1.	Bacopaside I (3-*O*-α-L-arabinofuranosyl-(1→2)-[6-*O*-sulphonyl-β-D-glucopyranosyl-(1→3)]-α-L-arabinopyranosyl pseudojujubogenin)	**59**, 60–64
2.	Bacopaside II (3-*O*-α-L-arabinofuranosyl-(1→2)-[β-D-glucopyranosyl-(1→3)]-β-D-glucopyranosyl pseudojujubogenin)	**59**, 60–68
3.	Bacopaside III (3-*O*-[6-*O*-sulfonyl-β-D-glucopyranosyl (1→3)]-α-L arabinopyranosyl pseudojujubogenin)	**69**
	(Zhou *et al.*, 2007[63] reported this molecule as bacopaside VI and claimed as a new molecule).	
	(Chakravarty *et al.*, 2003[70] reported 3-*O*-α-L-arabinofuranosyl-(1→2)-β-D-glucopyranosyl jujubogenin (a jujubogenin derivative) as bacopaside III).	
4.	Bacopaside V (3-*O*-β-D-glucopyranosyl-(1→3)-α-L-arabinopyranosyl psuedojujubogenin)	**70**, 61, 65, 67, 71
5.	Bacopaside XI (3-*O*-[α-L-arabinofuranosyl-(1→3)]-6-*O*-sulfonyl-β-D-glucopyranosyl pseudojujubogenin)	**71**
	(Zhou *et al.*, 2009[64] reported 3-*O*-[β-D-glucopyranosyl-(1→3)]-β-D-glucopyranosyl psuedojujubogenin as bacopaside XI).	
6.	Bacopaside N2 (3-*O*-[β-D-glucopyranosyl-(1→3)-β-D-glucopyranosyl] psuedojujubogenin)	**61**, 67, 72
	(Zhou *et al.*, 2009[64] isolated the same molecule and reported it as a new molecule, bacopaside XI).	
	(Bhandari *et al.*, 2009[71] reported 3-*O*-[α-L-arabinofuranosyl-(1→3)]-6-*O*-sulfonyl-β-D-glucopyranosyl pseudojujubogenin as bacopaside XI).	
7.	Bacopaside XII (3-*O*-[β-D-glucopyranosyl(1→3){-α-L-arabinofuranosyl(1→2)}-β-D-glucopyranosyl]-20-*O*-α-L arabinopyranosyl psuedojujubogenin)	**71**
8.	Bacopasaponin B (3-*O*-[α-L-arabinofuranosyl(1→2)-α-L-arabinopyranosyl] psuedojujubogenin)	**73**, 74
9.	Bacopasaponin C (3-*O*-[β-D-glucopyranosyl-(1→3){-α-L-arabinofuranosyl-(1→2)}-α-L-arabinopyranosyl] psuedojujubogenin)	**73**, 61–68, 74, 75
10.	Bacopasaponin D (3-*O*-[α-L-arabinofuranosyl(1→2)-β-D-glucopyranosyl] psuedojujubogenin)	**76**, 64

11.	Bacopasaponin H (3-*O*-[α-L-arabinopyranosyl] psuedojujubogenin)	**74**
12.	3-*O*-β-D-Glucopyranosyl pseudojujubogenin	67
13.	Bacoside A_2 (3-*O*-[α-L-arabinofuranosyl-(1→6)-*O*-{α-L-arabinopyranosyl-(1→5)}-*O*-α-D-glucofuranosyl] oxypseudojujubogenin)	77
14.	Bacopasaponin G (3-*O*-[β-D-glucopyranosyl(1→3)-β-D-glucopyranosyl]-20-deoxypseudojujubogenin)	**67**
	(Hou *et al.*, 2002[69] reported 3-*O*-[α-L-arabinofuranosyl-(1→2)]-α-L-arabinopyranosyl jujubogenin as bacopasaponin G).	
15.	Bacopasaponin I (3-*O*-[β-D-glucopyranosyl-(1→3)-*O*-{α-L-arabinofuranosyl-(1→2)}-*O*-β-D-glucopyranosyl]-20-deoxypseudojujubogenin)	72
[A2] Jujubogenin derivatives		
16.	Bacopaside III (3-*O*-α-L-arabinofuranosyl-(1→2)-β-D-glucopyranosyl-jujubogenin)	70
	(Hou *et al.*, 2002[69] reported 3-*O*-[6-*O*-sulfonyl-β-D-glucopyranosyl(1→3)]-α-L-arabinopyranosyl pseudojujubogenin (a pseudojujubogenin derivative) as bacopaside III).	
17.	Bacopaside IV (3-*O*-β-D-glucopyranosyl-(1→3)-α-L-arabinopyranosyl jujubogenin)	**70**, 61, 65, 67, 71
18.	Bacopaside VIII (3-*O*-[β-D-glucopyranosyl-(1→3){-α-L-arabinofuranosyl(1→2)}-β-D-glucopyranosyl]-20-*O*-α-L-arabinopyranosyl jujubogenin)	**63**
19.	Bacopaside IX (3-*O*-[β-D-glucopyranosyl-(1→4){-α-L-arabinofuranosyl(1→2)}-β-D-glucopyranosyl]-20-*O*-α-L-arabinopyranosyl jujubogenin)	**78, 64**
20.	Bacopaside X (3-*O*-[α-L-arabinofuranosyl-(1→2){-β-D-glucopyranosyl-(1→3)}-α-L-arabinopyranosyl] jujubogenin) (jujubogenin isomer of bacopasaponin C, bacopaside VII)	61, 63, 66, 68
	(Zhou *et al.*, 2009[64] isolated 3-*O*-[β-D-glucopyranosyl(1→2)]-α-L-arabinopyranosyl jujubogenin and named it as bacopaside X).	
21.	Bacopaside X (3-*O*-[β-D-glucopyranosyl(1→2)]-α-L-arabinopyranosyl jujubogenin)	**64**
	(Sivaramakrishna *et al.*, 2005[61] isolated the known molecule 3-*O*-[α-L-arabinofuranosyl-(1→2){-β-D-glucopyranosyl-(1→3)}-α-L-arabinopyranosyl] jujubogenin and gave the trivial name bacopaside X).	
22.	Bacopaside N1 (3-*O*-[β-D-glucopyranosyl-(1→3)-β-D-glucopyranosyl] jujubogenin)	**61**, 79, 72
23.	Bacopasaponin A (3-*O*-α-L-arabinopyranosyl-20-*O*-α-L-arabinopyranosyl jujubogenin)	73
24.	Bacopasaponin E (3-*O*-[β-D-glucopyranosyl(1→3){-α-L-arabinofuranosyl(1→2)}-α-L-arabinopyranosyl]-20-*O*-(α-L-arabinopyranosyl) jujubogenin)	**80**, 61

Table 8.1 (*Continued*)

No.	Compound	Reference[b]
25.	Bacopasaponin F (3-*O*-[β-D-glucopyranosyl(1 → 3){-α-L-arabinofuranosyl(1 → 2)}-β-D-glucopyranosyl]-20-*O*-(α-L arabinopyranosyl) jujubogenin)	**80**, 61
26.	Bacopasaponin G (3-*O*-[α-L-arabinofuranosyl-(1 → 2)]-α-L-arabinopyranosyl jujubogenin) (Pawar *et al.*, 2007[67] reported 3-*O*-[β-D-glucopyranosyl(1 → 3)-β-D-glucopyranosyl]-20-deoxypseudojujubogenin as bacopasaponin G).	**69**
27.	3-*O*-α-L-arabinopyranosyl jujubogenin	69
28.	Bacoside A_1 (3-*O*-[α-L-arabinofuranosyl-(1 → 3)-α-L-arabinopyranosyl] jujubogenin)	**81**
29.	Bacoside A_3 (3-*O*-[β-D-glucopyranosyl-(1 → 3)-*O*-{α-L-arabinofuranosyl-(1 → 2)}-*O*-β-D-glucopyranosyl] oxyjujubogenin)	**82**, 61, 65–69, 75
30.	Bacoside A_4 (3-*O*-α-L-arabinopyranosyl jujubogenin) (Mourya et. al., 1995[85] reported this compound as zizyotin from *Zizyphus oenoplea*).	83, 84
31.	Bacoside A_5 (20-*O*-α-L-arabinopyranosyl jujubogenin)	**83**
32.	Bacoside A_6 (3-*O*-[β-D-glucopyranosyl-(1 → 3)-*O*-{α-L-arabinofuranosyl-(1 → 2)}-*O*-β-D-glucopyranosyl]-20-deoxyjujubogenin)	**67**
33.	Bacomosaponin A (3-*O*-[β-D-glucopyranosyl(1 → 3){α-L-arabinofuranosyl(1 → 2)}-β-D-glucopyranosyl]-20-*O*-(3,4-diacetyl-α-L-arabinopyranosyl) jujubogenin)	**68**
34.	Bacomosaponin B (3-*O*-[β-D-glucopyranosyl(1 → 3){α-L-arabinofuranosyl(1 → 2)}-α-L-arabinopyranosyl]-20-*O*-(3,4-diacetyl-α-L-arabinopyranosyl) jujubogenin)	**68**
[B] Pentacyclic triterpenes		
35.	Ursolic acid	84
36.	Betulinic acid	72, 86, 87
37.	Lupeol	84
38.	28-Hydroxy lupeol	84
39.	Bacosine	**88**, 62, 89
40.	Ampelozigenin	84
[C] Phenylethyl glycosides		
41.	Bacopaside A ((3*R*)-1-octan-3-yl-(6-*O*-sulfonyl)-β-D-glucopyranoside)	**69**
42.	Bacopaside B (3,4-dihydroxy phenylethyl alcohol (2-*O*-feruloyl)-β-D-glucopyranoside)	**69**
43.	Bacopaside C (Phenylethyl alcohol [5-*O*-*p*-hydroxybenzoyl-β-D-apiofuranosyl-(1 → 2)]-β-D-glucopyranoside)	**69**

44.	Monnieraside I (α-*O*-[2-*O*-(4-hydroxybenzoyl)-β-D-glucopyranosyl]-4-hydroxyphenyl ethanol)	**90**, 72, 91
45.	Monnieraside II (α-*O*-[2-*O*-(3-methoxy-4-hydroxycinnamoyl)-β-D-glucopyranosyl]-3,4-dihydroxyphenyl ethanol)	**90**
46.	Monnieraside III (α-*O*-[2-*O*-(4-hydroxybenzoyl)-β-D-glucopyranosyl]-3,4-dihydroxyphenyl ethanol)	**90**, 68, 91
47.	Plantainoside A	69
48.	Plantainoside B (plantioside B)	68, 69, 90, 91
49.	Bacomoside A	**68**
50.	Bacomoside B_1[c]	**68**
51.	Bacomoside B_2[c]	**68**
52.	Calcerorioside B	69
[D] Phenylpropanoid glycosides		
53.	Martynoside	69
54.	Acetoside	69
[E] Cucurbitacins		
55.	Bacobitacin A (2,7β,16α,20*R*-tetrahydroxy-25-acetoxy-cucurbita-1,5,23*E*-triene-3,11,22-trione)	**91**
56.	Bacobitacin B (2,20*R*-dihydroxy-16α,25-diacetoxy-cucurbita-1,5,23*E*-triene-3,11,22-trione)	**91**
57.	Bacobitacin C (2-*O*-[{α-L-rhamnopyranosyl-(1 → 2)-α-L-arabinofuranosyl-(1 → 4)}-{α-L-rhamnopyranosyl-(1 → 2)-α-L-arabinopyranosyl}] cucurbitacin E)	**91**
58.	Bacobitacin D (2-*O*-[{α-L-rhamnopyranosyl-(1 → 2)-α-L-arabinofuranosyl-(1 → 4)}-{α-L-rhamnopyranosyl-(1 → 2)-α-L-arabinopyranosyl}]-7β,16α,20*R*-trihydroxy-25-acetoxy-cucurbita-1,15,23*E*-triene-3,11,22-trione)	**91**
59.	Cucurbitacin E	68, 91
60.	Cucurbitacin I	68
61.	Cucurbitacin E-2-*O*-β-D-glucopyranoside	68
[F] Steroids		
62.	Bacosterol	**88**, 62
63.	Stigmasterol	92

Table 8.1 (*Continued*)

No.	Compound	Reference[b]
64.	β-Sitosterol	92
65.	Stigmastanol	92
66.	Bacosterol-3-*O*-β-D-glucopyranoside	**62**
67.	Stigmasterol-3-*O*-β-glucopyranoside	84
68.	$\Delta^{5,24}$-Ergostadien-3β-ol	84
69.	β-Daucosterin	84
[G] Flavonoids		
70.	Luteolin	56, 69, 72, 84
71.	Luteolin-7-*O*-β-glucopyranoside	56, 62
72.	Luteolin-7-glucuronide	93
73.	Apigenin-7-glucuronide	93
74.	Quercetin	84
75.	Apigenin	69, 71, 84
[H] Alkaloids		
76.	Brahmine[d]	54
77.	Herpestine[d]	55
78.	Nicotine[d]	56
[I] Others		
79.	2,4,6-Trihydroxy-5-(3,3-dimethyl propenyl)-3-(4-hydroxy phenyl) propiophenone	94
80.	3-*O*-β-D-Glucopyranosyl (3*R*)-1-octen-3-ol	72
81.	*p*-Hydroxybenzoic acid	84, 72
82.	*p*-Hydroxybenzenemethanol	84
83.	3,4-Dimethoxy cinnamic acid	84
84.	Hersaponin[e]	55
85.	D-Mannitol	55, 92
86.	Feruloyl glucoside	84
87.	Rosavin	84
88.	Loliolide	84

89.	Monnierin[e]	95, 96
90.	Heptacosane	92
91.	Octacosane	92
92.	Nonacosane	92
93.	Triacontane	92
94.	Hentriacontane	92
95.	Dotriacontane	92
96.	3-Formyl-4-hydroxy-2*H*-pyran	56
97.	1-Linolenyl-3-*O*-β-D-galactopyranosyl-*sn*-glycerol	69
98.	β-(3,4-dihydroxyphenyl)ethyl-*O*-β-D-glucose	69

[a]Bacoside A, first isolated by Chatterji *et al.*, 1965,[57] was later identified as a mixture of four saponins (bacoside A_3 (**29**), bacopaside II (**2**), bacopasaponin C (**9**) and bacopaside X (**20**) (jujubogenin isomer of bacopasaponin C).[66] Bacoside B was reported as a mixture of four saponins (bacopaside N1 (**22**), bacopaside N2 (**6**), bacopaside IV (**17**) and bacopaside V (**4**)) (Sivaramakrishna *et al.*, 2005).[61] Deepak and Amit (2013)[79] mentioned that the details given by Sivaramakrishna and co-workers (2005)[61] are contradicting the original reports (Basu *et al.*, 1967;[97] Singh *et al.*, 1988[98]) describing bacoside B.

[b]References reporting new molecules are highlighted in bold.

[c]Ohta *et al.*, 2016[68] reported two molecules bacomoside B_1 and B_2 (**50, 51**) with optical rotation $[\alpha]_D^{27}$ +5.1 and $[\alpha]_D^{27}$ +22.0, respectively, but their structures are the same.

[d]Alkaloids, brahmine (**76**),[54] herpestine (**77**)[55] and nicotine (**78**)[56] reported in the early studies (1931-1972) were not isolated by other authors from *B. monnieri*.

[e]Hersaponin (**34**)[55] and monnierin (**89**)[95,96] were reported with doubtful structural details and not isolated by other authors.

extract.[59] Again the same authors isolated three new saponins [bacopaside III (**16**), IV (**17**) and V (**4**)] from its aerial parts in 2003.[70]

Bhandari and co-workers reported a new sterol glycoside, bacosterol-3-*O*-β-D-glucopyranoside (**66**) along with known compounds bacopaside I (**1**), bacopaside II (**2**), bacopasaponin C (**9**), bacosine (**39**), bacosterol (**62**) and luteolin-7-*O*-β-glucopyranoside (**71**) from *B. monnieri* aerial parts in 2006.[62] Again, in 2007, the same group isolated cucurbitacins, bacobitacin A-D (**55–58**) and cucurbitacin E (**59**), along with three known phenylethyl glycosides [monnieraside I (**44**), II (**45**), III (**46**)] and plantinoside B (**48**).[91] Zhou and co-workers isolated three new triterpene glycosides, bacopaside VI (**3**), VII (**20**) and VIII (**18**), along with known molecules, bacopaside I (**1**), bacopaside II (**2**) and bacopasaponin C (**9**), from *B. monnieri* whole plant[63] (Table 8.1).

Pawar and co-workers reported two dammarane glycosides, bacopasaponin G (**14**) and bacoside A_6 (**32**), together with 10 known compounds from *B. monnieri*.[67] Thereafter, bacopaside IX (**19**), X (**21**) and XI (**6**) along with known compounds bacopaside I (**1**), bacopaside II (**2**), bacopasaponin C (**9**) and bacopasaponin D (**10**) were isolated from *B. monnieri* whole plant.[64] A chalcone-type compound 2,4,6-trihydroxy-5-(3,3-dimethyl propenyl)-3-(4-hydroxy phenyl) propiophenone (**79**) was reported by Suresh *et al.*[94] Ohta and co-workers reported bacomosaponin A (**33**) and B (**34**) (dammarane-type oligoglycosides) and bacomoside A (**49**), B_1 (**50**) and B_2 (**51**) (phenylethyl glycosides) as new molecules from *B. monnieri* whole plant in 2016.[68] Dang and co-workers isolated a new saponin, bacopasaponin I (**15**), along with a few known molecules from *B. monnieri* in 2018[72] (Tables 8.1 and 8.2). There are several contradictory names in the compounds reported from *B. monnieri*, and these issues are addressed in this review [Table 8.1; examples, compounds (3), (5), (**6**), (**14**)].

A total of 98 secondary metabolites *viz.*, dammarane-type triterpenoid saponins (34, **1–34**), pentacyclic triterpenes (6, **35–40**), phenylethyl glycosides (12, **41–52**), phenylpropanoid glycosides (2, **53–54**), cucurbitacins (7, **55–61**), steroids (8, **62–69**), flavonoids (6, **70–75**), alkaloids (3, **76–78**) and other miscellaneous compounds (20, **79–98**) were isolated from *B. monnieri* (Table 8.1). Of these 98 compounds, 46 were reported as new molecules (Table 8.2).

The pharmacological effects of dammarane-type triterpenoid saponins (mainly bacosides) have been investigated and reviewed by various authors.[13,15,21,29–35,99,100] Other chemical entities (Table 8.1) in *B. monnieri* are relatively less explored.[34] In the context of this review, the chemical profiles and molecular entities of *B. monnieri* in Ayurvedic formulations are least investigated so far; and this is one of the priority areas toward justifying their efficacy.

8.4 Standardization of *Brahmi*-based and Other Ayurvedic Formulations

Phytochemical studies revealed dammarane-type triterpenoid saponins (**1–34**) as the major secondary metabolites in *B. monnieri*, of which bacosides

Table 8.2 Structures of new molecules reported from *B. monnieri*.[a]

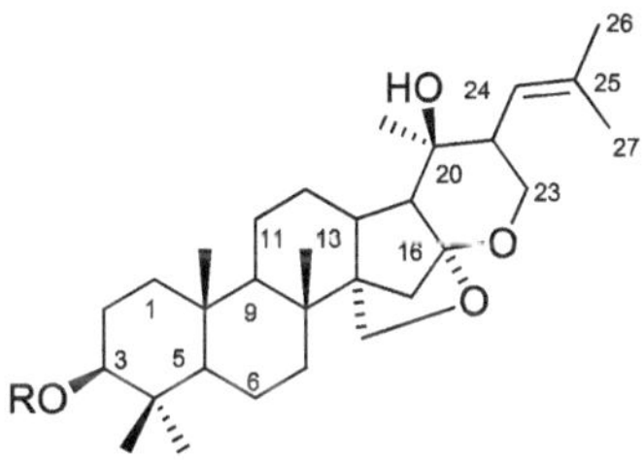

Pseudojujubogenin derivatives

No.	Compound	Substitution (R)
1.	Bacopaside I	α-L-araf-(1→2)-[6-*O*-sulfonyl-β-D-glc-(1→3)]-α-L-ara
2.	Bacopaside II	α-L-araf-(1→2)-[β-D-glc-(1→3)]-β-D-glc
3.	Bacopaside III	[6-*O*-Sulfonyl-β-D-glc-(1→3)]-α-L-ara
4.	Bacopaside V	β-D-glc-(1→3)-α-L-ara
5.	Bacopaside XI	[α-L-araf-(1→3)]-6-*O*-sulfonyl-β-D-glc
6.	Bacopaside N2	[β-D-glc-(1→3)-β-D-glc]
8.	Bacopasaponin B	[α-L-araf-(1→2)-α-L-ara]
9.	Bacopasaponin C	[β-D-glc-(1→3)-{α-L-araf-(1→2)}-α-L-ara]
10.	Bacopasaponin D	[α-L-araf-(1→2)-β-D-glc]
11.	Bacopasaponin H	[α-L-ara]
13.	Bacoside A_2	[α-L-araf-(1→6)-*O*-{α-L-ara-(1→5)}-*O*-α-D-glcf]

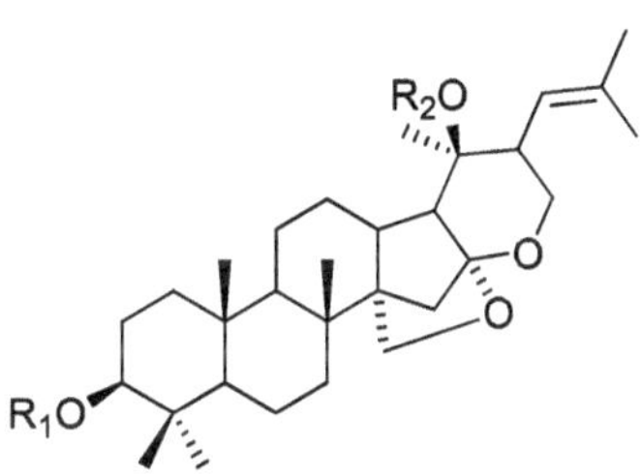

		R_1	R_2
7.	Bacopaside XII	[β-D-glc-(1→3)-{α-L-araf (1→2)}-β-D-glc]	α-L-ara

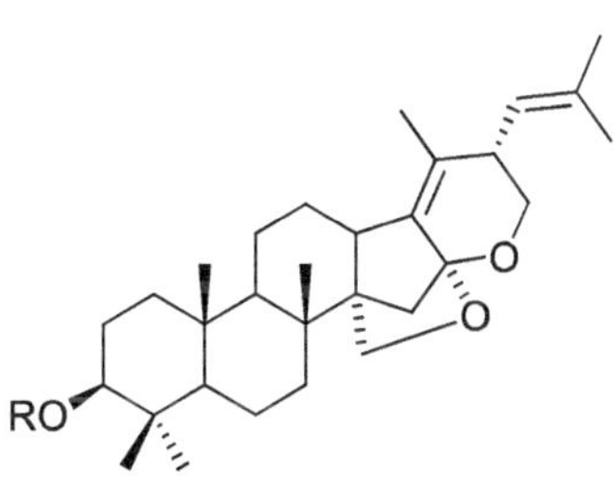

Table 8.2 (*Continued*)

		R
14.	Bacopasaponin G	[β-D-glc-(1→3)-β-D-glc]
15.	Bacopasaponin I	[β-D-glc-(1→3)-*O*-{α-L-araf-(1→2)}-*O*-β-D-glc]

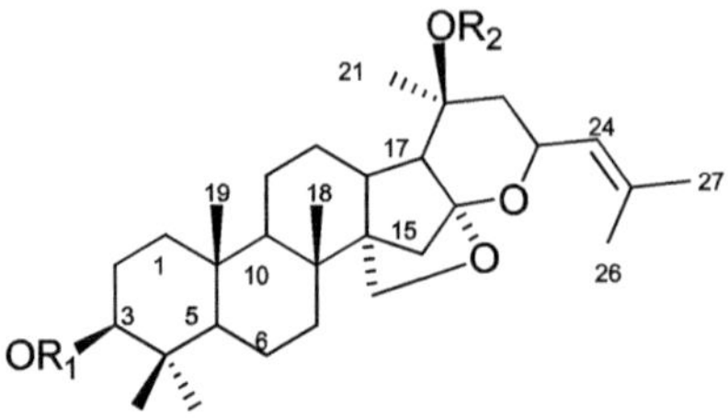

Jujubogenin derivatives

		R_1	R_2
16.	Bacopaside III	α-L-araf-(1→2)-β-D-glc	H
17.	Bacopaside IV	β-D-glc-(1→3)-α-L-ara	H
18.	Bacopaside VIII	[β-D-glc-(1→3)-{α-L-araf-(1→2)}-β-D-glc]	α-L-ara
19.	Bacopaside IX	[β-D-glc-(1→4)-{α-L-araf-(1→2)}-β-D-glc]	α-L-ara
21.	Bacopaside X	[β-D-glc-(1→2)]-α-L-ara	H
22.	Bacopaside N1	[β-D-glc-(1→3)-β-D-glc]	H
23.	Bacopasaponin A	α-L-ara	α-L-ara
24.	Bacopasaponin E	[β-D-glc-(1→3)-{α-L-araf-(1→2)}-α-L-ara]	α-L-ara
25.	Bacopasaponin F	[β-D-glc-(1→3)-{α-L-araf-(1→2)}-β-D-glc]	α-L-ara
26.	Bacopasaponin G	[α-L-araf-(1→2)]-α-L-ara	H
28.	Bacoside A_1	[α-L-araf-(1→3)-α-L-ara]	H
29.	Bacoside A_3	[β-D-glc-(1→3)-*O*-{α-L-araf-(1→2)}-*O*-β-D-glc]	H
31.	Bacoside A_5	H	α-L-ara
33.	Bacomosaponin A	[β-D-glc-(1→3)-{α-L-araf-(1→2)}-β-D-glc]	3,4-Diacetyl-α-L-ara
34.	Bacomosaponin B	[β-D-glc-(1→3)-{α-L-araf-(1→2)}-α-L-ara]	3,4-diacetyl-α-L-ara

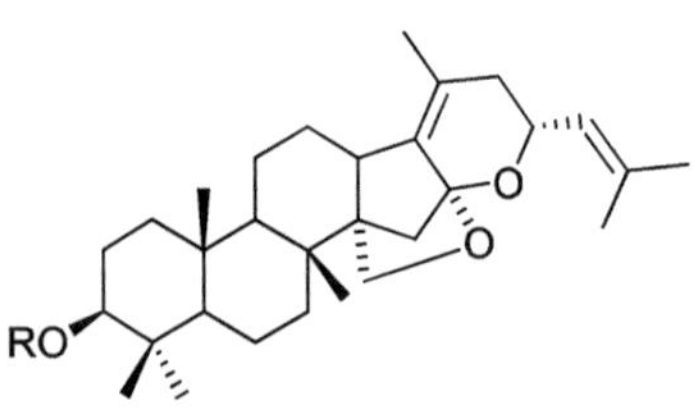

32. Bacoside A_6 (R = [β-D-glc-(1→3)-*O*-{α-L-araf-(1→2)}-*O*-β-D-glc])

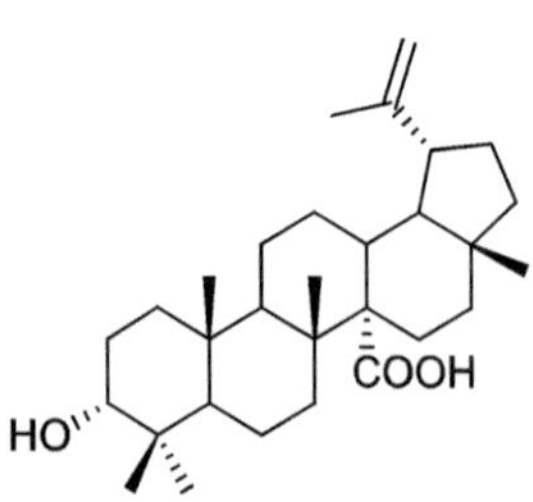

Pentacyclic triterpene

39. Bacosine

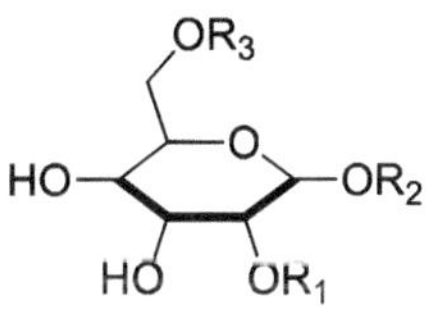

Phenylethyl glycosides		R₁	R₂	R₃
41.	Bacopaside A	H	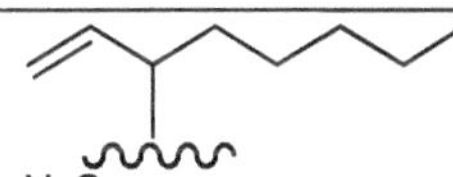	SO_3H
42.	Bacopaside B	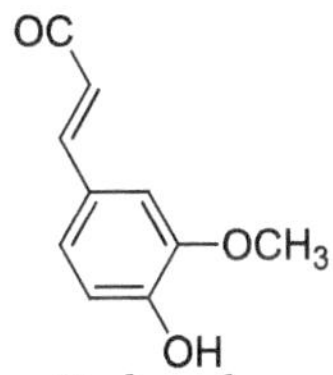	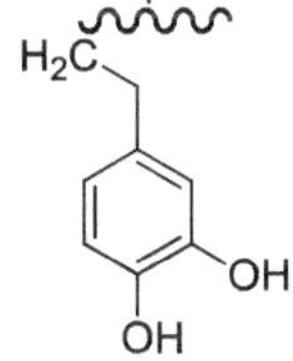	H
43.	Bacopaside C	*p*-Hydroxybenzoyl-β-D-apiofuranosyl	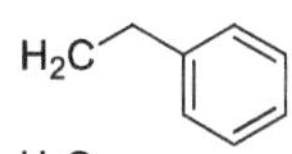	H
44.	Monnieraside I	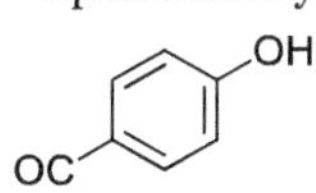	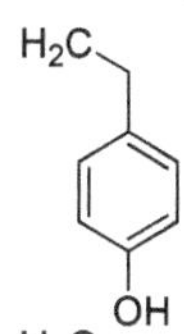	H
45.	Monnieraside II	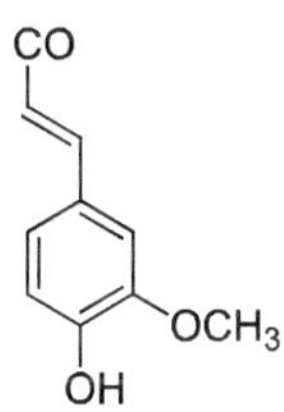	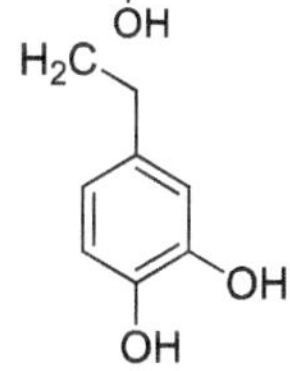	H
46.	Monnieraside III	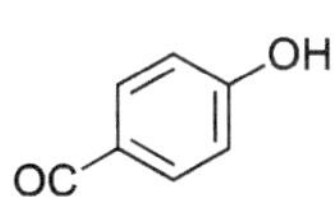	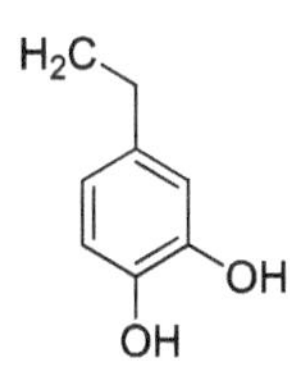	H
49.	Bacomoside A	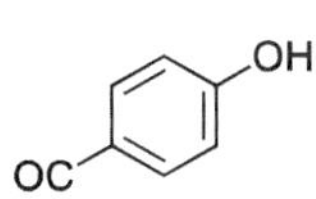	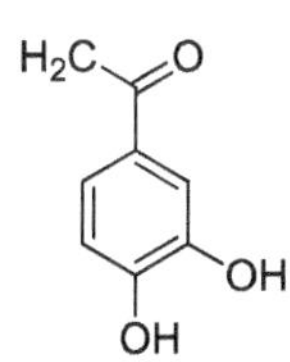	H

Table 8.2 (*Continued*)

50.	Bacomoside B_1[a]	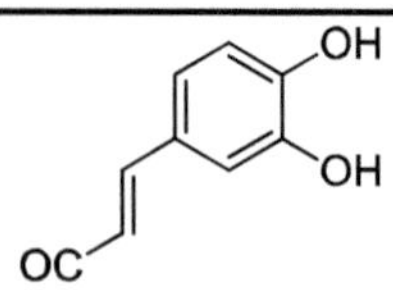	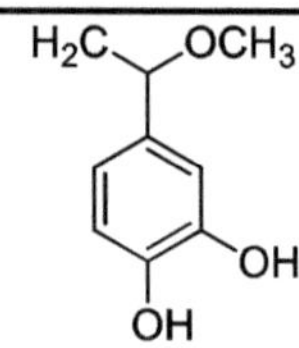	H
51.	Bacomoside B_2[a]	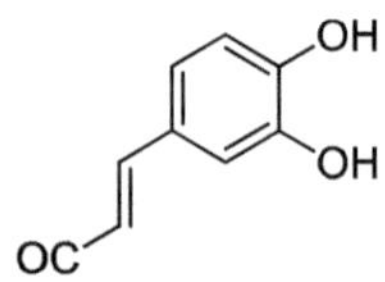	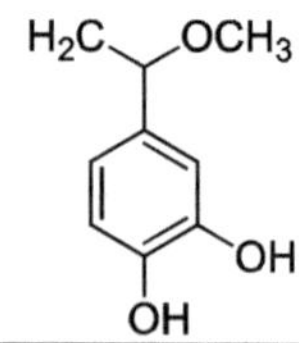	H

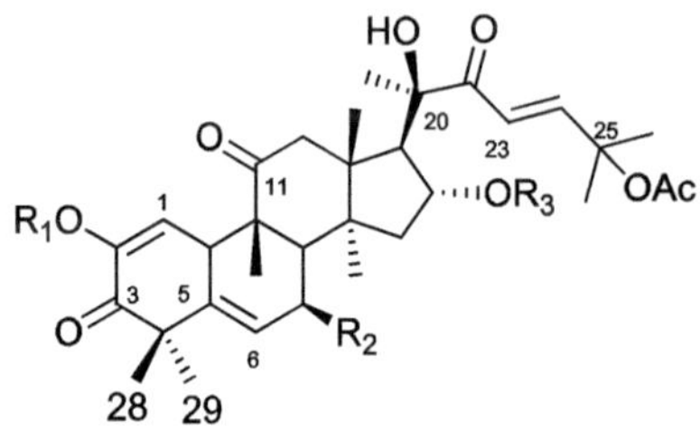

Cucurbitacins			
	R_1	R_2	R_3
55. Bacobitacin A	H	OH	H
56. Bacobitacin B	H	H	Ac
57. Bacobitacin C	[{α-L-rha-(1→2)-α-L-araf-(1→4)}-{α-L-rha-(1→2)-α-L-ara}]	H	H
58. Bacobitacin D	[{α-L-rha-(1→2)-α-L-araf-(1→4)}-{α-L-rha-(1→2)-α-L-ara}]	OH	H

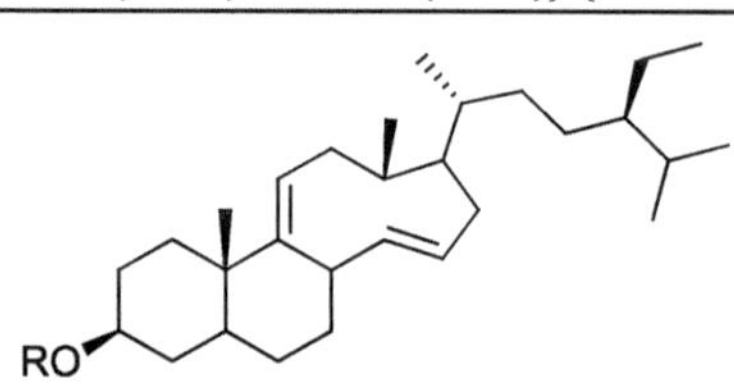

Steroids	
62.	Bacosterol (R = H)
66.	Bacosterol-3-*O*-β-D-glucopyranoside (R = glc)

[a]Ohta *et al.*,[68] reported two molecules bacomoside B_1 and B_2 (**50, 51**) with optical rotation $[\alpha]_D^{27}+5.1$ and $[\alpha]_D^{27}+22.0$, respectively, but their structures are the same. (ara: arabinopyranosyl; araf: arabinofuranosyl; glc: glucopyranosyl; glcf: glucofuranosyl; rha: rhamnopyranosyl.)

are the most studied biologically active entities (Table 8.1). The neuroprotective and memory-enhancing activities of bacosides are widely studied;[26,33] but the bioactive entities in most *Brahmi*-based Ayurvedic formulations are not elucidated. Despite a long history, one of the major drawbacks of Ayurvedic formulations is the lack of knowledge on specific chemical entities illustrating their biological activities. For example, this review discusses recent studies on the mechanism of neuroprotection and memory enhancement of *B. monnieri*-based formulations such as *Brahmi ghrita*, *Brahmi rasayana*, *Brahmi vati* and *Saraswata ghrita*, but their 'bioactive chemical entities' are not described; rather their activities are attributed to synergism of their herbal ingredients. Similarly, the efficacy of most Ayurvedic polyherbal formulations is credited to the synergistic effects of their multiple herbal ingredients.[26,31] Ayurvedic medicines/formulations offer a contrasting logic that they are multifarious integrated derivatives involving several distinct preparatory steps; and single structural entities may not provide full insight into their efficacy or mode of action.[8] However, in this advanced era, chemical fingerprinting and knowledge on individual bioactive structural entities of *Brahmi*-based and other Ayurvedic formulations could establish a strong evidence-based rationale for these ancient herbal formulations and practices.[9,34,38,48] Again, systematic studies are also essential on the bioavailability and toxicity of Ayurvedic herbal formulations.[4,35,101,102]

Ayurvedic medicine systems are not '*disease-centric*' but '*individual-centric*',[103] and this limits the applicability of medications on a general basis. Again, Ayurvedic formulations are prepared by traditional processing methods which involve the use of several herbs and minerals; and they are available in various forms such as decoction, juice, vati, oil, powder, clarified butter and alcoholic preparations.[48,104] Several studies have reported inconsistencies in the quantity and quality of these Ayurvedic formulations. In a recent example, Mishra and co-workers attempted standardization of various in-house and market samples of *Brahmi vati*, with bacoside A_3 (*B. monnieri*) and piperine (*P. longum*) as standards using HPLC and found inconsistencies in the contents of these standards, inferring non-uniformity of their therapeutic effects.[48] In a pharmacognostical and physicochemical analysis, Gubbannavar and co-workers detected bacoside in *Brahmi ghrita* by HPTLC-densitometry and observed consistency of the formulation.[34,105] Contrastingly, Saini and co-workers observed significant batch-to-batch variations in the quality and quantity of four *Brahmi*-based formulations marketed by different manufacturers. Bacoside A contents quantified by HPTLC in these formulations varied between batches of the same manufacturer and between manufacturers. The authors recommended stringent guidelines, strict implementation and awareness in consumers.[106]

The critically important parameters deciding the efficacy of these drugs are plant species identification, collection location, growth stage, adulteration, quality control standards and processing methods; and these multiple factors necessitate systematic chemical fingerprinting and quality control at various stages of production of Ayurvedic drugs.[12] Briefly, Ayurvedic drugs are to be regulated at three levels by standardization of (i) raw materials, (ii) methods of production and tools used and (iii) finished products.[9] The role of advanced phytochemical tools in quality control of traditional medicines has been reviewed recently.[107–109] Elite lines of *Brahmi* and other herbs used for Ayurvedic drugs with standardized agro-practices could offer consistency in herbal raw materials.[16,110–113] Moreover, the World Health Organization (WHO) recommends stringent assessment of heavy metals and pesticide residues in Ayurvedic products.[114] Medicinal plants having elevated levels of heavy metals may affect the central nervous system and various organs, and cause serious health hazards. Therefore, it must be mandatory to evaluate the heavy metal and pesticide residues in every batch of Ayurvedic drugs. This will help in consumer protection, and improve their export potentials.[115] Patwardhan *et al.*[116] and Rastogi[103] are recent articles dealing with the safety concerns in regard to Ayurvedic formulations. Genomic and metabolomic approaches could strengthen the scientific basis of Ayurvedic concepts and drugs.[11,108] So far, studies on human trials with strong experimental designs and models on Ayurvedic drugs are limited,[9,37,102] and efforts in these lines need to be intensified. These scientific approaches could unlock the possibility of integrating Ayurveda treatments/drugs with modern medicine.

8.5 Conclusions and Future Perspectives

B. monnieri, which has been recorded in the *Charaka Samhita* and *Susruta Samhita*, is now widely used in various Ayurvedic formulations. Its phytochemistry so far has led to the isolation of 98 secondary compounds, and the major bioactive constituents are bacosides (dammarane-type triterpenoid saponins). However, most *Brahmi*-based Ayurvedic medications and their biological effects are not defined based on structural entities. This is a major drawback of Ayurvedic medications. Chemical profiling (fingerprinting) of raw drugs and processed drugs is a priority, ensuring the quality and consistency of Ayurvedic drugs. The bioavailability and toxicity of Ayurvedic drugs must be systematically assessed. Similarly, their heavy metal and pesticide residues are to be estimated to avoid health risks. Briefly, modern scientific tools could play a crucial role in authenticating the long traditions of Ayurveda.

Abbreviations

ADHD	Attention deficit hyperactivity disorder
ASD	Autism spectrum disorder
BCE	Before the Common Era
DST	Department of Science and Technology
EHTN	Essential hypertension
GABA	Gamma amino butyric acid
HPLC	High-performance liquid chromatography

HPTLC	High-performance thin-layer chromatography
MES	Maximal electroshock
NMPB	National Medicinal Plant Board
TIFAC	Technology Information, Forecasting and Assessment Council
UFLC/MS	Ultra-fast liquid chromatography/mass spectrometry
WHO	World Health Organization

Acknowledgements

RK received Junior and Senior Research Fellowships from the Programme Support project (BT/PR12720/COE/34/21/2015/A 193 B1) funded by the Department of Biotechnology, Government of India.

References

1. B. Patwardhan, A. D. B. Vaidya and M. Chorghade, *Curr. Sci.*, 2004, **86**, 789.
2. C. P. Khare, *Encyclopedia of Indian Medicinal Plants: Rational Western Therapy, Ayurvedic and Other Traditional Usage, Botany*, Springer, Berlin, 2004.
3. M. M. Hossain, *Fitoterapia*, 2011, **82**, 102.
4. T. Dubey and S. Chinnathambi, *Arch. Biochem. Biophys.*, 2019, **676**, 108153.
5. S. A. Dahanukar and U. M. Thatte, *Phytomedicine*, 1997, **4**, 359.
6. S. Dev, *Environ. Health Perspect.*, 1999, **107**, 783.
7. P. K. Mukherjee, R. K. Harwansh, S. Bahadur, S. Banerjee, A. Kar, J. Chanda, S. Biswas, S. M. Ahmmed and C. K. Katiyar, *J. Ethnopharmacol.*, 2017, **197**, 10.
8. V. Dwivedi, E. M. Anandan, R. S. Mony, T. S. Muraleedharan, M. S. Valiathan, M. Matsuddi and S. C. Lakhotia, *PLoS One*, 2012, 7, e37113.
9. R. H. Singh, Exploring the medicinal value of green wealth with special reference to neuronutrient medhya rasayana plant drugs, in *New Age Herbals*, ed. B. Singh and K. Peter, Springer, Singapore, 2018.
10. A. Chopra and V. V. Doiphode, *Med. Clin. North Am.*, 2002, **86**, 75.
11. P. Govindaraj, S. Nizamuddin, A. Sharath, V. Jyothi, H. Rotti, R. Raval, J. Nayak, B. K. Bhat, B. V. Prasanna, P. Shintre, M. Sule, K. S. Joshi, A. P. Dedge, R. Bharadwaj, G. G. Gangadharan, S. Nair, P. M. Gopinath, B. Patwardhan, P. Kondaiah, K. Satyamoorthy, M. V. S. Valiathan and K. Thangaraj, *Sci. Rep.*, 2015, **5**, 15786.
12. Y. S. Jaiswal and L. L. Williams, *J. Tradit. Complementary Med.*, 2017, 7, 50–53.
13. V. C. Sekhar, G. Viswanathan and S. Baby, *Curr. Neuropharmacol.*, 2019, **17**, 438.
14. V. C. Sekhar, G. Viswanathan and S. Baby, *Phytomedicine Plus*, 2021, **1**, 100040.

15. A. Russo and F. Borrelli, *Phytomedicine*, 2005, **12**, 305.
16. C. Christopher, A. J. Johnson, P. J. Mathew and S. Baby, *Ind. Crops Prod.*, 2017, **98**, 76.
17. K. S. Chaudhari, N. R. Tiwari, R. R. Tiwari and R. S. Sharma, *Ann. Neurosci.*, 2017, **24**, 111.
18. M. D. Nemetchek, A. A. Stierle, D. B. Stierle and D. I. Lurie, *J. Ethnopharmacol.*, 2017, **197**, 92.
19. S. Chobe, S. K. Patra, M. Chobe and K. Metri, *J. Ayurveda Integr. Med.*, 2020, **S0975-9476**(20), 30106.
20. M. Dhanasekaran, B. Tharakan, L. A. Holcomb, A. R. Hitt, K. A. Young and B. V. Manyam, *Phytother. Res.*, 2007, **21**, 965.
21. S. Nandy, A. Mukherjee, D. K. Pandey and A. Dey, *Bacopa monnieri*: The neuroprotective elixir from the East – phytochemistry, pharmacology, and biotechnological improvement, in *Bioactive Natural Products in Drug Discovery*, ed. J. Singh, V. Meshram and M. Gupta, Springer, Singapore, 2020.
22. S. G. Silpa, G. R. Smitha and K. Ranjitha, *Ind. Crops Prod.*, 2021, **159**, 113064.
23. Anonymous, The Ayurvedic Pharmacopoeia of India, in *Ministry of Health and Family Welfare Department of Ayush*, Government of India, New Delhi, 2015, Part I, vol. II.
24. A. Karthikeyan, A. Madhanraj, S. K. Pandian and M. Ramesh, *Genet. Resour. Crop Evol.*, 2011, **58**, 769.
25. N. Tripathi, D. S. Chouhan, N. Saini and S. Tiwari, *3 Biotech*, 2012, **2**, 327.
26. T. Joshi, A. Gupta, P. Kumar, A. Singh and A. Kumar, *Bacopa monnieri* (Brahmi), in *Naturally Occurring Chemicals Against Alzheimer's Disease*, ed. T. Belwal, S. M. Nabavi, S. F. Nabavi, A. R. Dehpour and S. Shirooie, Academic Press, London, 2021, ch. 3.2.4.
27. M. Banerjee and S. Shrivastava, *World J. Microbiol. Biotechnol.*, 2008, **24**, 1355.
28. D. K. Ved and G. S. Goraya, *Demand and Supply of Medicinal Plants in India*, National Medicinal Plants Board, New Delhi & Foundation for Revitalization of Local Health Traditions, Bangalore, India, 2007.
29. K. Chowdhuri, D. Parmar, P. Kakkar, R. Shukla, P. K. Seth and R. C. Srimal, *Phytother. Res.*, 2002, **16**, 639.
30. S. Roodenrys, D. Booth, S. Bulzomi, A. Phipps, C. Micallef and J. Smoker, *Neuropsychopharmacology*, 2002, **27**, 279.
31. S. Aguiar and T. Borowski, *Rejuvenation Res.*, 2013, **6**, 313.
32. N. P. Sukumaran, A. Amalraj and S. Gopi, *Complementary Ther. Med.*, 2019, **44**, 68.
33. P. Bhandari, N. Sendri and S. B. Devidas, *Phytochemistry*, 2020, **172**, 112276.
34. S. Banerjee, U. Anand, S. Ghosh, D. Ray, P. Ray, S. Nandy, G. D. Deshmukh, V. Tripathi and A. Dey, *Phytother. Res.*, 2021, **35**, 5668.

35. V. T. Shalini, S. J. Neelakanta and J. S. Sriranjini, *Mol. Biol. Rep.*, 2021, **48**, 2653.
36. R. H. Singh, K. Narsimhamurthy and G. Singh, *Biogerontology*, 2008, **9**, 369.
37. M. S. Baliga, S. Meera, A. R. Shivashankara, P. L. Palatty and R. Haniadka, The health benefits of Indian traditional Ayurvedic rasayana (anti-aging) drugs: A review, in *Foods and Dietary Supplements in the Prevention and Treatment of Disease in Older Adults*, ed. R. R. Watson, Elsevier, Amsterdam, 2015.
38. G. M. Doshi, H. D. Une and P. P. Shanbhag, *Pharmacogn. Rev.*, 2013, **7**, 92.
39. R. H. Singh and S. Rastogi, Rasayana therapy and rejuvenation, in *Evidence-Based Practice in Complementary and Alternative Medicine: Perspectives, Protocols, Problems, and Potential in Ayurveda*, ed. S. Rastogi, F. Chiappelli, M. H. Ramchandani and R. H. Singh, Springer-Verlag, Berlin, 2012, ch. 10.
40. K. D. Yadav, K. R. C. Reddy and V. Kumar, *Ayu*, 2014, **35**, 325.
41. G. S. Achliya, U. Barabde, S. Wadodkar and A. Dorle, *Indian J. Pharmacol*, 2004, **36**, 159.
42. S. Bhalerao, R. Munshi, T. Nesari and H. Shah, *Anc. Sci. Life*, 2013, **33**, 123.
43. K. D. Yadav, K. Reddy and V. Kumar, *Anc. Sci. Life*, 2012, **32**, 11.
44. Y. S. Deole and H. M. Chandola, *Ayu*, 2008, **29**, 207.
45. H. Joshi and M. Parle, *Evidence-Based Complementary Altern. Med.*, 2006, **3**, 79.
46. P. Jain, N. K. Khanna, N. Trehan, V. K. Pendse and J. L. Godhwani, *Indian J. Exp. Biol.*, 1994, **32**, 633.
47. B. Shukia, N. K. Khanna and J. L. Godhwani, *J. Ethnopharmacol.*, 1987, **21**, 65.
48. A. Mishra, A. K. Mishra, O. P. Tiwari and S. Jha, *J. Young Pharm.*, 2013, **5**, 77.
49. A. Mishra, A. K. Mishra and S. Jha, *Epilepsy Behav.*, 2018, **80**, 144.
50. D. Mishra and B. R. Tubaki, *J. Ayurveda Integr. Med.*, 2019, **10**, 269.
51. M. P. Mamidala, N. Rajesh and V. Rajesh, *Rapid Commun. Mass Spectrom.*, 2016, **30**, 1413.
52. R. R. Parekar, K. S. Jadhav, P. A. Marathe and N. N. Rege, *J. Ayurveda Integr. Med.*, 2014, **5**, 141.
53. S. A. Giramkar, O. P. Kulkarni, S. D. Jagtap, A. A. Kuvalekar, S. Mukherjee, R. R. Jagtap, A. R. Wagh, D. D. Bandawane and S. S. Nipate, *J. Pharm. Res.*, 2013, 7, 787.
54. K. C. Bose and N. K. Bose, *J. Indian Med. Assoc.*, 1931, **1**, 60.
55. M. S. Sastri, N. S. Dhalla and C. L. Malhotra, *Indian J. Pharmacol.*, 1959, **21**, 303.
56. K. E. Schulte, G. Ruecker and M. El-Kersch, *Phytochemistry*, 1972, **11**, 2649.
57. N. Chatterji, R. P. Rastogi and M. L. Dhar, *Indian J. Chem.*, 1965, **3**, 24.

58. K. I. Kawai and S. Shibata, *Phytochemistry*, 1978, **17**, 287.
59. A. K. Chakravarty, T. Sarkar, K. Masuda, K. Shiojima, T. Nakane and N. Kawahara, *Phytochemistry*, 2001, **58**, 553.
60. A. K. Chakravarty, T. Sarkar, K. Masuda, K. Shiojima, T. Nakane and N. Kawahara, *Phytochemistry*, 2002, **59**, 365.
61. C. Sivaramakrishna, C. V. Rao, G. Trimurtulu, M. Vanisree and G. V. Subbaraju, *Phytochemistry*, 2005, **66**, 2719.
62. P. Bhandari, N. Kumar, B. Singh and V. K. Kaul, *Chem. Pharm. Bull.*, 2006, **54**, 240.
63. Y. Zhou, Y. H. Shen, C. Zhang, J. Su, R. H. Liu and W. D. Zhang, *J. Nat. Prod.*, 2007, **70**, 652.
64. Y. Zhou, L. Peng, W. D. Zhang and D. Y. Kong, *Planta Med.*, 2009, **75**, 568.
65. M. Ganzera, J. Gampenrieder, R. S. Pawar, I. A. Khan and H. Stuppner, *Anal. Chim. Acta*, 2004, **516**, 149.
66. M. Deepak, G. K. Sangli, P. C. Arun and A. Amit, *Phytochem. Anal.*, 2005, **16**, 24.
67. R. S. Pawar, S. I. Khan and I. A. Khan, *Planta Med.*, 2007, **73**, 380.
68. T. Ohta, S. Nakamura, S. Nakashima, Y. Oda, T. Matsumoto, M. Fukaya, M. Yano, M. Yoshikawa and H. Matsuda, *J. Nat. Med.*, 2016, **70**, 404.
69. C. Hou, S. J. Lin, J. T. Cheng and F. L. Hsu, *J. Nat. Prod.*, 2002, **65**, 1759.
70. A. K. Chakravarty, S. Garai, K. Masuda, T. Nakane and N. Kawahara, *Chem. Pharm. Bull.*, 2003, **51**, 215.
71. P. Bhandari, N. Kumar, B. Singh and I. Kaur, *Can. J. Chem.*, 2009, **87**, 1230.
72. P. H. Dang, H. K. T. Pham, T. H. N. Phan, K. D. H. Nguyen, M. T. T. Nguyen and N. T. Nguyen, *Chem. Nat. Compd.*, 2018, **54**, 124.
73. S. Garai, S. B. Mahato, K. Ohtani and K. Yamasaki, *Phytochemistry*, 1996, **42**, 815.
74. S. Mandal and S. Mukhopadhyay, *Indian J. Chem.*, 2004, **43B**, 1802.
75. R. Pawar, C. Gopalakrishnan and K. K. Bhutani, *Planta Med.*, 2001, **67**, 752.
76. S. Garai, S. B. Mahato, K. Ohtani and K. Yamasaki, *Phytochemistry*, 1996, **43**, 447.
77. S. Rastogi and D. K. Kulshreshtha, *Indian J. Chem.*, 1999, **38B**, 353.
78. Y. Zhou, L. Peng and W. D. Zhang, *Chin. Chem. Lett.*, 2009, **20**, 569.
79. M. Deepak and A. Amit, *Fitoterapia*, 2013, **87**, 7.
80. S. B. Mahato, S. Garai and A. K. Chakravarty, *Phytochemistry*, 2000, **53**, 711.
81. P. Jain and D. K. Kulshreshtha, *Phytochemistry*, 1993, **33**, 449.
82. S. Rastogi, R. Pal and D. K. Kulshreshtha, *Phytochemistry*, 1994, **36**, 133.
83. R. S. Pawar and K. K. Bhutani, *Indian J. Chem.*, 2006, **45B**, 1511.
84. Y. Zhou, Y. H. Shen, C. Zhang and W. D. Zhang, *Chem. Nat. Compd.*, 2007, **43**, 355.
85. S. K. Maurya, D. P. Pandey, J. P. Singh and V. B. Pandey, *Pharmazie*, 1995, **50**, 372.

86. P. K. Chaudhuri, R. Srivastava, S. Kumar and S. Kumar, *Phytother. Res.*, 2004, **18**, 1147.
87. V. Viji, B. Shobha, S. K. Kavitha, M. Ratheesh, K. Kripa and A. Helen, *Int. Immunopharmacol.*, 2010, **10**, 843.
88. B. Ahmed and A. Rahman, *Indian J. Chem.*, 2000, **39B**, 620.
89. L. Kishore, N. Kaur and R. Singh, *J. Funct. Foods*, 2017, **34**, 237.
90. A. K. Chakaravarty, T. Sarkar, N. Takahisa, N. Kawahara and K. Masuda, *Chem. Pharm. Bull.*, 2002, **50**, 1616.
91. P. Bhandari, N. Kumar, B. Singh and V. K. Kaul, *Phytochemistry*, 2007, **68**, 1248.
92. Anonymous, *The Wealth of India: Raw Materials*, Council of Scientific and Industrial Research, New Delhi, 1988, vol. 2B.
93. *Quality Standards of Indian Medicinal Plants*, ed. T. Neeraj and S. Madhu, Indian Council of Medical Research, New Delhi, 2010, vol. 8.
94. A. Suresh, X. Q. R. Sheela, R. Kanmani, C. Mani, E. Lalith, A. L. Stanley and V. A. Ramani, *Asian J. Chem.*, 2010, **22**, 965.
95. T. Dutta and U. P. Basu, *Indian J. Chem.*, 1963, **1**, 400.
96. U. P. Basu and T. Dutta, *Tetrahedron Lett.*, 1967, **8**, 2937.
97. N. Basu, R. P. Rastogi and M. L. Dhar, *Indian J. Chem.*, 1967, **5**, 84.
98. H. K. Singh, R. P. Rastogi, R. C. Srimal and B. N. Dhawan, *Phytother. Res.*, 1988, **2**, 70.
99. K. Mathur, V. Goyal, Koul and A. Anand, *Front. Pharmacol.*, 2016, **7**, 44.
100. S. Ghosh, R. Khanam and A. A. Chowdhury, *Nutr. Cancer*, 2021, **73**, 2166.
101. A. A. Farooqui, T. Farooqui, A. Madan, J. H.-J. Ong and W.-Y. Ong, *Evidence-Based Complementary Altern. Med.*, 2018, **2018**, 2481076.
102. A. Ranade, M. Surana, S. V. Dhokne, S. Gaidhani and S. D. Pawar, Ayurvedic ideology on Rasapanchak-based cognitive drug intervention, in *Medicinal Herbs and Fungi*, ed. D. C. Agrawal and M. Dhanasekaran, Springer, Singapore, 2021.
103. S. Rastogi, *J. Ayurveda Integr. Med.*, 2018, **9**, 81.
104. S. Parasuraman, G. S. Thing and S. A. Dhanaraj, *Pharmacogn. Rev.*, 2014, **8**, 73.
105. J. S. Gubbannavar, H. Chandola, C. R. Harisha, R. Kalyani and V. J. Shukla, *Ayu*, 2012, **33**, 289.
106. N. Saini, R. Mathur and S. S. Agrawal, *Indian J. Pharm. Sci.*, 2012, **74**, 24.
107. G. A. Cordell and M. D. Colvard, *J. Nat. Prod.*, 2012, **75**, 514.
108. G. A. Cordell, *Phytochem. Lett.*, 2014, **10**, xxviii–xl.
109. G. A. Cordell, *Nat. Prod. Comm.*, 2017, **12**, 1371.
110. M. T. Thomas, R. Kurup, A. J. Johnson, S. P. Chandrika, P. J. Mathew, M. Dan and S. Baby, *Ind. Crops Prod.*, 2010, **32**, 545.
111. R. Rajan, S. C. Varghese, R. Kurup, R. Gopalakrishnan, R. Venkataraman, K. Satheeshkumar and S. Baby, *Ind. Crops Prod.*, 2013, **43**, 472.
112. R. Rajan, S. C. Varghese, R. Kurup, R. Gopalakrishnan, R. Venkataraman, K. Satheeshkumar and S. Baby, *Cogent Chem.*, 2016, **2**, 1275408.

113. H. Pulikkalpura, R. Kurup, P. J. Mathew and S. Baby, *Sci. Rep.*, 2015, **5**, 11078.
114. World Health Organization, *WHO guidelines for assessing quality of herbal medicines with reference to contaminants and residues*, World Health Organization, Geneva, 2007.
115. A. Mishra, A. K. Mishra, O. P. Tiwari and S. Jha, *J. Integr. Med.*, 2016, **14**, 44.
116. K. Patwardhan, J. Pathak and R. N. Acharya, *J. Ayurveda Integr. Med.*, 2017, **8**, 279.

CHAPTER 9

Therapeutic Potential of Triphala in Ayurveda and Its Phytochemical Studies

A. WELE ASMITA,*[a] B. KOPPIKAR SOUMYA,[b] NIMKAR GAURI[a] AND BENDALE RAMA[a]

[a] Bharati Vidyapeeth (Deemed to be University), College of Ayurveda, Katraj-Dhankawadi Campus, Pune-Satara Road, Pune 411043, India;
[b] Independent Researcher, Santa Clara, California, USA
*Email: asmitawele.ayurveda@unideb.hu

9.1 Triphala: The Three Myrobalans

Triphala, recently included in the American Herbal Pharmacopoeia and Therapeutic Compendium, is a well-known indigenous poly-herbal formulation consisting of three myrobalans, namely *Amalaki, Bibhitak* and *Haritaki*.[1] In the *Sanskrit* language *tri* means three and *phala* means fruit. It is widely used for the maintenance of health and longevity. It is classified as a *tridoshic rasayana*[1] in Ayurveda as it endorses rejuvenation in patients of all constitutions and ages. Triphala is comprised of the dried pericarp of the fruits of three plants, *Phyllanthus emblica* L. (Euphorbiaceae), *Terminalia bellirica* (Gaertn.) Roxb. (Combretaceae) and *Terminalia chebula* Retz. (Combretaceae) mixed in equal quantities (1 : 1 : 1).[2] The traditional treatises further specify that for curative purposes the proportion is 1 : 1 : 1 (*Charak, Sushruta, Bhavamishra*), while for *rasayana* purposes it is 1 : 2 : 4 (*Charak, Sharangdhar, Yogaratnakara*). Triphala is an easily available, uncomplicated, multitargeted, efficacious formulation with few side effects,

Chemistry, Biological Activities and Therapeutic Applications of Medicinal Plants in Ayurveda
Edited by Augustine Amalraj, Sasikumar Kuttappan and Karthik Varma A. C.

Published by the Royal Society of Chemistry, www.rsc.org

and has been thoroughly studied by Indian and Western scientists alike. The epistemological differences of indigenous medical science Ayurveda and biomedicine, at times, make it difficult to translate the exact mode of all attributes and activities of Triphala; however the effort is worthwhile. In this chapter, the Ayurvedic perspectives of the three plants, known bioactive components of the formulation, *in vitro* and *in vivo* experiments, and clinical trials oriented to specific diseases are discussed in brief.

9.2 Ayurvedic Perspective on Triphala

Ayurveda pharmacology *Dravyagunavigyan* explains the tastes (*rasa*), potency (*virya*), post-digestive effect (*vipaka*), attributes (*guna*) and actions (*karma*) of each plant, and of some prominent herbal formulations like Triphala. The functions are inferred based on the rasa and attributes. Triphala is a *pancharasa* formulation, having five tastes, *viz.* sour, astringent, pungent, bitter and sweet. The only missing taste is salt. Its post-digestive effect is sweet and potency neutral.[1] It is a perfect balance of heavy and light along with dry attributes. The formulation balances, restores and rejuvenates all three *doshas*, namely *vata, pitta* and *kapha*, wherein *Amalaki* works on *Pittadosha*, *Bibhitaka* on *Kaphadosha*, and *Haritaki* on *Vatadosha*.[3,4] The synergistic action of these three ingredients is unique to Triphala.

Both conventional medicine and Ayurveda have endorsed that health and disease initiate in the gut.[5,6] Triphala promotes absorption, digestion, elimination and rejuvenation. Numerous references in well-respected Ayurvedic medical texts make clear that Triphala is well-regarded as a multiuse therapeutic and perhaps the best remedy traditionally.[5,7] Triphala is a well-known Ayurvedic herb constituent of about 1500 Ayurveda formulations and is widely used in several diseases, especially for anaemia, abdominal pain, blood purification, cough, upper respiratory diseases, dental, skin, constipation, malignancies, orbital oedema, liver disorders, peptic and mouth ulcers, obesity, wound healing and toxic conditions[2] (Figure 9.1 and Table 9.1).

It has been used as a stand-alone formulation in powder, decoction, oil, ghee and *mashi* forms.[10] Triphala decoction and ghee are used as accompaniments (*anupan*) to the main drug and have interesting, targeted effects. Triphala powder with sugar works on acid peptic disease, while with hot water it is used as a mild laxative and with oil in non-healing ulcers.[11] The most widely used formulations containing Triphala are *Triphala Ghrita* for vision loss or retinopathy, and *Triphala Guggul* for joint diseases, haemorrhoids/fistula, sciatica and other musculoskeletal/neurovegetative diseases (Figure 9.2). *Triphala Guggul* inhibits the hyaluronidase and collagenase enzymes that degrade cartilage matrix, and also inhibits protein fibrillation responsible for neurodegeneration.[12,13] Another important formulation is *Triphala mashi* wherein Triphala powder is carbonised at 450 °C and is

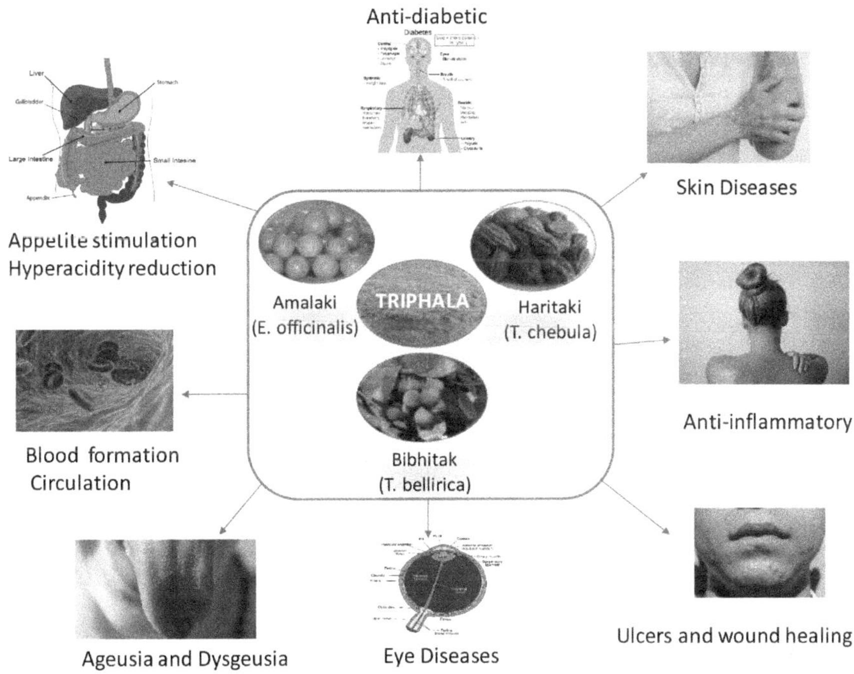

Figure 9.1 Therapeutic uses of Triphala.

traditionally used to treat hiccups and dyspnoea. It has been reported to be non-toxic and has antimicrobial and antidiarrhoeal activities.

It is used to process inorganic metals and minerals such as iron, zincite, biotite mica, *etc.*, which belong to the *Rasaushadhi* class of Ayurvedic drugs (Table 9.2). The conversion of inorganic metals/minerals into biologically active nanomolecules with a therapeutic effect is brought about by various processes including quenching, amalgamation, trituration and heating. It has been reported that during the specific processes various phytochemicals find their way to bind at different sites of the metal molecule which can be traced using FTIR-like techniques.[14–16]

Triphala has been studied for its biological activity through cellular, molecular and animal experiments (Figure 9.3). It is reported to possess anti-arthritic, antibacterial, analgesic, antiaging, antiallergic, anti-inflammatory, antimicrobial, anticancer, antifungal, antiviral, apoptogenic, cardiotonic, hypoglycaemic, immunomodulatory, antimutagenic, antimalarial, hepatoprotective and capillary-strengthening activities.[18–30]

9.3 Bioactive Components of Triphala

The bioactive herbs are commonly found under the Universal Natural Products Database (UNPD). According to the UNPD, Triphala herbs contain 114 bioactives from *Phyllanthus emblica* L., 25 from *Terminalia bellirica* and

Table 9.1 Therapeutic uses of Triphala from Ayurvedic treatises.[a]

No.	*Phyllanthus emblica* L.[8]	*Terminalia bellirica*[9]	*Terminalia chebula*[9]
1	Impaired digestion (ACB-23) (*Agnimandya*)	Cough (EA-3) and dyspnoea (EA-4) (*Kasa, shwasa*)	Intermittent pyrexia (EC-3.0) (*Visham jwara*)
2	Anorexia (EC-7) (*Aruchi*)	Diarrhoea (EB-2) (*Atisara*)	Filariasis (ED-13) (*Shipada*)
3	Constipation (AAC-12.4) (*Vibandha*)	Oedema (EK-3) (*Shotha*)	Diarrhoea (EB-2) (*Atisara*)
4	Liver disorders (Yakrut Vikar)	Vitiligo (ED-5) (*Shwitra*)	Dysentery (EB-9) (*Pravahika*)
5	Hyperacidity (EB-4) (*Amlapitta*)	Lithiasis (EJ-2) (*Ashmari*)	Constipation (AAC-12.4) (*Vibandha*)
6	Ascites (EK-4) (*Udara*)	Acute spreading eruptions (ED-10) (*Visarpa*)	Impaired digestion (ACB-23) (*Agnimandya*)
7	Obstipation (EB-11) (*Udavarta*)	Corneal ulcer (HE-1) (*Savrana shukla*)	Indigestion (EB-3) (*Ajirna*)
8	Haemorrhoid (EE-3) (*Arsha*)	—	Vomiting (EB-6) (*Chardi*)
9	Heart diseases (EC-2) (*Hridrog*)	—	Hyperacidity (EB-4) (*Amlapitta*)
10	Epistaxis/bleeding disorder (ED-7) (*Raktapitta*)	—	Obstipation (EB-11) (*Udavarta*)
11	Anaemia (EC-5) (*Pandu*)	—	Ascites (EK-4) (*Udara*)
12	Cough (EA-3) and dyspnoea (EA-4) (*Kasa, Shwasa*)	—	Disorders of lower G.I.T. (EB-7) (*Grahani*)
13	Consumption (BI) (*Rajayakshma*)	—	Duodenal ulcer (EB-10.10) (*Parinamshool*)
14	Improves memory (*Smriti and medha vardhak*)	—	Lump (EB-8) (*Gulma*)
15	Menorrhagia (EL-4) (*Raktapradara*)	—	Hepatomegaly (EK-4.6) (*Yakruddaludar*)
16	Infertility (EI-1) (*Garbhashaya daurbalya*)	—	Anaemia (EC-5) (*Pandu*)
17	Spermaturia (EF-2.2.6) (*Shukrameha*)	—	Jaundice (ED-3) (*Kamala*)
18	Dysuria (EJ-4) (*Mutrakruchchra*)	—	Obesity (EF-3) (*Medoroga*)
19	Diabetes (EF-2.3) (*Pittaj Prameha*)	—	Diabetes (EF-2) (*Prameha*)
20	Acute spreading eruptions (ED-10) (*Visarpa*)	—	Rheumatism (EC-6) (*Amavata*)
21	Integumentary diseases (ED-4) (*Kushtha*)	—	Sciatica (AAC-20) (*Grudhrasi*)
22	Chronic fever (*Jeerna jwara*)	—	Rheumatism (ED-8) (*Vatarakta*)
23	Polidypsia (EK-2) (*Trishna*)	—	Oedema (EK-3) (*Shotha*)
24	Burning sensation (ED-20) (*Daha*)	—	Integumentary diseases (ED-4) (*Kushtha*)
25	Rejuvenation (*Rasayana*)	—	Syphilis (EI-5) (*Upadansh*)

Table 9.1 (*Continued*)

No.	*Phyllanthus emblica* L.[8]	*Terminalia bellirica*[9]	*Terminalia chebula*[9]
26	—	—	Pytiriasis capitis (FA-4) (*Darunak*)
27	—	—	Rejuvenation (*Rasayana*)

[a]The letters and numbers in brackets show the NAMC (National Ayurveda Morbidity Code).

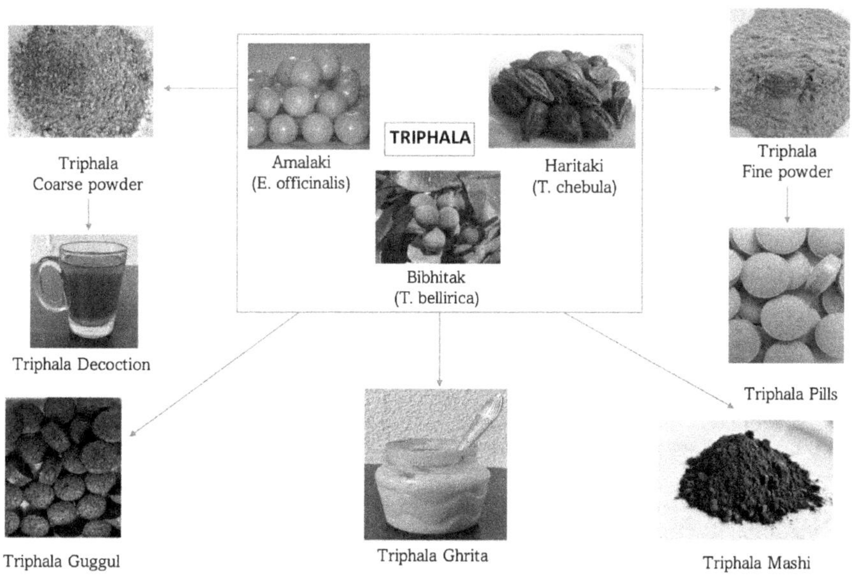

Figure 9.2 Different dosage forms of Triphala. All images are taken from pixabay.com.

63 from *Terminalia chebula*. A few bioactives are common to all three herbs. Altogether, the Triphala formulation comprises of 177 bioactive compounds. These bioactive data further help to build human and microbial proteome targeting networks. The bioactives are scored from 0.9–1, based on the maximum binding closeness between the ligand and target protein, where score 1 denotes a higher binding database (DB). According to June 2015 binding DB search, Triphala has 36 bioactives scoring 1, as depicted in Figures 9.4–9.6[31] and Table 9.3.

9.4 Effects of Triphala on Various Diseases

9.4.1 Anticancer Activity and Triphala

Triphala has been used for centuries to treat different ailments. Numerous reports of recent *in vitro*, *in vivo* and clinical studies on Triphala have

Table 9.2 Triphala as a processing formulation for inorganic drug sources.[17]

No	Name of drug	Name of the process	Form of Triphala as a processing ingredient
For processing			
1	*Abhrak* (biotite mica)	*Shodhana* (detoxification) by *Nirvap* (quenching)	*Triphala quath*
2	*Abhrak* (biotite mica)	*Maran* (incineration)	*Triphala quath*
3	*Vaikrant* (tourmaline)	*Satwapatan* (extraction)	*Triphala churna*
4	*Shilajeet* (black bitumen)	Detoxification	*Triphala quath*
5	*Rasak* (zinc carbonate)	Extraction	*Triphala churna*
8	*Loha* (iron)	Detoxification by Quenching followed by *bhavana* (trituration)/boiling/sun drying followed by incineration	*Triphala quath*
9	*Mandoor* (iron oxide)	Detoxification by quenching	*Triphala quath*
For internal use			
1	*Shilajeet* (black bitumen)	Mixture	*Triphala churna*
2	*Rasak* (zinc carbonate)	Mixture	*Triphala quath*
3	*Gandhak* (sulfur)	Mixture	*Triphala churna*
4	*Gandhak* (sulfur) (*Kushthahar yog*)	Mixture	*Triphala churna*
5	*Kasis* (ferrous sulfate)	Mixture	*Triphala churna*
6	*Rajat* (silver)	Mixture	*Triphala churna*
7	*Nag*	Mixture	*Triphala quath*

demonstrated its protective and therapeutic efficacy against malignancies such as melanoma, breast, colon, cervical, endometrial, pancreatic, prostatic, ovarian and lymphatic cancers (Table 9.4).[13,32–41] On the other hand, experimental studies have also revealed that through modulation of various signalling molecules, Triphala displays antineoplastic, chemoprotective and radioprotective effects.[42]

Different cancer cell lines have been tested using Triphala where it has shown an anticancer effect. The chemical analysis of Triphala (aqueous extract) showed that it was rich in gallic acid (a polyphenol), which is considered as a major factor in inducing cytotoxicity of breast and prostate cancer cells.[40] Sandhya *et al.*[30] presented that the increasing concentrations of Triphala consistently decreased the viability and induced apoptosis in treated breast cancer cells. Interestingly, normal breast epithelial cells, human peripheral blood

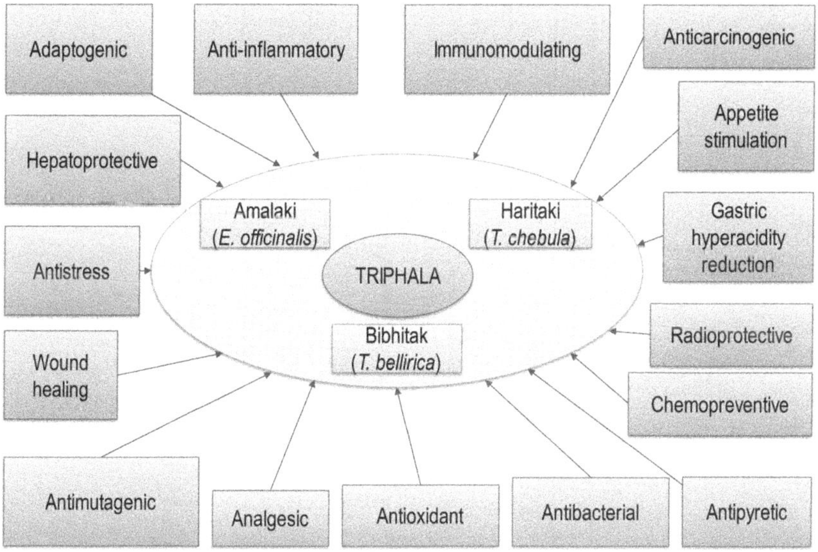

Figure 9.3 Biological activities of Triphala.

mononuclear cells, and mouse liver and spleen cells did not show a cytotoxic effect or DNA damage with the same concentrations of Triphala. Hence, this study revealed that Triphala was selectively cytotoxic to cancer cells.[30]

Triphala has cancer chemo-preventive potential as shown in animal studies. This effect was elucidated by the molecular mechanism against human pancreatic cancer in the cellular and *in vivo* model. Oral administration of Triphala (50–100 mg kg^{-1}) significantly reduces the growth of Capan-2 pancreatic tumour-xenograft. It was found that a decrease in tumour growth by Triphala in mice was due to increased apoptosis in tumour cells, which was associated with elevated stimulation of p53 gene and extracellular-signal-regulated kinase (ERK).[43] Another *in vivo* study demonstrated that the tumour inhibitory effects of Triphala or its active ingredients were enabled through the suppression of vascular endothelial growth factor (VEGF). Triphala and one of its active ingredients, chebulinic acid, specifically prevent VEGF-induced angiogenesis by suppressing VEGF receptor-2 (VEGFR-2) phosphorylation and thus reduce tumour growth and metastasis.[44] Triphala treatment in a zebrafish xenograft model inhibited the growth and metastasis of transplanted gastric carcinoma cells. The antineoplastic effect of Triphala was assessed by western blotting and the results demonstrated that it inhibited phosphorylation of epidermal growth factor receptor (EGFR), protein kinase B (PKB, also known as AKT) and ERK.[38]

9.4.2 Immunomodulatory Effect and Triphala

Triphala has been shown to strengthen the immune system and act as an immunomodulatory agent. A review of the literature suggests that the

Table 9.3 The chemical constituents, score-1 bioactives and their corresponding protein targets of Triphala.

Chemical constituents of Triphala Ingredients		Score-1 bioactives of Triphala	Protein targets of Triphala
P. emblica[32,33]	Ascorbic acid	Catechin	Alkaline phosphatase (ALPL)
	Catechol	Chebulanin	Beta-secretase 1 (BACE1)
	Ellagic acid	Ellagic acid	Carbonic anhydrase 7 (CA7)
	Flavonoids	Epicatechin	Coagulation factor X (F10)
	Gallic acid	Gallocatechin	DNA repair protein RAD51 homolog 1 (RAD51)
	Pectin	Gallic acid	GSTM1 protein (GSTM1)
	Phyllaemblic compounds	Kaempferol	Plasminogen activator inhibitor 1 (SERPINE1)
	Quercetin	Methyl gallate	Prothrombin (F2)
	Rutin	Quercetin	Regulator of G-protein signalling (RGS) 4, 7, and 8, tissue-type plasminogen activator (PLAT)
	Tannins	Tannic acid	Tyrosine–protein phosphatase nonreceptor type 2 (PTPN2)
	Vitamin C	1,2,3,6-Tetrakis-(*O*-galloyl)-beta-D-glucose	
		1,2,3,4,6-Pentagalloylglucose	
		1,6-Digalloyl-beta-D-glucopirano-side	
		Trans- 3,3′,4′,5,7-pentahydroxylflavane	
T. bellirica[34]	Ellagic acid		
	Ethyl acid		
	Ethyl gallate		
	Gallic acid		
	Galloyl glucose		
	Chebulaginic acid		
	β-Sitosterol		

T. chebula(35–37)	1,2,3,4,6-Penta-*O*-galloyl-β-D-glucose
	1,6-Di-*o*-galloyl-D-glucose
	3,4,6-Tri-*o*-galloyl-D-glucose
	Anthraquinones
	Casuarinin
	Chebulagic acid Punicalagin
	Chebulanin
	Chebulinic acid
	Corilagin
	Ellagic acid
	Gallic acid
	Galloyl glucose
	Maslinic acid
	Neochebulinic acid
	Punicalagin
	Terchebulin
	Terflavin A

1.

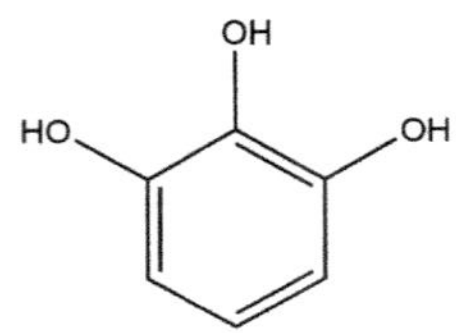

benzene-1,2,3-triol;methanesulfonic acid (1,2,3-benzenetriol)

2.

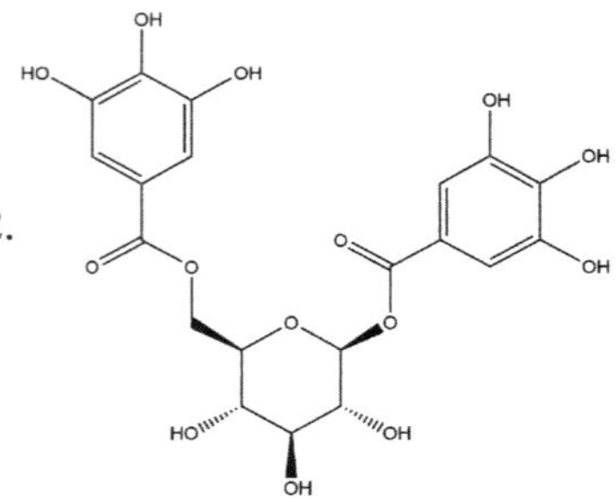

[(2R,3S,4S,5R,6S)-3,4,5-trihydroxy-6-(3,4,5-trihydroxybenzoyl)oxyoxan-2-yl]methyl 3,4,5-trihydroxybenzoate (1,6-digalloyl-beta-Dglucopirano-side)

3.

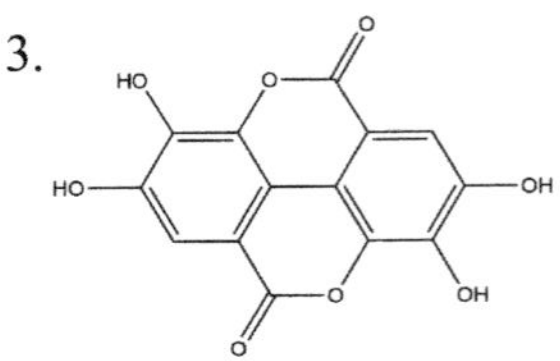

2,3,7,8-Tetrahydroxy[1]benzopyrano[5,4,3-cde][1]benzopyran-5,10-dione (ellagic acid)

4.

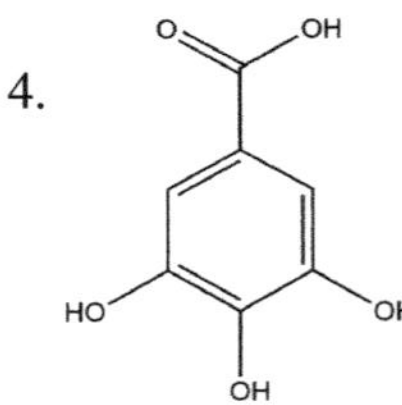

3,4,5-trihydroxybenzoic acid (gallic acid)

5.

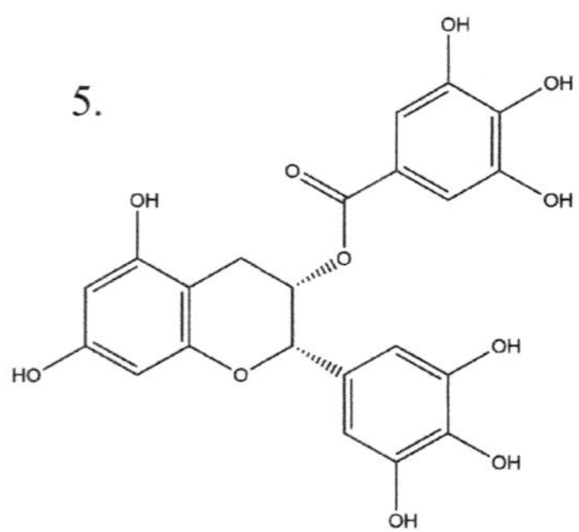

(2R,3R)-5,7-Dihydroxy-2-(3,4,5-trihydroxyphenyl)-3,4-dihydro-2H-1-benzopyran-3-yl 3,4,5-trihydroxybenzoate (epigallocatechin-3-gallate)

6.

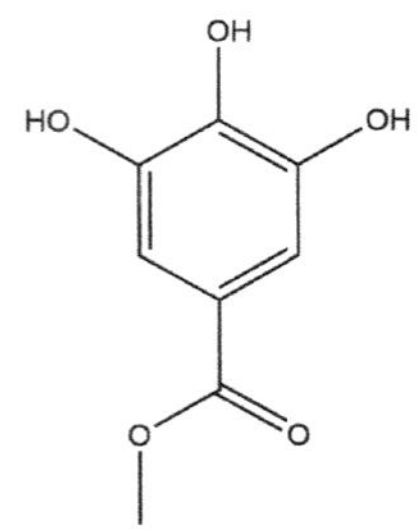

methyl 3,4,5-trihydroxybenzoate (methyl gallate)

7.

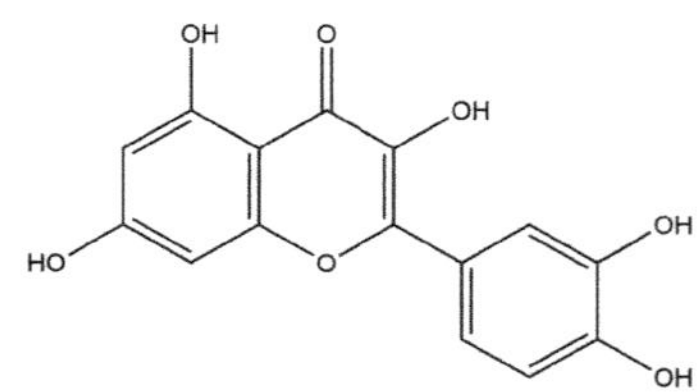

2-(3,4-dihydroxyphenyl)-3,5,7-trihydroxychromen-4-one (quercetin)

8.

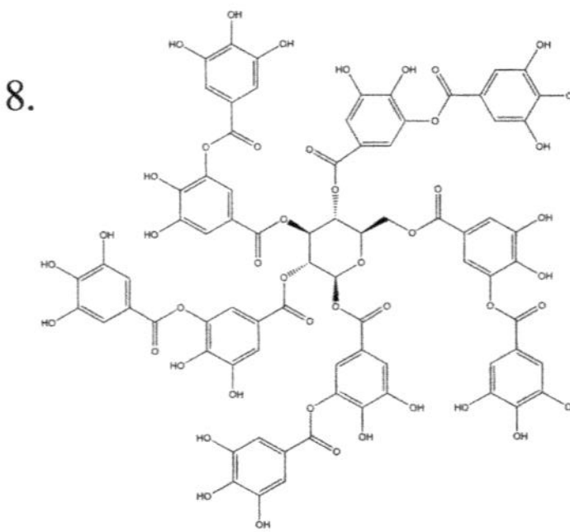

[2,3-dihydroxy-5-[[(2R,3R,4S,5R,6S)-3,4,5,6-tetrakis[[3,4-dihydroxy-5-(3,4,5-trihydroxybenzoyl)oxybenzoyl]oxy]oxan-2-yl]methoxycarbonyl]phenyl] 3,4,5-trihydroxybenzoate (tannic acid)

Figure 9.4 Chemical structures of active compounds of Triphala, part I.

1.
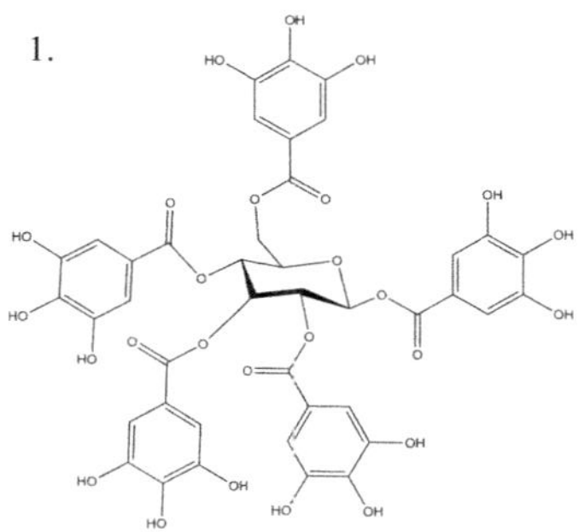
[(2R,3R,4S,5R,6S)-3,4,5,6-tetrakis[(3,4,5-trihydroxybenzoyl)oxy]oxan-2-yl]methyl 3,4,5-trihydroxybenzoate (1,2,3,4,6-pentagalloylglucose)

2.
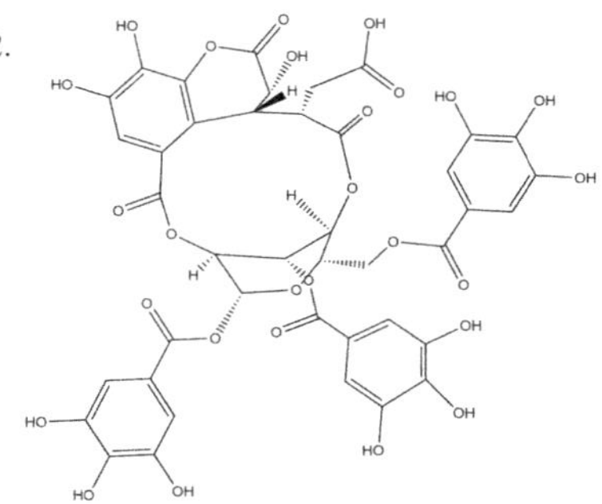
2-[(4R,5S,7R,8R,11S,12S,13S,21S)-13,17,18-trihydroxy-2,10,14-trioxo-5,21-bis[(3,4,5-trihydroxybenzoyl)oxy]-7-[(3,4,5-trihydroxybenzoyl)oxymethyl]-3,6,9,15-tetraoxatetracyclo[10.7.1.14,8.016,20]henicosa-1(19),16(20),17-trien-11-yl]acetic acid (chebulinic acid)

3.
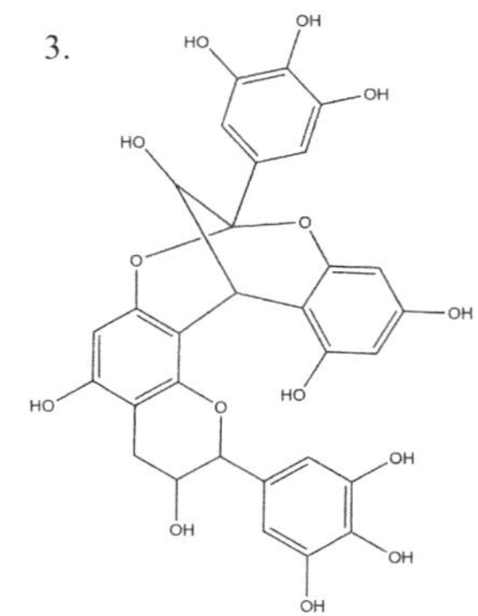
5,13-bis(3,4,5-trihydroxyphenyl)-4,12,14-trioxapentacyclo[11.7.1.02,11.03,8.015,20]henicosa-2(11),3(8),9,15,17,19-hexaene-6,9,17,19,21-pentol (prodelphinidin A1)

4.
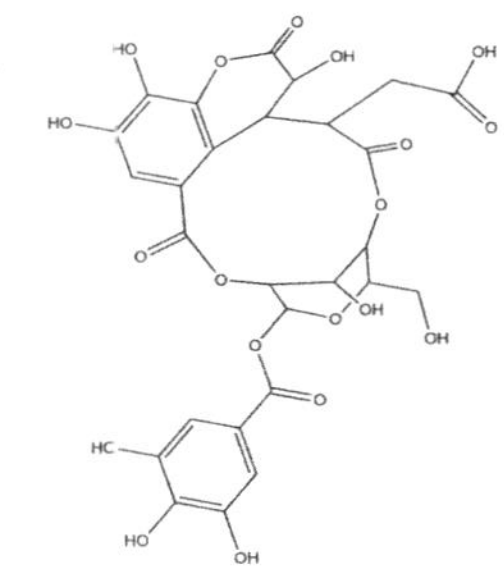
2-[13,17,18,21-tetrahydroxy-7-(hydroxymethyl)-2,10,14-trioxo-5-(3,4,5-trihydroxybenzoyl)oxy-3,6,9,15-tetraoxatetracyclo[10.7.1.14,8.016,20]henicosa-1(19),16(20),17-trien-11-yl]acetic acid (chebulanin)

5.
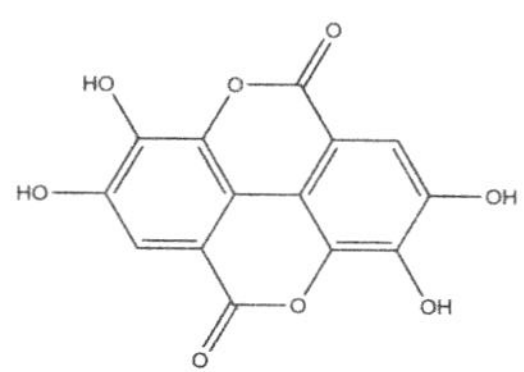
2,3,7,8-Tetrahydroxy[1]benzopyrano[5,4,3-cde][1]benzopyran-5,10-dione (ellagic acid)

6.
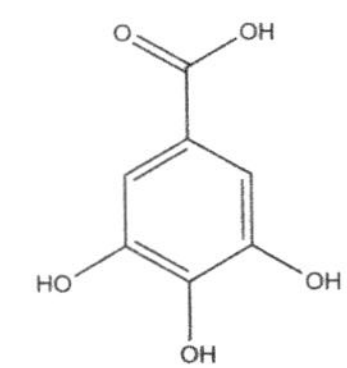
3,4,5-trihydroxybenzoic acid (gallic acid)

7.
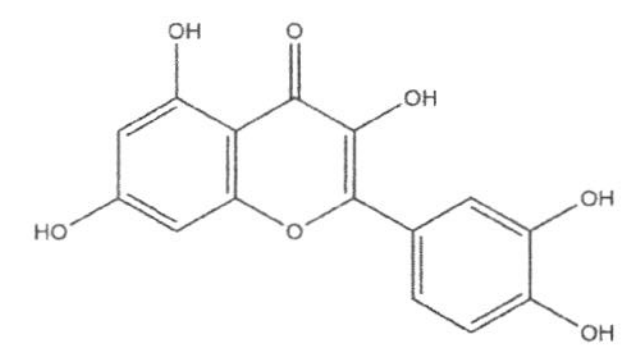
2-(3,4-dihydroxyphenyl)-3,5,7-trihydroxychromen-4-one (quercetin)

8.
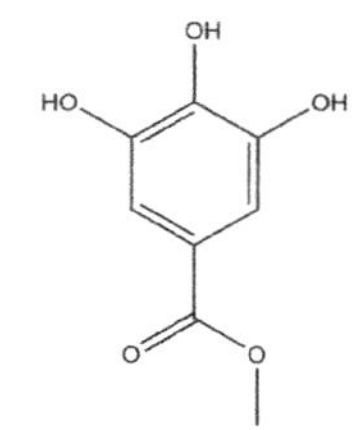
methyl 3,4,5-trihydroxybenzoate (methyl gallate)

Figure 9.5 Chemical structures of active compounds of Triphala, part II.

1.

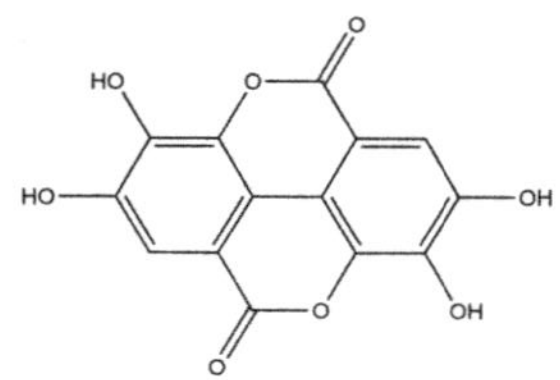

2,3,7,8-Tetrahydroxy[1]benzopyrano [5,4,3-cde][1]benzopyran-5,10-dione (ellagic acid)

2.

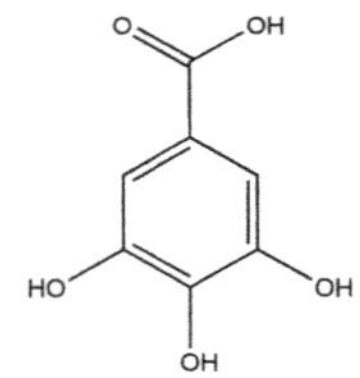

3,4,5-trihydroxybenzoic acid (gallic acid)

3.

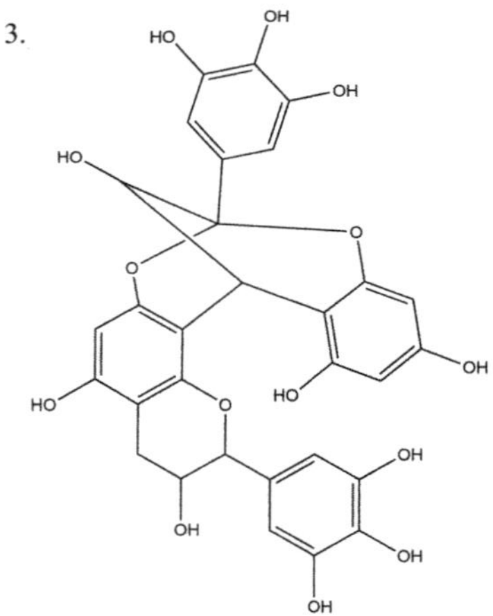

5,13-bis(3,4,5-trihydroxyphenyl)-4,12,14-trioxapentacyclo[11.7.1.02,11.03,8.015,20]henicosa-2(11),3(8),9,15,17,19-hexaene-6,9,17,19,21-pentol (prodelphinidin A1)

4.

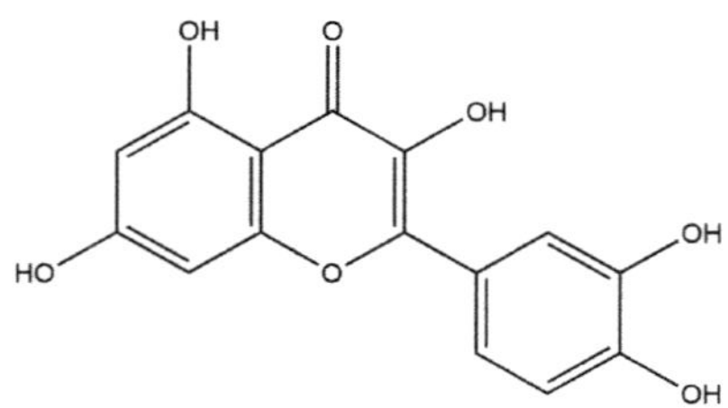

2-(3,4-dihydroxyphenyl)-3,5,7-trihydroxychromen-4-one (quercetin)

5.

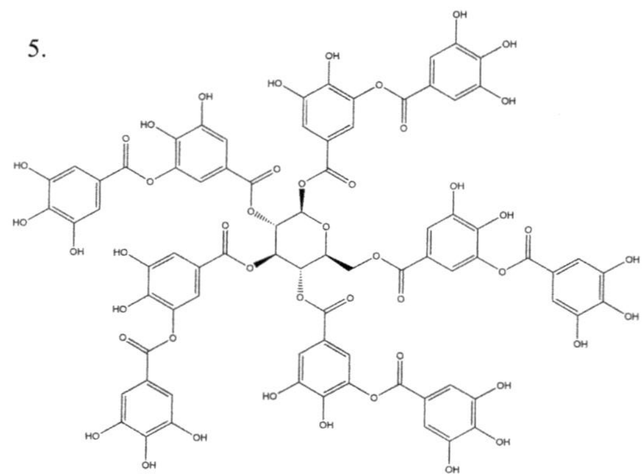

[2,3-dihydroxy-5-[[(2R,3R,4S,5R,6S)-3,4,5,6-tetrakis[[3,4-dihydroxy-5-(3,4,5-trihydroxybenzoyl)oxybenzoyl]oxy]oxan-2-yl]methoxycarbonyl]phenyl] 3,4,5-trihydroxybenzoate (tannic acid)

Figure 9.6 Chemical structures of active compounds of Triphala, part III.

Table 9.4 Triphala's significant bioactives and therapeutic targets involved in various types of cancers.[31,45]

Therapeutic target and bioactives of Triphala in cancer		
Organ types	Bioactives of Triphala	Protein targets
Breast	1,2,3-Benzenetriol	BACE1
Colorectal	Ellagic acid	Cellular antigen p53 (TP53)
Gastric	Quercetin	EGFR
Lung		F2
Pancreatic	Prodelphinidin A1	Insulin-like growth factor-1 receptor (IGF1R)
Prostate	—	Interleukin 6 (IL-6)
—	—	Matrix metalloproteinase-2 (MMP2)
—	—	MMP-9
—	—	RAD51

Table 9.5 The protein targets and important bioactives of Triphala involved in immunomodulatory disorders.[51]

Immunomodulatory disorders	Important bioactives	Protein targets
Autoimmune diseases[(51)]	4-*O*-Methygallic acids	Cyclooxygenase-2 (COX-2)
—	Bellericanin	IL-1β
—	Chebulagic	IL-2
—	Ellagic acid	IL-4
—	Gallic acid	IL-6
—	—	Interferon gamma (IFN-γ)
—	—	Tumour necrosis factor alpha (TNF-α)

immunomodulatory activity of Triphala was elucidated by testing the various functions of neutrophil-like adherence, and avidity index in animal studies (Table 9.5). The activity index was found to be amplified after Triphala treatment in *in vivo* studies. Moreover, neutrophil function was also increased, which reduced the corticosterone level in the Triphala-treated group.[26] Therefore, Triphala shows strong neutrophil stimulation and prevents stress induction in neutrophils in the Triphala-supplemented group. In another report, the Triphala-immunised group prevented noise/stress-induced changes in the cell-mediated immune response in rats.[46] In a preclinical study, the immunostimulatory action of Triphala was evaluated. Consumption of Triphala by healthy volunteers validated significant immunostimulatory effects on cytotoxic T cells ($CD3^{-}$ $CD8^{+}$) and natural killer cells ($CD16^{+}$ $CD56^{+}$). However, Triphala did not affect the cytokine level in volunteers.[47] On the other hand, the individual Triphala components have also exhibited immunomodulatory action. *T. chebula* fruit extract demonstrated an increase in rat spleen lymphocyte proliferation and boosted the cytokines such as IL-2, IL-10 and TNF-α in an animal study.[48] The methanolic extract of *T. bellirica* affected the mouse immune system, specifically both the cellular and humoral immune response in *in vitro*. This extract

Table 9.6 Tabular data representing the therapeutic linkage between score-1 bioactives and the target group of Triphala in cardiovascular diseases.[31]

Cardiovascular diseases	Important bioactives	Protein targets
Atherosclerosis	1,2,3,4,6-Penta-galloylglucose	F2
Cerebral vasospasm	1,2,3,6-Tetrakis-(*O*-galloyl)-beta-D-glucose	F10
Hypertension	Bellericagenin A1	FABP4
Infarction	Quercetin	SERPINE1
Myocardial ischaemia	Prodelphinidin A1	—
Thrombosis	—	—

stimulated phagocytic activity and T-lymphocyte proliferation.[49] Suja *et al.*[50] demonstrated the immunostimulatory action of *E. officinalis* fruit extract on humoral and cell-mediated immune responses with macrophages and phagocytes.[50] In a nutshell, the above studies illustrate that Triphala and its three individual components have the potential to stimulate immune systems.

9.4.3 Cardiovascular Health and Triphala

Cardiovascular disease is a major health condition affecting the heart and blood vessels, and often elevated cholesterol levels (hypercholesterolaemia) are the leading risk factor.[52] Animal experiments have shown that hypercholesteremia can be managed by Triphala (Table 9.6). In a study, rats were fed low-density and very low-density lipoprotein, high-density lipoprotein and free fatty acid diets for 48 days. The results demonstrated that total cholesterol was decreased in the Triphala-administered group.[25] In another study, an atherogenic diet along with *Haritaki* was fed to animals. The results showed a significant decrease in total cholesterol, triglycerides and total protein, and increased high-density lipoprotein cholesterol was found in the Haritaki-treated group as compared to the control group.[53] Triphala addresses disparities in the gastrointestinal tract and cardiovascular system and should be extensively reviewed in the context of these common diseases.

9.4.4 Nervous System and Triphala

Chronic stress or depression and anxiety cause serious effects on the body and mind, and are identified as major causes of disability worldwide.[54] Stress is an imbalance caused by apparent threat that is counteracted by an adaptive response to re-establish homeostasis and is associated with many chronic disorders. *In vivo* studies have reported that Triphala provides protection against cold-induced stress and reversed stress-induced behavioural and biochemical alterations such as elevated lipid peroxidation and corticosterone levels.[55] Triphala also prevented noise-induced stress in animals.[46] Triphala successfully prevented noise-induced metabolic changes by facilitating the cell-mediated immune and antioxidant status.[46,56–59] Thus,

Table 9.7 The therapeutic network addressing the nervous system of Triphala's bioactives and protein targets.[31]

Nervous system disorders	Important bioactives	Protein targets
Alzheimer's	1,2,3,4,6-Pentagalloylglucose	ALDR
Diabetes	1,2,3,6-Tetrakis-(*O*-galloyl)-beta-D-glucose	BACE1
Parkinson's	Epigallocatechin-3-gallate	CA2
Retinopathy	Quercetin	PLAT
—	—	SERPINE1

high levels of stress or depression in humans may be tackled with Triphala treatments in clinical practice, based on its reported adaptogenic activity (Table 9.7).

9.4.5 Metabolic Syndrome and Triphala

Metabolic syndrome is a pathological condition characterised by central obesity, elevated blood pressure, increased lipid levels, and high blood glucose.[60–63] The overall prevalence of metabolic syndrome is estimated to be one-fourth of the global adult population.[61] The scientific literature has also endorsed the potential of Triphala as a therapeutic agent for weight loss and reduction of body fat (Table 9.8). In an animal study, diet-induced obese mice were fed with Triphala for 10 weeks.[64] Triphala administration subsequently decreased the percentage of body fat, body weight and energy intake. Triphala also decreased total cholesterol, triglycerides and low-density lipoprotein cholesterol in the experimental group compared with the controls.[65] In a clinical study ($n = 749$), Triphala-treated groups showed significantly reduced low-density lipoprotein-cholesterol, total cholesterol and triglyceride. Five RCTs have shown that Triphala administration significantly reduces body weight, body mass index (BMI) and waist circumference in obese patients. Moreover, Triphala significantly reduces the fasting blood glucose level in diabetic patients, but not in healthy individuals. Given the global obesity epidemic, more treatment options are necessary to reduce the associated healthcare burden.[65]

Triphala plays a key role in inhibiting starch digestion and absorption, thus reducing postprandial hyperglycaemia. A similar process is carried out by miglitol and acarbose (diabetes pharmaceutical drugs), which correspondingly target these glycolytic enzymes. In addition, Patel *et al.*[66] discovered that Triphala decreases the blood glucose level in alloxan-induced diabetic as well as normal rats.[66] Triphala powder (5 g) was supplemented to diabetes mellitus patients for 45 days in a clinical study. The results elucidated a significant decrease in blood glucose level.[67] Interestingly, both fasting and postprandial blood glucose were reduced, which may be due to active ingredients such as menthol and sorbitol. However, ellagitannins and gallotannins, the other constituents of Triphala, also improve the immune responses for glucose uptake by enhancing both PPAR-alpha and -gamma,

Table 9.8 The relationship between metabolic disorders and the Triphala network.[31]

Therapeutic target and bioactives of Triphala in metabolic syndrome		
Metabolic diseases	Important bioactives	Protein targets
Diabetes	Ellagic acid	ALDR
Endometriosis	Quercetin	AKR1B1
Hypercholesterolaemia	Bellericagenin A1	FABP5
Hyperlipidaemia	—	IGF1R
Gout	—	—
Psoriasis	—	—
Lung fibrosis	—	—
Obesity	—	—
Osteoarthritis	—	—
Ulcerative colitis	—	—

without inducing adipogenesis.[68] Hence, these polyphenols may also be promoted to reduce blood glucose and insulin levels in diabetic patients.

9.5 The Way Ahead for Triphala

The major diseases in the present era are more like syndromes rather than simple diseases. Synthetic molecules are double-edged swords, offering benefits laced with adverse effects. To treat these patients, pharmacy practice is experiencing a change from synthetic chemicals to refined single herbs and polyherbal compounds. Eventually it will find its way to herbo-mineral compounds. The knowledge of *Ayurvedic Rasashastra* (Indian Iatrochemistry) based metal nanomolecules made using Triphala and other herbs will pave the way to a new age of drugs.

The total of 177 bioactives of the formulation Triphala act in unison to eradicate disease, and do not harm healthy tissues at the same time, which is unique to this formulation. Whether or not the change in the proportion of the three plants reflects making it a therapeutic, antiaging, or immunomodulator substance remains to be explored. Apart from powder-filled capsules or compressed tablets and decoctions (*quath*), other dosage forms like *Triphala-ghrita*, oil and *guggul* may be explored for targeted drug delivery for complex and morbid diseases. The *Mashi* dosage form can be developed further using various carbon nanotube development methods for diagnostic and therapeutic purposes. The Ayurveda-described Triphala bio-iron, bio-zinc and bio-copper are a challenging area of drug development for the anti-inflammatory, anticancer, anti-Alzheimer, antiaging, *etc.* activities. The known phytoconstituent profile of Triphala may prove to be beneficial for exploration of its genetic effects in non-communicable diseases. Simple *in vitro* experiments, preclinical and clinical trials of Triphala, and its advanced dosage forms such as Triphala Guggul with various accompaniments including water, honey, sugar, ghee and oil for indicated disease conditions would substantiate the therapeutic potential of this formulation.

Abbreviations

AKT	Ak strain transforming
BMI	Body mass index
CD-3	Cluster of differentiation-3
DB	Database
Dravyagunavigyan	Indian Pharmacology
EGFR	Epidermal growth factor receptor
ERK	Extracellular signal-regulated kinase
FTIR	Fourier transform infrared spectroscopy
IL-2	Interleukin-2
P53	Tumour protein P53
PKB	Protein kinase B
PPAR	Peroxisome proliferator-activated receptor
Rasa shastra	Indian iatrochemistry
RCT	Randomised control trial
TNFα	Tumour necrosis factor alpha
UNPD	Universal Natural Products Database
VEGF	Vascular endothelial growth factor
VGEFR	Vascular endothelial growth factor receptor

Acknowledgements

The authors acknowledge all the known and unknown sources used in this chapter.

References

1. AHP Monographs – Triphala – American Herbal Pharmacopoeia. Available from: https://herbal-ahp.org/ahp-monographs-triphala/ (accessed 2021 Aug 3).
2. A. Manoraj, V. Thevanesam, B. M. R. Bandara, A. Ekanayake and V. Liyanapathirana, *BMC Complementary Altern. Med.*, 2019, **19**, 191.
3. C. T. Peterson, K. Denniston and D. Chopra, *J. Altern. Complement. Med.*, 2017, **23**, 607.
4. S. Ahmed, X. Ding and A. Sharma, *J. Ethnopharmacol.*, 2021, **273**, 113829.
5. R. K. Sharma and B. Dash, *Carak Samhita*, Chaowkhamba Sanskrit Studies, Varanasi, 2009, pp. 1–549.
6. G. E. R. Lloyd, J. Chadwick and W. N. Mann, *Hippocratic writings*, Harmondsworth, New York, Penguin, 1983.
7. P. V. Sharma, *Susruta Samhita*, Chaukhamba Visvabharati, Varanasi, 2010, pp. 1–568.
8. V. M. Gogte, *Ayurvedic pharmacology and therapeutic uses of medicinal plants*, Chaukhambha Publication, New Delhi, 2009, pp. 1–841.

9. *Safety reviews on selected Indian medicinal plants*, ed. S. S. Yadav and N. Tandon, Indian Council of Medical Research, New Delhi, 2018, pp. 1–1421.
10. M. S. Baliga, S. Meera, B. Mathai, M. P. Rai, V. Pawar and P. L. Palatty, *Chin. J. Integr. Med.*, 2012, **18**, 946.
11. U. Chandran, N. Mehendale, G. Tillu and B. Patwardhan, *Comb. Chem. High Throughput Screening*, 2015, **18**, 846.
12. V. N. Sumantran, A. A. Kulkarni, A. Harsulkar, A. Wele, S. J. Koppikar, R. Chandwaskar, V. Gaire, M. Dalvi and U. V. Wagh, *J. Biosci.*, 2007, **32**, 755.
13. S. Cheriyamundath, T. Mahaddalkar, S. N. Save, S. Choudhary, R. V. Hosur and M. Lopus, *Biomed. Pharmacother.*, 2018, **98**, 76.
14. R. M. Amir, F. M. Anjum, M. I. Khan, M. R. Khan, I. Pasha and M. Nadeem, *J. Food Sci. Technol.*, 2013, **50**, 1018.
15. S. Kantak, N. Rajurkar and P. Adhyapak, *J. Ayurveda Integr. Med.*, 2020, **11**, 236.
16. A. Wele, S. De, M. Dalvi, N. Devi and V. Pandit, *J. Ayurveda Integr. Med.*, 2021, **12**, 269.
17. V. A. Dole, *Rasaratnasamuccaya of Sri Vagbhatacharya*, Chowkhambha Sanskrit Series Office, Varanasi, UP, 1st edn, 2006, p. 420.
18. Y. S. Biradar, S. Jagatap, K. R. Khandelwal and S. S. Singhania, *Evidence-Based Complementary Altern. Med.*, 2008, **5**, 107.
19. T. Vani, M. Rajani, S. Sarkar and C. J. Shishoo, *Int. J. Pharmacogn.*, 1997, **35**, 313.
20. S. Kaur, S. Arora, K. Kaur and S. Kumar, *Food Chem. Toxicol.*, 2002, **40**, 527.
21. M. Nariya, V. Shukla, S. Jain and B. Ravishankar, *Phytother. Res.*, 2009, **23**, 1092.
22. E. P. Sabina and M. Rasool, *Vasc. Pharmacol.*, 2008, **48**, 14.
23. B. Hazra, R. Sarkar, S. Biswas and N. Mandal, *BMC Complementary Altern. Med.*, 2010, **10**, 20.
24. E. Jayajothi, T. Elavarasu, M. Hamsaveni and S. K. Sridhar, *Nat. Prod. Sci.*, 2004, **10**, 16.
25. S. Saravanan, R. Srikumar, S. Manikandan, N. Jeya Parthasarathy and R. Sheela Devi, *Yakugaku Zasshi*, 2007, **127**, 385.
26. R. Srikumar, N. J. Parthasarathy and R. S. Devi, *Biol. Pharm. Bull.*, 2005, **28**, 1398.
27. S. Girdhani, S. M. Bhosle, S. A. Thulsidas, A. Kumar and K. P. Mishra, *J. Cancer Res. Ther.*, 2005, **1**, 129.
28. S. Kaur, H. Michael, S. Arora, P. L. Härkönen and S. Kumar, *J. Ethnopharmacol.*, 2005, **97**, 15.
29. S. Kaur and R. K. Jaggi, *Indian J. Exp. Biol.*, 2010, **48**, 925.
30. T. Sandhya, K. M. Lathika, B. N. Pandey and K. P. Mishra, *Cancer Lett.*, 2006, **231**, 206.
31. U. Chandran, N. Mehendale, S. Patil, R. Chaguturu and B. Patwardhan, *Innovative Approaches in Drug Discovery*, 2017, p. 127.

32. Habib-ur-Rehman, K. A. Yasin, M. A. Choudhary, N. Khaliq, Atta-ur-Rahman, M. I. Choudhary and S. Malik, *Nat. Prod. Res.*, 2007, **21**, 775.
33. A. K. Chahal, G. Chandan, R. Kumar, A. K. Chhillar, A. K. Saini and R. V. Saini, *J. Food Biochem.*, 2020, **44**, e13115.
34. N. Kumar and S. M. P. Khurana, *Indian J. Nat. Prod. Resour.*, 2018, **9**, 97.
35. A. Bag, S. K. Bhattacharyya and R. R. Chattopadhyay, *Asian Pac. J. Trop. Biomed.*, 2013, **3**, 244.
36. L. J. Juang, S. J. Sheu and T. C. Lin, *J. Sep. Sci.*, 2004, **27**, 718.
37. E. M. Williamson, *Major herbs of ayurveda*, Churchill Livingstone, Edinburgh, New York, 2002.
38. J. Tsering and X. Hu, *BioMed Res. Int.*, 2018, **2018**, 7046927.
39. H. Z. Huang, S. Y. Zhao, X. M. Ke, J. Z. Lin, S. S. Huang, R. C. Xu, H. Y. Ma, Y. Zhang, L. Han and D. K. Zhang, *J. Pharm. Biomed. Anal.*, 2018, **158**, 247.
40. S. Kaur, H. Michael, S. Arora, P. L. Härkönen and S. Kumar, *J. Ethnopharmacol.*, 2005, **97**, 15.
41. N. Birla and P. K. Das, *Pharm. Biol. Eval.*, 2016, **3**, 366.
42. S. Prasad and S. K. Srivastava, *Antioxidants*, 2020, **9**, 72.
43. Y. Shi, R. P. Sahu and S. K. Srivastava, *BMC Cancer*, 2008, **8**, 294.
44. K. Lu, D. Chakroborty, C. Sarkar, T. Lu, Z. Xie, Z. Liu and S. Basu, *PLoS One*, 2012, 7, e43934.
45. W. Wang, T. Liu, L. Yang, Y. Ma, F. Dou, L. Shi, A. Wen and Y. Ding, *Biomed. Pharmacother.*, 2019, **116**, 108994.
46. R. Srikumar, N. J. Parthasarathy, S. Manikandan, G. S. Narayanan and R. Sheeladevi, *Mol. Cell. Biochem.*, 2006, **283**, 67.
47. P. Phetkate, T. Kummalue, Y. U-pratya and S. Kietinun, *Evidence-Based Complementary Altern. Med.*, 2012, **2012**, 239856.
48. V. Aher and A. Wahi, *Trop. J. Pharm. Res.*, 2011, **10**, 567.
49. A. Saraphanchotiwitthaya, P. Sripalakit and K. Ingkaninan, *Maejo Int. J. Sci. Technol.*, 2008, **2**, 400.
50. R. S. Suja, A. M. C. Nair, S. Sujith, J. Preethy and A. K. Deepa, *Indian J. Anim. Res.*, 2009, **43**, 103.
51. A. Tarasiuk, P. Mosińska and J. Fichna, *Chin. Med.*, 2018, **13**, 39.
52. T. Basu, S. Panja, N. B. Ghate, D. Chaudhuri and N. Mandal, *Cytotechnology*, 2017, **69**, 201.
53. V. Maruthappan and K. S. Shree, *J. Adv. Pharm. Technol. Res.*, 2010, **1**, 229.
54. H. A. Whiteford, A. J. Ferrari, L. Degenhardt, V. Feigin and T. Vos, *PLoS One*, 2015, **10**, e0116820.
55. S. Dhanalakshmi, R. S. Devi, R. Srikumar, S. Manikandan and R. Thangaraj, *Yakugaku Zasshi*, 2007, **127**, 1863.
56. P. Belapurkar, P. Goyal and P. Tiwari-Barua, *Indian J. Pharm. Sci.*, 2014, **76**, 467.
57. M. H. Carlsen, B. L. Halvorsen, K. Holte, S. K. Bøhn, S. Dragland, L. Sampson, C. Willey, H. Senoo, Y. Umezono, C. Sanada, I. Barikmo,

N. Berhe, W. C. Willett, K. M. Phillips, D. R. Jacobs Jr and R. Blomhoff, *Nutr. J.*, 2010, **9**, 3.
58. N. Kumari, P. Kumar, D. Mitra, B. Prasad, B. N. Tiwary and L. Varshney, *J. Food Sci.*, 2009, **74**, M109.
59. M. Rasool and E. P. Sabina, *Phytother. Res.*, 2007, **21**, 889.
60. K. G. Alberti, R. H. Eckel, S. M. Grundy, P. Z. Zimmet, J. I. Cleeman, K. A. Donato, J. C. Fruchart, W. P. James, C. M. Loria, S. C. Smith Jr, International Diabetes Federation Task Force on Epidemiology and Prevention, Hational Heart, Lung, and Blood Institute, American Heart Association, World Heart Federation, International Atherosclerosis Society and International Association for the Study of Obesity, *Circulation*, 2009, **120**, 1640.
61. H. Beltrán-Sánchez, M. O. Harhay, M. M. Harhay and S. McElligott, *J. Am. Coll. Cardiol.*, 2013, **62**, 697.
62. A. Scuteri, S. Laurent, F. Cucca, J. Cockcroft, P. G. Cunha, L. R. Mañas, F. C. Mattace Raso, M. L. Muiesan, L. Ryliškytė, E. Rietzschel, J. Strait, C. Vlachopoulos, H. Völzke, E. G. Lakatta and P. M. Nilsson, *Eur. J. Prev. Cardiol.*, 2015, **22**, 486.
63. R. Li, W. Li, Z. Lun, H. Zhang, Z. Sun, J. S. Kanu, S. Qiu, Y. Cheng and Y. Liu, *BMC Public Health*, 2016, **16**, 296.
64. S. Gurjar, A. Pal and S. Kapur, *Altern. Ther. Health Med.*, 2012, **18**, 38.
65. W. Phimarn, B. Sungthong and H. Itabe, *J. Evid. Based Integr. Med.*, 2021, **26**, DOI: 10.1177/2515690X211011038.
66. D. Patel, R. Kumar, D. Laloo and S. Hemalatha, *Asian Pac. J. Trop. Biomed.*, 2012, **2**, 411.
67. S. S. Rajan and S. Antony, *Anc. Sci. Life*, 2008, **27**, 45.
68. M. H. Yang, Y. Vasquez, Z. Ali, I. A. Khan and S. I. Khan, *J. Ethnopharmacol.*, 2013, **149**, 490.

CHAPTER 10

Pharmacological Potential and Phytochemical Evaluation of **Emblica officinalis**: *A Wonder Herb in Ayurveda*

GURMEET SINGH,* SHRIDEVI GOTHE, NADANA VINAYAGAN AND SMRITI CHAWALA

Centre for Ayurveda Biology and Holistic Nutrition, The University of Trans-Disciplinary Health Sciences and Technology (TDU), Bengaluru, Karnataka 560064, India
*Email: gurmeet.singh@tdu.edu.in

10.1 Introduction

Emblica officinalis (*E. officinalis*) is conventionally called as Amla or Indian gooseberry, a superfood of Ayurveda. Its use as a medicine predates Ayurveda as it finds a prominent place for its health-promoting effects in Indian mythological literature such as Vedas, Upanishads and Puranas. These references suggest that because of its medicinal importance, the amla tree had wide cultural and religious significance as well as being revered, with its fruit considered sacred and it being attributed with the quality of giving longevity to those who consumed it regularly. Amla is considered to be the beloved tree of Lord Vishnu and so is worshipped. The Indian mythological epic Mahabharat mentions that the Amla berry originated from a drop of Amrit, the heavenly nectar of immortality, that fell from the home of the gods.

Chemistry, Biological Activities and Therapeutic Applications of Medicinal Plants in Ayurveda
Edited by Augustine Amalraj, Sasikumar Kuttappan and Karthik Varma A. C.

Published by the Royal Society of Chemistry, www.rsc.org

Hence, it is believed that the fruit is bestowed with the power to cure all kinds of health ailments.

The Sanskrit name 'Amalaki', which means 'prosperity' or 'sustained', itself points to its medicinal significance. Amla is an integral part of Ayurvedic medicinal preparations. Through the curation of all references to Amla in the nine key texts of Ayurveda, from Charaka Samhita to Bhavaprakasha, we have obtained 892 references and 354 formulations which were prescribed for over 40 different disease conditions. In Ayurveda, besides its use as an important botanical in various medicinal formulations for the treatment of an array of ailments of the body, Amla is described as a **"Rejuvenator (Rasayana)", and wholesome daily consumable food (Nitya sevaniya)** for boosting the immune system, preventing disease, enhancing good health, as well as revitalising the body and mind.[1,2] Acharya Charaka mentions it as **"Vayasthapananam shrestham"** which halts ageing.[2] It is also described as **"Sada pathyam"**, a health-promoting food which is the all-time most wholesome food.

Amla's use is not restricted to Ayurveda pharmacological Materia Medica or other traditional medicinal systems as it is also a part of Indian cuisine. Even in Charaka Samhita, Amla is included in the list of 11 nitya sevaniya aahara – the food ingredients that can be consumed daily throughout the year by all. Amla is also included in the 'National Institute of Nutrition', Indian Food Composition Tables. From Charaka Samhita to Indian Food Composition Tables, the documented use of Amla for culinary purposes dates back over 3000+ years. Together, the ancient culinary importance of Amla, its acknowledged nutritive value in modern nutrition science, the vast number of ethnomedicinal formulations that include Amla, and the wide range of diseases in which it is used for preventive or curative purposes, gives Amla a wonderful botanical status.

10.2 General Description of *E. officinalis*

10.2.1 Scientific Classification and Morphological Description

Amla belongs to the family Phyllanthaceae (Table 10.1). It is a deciduous plant of moderate size. It has a greyish, smooth stem bark. The wood is red and hard, and the leaf is divided into small, narrow leaflets. A group of leaves

Table 10.1 Scientific classification of Amla (*Emblica officinalis*).

Kingdom	Plantae
Botanical name	*Emblica officinalis* Gaertn. / *Phyllanthus emblica Linn. Phyllanthus emblica*
Genus	Phyllanthus
Species	Emblica
Family	Euphorbiaceae
Order	Malpighiales
Clade	Rosids

resemble a feather. The flowers are unisexual. The two types of flowers are found arranged in dense clusters at the axis of the leaves. In the flowering axis, male flowers are seen at the apex, while female flowers are at the base. The female flowers are fewer in number. The fruits are light green drupes, which are compressed, spherical, smooth, glossy, fleshy, and juicy. There are six inconspicuous vertical furrows. The drupe has three crustaceous compartments which contain two seeds each.

The fruit, both fresh and dried, is used as a medicine in Ayurveda. The fresh fruit is globular, about 1.5″ in diameter, six-lobed, yellowish-green, with a smooth surface. The fruit tastes sour and astringent, with bitter, sweet, and pungent notes. The dried fruit breaks easily by hand, exposing the dried pulp and endocarps which contain the triangular seeds. Nearly 17 000 tons of fruit were consumed by the Indian domestic herbal manufacturing units in 2005,[3] making it one of the most used Ayurveda botanicals.

The Amla tree is found growing wild in subtropical and tropical regions of India and around the world. It is common in mixed deciduous forests and exposed slopes of hills. While historically Amla was gathered from the wild, today it is cultivated in many parts of the country.

10.2.2 Vernacular Names

More than 200 Sanskrit and vernacular names have been recorded for *E. officinalis*, underpinning its popularity and superfood status (FRLHT's Clinically Important plants of Ayurveda) (Table 10.2).

Sanskrit synonyms such as Dhatriphalam, Sriphalam, Vayastha, Adiphala, Amritaphala, Kayastha, Vrishya, Karshaphala, Vrittaphala, Kayastha, Parvakeeta, Tishya, Shadrasa, Shriphala, Rochani, Shiva, Pancharasa, Jatiphala, and Sita, are suggestive of its extensive use and multifaceted pharmacological health benefits. Most of these synonyms refer to the antiaging, antioxidant, rejuvenating, and aphrodisiac activities of Amla.

A select few interesting synonyms of Amla and their meanings are listed below.

- Dhatri – one who nourishes the child, owing to the medicinal properties of Amla. It is believed that regular consumption of Amla protects the body like a mother protects her child.
- Dhatriphalam – fruit that nourishes the human body like a mother.
- Vayastha – which seizes the youth; rejuvenator.
- Adiphala – the first fruit originated on Earth.
- Amritaphala – fruit that is as sweet/beneficial as nectar, or that which ensures the longevity of humans.
- Amalaki – one which retains health because of its rejuvenating properties.
- Kayastha – stabilises the body.
- Vrishya – aphrodisiac.
- Rochan – taste enhancer.

Table 10.2 Vernacular names of Amla.[8,34,74]

Language	Names	No. of names
Arabic	Amla khushk, amlag, amlaj	3
Hindi	Ambala, amla, amlika, anvurah, anwala, aonla, aonta, aoula, auna, amalaci, aamla, anola, anuli, amalak, anwiya, anvula, aola, aungra, awala, anwerd, amvala, aunra, daula	23
Kannada	Dadi, dhanya, nelli, nellikai, aamalaka, amalaka, aamalakee, ambla, bettada nelli, bettada nelli kaayi, betta nelli, cattu, chattu daadi kaayi, chattu, dodda nelli, dhatri, nellka, nelli chattu, nilika, sudhe, perunelli	21
Malayalam	Nelli, amalakam, nellikka, nellimaram, nellikkaya, nilica, nilika, nilicamaram, nellikai	9
Marathi	Aouli, avalkati, avil, avla, awla, aamlee, anvala, aonli, arola, avala, bhuiawali	11
Sanskrit	Amruthaphala, amala, amulki, akara, amrita, amalaka, adiphala, amalakah, amlika, amalakam, amalaki, amamalakam, amritaphala, amraphala, amraphalam, bahuphali, dhatri, dhatri-phala, dhatrika, dhatriphala, dhatriphalam, jatiphala, karshaphala, kamu, kayastha, parvakeeta, pancharasa, sriphalam, shriphali, shanta, Sita, shiva, shriphala, shadarasa, rochani, vrishya, Vayasya, Vrsyam, vayastha, vrittaphala, umrita, triphala, tishya, tisya	44
Tamil	Alakam, alakam, alikam, amalakamalam, amalai, amalakam, amalaki, amalakki, amalam, amirtai, amalagam, ampal, ampalkamaram, amutai, amukam, amuttam, antakolam amirtapalam, anantai, andakaram, antakaramaram, amirtapalai, antor, antakolam, atikoram, attakolam, attakoram, antakoram, antakaram, avutanci, amalakam, amarikam, ampal, korankam, mintu, mirutupala, nelli, nelli mulli, nellikkaai, nellikkai, nellikkay, cilata, cinkitakamaram, civai, cirottam, civam, indul, citupalam, cinkitakam, intulam, intul, intulapam, intulix, intulia, ciripalam, intuli, iratai, kaiyirkani, kapitattirimaram, kapitattiri, kantamulakam, kanakam, kantumalalam, kantattiri, kapi, kayastam, kokkam, kattunelli, konkal, konkam, korankamikam, kotam, korankamikamaram, kotimukakkini, kotimukakkinimaram, makanti, makantika, makantikam, mankantikamaram, malainelli, mirutupala, mamalakam, mamalakkay, miruntikam,mirutupalamaram, mitintu, miruntikamaram, miruntu, mirutupala, mitintu, mituntumaram, nelli-kay, nellie poo, nellikai, nellie kai, pancatcaravaruni, parainatavanci, perunelli, tattari, tamalaikkay, tattili, parainatavancimaram, tattinimaram, tattirimaram, tattiripalai, tattiripalam, tecomantaram, tattiri, tecomantiram, toppunelli, tantiri, tecomaram, tiriciyapalam, tiripalati, toppi, tiriciyam, toppi, tottiki, toppi, totti, tottikimaram, tuttarikam, tuppunelli, vanamalakam, yantikoram, yankoram	127

Table 10.2 (*Continued*)

Language	Names	No. of names
Telugu	Aamalaki, amalakamu, aamalakamu, amalaki, assereki, asereki, pedda usiri, peddavusirika, putla nelli, oosree, pullayusirika, pedda, nelli kaaya, usirika, triphalamu, usiri, usirikai, usirika, usari, usarika, usirikaaya, usrikayi, triphalam, useri, usirikaya, usiriki, userikai usiuka, userakee, usri, vusirikaya, vusirika, yusirika	33
Tibetan	Skyu-ru-ra, skyu ru ra, sk yu ru ra	3
Urdu	Aamla, amal khusk, amala khushk, amal khushk, baobarang,amla khushk, amla, amla khusk, amla sabz, guthli nikalal hua, guthli, khushbu amla, guthli nikala hua, amlay taza aur farba, tirphala, aonla, murraba-i-amla, triphala	18
English	Indian gooseberry, Emblica myroblan	2
Persian	Amelah, ameleh, amla, amuleh	4
Italian	Mirabolano emblico	1
German	Amla	1
French	Phyllanthe emblica	1
Nepalese	Amba	1
Chinese	Anmole	1
Malaysian	Popok Melaka	1
Portuguese	Mirabolano emblico	1
Lao	Mak kham bom Nellikka	1
Thai	Ma kham pom	1
Assamese	Amlakhi	1
Bengali	Amlaki	1
Gujarati	Amla	1
Konkani	Aavalo	1
Manipuri	Heikru	1
Odiya	Aanla	1
Punjabi	Olay	1

10.3 Amla in Ayurveda Classical Text

Amla is used in a wide range of Ayurveda formulations for the treatment of different health-related issues. As noted above, Amla is mentioned 892 times in the seven key texts of Ayurveda (Table 10.3).

Charaka Samhita, being a text that is more focused on Rasayana and general medicine, has the most references at 183. Susruta Samhita mostly explains surgical conditions, hence it has 91 Amla references, which is significantly less compared to Charaka Samhita. Astanga Sangraha is a compilation of both Charaka and Susruta, and contains 204 Amla references. Other books, Astanga Hrdaya, Chakradatta Sarangadhara, and Bhavaprakash have 108, 131, 39, and 136 Amla references, respectively. The extensive references to Amla from Charaka Samhita, the first book of Ayurveda, down to 16th-century texts such as Bhavaprakasha gives an indication of the strength of its medicinal merits besides its continuing popularity.

Table 10.3 Number of references and formulations in different Ayurveda classics.

S. no.	Name of the book	Time period of the book	Number of references	Number of available formulations
1	Charaka Samhita	15th century BCE–4th century CE	183	52
2	Sushruta Samhita	15th century BCE–5th century CE	91	33
3	Ashtanga Sangraha	4th–5th centuries CE	204	65
4	Ashtanga Hridaya	6th century CE	108	47
5	Sharangadhara Samhita	8th century CE	39	19
6	Chakradatta	11th century CE	131	65
7	Bhavaprakasha Samhita	16th century CE	136	53
8	Bhaishajya ratnavali	18th century CE		8
9	Sahasrayoga	19th century CE		12
			892	**354**

Due to its continued use for its medicinal properties over centuries, Amla's classical properties, referred to as Ayurveda pharmacology, have been documented extensively in many of the key texts. These properties include rasapanchaka – rasa, guna, virya, vipaka, and karma. These terms are explained in the Dravya guna text by J. L. N. Shastri[4] or P. V. Sharma.[5] All the classical texts agree that Amla fruit has five of the six rasas – sour (Amla), sweet (Madhura), astringent (Kashaya), bitter (Tikta), and pungent (Katu). Ayurveda recommends that our diet should have all six rasas at every meal. Hence botanicals that have more rasas are highly valued. Very few botanicals in the Ayurveda Materia Medica (comprising of 1200+ botanicals) have such a diverse rasa profile as Amla. This rasa profile gives amla a hallowed position in the Ayurveda Materia Medica. The effect after digestion (vipaka) is sweet (Madhura). In terms of its Gunas (properties) the dominant ones are Laghu (light), Ruksha (dry), and mridu (soft). Equally important is Amla's tridoshara property, which implies that it can reduce aggravated levels of all the three dosha – vata, pitta and kapha (for detailed information on doshas refer to the Roga Vijnana and Vikriti Vijnana text by Dr Manoj Sankaranarayana).[6] Together the rasapanchaka and doshakarma define the pharmacological action of Amla in Ayurveda.[7]

According to Charaka Samhita, an imbalance or vitiation of the three doshas in the body is the cause of disease.[2] Herbs have been valued for their ability to reduce the different doshas. Most herbs balance vitiation of one or two doshas. More importantly, while doing so, they may disturb the balance of the remaining doshas. There are only a few herbs that balance all three doshas. Amla is one of those rare botanicals that balances all three doshas. It reduces Vata because of the sour taste (Amlata), Pitta because of its sweet taste (Madhurata) and cooling properties (sheeta virya), and Kapha because of its astringent taste (kashaya) and dry property (ruksha guna).[7]

The Karma or pharmacological action of Amla is dictated by the properties described above. Amla sharpens the sense of taste, thus it stimulates the first step of digestion, the taste experience. Amla also improves appetite because of its sour taste that in turn kindles Agni (the digestive fire), which is at the core of healthy digestion. While doing so it doesn't aggravate pitta because of its bitter, astringent taste and cold property. Thus, Amla maintains a healthy and balanced digestive system, in turn helping in the conversion of food into a bio-assimilable form.

According to Ayurveda, the bitter property of Amla is responsible for better circulation of Rasa dhatu or the digested essence of food, and this, in turn, contributes to its use as a hepatotonic. Sara guna (a property that promotes movements) helps in the excretion of metabolic waste and other toxins out of the body. This provides a boost and protection for the body's natural defence system. The multiple points of indigestion and toxin elimination that are influenced positively by Amla make it one of the most important botanicals for maintaining overall healthy digestion and for nourishing all body tissues.

Amla, being a tridoshahara medicine, helps to normalise the metabolism, thus preventing the accumulation of fat (Medha dhatu). It helps to manage body weight and aids in weight loss. It also slows down the ageing process and boosts the body's immunity. According to Ayurveda, Amla improves memory and cognition, and is also a neurotonic. It improves sense organs by enhancing their sensitivity. Amla acts as an aphrodisiac and strengthens Shukra dhatu and rejuvenates the male reproductive system and increases male fertility, for this it is included in Vajikarana therapy as an Ayurvedic herb for treating erectile dysfunction.

These karmas (pharmacological effects) are captured in the many shlokas in classical Ayurvedic texts. Two such representative shloka from Bhavaprakasha nighantu are reproduced below. The first emphasises that though the properties are similar to *Terminalia chebula* Retz (Haritaki), it is specially used for bleeding disorders (raktapitta), diabetes (prameha) treatment, and as an aphrodisiac and rejuvenator. The second shloka lists several benefits and states that Amla is good for the eyes, enhances a healthy life span, nourishes the body, and regulates the movement of urine and faeces and is also claimed to be beneficial in respiratory disorders, diabetes, haemorrhoids, skin diseases, inflammation, fever, hoarseness of voice, IBS (irritable bowel syndrome), constipation, abdominal masses, abdominal distention, nausea and vomiting, hiccup, eczema, cardiac diseases, jaundice, abdominal pain, liver and spleen disorders, renal calculi, and urinary disorders.[7]

हरीतकीसमं धात्रीफलं किन्तु विशेषतः ।
रक्तपित्तप्रमेहघ्नं परं वृष्यं रसायनम् ॥३६॥

(Bhavaprakasha nighantu 2/36)

चक्षुष्या लघुरायुष्या बृंहणी चानुलोमिनी ॥१८॥
श्वासकासप्रमेहार्शःकुष्ठशोथोदरक्रिमीन् ।
वैस्वर्यग्रहणीरोगविबन्धविषमज्वरान् ॥१९॥
गुल्माध्मानतृषाछर्दिहिक्काकण्डूहृदामयान् ।
कामलां शूलमानाहं प्लीहानञ्च यकृत्तथा ।
अश्मरीमूत्रकृच्छ्रं च मूत्राघातं च नाशयेत् ॥२०॥

(Bhavaprakasha nighantu 2/18–20)

If the importance of medicinal herbs is defined by their usage in different formulations and the number of diseases in which they are indicated, then Amla is indeed a wonder drug. Amla's use has been documented for a wide range of disease indications and it is present in more than 1000 formulations in Ayurvedic classics. The in-house CIPA database of FRLHT (Clinically Important Plants of Ayurveda) which has curated all formulations given in the nine classic Ayurveda texts (Table 10.3) lists 354 Amla formulations. These formulations range from a simple powder (Choorna) to more complex jams (Avaleha kalpana) like Chyawana Prasha.

Amla is indicated for treating a wide array of acute to chronic major health ailments and also for the maintenance of wellness of health. An analysis of the 354 Amla formulations given in the CIPA reveals that they have been prescribed for the treatment of 40 different disease conditions. A frequency plot of the number of formulations for each disease condition is given in Figure 10.1. As per this analysis, the top diseases and medical conditions in which formulations containing Amla have been recommended are fever (Jvara) (68 formulations), skin diseases (Kusta) (34 formulations), diabetes (Prameha) (27 formulations), cough (Kasa) (27 formulations), rejuvenation (Rasayana) (25 formulations), eye diseases (Akshi roga) (20 formulations), and anaemia (Pandu) (19 formulations).

As depicted in Figure 10.2, Amla formulations range from simple formulations of 2–3 ingredients to complex formulations with scores of ingredients. Simple formulations are most often indicated in the treatment of one or two symptoms or disease conditions that are acute in origin and/or easily curable (sadhya vyadhi). However, complex formulations like Chavyanaprash, containing more than 20 ingredients, are used as tonics or in the treatment of multiple disease conditions or syndromes, diseases involving multiple systems, and chronic and/or difficult to cure or incurable diseases (kricchrasadhya/asadhya vyadhi).

An analysis of the complexity of these formulations shows that in these 354 Amla formulations, it has been paired with more than 30 different ingredients (Figure 10.3). Among these ingredients the most used are Haritaki (*Terminalia chebula*), Pippali (*Piper longum*), Vibhitaki (*Terminalia bellirica*),

Figure 10.1 Frequency chart of Amla indications in different diseases.

Shunti (*Zingiber Officinale*), Musta (*Cyperus rotundus*), Guduchi (*Tinospora cordifolia*), Draksha (*Vitis vinifera*), Patola (*Trichosanthes dioica*), Madhu (Honey), and Haridra (*Curcuma longa*) (95, 84, 62, 55, 49, 34, 31, 28, and 27 times, respectively). Meanwhile, in general, it is believed that the pairings of herbs represent the synergistic effects on a wide range of treatment areas, a network analysis of such pairing datasets may reveal insights for new and more effective formulations.

Of these pairings, a few deserve special mention due to their popularity. One such pairing is its combination with *Terminalia chebula* and *Terminalia bellirica*, known as Triphala, a well-known Ayurvedic formulation. Triphala is an appetiser and carminative, considered beneficial for the eyes, that alleviates Kapha pitta dosha, and cures leprosy (kushta), constipation (vibandha), anomalies of urinary system/diabetes (prameha), eye disease (akshiroga), and all types of fever (jvara). It also enhances taste and digestion.[7] Apart from Ayurvedic literature, some research studies have indicated triphala as an appetiser and antacid. It is also known for its anti-inflammatory, immunomodulating, antibacterial, antimutagenic, antioxidant, chemoprotective,

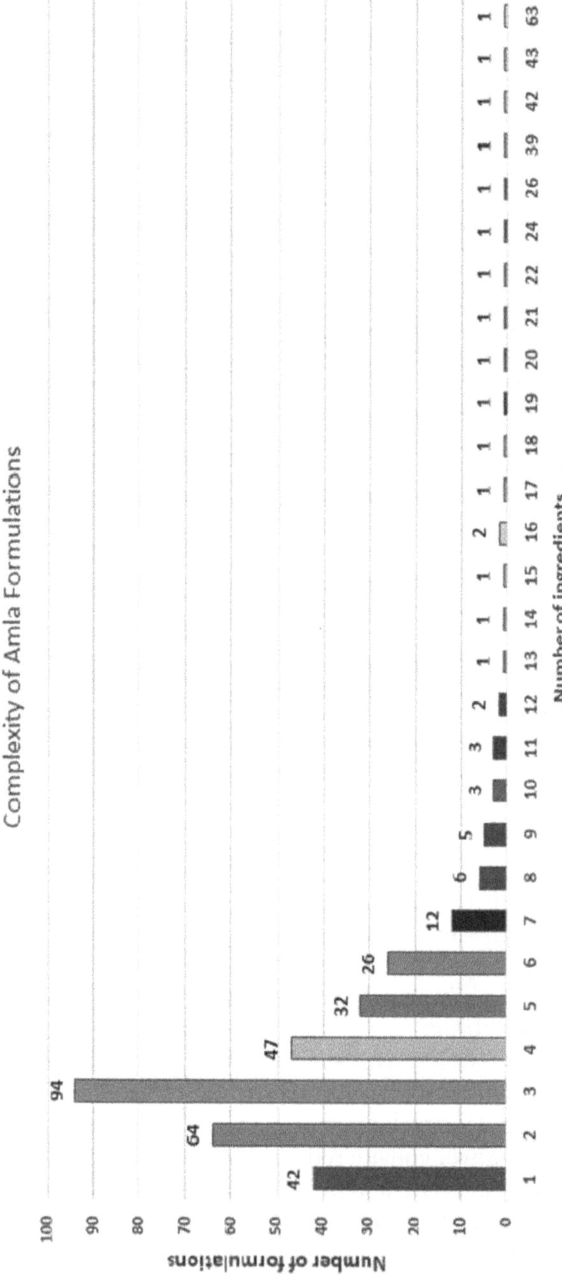

Figure 10.2 Frequency chart of the complexity of Amla formulations.

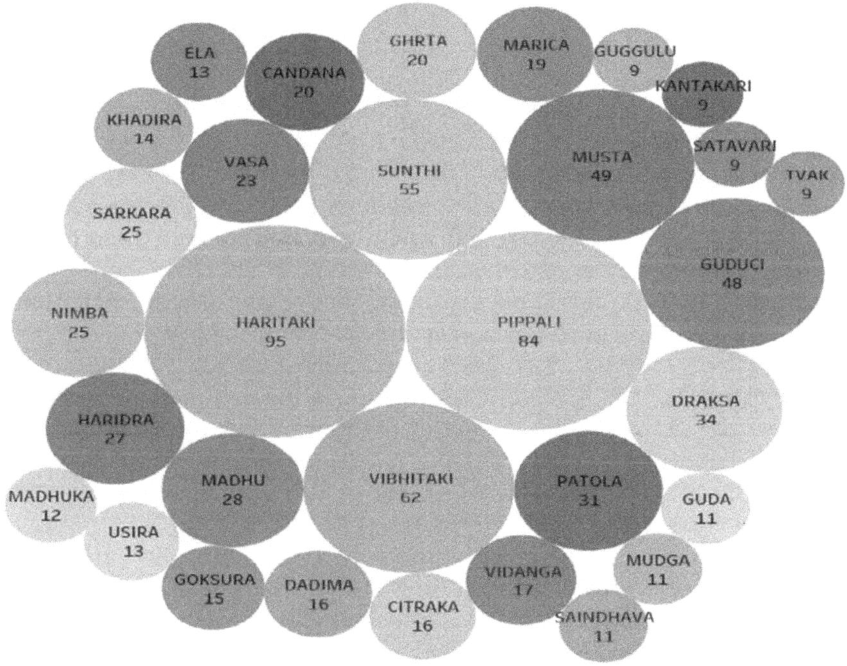

Figure 10.3 Frequency of Amla pairings with other ingredients.

antidiabetic, hypoglycaemic, antineoplastic, and radioprotective effects.[8] Another popular pairing of Amla is with turmeric (Haridra), which forms the renowned formulation Nishamalaki. Nishamalaki is mainly indicated in the treatment of prameha, a disease condition symptomatically similar to diabetes. Amla when paired with iron (loha), yashti (*Glycyrrhiza glabra*), and guduchi (*Tinospora cordifolia*) forms an esteemed Ayurveda formulation known as Dhatri loha. Dhatri loha is used in the treatment of iron deficiency anaemia. Amla is one of the most popular Rasayanas for improving appetite, nutrient absorption, and assimilation. This property in turn helps in iron absorption and hepatotonic activity. When paired with Shunti, Guduchi, Musta, Draksha, and Patola, it acts as a pittahara and is indicated in fever (jvara), inflammation (shopha), bleeding disorders (raktapitta), diabetes (prameha), and eczema (shitapitta).

10.4 Phytochemistry of *E. officinalis*

There has been tremendous interest in unravelling the molecular profile of *E. officinalis* and linking the pharmacological action of *E. officinalis* reported in traditional knowledge systems to its nutritional, phytochemical, and molecular profile. *E. officinalis* fruit has been well characterised from a nutritional perspective as 180 nutritional components of *E. officinalis* have been reported by the National Institute of Nutrition in the Food Composition Tables. According to these tables, *E. officinalis* is a reservoir of various vitamins, flavonoids, and tannins. The nutritional data of Amla from NIN IFCT are listed in Table 10.4. One hundred grams of *E. officinalis* has 300 mg of vitamin C, which is almost three times the RDA for adults (65–90 mg per day).[9]

Molecules such as citric acid, gallic acid, ascorbic acid, ellagic acid, chebulinic acid, quercetin, pectin, emblicanin-A, trigallayl glucose, pedunculagin, ellagotannin, chebulagic acid, emblicanin-B, and punigluconin have been reported in *E. officinalis*. However, good characterisation of the phytochemicals of *E. officinalis* has been challenging due to the reactive nature of many of its phytochemicals and their binding to the *E. officinalis* insoluble matrix. This has been especially true of phenolic acids and polyphenols of *E. officinalis* which form a major functional group of *E. officinalis*'s phytochemicals. One of the approaches to studying phytochemistry has been by characterising the phenolic acid fractions as free and bound, such as in the work of Kumar *et al.*[10] These authors used 70% ethanol to extract the phenolic acid-free and the phenolic acid-bound fractions, subsequently extracted under a nitrogen atmosphere using 1 M sodium hydroxide with 0.5% sodium borohydride.[10] They attributed the antioxidant activity to phytochemicals such as gallic acid and ascorbic acid, but did not characterize it further. Other studies have progressed with the isolation and identification of polyphenols and organic acid gallates from *E. officinalis* fruit juice.[11] Along with hydrolysable tannins, *E. officinalis*'s other major phenolic constituents are L-malic acid 2-*O*-gallate, organic acids gallates,[12] mucic acid 2-*O*-gallate,[13] chebulagic acid, 1-*O*-galloyl-α-D-glucose,[14] and corilagin.[14,15] Bansal *et al.*[16] reported that fruit juice contains major ellagitannins like *E. officinalis* carpusin and putranjivain A.[16] Our analysis of the literature for phytochemicals of *E. officinalis* resulted in a list of 200 molecules, of which we were able to establish a PubChem ID for 123.

An analysis of these 123 molecules with PubChem ID's using the ChemMine tool clustered them into 19 functional groups, as depicted in Figure 10.4. The top four functional groups are flavonoids, tannins, organooxygen compounds, and prenol lipids. Flavonoids and tannins together are part of a larger polyphenol group, and this forms the major section of the *E. officinalis* phytochemical profile. Polyphenols are being studied by scientists for their wide-ranging biological activity and therapeutic potential.[17] Various polyphenols may play a pivotal role in metabolic pathways as they interact in biological activities related to multiple signalling pathways involved in the homeostasis of protein, DNA repair, metabolism regulation, and antioxidant mechanisms.[18] Williams *et al.*[19] reported that dietary flavonoids have pharmacological neuroprotective, cardioprotective, and chemopreventive actions.[19] As mentioned by Hussain *et al.*,[20] phytomolecules belonging to tannins possess antioxidant, antiaging, and antiapoptotic activities, and prevent neurodegenerative diseases and neuropsychiatric disorders.[20]

Table 10.4 Nutritional components of *Emblica officinalis*.

Nutrition group	Components	Unit	Content	Standard deviation
Proximate principles and dietary fibre	Moisture	g/100 g	87.02	0.78
	Protein	g/100 g	0.34	0.03
	Ash	g/100 g	0.34	0.05
	Total fat	g/100 g	0.16	0.04
	Dietary fibre: total	g/100 g	7.75	0.64
	Dietary fibre: insoluble	g/100 g	6.2	0.43
	Dietary fibre: soluble	g/100 g	1.55	0.43
	Carbohydrate	g/100 g	4.39	1.09
	Energy	kJ/100 g	99	19
Water-soluble vitamins	Thiamine (B1)	mg/100 g	0.01	0.009
	Riboflavin (B2)	mg/100 g	0.03	0.012
	Niacin (B3)	mg/100 g	0.12	0.02
	Pantothenic acid (B5)	mg/100 g	0.35	0.04
	Total B6	mg/100 g	0.27	0.054
	Biotin (B7)	µg/100 g	1.42	0.35
	Total folates (B9)	µg/100 g	7.86	1.27
	Total ascorbic acid	mg/100 g	252	30.4
	Retinol	µg/100 g		
Fat-soluble vitamins	Ergocalciferol (D2)	µg/100 g	0.27	0.02
	Tocopherols: alpha	mg/100 g	0.11	0.01
	Tocopherols: gamma	mg/100 g	0.01	0
	Tocopherols: delta	mg/100 g	0.06	0.01
	Tocotrienols: alpha	mg/100 g	0.05	0.01
	α-Tocopherol equivalent	mg/100 g	0.12	0.02
	Phylloquinones (K1)	µg/100 g	1.64	0.31
Carotenoids	Lutein	µg/100 g	38.7	9.78
	Zeaxanthin	µg/100 g	2.86	1.61
	β-Carotene	µg/100 g	1.58	0.09
	Total carotenoids	µg/100 g	62.01	6.45
Minerals and trace elements	Aluminium (Al)	mg/100 g	0.08	0.02
	Calcium (Ca)	mg/100 g	20.14	2.51
	Magnesium (Mg)	mg/100 g	6.5	0.93
	Copper (Cu)	mg/100 g	0.12	0.08
	Chromium (Cr)	mg/100 g	0.007	0.003
	Lead (Pb)	mg/100 g	0.004	0.002
	Iron (Fe)	mg/100 g	1.25	0.08
	Molybdenum (Mo)	mg/100 g	0.001	0
	Manganese (Mn)	mg/100 g	0.11	0.02
	Phosphorus (P)	mg/100 g	21.85	1.87
	Nickel (Ni)	mg/100 g	0.01	0.003
	Lithium (Li)	mg/100 g	0.001	0.001
	Potassium(K)	mg/100 g	223	12.6
	Sodium (Na)	mg/100 g	1.37	0.19
	Zinc (Zn)	mg/100 g	0.05	0.01
Starch and individual sugars	Total available CHO	g/100 g	3.95	0.05
	Total starch	g/100 g	0.56	0.03
	Fructose	g/100 g	2.1	0.02
	Glucose	g/100 g	1.29	0.02
	Total free sugars	g/100 g	3.39	0.02

The third largest group in *E. officinalis* is organo-oxygen compounds which are aliphatic and aromatic, a combination of C–O bonds, including alcohols and aldehydes that have been used in gene-delivery systems to enhance the efficacy of molecular studies.[21] Another group of *E. officinalis* phytochemicals, the prenol lipids, may also have therapeutic properties through their involvement in multiple cellular processes and metabolic lipid pathways that are interconnected with obesity, cancer, cell homeostasis, infectious, inflammatory, and neurodegenerative diseases.[22] Thus, *E. officinalis* has potential therapeutic properties that are driven by the above functional groups.

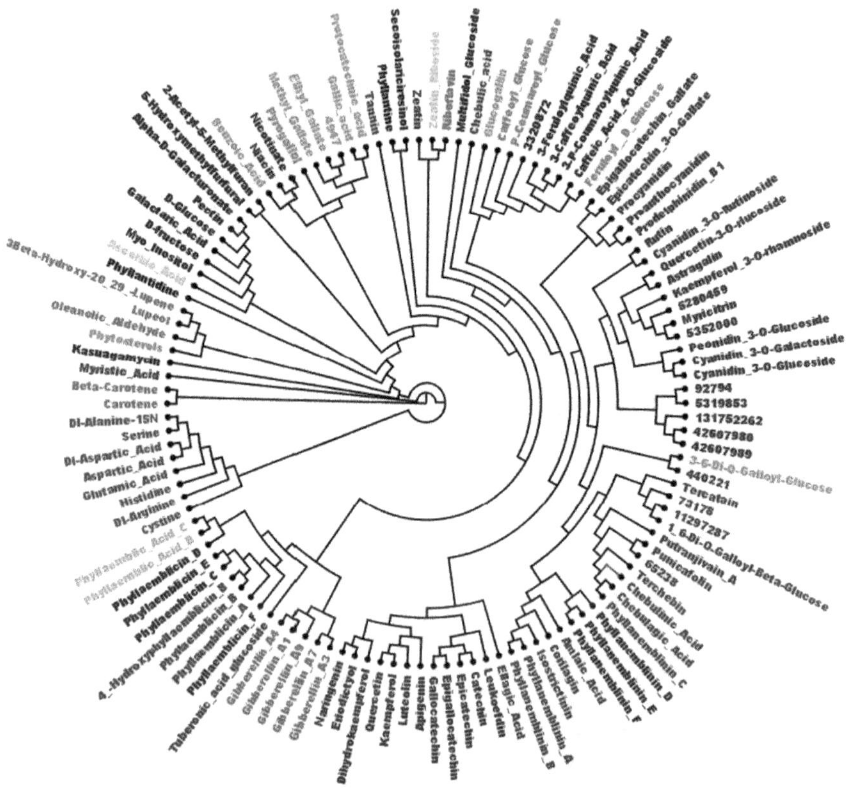

Figure 10.4 Amla molecules class clustering by ChemMine tools. Note: Each group is distinguished as a different colour, namely: polyphenols are purple, tannins are dark green, organooxygen compounds are blue, prenol lipids are light lavender, carboxylic acids and derivatives are deep moss green, benzene and substituted derivatives are rama green, fatty acyls are red, saccharolipids are navy blue, cinnamic acids and derivatives are peach, dibenzylbutane lignans are dark purple, pyridines and derivatives are dark brown, dihydrofurans are light parrot green, imidazopyrimidines are dark lavender, oxazinanes are chocolate, phenols are dark parrot green, pteridines and derivatives are pink, purine nucleosides are golden yellow, steroids and steroid derivatives are baby pink.

10.5 Pharmacological Activity of *E. officinalis*

As mentioned above, in Ayurveda the top diseases and medical conditions in which formulations containing *E. officinalis* have been recommended are fever (Jvara), skin diseases (Kushta), diabetes (Prameha), cough (Kasa), rejuvenation (Rasayana), eye diseases (Akshi roga), and anaemia (Pandu) (Figure 10.1). Its use has also been described in Ayurveda for respiratory disorders, haemorrhoids, inflammation, IBS, constipation, abdominal masses, abdominal distention, nausea and vomiting, hiccup, eczema, cardiac diseases, jaundice, abdominal pain, liver and spleen disorders, renal calculi, and urinary disorders. This breadth of therapeutic properties has to be derived from a range of pharmacological activities. Some of the most relevant are antioxidant, glycaemic modulation, lipemic modulation, immunomodulation, anti-inflammatory, blood circulation, anticancer and antimutagenic, hepatoprotective, gastroprotective functionalities, analgesic and antipyretic, antimicrobial, antitussive, neuroprotective, and antiaging activities.[23]

10.5.1 Analgesic, Antipyretic, and Antimicrobial Activities

E. officinalis-based formulations are used for Jwara (fever) in Ayurveda. This points to its analgesic, antipyretic, and antimicrobial activity. As expected, extracts of *E. officinalis* fruits have shown antipyretic and analgesic activities in various studies.[24–27] These activities have been ascribed to its bioactive phytochemicals such as carbohydrates, alkaloids, amino acids, phenolates, and tannoids. An experimental study on rats with *E. officinalis* extracts reported that these extracts notably reduced hypersensitivity-related neuropathic pain in rats.[24] An animal model study concluded that the animals induced by brewer's yeast showed a reduction in hyperthermia.[8] Furthermore, the writhing response induced by acetic acid in mice was inhibited by extracts of *E. officinalis*.[27] It has been suggested that the analgesic activity of *E. officinalis* is because of its ability to inhibit nociceptors and pain mediator synthesis.[26] An *in vivo* pain model study by Lim *et al.*[24] claimed that extract of *E. officinalis* has a potent analgesic agent in both neuropathic and postoperative pain.

The antimicrobial activity of *E. officinalis* extracts may also be a reason for the use of its extract in fever (jwara) treatment in Ayurveda. Many of the phytochemicals that contribute to its analgesic activity also contribute to its antimicrobial activity such as alkaloids, flavonoids, and terpenoids, which are present at substantial levels in *E. officinalis*.

The antimicrobial activity of *E. officinalis* extracts has been studied in many *in vitro* model systems involving bacteria, fungi, viruses, and protozoa. In an extensive study, the alcoholic extract of *E. officinalis* was shown to have an excellent antibacterial effect on a broad range of pathogenic Gram-positive and -negative bacteria. The extracts also showed potent inhibitory action on fungi such as *S. cerevisiae, C. albicans*, and *A. niger*.[28,29] A study observed that the chloroform fractionated methanolic extract of *E. officinalis* had potent cytotoxicity with an LC50 of 10.257 ± 0.770 μg mL^{-1} [28] against pathogenic Gram-positive and -negative bacteria and it also exhibited inhibitory activity on dermatophytes.[29] The antiviral effect of *E. officinalis* against herpes virus was ascribed to the polyphenols that inactivated viral particles in the extracellular area and inhibited biosynthesis of virus in the host.[30] Furthermore, Bagavan *et al.*[31] evaluated the antimalarial activity of leaf extracts of *E. officinalis* against chloroquine-sensitive and -resistant strains of *Plasmodium falciparum* and found considerable antimicrobial activity against both strains.[31]

Experiments conducted on mice showed that dietary supplementation with *E. officinalis* proved to be effective against the respiratory tract infection induced by *K. pneumoniae*.[32] *In vitro* experiments carried out by Thaweboon *et al.*[33] also demonstrated antimicrobial activity of ethanolic extracts of *E. officinalis* and reported the alcoholic extract inhibits *Candida albicans* on human oral epithelial cells and denture acrylic surfaces. Antiplasmodial activity has been reported in *E. officinalis* extracts using various *in vivo* and *in vitro* models. *C. albicans* is an opportunistic pathogen found on human oral epithelial cells and denture acrylic surfaces which causes oral infection; its growth and adhesion were significantly inhibited by ethanolic extracts of *E. officinalis*.[33]

10.5.2 Antioxidant Activity

E. officinalis and *E. officinalis*-based formulations have been studied extensively for their metal ion chelation, antioxidant, and prooxidant activities.[34] Metal ion chelation prevents the generation of free radicals by the oxidative chemistry induced by metal ions in the gut and elsewhere in the body. Antioxidants bind to the reactive oxygen and reactive nitrogen species generated in the body as a result of bodily functions. Pro-oxidants do the opposite and promote the generation of oxygen species, thereby killing the cells in which pro-oxidants are active. Metal ion chelation and antioxidants play a role in chemoprevention, while pro-oxidant properties can be exploited for chemotherapeutic roles. *E. officinalis* appears to have all three properties as shown by a number of studies. Polatnov *et al.*[35] evaluated the antioxidant activity of four different commercially available *E. officinalis* extracts and evaluated their antioxidant activity using the 2,2-diphenyl-1-picryl-hydrazyl-hydrate (DPPH) and superoxide scavenging assay.[35] They concluded that all the extracts had antioxidant properties in these assays at gastric concentrations and pH, and thus by stabilising reactive species, *E. officinalis* extracts prevented the initiation and propagation of a free radical-derived chain reaction, *e.g.* hydrogen abstraction from susceptible polyunsaturated fatty acids. An *in vitro* study carried out on a HepG2 cell line (hepatocyte cell line) proved the antioxidant effect of *E. officinalis* extract by demonstrating a reduction in the level of molecules, namely lipid hydroperoxide reactive oxygen species (ROS) that causes oxidation induced by stress and an increment in the level of antioxidant enzymes such as glutathione reductase (GSR), glutathione *S*-transferase, and glutathione peroxidase.[36] A study on human dermal fibroblasts showed that *E. officinalis* was effective in repairing collagen damage and ultraviolet-B (UVB) irradiation-induced ROS damage.[37]

Several researchers have studied the antioxidant effects of *E. officinalis* in animal model systems wherein oxidative stress or cell damage was created by metal ions, radiation, and chemicals/drugs. In one study, the oral administration of *E. officinalis* juice demonstrated antioxidant activity by preventing brain damage in a radiation and cadmium-induced rat model.[38] Furthermore, Singh *et al.*[39] illustrated the antioxidant effect of *E. officinalis* in hepatocytes of mice by preventing oxidative damage, necrosis, and apoptosis caused by arsenic.[39] It has also been reported that the antioxidant potential and cytoprotective action of *E. officinalis* was effective in protection from non-steroidal anti-inflammatory drug (NSAID)-induced gastric lesions.[40] A recent study carried out on lead and sodium arsenite-induced rats reported that administration of *E. officinalis* leaf extracts rich in antioxidants was effective in repairing renal tissue damage by restoring the abnormal concentration of urea and creatinine.[41]

Mehrotra *et al.*[42] reported that fruit pulp alcoholic extract of *E. officinalis* also has free radical scavenging and antioxidant activities.[42] Bhattacharya *et al.*[11] proved the potential antioxidant effect of *E. officinalis*, through an increase in the free radical scavenging enzyme level, on rat brain striatal and frontal cortical. This property may be ascribed to tannins, *viz.* emblicanin B, emblicanin A, pedunculagin, and punigluconin, which have vitamin C-like properties.[11]

While much of the literature has focused on the antioxidant properties of *E. officinalis*, there has been mention of its role as a pro-oxidant in biological systems also. Studies have reported that Triphala, a formulation containing 33% *E. officinalis*, exhibits a pro-oxidant nature by inducing ROS production in the murine spleen, normal breast MCF-10F cells, and liver cells.[43] Studies have shown that the higher level of ROS leads to increased lethality in cancer cells.[44,45] Based on the above report, it was concluded that the pro-oxidative nature and anticancer property of Triphala were due to ROS production, which furthermore inhibits proliferation and induces apoptosis in cancer cells.[46]

10.5.3 Immunomodulatory and Antiinflammatory Activities

E. officinalis extracts have been studied for their ability to regulate numerous components of inflammation and to trigger anti-inflammatory mechanisms, including gene expression, regulation of immune cell, and pro-inflammatory cytokine synthesis. The action of *E. officinalis* on the immune system is attributed to biologically active constituents that could potentially protect from various chronic inflammatory diseases like cancers, obesity, degeneration of neurons, type 2 diabetes, and cardiovascular diseases, among other conditions.

Much of the work has been conducted on animal and cell-based models. Treatment of albino mice with *E. officinalis* showed a dose-dependent elevation in the macrophage emigration index, haemagglutination antibody titre, allergic reaction, total leukocyte count, respiratory burst activity of the peritoneal macrophages, distribution of lymphocyte, serum globulin, and weight of lymphoid organ. This explains the immunomodulatory effect of *E. officinalis*.[47] Suresh and Vasudevan (1994) demonstrated the immunomodulatory effect in *E. officinalis*-treated BALB/c mice due to a two-fold increment in the activity of antibody-mediated cellular cytotoxicity and natural killer cell activity. This activity seemed to be because of ascorbic acid.[48] It has been reported that *E. officinalis* treatment attenuated oxidative stress and immunosuppression induced by chromium. It also restored lymphocyte proliferation caused by intoxication of lipopolysaccharide and concanavalin-A. *E. officinalis* has been reported to inhibit DNA fragmentation and is useful in restoring the interferon-γ and IL-2 levels.[49]

Santoshkumar *et al.*[50] reported that *E. officinalis* fruit powder showed a strong anti-inflammatory effect on both acute and chronic inflammation models in comparison with the standard drug, diclofenac sodium.[50] Asmavi *et al.*[51] observed significant anti-inflammatory activity of *E. officinalis* aqueous extract on a carrageenan-induced rat model at 200 mg per kg per day dosage.[51]

A clinical study carried out on a small population of diabetes type 2 patients illustrated that *E. officinalis* fruit extract in single or multiple doses decreases the platelet aggregation at a dose of 500 mg per day.[34] In a study undertaken by Ihantola-Vormisto *et al.*[52] it was found that various extracts of *E. officinalis* leaves attenuated TXB2 produced in platelets during clotting, LTB4 secretion from human PMNs caused by calcium ionophore A23187, platelet aggregation due to adrenaline, emigration of human polymorphonuclear leukocytes induced by leukotriene B4, and degranulation triggered by FMLP.[52] The anti-inflammatory property of *E. officinalis* can be ascribed to the presence of gallic acid, rutin, and ellagic acid. Srikumar *et al.*[53] described an *E. officinalis*-containing Ayurvedic formulation called Triphala that exhibited a stimulative effect on neutrophil functions in albino rats.[53] An experimental study reported that the inflammatory effect of *E. officinalis* extracts can be attributed to attenuation of synthesis and release of inflammatory regulators such as prostaglandins, serotonin, and histamine.[54] In addition, this study was also employed on a cotton pellet granuloma model in rats to evaluate the effect of the hydroalcoholic extract of *E. officinalis* against chronic inflammation. The study reported a significant reduction of granuloma in the treated group, with a 52.36% reduction compared to the control group.[54] Moreover, it has been observed that *E. officinalis* extract inhibited a key transcriptor NF-κB activation which is involved in ageing and chronic inflammatory response.[55] *In vitro* studies on cystic fibrosis extracted bronchial epithelial cells illustrated that *E. officinalis* extract treatment led to the inhibition of pro-inflammatory cytokine expression.[56]

The wide range of anti-inflammatory mechanisms that *E. officinalis* extracts can trigger hold promise of newer formulations and need to be further researched with well-characterised extracts and taken forward to clinical studies.

10.5.4 Antiproliferative, Antitumour, and Antimutagenic Activities

E. officinalis extracts have been studied for their effects on several pathways that play a role in controlling cancer cell proliferation and mutagenesis, such as enhancement of the apoptosis rate, DNA damage, and synthesis of a pro-inflammatory marker.[34] One of the best approaches to controlling cancer is to combine food phytomolecules and chemotherapy. *E. officinalis* could be the food source for the phytochemicals that complement chemotherapy. A study conducted on high-grade serous cancer of ovary cells with p53 and BRCA1/2 mutations which were carboplatin- and taxol-resistant reported that treatment with *E. officinalis* extracts significantly reduced the invasiveness and emigration of cells. In this study, it was also observed that there was a reduction of IGFIR, SNAIL, and HIF-IL expression, and enhancement of E-cadherin.[57]

An *E. officinalis* aqueous extract study showed a dose-dependent size reduction of the solid tumour through inhibition of Cdc25 phosphatase, a cell cycle regulator. This indicated the antiproliferative effect of the *E. officinalis* extract.[58] The anticancer effect of *E. officinalis* has been attributed to multiple phytomolecules such as gallic acid, proanthocyanidin polymers, norsesquiterpenoids and phenolic compounds.[59] Ngamkitidechakul *et al.*[60] illustrated the antitumour capacity of *E. officinalis* extract by an *in vitro* study carried out on various human cancer cell lines, namely SW620 (colorectal), A549 (lung), MDB-MB-231 (breast, triple-negative), HepG2 (hepatocytes), SK-OV3 (ovarian), and HeLa (cervical).[60] An *in vivo* study carried out on DMBA (7,12-dimethylbenz[*a*]anthracene)/TPL (12-*O*-tetradecanoyl-phorbol-13-acetate) induced skin tumours in mice.[60] The antiproliferative effect of *E. officinalis* was demonstrated by a study carried out on four different bone marrow cell lines, Jurkat (T-lymphoid), K562 (erythromyeloid), HEL (leukaemic erythroblast), and Raji (B-lymphoid), which can be ascribed to a potent tumour cell growth inhibitor pyrogallol, present in *E. officinalis.* It has been suggested that *E. officinalis* fruit extract prevented chromosome damage in 3,4-benzo(*a*)pyrene-induced mice.[61] An experimental study illustrated the anticancer activity of *E. officinalis* in tumour-bearing mice, leading to a 35% increase in life expectancy. Another *in vitro* study of *E. officinalis* extract on human ovary cancer cells showed suppression of the angiogenic gene and stimulated expression of autophagic proteins.[62] In an animal experimental model, extract of *E. officinalis* exhibited protection against lung lesions induced by benzopyrene through the suppression of interleukin 1β.[58] The antimutagenic activity of *E. officinalis* was illustrated by different studies which can be ascribed to various secondary metabolites such as terpenoids, phenolates, flavonoids, and glycosides. Jose *et al.*[63] reported that *E. officinalis* extract treatment led to suppression of the activation and mutation of 2-acetamidoflurene and sarcoma induced by 20-methylcholanthrene.[63] This antimutagenic activity was attributed to its ability to inhibit phase I enzymes and free radical scavenging properties.[34] Protection against mutations induced by cyclophosphamide in mice was illustrated by Haque *et al.*[64]

10.5.5 Neuroprotective and Antiaging Activity

Currently, very few curative or preventive agents are available for the treatment of brain health disorders such as Parkinsonism and Alzheimer's disease. *E. officinalis*'s use in traditional medicine as a neuroprotective has triggered scientific interest in the use of *E. officinalis* extracts for their effects on various pathways linked with brain and neural function.

Studies have reported that the antidepressant-like effect of *E. officinalis* extracts can be attributed to attenuation of neurotransmitters and certain enzymes such as gamma-aminobutyric acid (GABA) and monoamine oxidase-A (MAO-A).[65] Studies have also illustrated that *E. officinalis* powder produced dose-dependent enhancement of memory, which indicates that *E. officinalis* can be used as a therapeutic agent for treating Alzheimer's disease because of its multitudinous benefits in memory loss reversal and memory enhancement.[66] *E. officinalis* has also been useful in curing epilepsy. *E. officinalis* hydroalcoholic extract was observed to be beneficial against status epilepticus and generalised tonic convulsions induced by kainic acid and pentylenetetrazole.[67] Justin Thenmozhi *et al.*[67] evaluated the nootropic activity of *E. officinalis* on aluminium chloride induced. *E. officinalis* treatment with 200 mg per kg dosage yielded a significant decrease of hippocampal and cortical enzymes by inhibiting amyloid precursor protein (APP), acetylcholinesterase, and A-beta, beta and gamma secretase.[67] Damodar Reddy *et al.*, demonstrated the effect of *E. officinalis* against alcohol-induced brain mitochondrial dysfunction which revealed that treatment with *E. officinalis* improved the activity of cytochrome C oxidase and endogenous antioxidant system.[68] In an experimental study on stress-induced mice, it was observed that *E. officinalis* extract treatment enhanced mental health by reducing corticosterone levels.[69] Furthermore, *E. officinalis* extract was found to exhibit an antiaging effect that benefits collagen synthesis and prevents its degradation.[70] An *in vitro* study reported that one of the phytochemical chebulagic acids that exist in *E. officinalis*, ameliorates intracellular organelles and proteins autophagy. This process leads to protection from conditions mimicking Parkinson's disease induced by 1methyl-4-phenylpyridinium (MPP+), a neuronal toxin.[69] A study carried out on a scopolamine-induced amnesia mice model to assess the therapeutic efficacy of hydroalcoholic *E. officinalis* extract also gave promising results and reported that the extract has the potential to inhibit cholinergic dysfunction, in turn leading to a reduction in cognitive deficits.[71]

10.5.6 Hepatoprotective Activity

Studies have shown that *E. officinalis* possesses hepatoprotective effects.[72] Many studies have used model systems wherein acute hepatic damage was induced by using chemicals such as CCl_4 and thioacetamide (an organosulphuric drug that leads to hepatocarcinoma and related disorders) and alcohol. *E. officinalis* extract was useful in decreasing the graveness of hepatic fibrosis caused by carbon tetrachloride and thioacetamide,[8] while in other studies the extracts reduced the percentage of liver infiltration and focal necrosis,[34] reduced the synthesis of DNA and thioacetamide-augmented ornithine decarboxylase, and balanced the levels of GSH, SGOT, SGPT, and glutathione peroxidase G6PD. Pramyothin *et al.*[73] also showed that *E. officinalis* treatment has a potential hepatoprotective effect by balancing serum SGOT, SGPT, TNF, TG, and IL-1β on a primary hepatocyte culture of rats intoxicated with alcohol.[73] The protective activity of aqueous extract on hepatic cells in non-alcoholic fatty liver disease[58] was shown in an animal model and attributed to the phytochemical quercetin present in *E. officinalis* fruit. The wide-ranging hepatoprotective activity of *E. officinalis* may be due to its property to inhibit CYP 2E1, and its antioxidant and membrane-stabilising activities.[74]

Another area of research for *E. officinalis* extracts has been hepatotoxicity caused by iron overload.[75] From a therapeutic point of view, chelation has been a very important method in the treatment of iron overload and deferoxamine (Desferal), being a classical chelator of iron, is useful in the reduction of pathogenesis induced by iron overload but it cannot be consumed orally and has certain side effects, and hence there is a need to develop new effective iron chelators for oral consumption. An *E. officinalis*-based chelator may address this need. Preclinical studies carried out in rats have reported that a emblicanin A- and B-rich *E. officinalis* prophylactic treatment reduced ALT, LDH, AST, and lipid peroxidation induced by iron overload.[76] Also, studies have demonstrated the benefits of *E. officinalis* extract against iron load-induced liver damage by chelating the labile iron due to its properties to reduce ferritin release, DNA break down due to ferrous, and it also enhances antioxidant enzyme levels.[77] *E. officinalis* thus is a potential hepatoprotective agent which protects against a broad range of hepatotoxins such as heavy metals, ethanol, ochratoxins, paracetamol, hexachlorocyclohexane, antitubercular drugs, and carbon tetrachloride,[8] and in turn holds promise for developing a new range of iron chelators.

10.5.7 Gastroprotective Activity

Many Ayurvedic formulations containing *E. officinalis* are effective in ulcer relief. Studies carried out in animal models using rats have shown that *E. officinalis* alcoholic extract prevents ulcerative colitis induced by acetic acid.[34] Extracts of *E. officinalis* demonstrated a significant reduction in ethanol-induced gastric ulcers. Pretreatment of the rats with butanol extracts of water fraction of *E. officinalis* significantly reduced the severity of indometacin-induced stomach mucous layer damage.[78] Mehmood *et al.*[79] demonstrated that extract of *E. officinalis* is a potential inhibitor of Ca^{2+} channels and muscarinic receptors mediated antidiarrheal and spasmolytic activities.[79] A clinical study reported that oral administration of *E. officinalis* crude powder for 7 days at 3 g thrice daily reduced gastritis syndrome in 85% of cases.[80] A study illustrated the protective effect of *E. officinalis* alcoholic extract against ulcers induced by NSAIDs.[58] Another study reported the *H. pylori* inhibition effect of *E. officinalis* alcoholic extract at a minimum concentration ranging from 0.91 to 1.87 $\mu g\, \mu l^{-1}$.[58] In a study carried out in rats treatment of *E. officinalis* extracts at 250 $mg\, kg^{-1}$ dosage remarkably decreased the stomach mucosal layer, secretions from the stomach, and titrable acidity.[23]

E. officinalis as a whole improves gut health, thereby improving nutrient absorption which is also a part of its Rasayana activity. Thus, *E. officinalis* helps in the treatment of malabsorption and nutrition deficiency disorders. Among the 354 formulations of *E. officinalis*, 19 are widely used in the treatment of iron deficiency anaemia (Pandu roga), which probably improves iron uptake from the gut. Dietary iron is less soluble and poorly absorbed.[81] Duodenal villi contain Dcytb reductase, which converts ferric iron into the ferrous state. The ferrous state is easily taken up by a divalent metal transporter into the cell. This iron is further stored as ferritin and released as and when required.[82]

10.5.8 Serum Glycaemic Modulation

Among the 354 *E. officinalis* formulations, about 27 formulations are used for the treatment of Prameha (diabetes mellitus and similar disorders), this has additionally advanced several non-clinical and clinical researches on the potential of *E. officinalis* in glycaemic modulation.[34] A clinical study carried out on human diabetic volunteers administered with *E. officinalis* extract for 21 days showed a significant reduction in both fasting and postprandial glucose levels. A clinical study on streptozotocin-induced diabetic rat models to understand the hypoglycaemic activity of *E. officinalis* fruit extract 200 $mg\, kg^{-1}$ body weight revealed that oral extracts showed a time-bound reduction in blood glucose levels in the diabetic rats.[61] Serum adiponectin is a protein hormone involved in regulating blood glucose and the breakdown of glycosylated protein. Rao *et al.*[83] concluded that the administration of *E. officinalis* upregulated the serum adiponectin levels and thus aided in a reduction of glycosylated protein.[83] In another study carried out on BRIN-BD11, a clonal pancreatic cell line, to study insulin output it was proved that aqueous extract of *E. officinalis* enhanced glucose-dependent insulin secretion alongside basal insulin output.[34] Conclusively, the hypoglycaemic effect of *E. officinalis* is directly attributed to its polyphenol-gallic acid and ellagic acid[83] and other 'organic acid constituents' ability to inhibit gluconeogenesis and glycogenolysis.[61]

E. officinalis has also been shown to improve glucose utilisation and maintain glucose homeostasis. A preclinical study concluded that *E. officinalis*'s pivotal role in glucose homeostasis was by maintaining the equilibrium of important enzymes involved in glycolysis, glycogenesis, and gluconeogenesis.[84] A study involving 3T3L1 adipocyte cell lines and placebo-treated cohorts proved that *E. officinalis* extract-pretreated cell lines showed a concentration-dependent effect in glucose uptake.[21] A study carried out on an arsenic-induced hyperglycaemic mice model revealed that coadministration of arsenic and *E. officinalis* (500 $mg\,kg^{-1}$ body weight) significantly reduced the concentration of phosphoenolpyruvate carboxykinase, glucose-6 phosphatase, and pancreatic inflammation markers like IL-1β, and accordingly regulated the hepatic glucose regulatory enzyme levels and blood glucose levels.[85,86] Unequivocally, the above observations emphasise the hypoglycaemic properties and antihyperglycaemic potential of *E. officinalis* and its extracts.

10.5.9 Cardioprotective Activity

Several cell and animal studies have illustrated *E. officinalis*'s cardioprotective potential with regards to its ability to inhibit oxidised low-density lipoprotein-induced vascular cell proliferation and thus prevent atherosclerosis.[62,85] An experimental study by Duan *et al.*[59] on *E. officinalis*'s extract containing corilagin and its analogue 1,6-di-*O*-galloyl-D-glucose against vascular smooth muscle cells of rat and endothelial cells of human umbilical vein proved a positive decrease in oxidative stress and also inhibited oxidised low-density lipoprotein-cholesterol induced vascular cell proliferation.[59] Additionally, studies have also proved *E. officinalis* 'fruit extracts' efficacy in preventing diabetes-induced myocardial dysfunction[62] alongside reducing hypertension, bradycardia, hypertrophy, and myocardial functional alterations.[84] The administration of *E. officinalis* extracts on high-fat diet-fed rats altered the heart rate and sympathovagal balance and also resulted in a progressive improvement of cardiac autonomic malfunctions.[85] On the whole, the cardioprotective effect of *E. officinalis* is generally ascribed to its constituents which include tannoids, corilagin, gallic acid, ellagic acid, and emblicanin-A and B.[34]

10.5.10 Lipid Profile Modulation

Oxidised-LDL production is considered to be one of the risk factors of atherosclerosis. Several studies have proven *E. officinalis*'s prophylactic property for atherosclerosis. A clinical study on subjects administered a high-fat diet attributed the antioxidant property and hypolipidemic activity of *E. officinalis* for the dose-dependent reduction of oxidised low-density lipoprotein levels and amelioration of atherosclerosis in the subjects.[34] A clinical study comprising subjects with Class I obesity with a BMI of 25–35 showed a significant reduction in levels of LDL-cholesterol, high-sensitive C-reactive protein, and total cholesterol in subjects fed with extracts of *E. officinalis* (500 mg) for 12 weeks.[34] Hepatic PPAR alpha protein, a protein that regulates the gene transcription involved in cholesterol and lipid metabolism was upregulated significantly in rats that were orally administered extracts of *E. officinalis* over aged control rats.[62] A study on a high-cholesterol diet-fed rabbit model showed a remarkable reduction in serum triglyceride, hepatic cholesterol content, LDL cholesterol, and phospholipid levels.[34]

Clinical research to evaluate the hypolipidaemic activity of *E. officinalis* in type II hyperlipidaemic patients in comparison with a simvastatin standard group reported that groups treated with *E. officinalis* extracts showed a significant decrease in total cholesterol, triglycerides, VLDL, and LDL, and a rise in HDL levels over the simvastatin-treated group.[86] A recent study has revealed the potential of *E. officinalis* to enhance peroxisome lipid oxidation and proliferator-activated receptors-α (PPARα) expression and reduce the lipogenic activity in the liver.[87]

10.5.11 Antitussive and Other Respiratory Problems

E. officinalis as a fruit in combination with other medicinal herbs has been extensively used in managing respiratory complications associated with inflammation, such as cough and asthma. Nosal'ova *et al.*[88] evaluated the antitussive potential of *E. officinalis* on conscious cats which was stimulated for tracheobronchial and laryngopharyngeal mucosal secretions. A dose-dependent cough-suppressing effect was observed on the mucous secretion pathway.[88] A clinical trial undertaken on 30 subjects with chronic smoking habits alongside a double-blind placebo control reported that the subjects consuming the fruit extract of *E. officinalis* (250 mg) consisting of hydrolysable tannins like emblicanin-A and -B, pedunculagin, and punigluconin for 2 months, twice a day showed a good prognostication over the placebo-treated control.[89] Studies have revealed that an *E. officinalis*-supplemented diet modulated proinflammatory cytokine (TNF-α) induction and thus acts as a prophylactic measure against pneumonia-initiated respiratory tract infection.[32] A study on anaesthetised male dogs by Geer *et al.*[90] revealed *E. officinalis*'s hypotensive effect, which was mediated by synergistic cholinergic and histaminergic actions on arterial blood pressure.[90] In some parts of the world, *E. officinalis* as a fruit is used as a cure for lung inflammation. Additionally, *E. officinalis* juice mixed with honey and pippali (long pepper) is used as a remedy for asthma, cough, hiccough, and other respiratory-related diseases.[91] *E. officinalis* fruit extract as a good source of vitamins, antioxidants, tocopherol, and beta-carotene has been proven to have inhibitory effects on chronic cough.[92]

10.6 Conclusion

Emblica officinalis is one of the most widely used plants in the Ayurveda system of medicine of India, wherein *E. officinalis*-based formulations are used as therapeutics in the treatment of a wide spectrum of diseases. *E. officinalis* is also a part of many food supplements such as Chyawanprasha. It is also consumed as a food and juice because of its various potential health benefits. This diverse therapeutic activity suggests that phytochemicals of *E. officinalis* play a role in many pharmacological functions. Research using cell and animal model systems is providing evidence of this wide-ranging pharmacological activity of *E. officinalis* extracts. A few clinical studies conducted using *E. officinalis* extracts that build on therapeutic benefits linked to pharmacological properties have shown promising results.

One challenge that research studies based on *E. officinalis* face is its complex phytochemistry. Much of the pharmacological activity of *E. officinalis* comes from its polyphenols. Nearly 200 polyphenols were curated from literature for this work. Only 130 of these could be linked to PubChem IDs. This in itself is indicative of the characterisation challenge that *E. officinalis* extracts face. Polyphenols are also a very reactive group of molecules that oxidise easily. This further adds to the complexity of characterising extracts used in preclinical and clinical studies. Poltanov *et al.*[35] studied four commercially available extracts of *E. officinalis* and found huge differences in their composition. The pharmacopoeias require only a simple specification – gallic acid and catechins. This does not capture the functionality of *E. officinalis*. To be able to study *E. officinalis* pharmacology to create quality functional products will require more detailed characterisation. With the growing scientific interest in polyphenols, our ability to characterise this group is also improving. Studies using *E. officinalis* will gain from this. Only through a strong link between phytochemistry and pharmacology will the true potential of *E. officinalis* be realised.

Abbreviations

ALT	Alanine aminotransferase
AST	Aspartate aminotransferase
FRLHT CIPA	Foundation for revitalisation of local health traditions – clinically important plants of Ayurveda
G6PD	Glucose-6-phosphate dehydrogenase
HDL	High-density lipoprotein
IL	Interleukin
LDH	Lactate dehydrogenase
LDL	Low-density lipoprotein
NIN IFCT	National Institute of Nutrition – Indian Food Composition Tables
SGOT	Serum glutamic-oxaloacetic transaminase
SGPT	Serum glutamic pyruvic transaminase
STZ	Streptozotocin
TG	Triglycerides
TNF	Tumour necrosis factor
VLDL	Very low-density lipoprotein

Acknowledgements

The authors wish to acknowledge the support of Rural India Support Trust for funding the research on Amla.

References

1. P. V. Sharma, *Charaka Samhita*, Chowkhamba Sanskrit series office, Varanasi India, 1st edn, 2007.
2. P. V. Sharma, *Charaka Samhita*, Chowkhamba Sanskrit Series office Varanasi, Varanasi India, 1st edn, 2007.
3. S. N. B. K. Ravikumar, D. K. Ved, J. R. Bhatt and G. S. Goraya, *Compendium of Traded Indian Medicinal Plants*, Foundation for Revitalization of local Health Traditions, Banglore India, 2018.
4. D. J. L. N. Sastry, *Dravyaguna Vijnana*, Chaukhambha Orientalia, Varanasi India, 2012.
5. P. P. V. Sharma, *Dravyaguna Vijnana*, Chaukhambha Bharati Academy, Varanasi India, 1st edn, 2007.
6. D. M. Sankaranarayana, *Roga vijnana and Vikriti vijnana*, Chowkhambha Sanskrit series office, Varanasi India, 1st edn, 2007.
7. B. Mishra, *Bhavaprakash*, Chowkamba krishnadas Academy, Varanasi India, 1st edn, 2004.
8. K. H. Khan, *Bot. Res. Int.*, 2009, **2**, 218.
9. T. Longvah, I. Anantan, K. Bhaskarachary, K. Venkaiah and T. Longvah, *Indian food composition tables*, National Institute of Nutrition, Indian Council of Medical Research Hyderabad, 2017.
10. G. S. Kumar, H. Nayaka, S. M. Dharmesh and P. V. Salimath, *J. Food Compos. Anal.*, 2006, **19**, 446.
11. A. Bhattacharya, A. Chatterjee, S. Ghosal and S. K. Bhattacharya, *Indian J. Exp. Biol.*, 1999, **37**, 676–680.
12. S. S. Yadav, M. K. Singh, P. K. Singh and V. Kumar, *Biomed. Pharmacother.*, 2017, **93**, 1292.
13. R. Soman and P. P. Pillay, *Curr. Sci.*, 1962, **31**, 13.
14. K. S. Jamwal, I. P. Sharma and L. Chopra, *J. Sci. Ind. Res.*, 1959, **18**, 180.
15. S. M. J. S. Quadry, S. S. Banga and C. K. Atal, *Ind. J. Pharm.*, 1962, **24**, 2.
16. V. Bansal, A. Sharma, C. Ghanshyam and M. L. Singla, *J. Liq. Chromatogr. Relat. Technol.*, 2015, **38**, 619.
17. M. Leri, M. Scuto, M. L. Ontario, V. Calabrese, E. J. Calabrese, M. Bucciantini and M. Stefani, *Int. J. Mol. Sci.*, 2020, **21**, 1250.
18. B. Halliwell, *Drugs Aging*, 2001, **18**, 685.
19. R. J. Williams, J. P. Spencer and C. Rice-Evans, *Free Radic. Biol. Med.*, 2004, **36**, 838.

20. G. Hussain, J. Huang, A. Rasul, H. Anwar, A. Imran, J. Maqbool, A. Razzaq, N. Aziz, E. U. H. Makhdoom, M. Konuk and T. Sun, *Molecules*, 2019, **24**, 2213.
21. S. J. Hong, M. H. Ahn, J. Sangshetti and R. B. Arote, *Acta Biomater.*, 2019, **97**, 105.
22. M. Markovic, S. Ben-Shabat, A. Aponick, E. M. Zimmermann and A. Dahan, *Int. J. Mol. Sci.*, 2020, **21**, 3248.
23. A. Charu, B. Dipti, A. Dhruv and T. Vinita, *Ann. Hortic.*, 2018, **11**, 1–12.
24. D. W. Lim, J. G. Kim and Y. T. Kim, *Nutrients*, 2016, **8**, 760.
25. B. Goel, N. Pathak, D. K. Nim, S. K. Singh, R. K. Dixit and R. Chaurasia, *Int. J. Basic Clin. Pharmacol.*, 2014, **3**, 365.
26. K. Jaijoy, N. Soonthornchareonnon, A. Panthong and S. Sireeratawong, *Int. J. Appl. Res. Nat. Prod.*, 2010, **3**, 28.
27. J. B. Perianayagam, S. Sharma, A. Joseph and A. Christina, *J. Ethnopharmacol.*, 2004, **95**, 83.
28. S. Rahman, M. Akbor, A. Howlader and A. Jabbar, *Pak. J. Biol. Sci.*, 2009, **12**, 1152.
29. B. Dutta, I. Rahman and T. Das, *Mycoses*, 1998, **41**, 535.
30. Y. Xiang, Y. Pei, C. Qu, Z. Lai, Z. Ren, K. Yang, S. Xiong, Y. Zhang, C. Yang and D. Wang, *Phytother. Res.*, 2011, **25**, 975.
31. A. Bagavan, A. A. Rahuman, N. K. Kaushik and D. Sahal, *Parasitol. Res.*, 2011, **108**, 15.
32. A. Saini, S. Sharma and S. Chhibber, *Indian J. Med. Res.*, 2008, **128**, 188.
33. B. Thaweboon and S. Thaweboon, *Asian Pac. J. Trop. Med.*, 2011, **4**, 41.
34. B. C. Variya, A. K. Bakrania and S. S. Patel, *Pharmacol. Res.*, 2016, **111**, 180.
35. E. A. Poltanov, A. N. Shikov, H. J. Dorman, O. N. Pozharitskaya, V. G. Makarov, V. P. Tikhonov and R. Hiltunen, *Phytother. Res.*, 2009, **23**, 1309.
36. M. M. Shivananjappa and M. K. Joshi, *Pharm. Biol.*, 2012, **50**, 497.
37. M. Majeed, B. Bhat, S. Anand, A. Sivakumar, P. Paliwal and K. G. Geetha, *J. Cosmet. Sci.*, 2011, **62**, 49.
38. R. Purohit, P. K. Joshi, A. Basu, M. Agarwal, A. Chakrawarti and K. Bhartiya, *Pharmacologyonline*, 2010, **3**, 304.
39. M. K. Singh, S. Dwivedi, S. S. Yadav, P. Sharma and S. Khattri, *Indian J. Clin. Biochem.*, 2014, **29**, 29.
40. A. Chatterjee, S. Chattopadhyay and S. K. Bandyopadhyay, *Evidence-Based Complementary Altern. Med.*, 2011, **2011**, 146808.
41. M. Fazal and R. Gayathri, *Bioinformation*, 2021, **17**, 295.
42. S. Mehrotra, R. Jamwal, R. Shyam, D. K. Meena, K. Mishra, R. Patra, R. De, A. Mukhopadhyay and A. K. Srivastava, *J. Med. Plant Res.*, 2011, **5**, 2577.
43. T. Sandhya, K. M. Lathika, B. N. Pandey and K. P. Mishra, *Cancer Lett.*, 2006, **231**, 206.
44. J. Wang and J. Yi, *Cancer Biol. Ther.*, 2008, 7, 1875.

45. L. Wang, R. Leite de Oliveira, S. Huijberts, E. Bosdriesz, N. Pencheva, D. Brunen, A. Bosma, J. Y. Song, J. Zevenhoven, G. T. Los-de Vries, H. Horlings, B. Nuijen, J. H. Beijnen, J. H. M. Schellens and R. Bernards, *Cell*, 2018, **173**, 1413.
46. T. Sandhya and K. P. Mishra, *Cancer Lett.*, 2006, **238**, 304.
47. R. Suja, A. Nair, S. Sujith, J. Preethy and A. Deepa, *Indian J. Anim. Res.*, 2009, **43**, 103.
48. K. Suresh and D. Vasudevan, *J. Ethnopharmacol.*, 1994, **44**, 55.
49. M. S. Ram, D. Neetu, B. Yogesh, B. Anju, P. Dipti, T. Pauline, S. Sharma, S. Sarada, G. Ilavazhagan and D. Kumar, *J. Ethnopharmacol.*, 2002, **81**, 5.
50. J. Santoshkumar, M. S. Devarmani, M. Sajjanar, M. Pranavakumar and P. Dass, *Med. Innov.*, 2013, **2**, 17.
51. M. Asmawi, H. Kankaanranta, E. Moilanen and H. Vapaatalo, *J. Pharm. Pharmacol.*, 1993, **45**, 581.
52. A. Ihantola-Vormisto, J. Summanen, H. Kankaanranta, H. Vuorela, Z. M. Asmawi and E. Moilanen, *Planta Med.*, 1997, **63**, 518.
53. R. Srikumar, N. J. Parthasarathy and R. S. Devi, *Biol. Pharm. Bull.*, 2005, **28**, 1398.
54. M. Golechha, V. Sarangal, S. Ojha, J. Bhatia and D. S. Arya, *Int. J. Inflamm.*, 2014, **2014**, 178408.
55. T. Yokozawa, H. Y. Kim, H. J. Kim, T. Okubo, D.-C. Chu and L. R. Juneja, *Br. J. Nutr.*, 2007, **97**, 1187.
56. S. Lanka, *World J. Pharm. Res.*, 2018, **7**, 380.
57. A. De, B. Powers, A. De, J. Zhou, S. Sharma, P. Van Veldhuizen, A. Bansal, R. Sharma and M. Sharma, *Oncotarget*, 2016, **7**, 31484.
58. M. Fazil and S. Nikhat, *European J. Med. Plants*, 2019, **30**, 1.
59. W. Duan, Y. Yu and L. Zhang, *Yakugaku Zasshi*, 2005, **125**, 587.
60. C. Ngamkitidechakul, K. Jaijoy, P. Hansakul, N. Soonthornchareonnon and S. Sireeratawong, *Phytother. Res.*, 2010, **24**, 1405.
61. S. Mirunalini and M. Krishnaveni, *J. Basic Clin. Physiol. Pharmacol.*, 2010, **21**, 93.
62. P. R. Bhandari and M. A. Kamdod, *Int. J. Green Pharm.*, 2012, **6**, 257.
63. J. K. Jose, G. Kuttan, J. George and R. Kuttan, *J. Clin. Biochem. Nutr.*, 1997, **22**, 171.
64. R. Haque, B. Bin-Hafeez, I. Ahmad, S. Parvez, S. Pandey and S. Raisuddin, *Hum. Exp. Toxicol.*, 2001, **20**, 643.
65. D. Dhingra, P. Joshi, A. Gupta and R. Chhillar, *CNS Neurosci. Ther.*, 2012, **18**, 419.
66. M. Vasudevan and M. Parle, *Physiol. Behav.*, 2007, **91**, 46.
67. A. Justin Thenmozhi, M. Dhivyabharathi, T. R. William Raja, T. Manivasagam and M. M. Essa, *Nutr. Neurosci.*, 2016, **19**, 269.
68. V. Damodara Reddy, P. Padmavathi, S. Gopi, M. Paramahamsa and N. Ch. Varadacharyulu, *Indian J. Clin. Biochem.*, 2010, **25**, 419.
69. H. J. Kim, J. Kim, K. S. Kang, K. T. Lee and H. O. Yang, *Biomol. Ther.*, 2014, **22**, 275.

70. P. Chanvorachote, V. Pongrakhananon, S. Luanpitpong, B. Chanvorachote, S. Wannachaiyasit and U. Nimmannit, *J. Cosmet. Sci.*, 2009, **60**, 395.
71. I. Husain, S. Zameer, T. Madaan, A. Minhaj, W. Ahmad, A. Iqubaal, A. Ali and A. K. Najmi, *Metab. Brain Dis.*, 2019, **34**, 957.
72. K. R. Thilakchand, R. T. Mathai, P. Simon, R. T. Ravi, M. P. Baliga-Rao and M. S. Baliga, *Food Funct.*, 2013, **4**, 1431.
73. P. Pramyothin, P. Samosorn, S. Poungshompoo and C. Chaichantipyuth, *J. Ethnopharmacol.*, 2006, **107**, 361.
74. S. Talreja, S. Kumari, P. Srivastava and S. Pandey, *World J. Pharm. Pharm. Sci.*, 2019, **8**, 622.
75. E. Prus and E. Fibach, *J. Biol. Regul. Homeost. Agents*, 2012, **26**, 203.
76. A. Bhattacharya, M. Kumar, S. Ghosal and S. Bhattacharya, *Phytomedicine*, 2000, **7**, 173.
77. R. Sarkar, B. Hazra and N. Mandal, *Toxicol. Ind. Health*, 2015, **31**, 656.
78. A. Al-Rehaily, T. Al-Howiriny, M. Al-Sohaibani and S. Rafatullah, *Phytomedicine*, 2002, **9**, 515.
79. M. H. Mehmood, H. S. Siddiqi and A. H. Gilani, *J. Ethnopharmacol.*, 2011, **133**, 856.
80. M. Bhagat, Indian gooseberry (*Emblica officinalis*): Pharmacognosy review, in *Utilisation and management of medicinal plants*, ed. A. Gupta and V. K. Kaul, Daya Publishing House, New Delhi, 2014, pp. 471–487.
81. K. J. Wienk, J. J. Marx and A. C. Beynen, *Eur. J. Nutr.*, 1999, **38**, 51.
82. P. Sharp and S. K. Srai, *World J. Gastroenterol.*, 2007, **13**, 4716.
83. T. Rao, N. Sakaguchi, L. Juneja, E. Wada and T. Yokozawa, *J. Med. Food*, 2005, **8**, 362.
84. J. J. D'souza, P. P. D'souza, F. Fazal, A. Kumar, H. P. Bhat and M. S. Baliga, *Food Funct.*, 2014, **5**, 635.
85. P. S. Kanthe, B. S. Patil, S. C. Bagali, R. C. Reddy, M. R. Aithala and K. K. Das, *J. Clin. Diagn. Res.*, 2017, **11**, CC05.
86. S. A. Almatroodi, M. A. Alsahli, A. Almatroudi, K. Dev, S. Rafat, A. K. Verma and A. H. Rahmani, *Gene Rep.*, 2020, **21**, 100820.
87. B. C. Variya, A. K. Bakrania, Y. Chen, J. Han and S. S. Patel, *Biomed. Pharmacother.*, 2018, **108**, 1274.
88. G. Nosal'ova, J. Mokrý and K. T. Hassan, *Phytomedicine*, 2003, **10**, 583.
89. T. K. Biswas, S. Chakrabarti, S. Pandit, U. Jana and S. K. Dey, *J. Herb. Med.*, 2014, **4**, 188.
90. M. I. Geer, U. A. Zia, B. Dilawer, G. Moinuddin, A. K. Anees and D. Kshama, *JK-Pract.*, 2005, **12**, 213.
91. D. Jayaweera, *Medicinal Plants (indigenous and Exotic) Used in Ceylon: Cactaceae-Fagaceae*, National Science Council of Sri Lanka, 1980.
92. Md. D. Khurshid, V. Shukla, B. Kumar and A. Singh, *Int. J. Pharm. Biol. Sci.*, 2020, **10**, 102.

CHAPTER 11

Galangal – a Promising Herb in Ayurvedic Formulations: Phytochemistry and Pharmacological Evaluations

GURUPRASAD C. NILLE,*[a,†] SHARDENDU KUMAR MISHRA,[b,†] OMKAR S. NILLE,[c] ANGAD KUMAR SINGH[d] AND ANAND KUMAR CHAUDHARY[a]

[a] Department of Rasa Shastra & Bhaishajya Kalpana, Faculty of Ayurveda, Institute of Medical Sciences, Banaras Hindu University, Varanasi, India; [b] Department of Pharmacology, KIET School of Pharmacy, KIET Group of Institutions, Ghaziabad, India; [c] Fluorescence Spectroscopy Research Laboratory, Department of Chemistry, Shivaji University, Kolhapur, India; [d] Department of Chemistry, Central University of South Bihar, Gaya, India
*Email: drguruprasadnille0412@gmail.com

11.1 Introduction

Ayurveda (a traditional Indian system of medicine) uses various medicinal herbs to treat different health ailments. The practice of Ayurveda includes thousands of medicinal herbs with different medicinal preparations, either in the form of a single herb or polyherbal formulations. Galangal, a common designation used for the plant species in Ayurveda, has controversial status amongst similar herbs of the Zingiberaceae family. The term galangal is used

†These authors have contributed equally to this work and share the first authorship.

Chemistry, Biological Activities and Therapeutic Applications of Medicinal Plants in Ayurveda
Edited by Augustine Amalraj, Sasikumar Kuttappan and Karthik Varma A. C.

Published by the Royal Society of Chemistry, www.rsc.org

to denominate *Alpinia galanga* (L.) Willd., known as greater galangal, and two other closely allied varieties, *viz. Alpinia officinarum* Hance and *Alpinia calcarata* Roscoe, known as lesser galangal.[1,2] Similarly, the word galanga is used for *Kaempferia galanga* L., which has no relation to any of the above species.[1,2] Further, the confusion is not only over with the species name but also the traditional plant "*Rasna*" in Ayurveda, which is used as galangal with different botanical names as per its geographical origin and following the local tradition.[3] In South India, *Alpinia galanga* (L.) Willd. and *Alpinia calcarata* Roscoe are used as *Rasna* whereas, in North India, *Pluchea lanceolata* C.B. Clarke is used as the source plant of *Rasna*.[3–5] The leaves of *Rasna* and the *Alpinia galanga* (L.) Willd. resemble the leaves of *Ela* (*Elettaria cardamomum* Maton), and the rhizomes of both plants possess a strong aroma or aromatic odour. Therefore, the Ayurvedic physicians of South India accept *Alpinia galanga* (L.) Willd. as *Rasna*.[5] Besides, in Ayurveda *Elaparni* and *Sugandha*, the synonyms of *Rasna*, also support the above findings between *Rasna* and *Alpinia galanga* (L.) Willd. However, *Nakuli* and *Gandhanakuli* have commonly used Sanskrit synonyms of *Rasna* in ancient texts of Ayurveda,[5] whereas *Cittaratta* and *Peraratta* are the Malayalam terms used for *Nakuli* and *Gandhanakuli*, respectively, in *Osadhinigantu* and *Ayurvedavisvakosam*, texts of the Keralian tradition of Ayurveda.[6] In South Indian markets, *Alpinia galanga* (L.) Willd. is called *Peraratta*, and *Alpinia calcarata* Roscoe is called *Cittaratta*, while both are available at the local markets as *Rasna*. However, the rhizomes of *Alpinia galanga* (L.) Willd. possesses a more aromatic odour than *Alpinia calcarata* Roscoe.[7] Although there remains controversy, *Alpinia galanga* (L.) Willd. is accepted and widely used as *Rasna* and galangal in most Ayurvedic formulations.

11.1.1 *Rasna* in Ayurveda

The first references to *Rasna* date back to the Vedic period. It has been used as a medicinal plant alone and as a principal ingredient in many compound formulations. *Charaka* (300 BC) has included *Rasna* in "*Vayasthapaka Gana*", a group of medicinal plants with antiageing properties that can maintain vigour and strength in the body.[8] Further, *Charaka* mentioned that *Rasna* is the best drug for pacifying the *Vata dosha* in the human body.[9] It supports the traditional claims and its medicinal uses in all kinds of degenerative and chronic inflammatory conditions, such as arthritis and neurodegenerative diseases. It was also indicated in rheumatism, intermittent fevers, dyspepsia, cough, bronchitis, asthma, and other respiratory ailments. In Ayurveda, *Rasna* has many synonyms such as *Rushna*, *Rasana*, *Rasya*, *Elaparni*, *Sugandha*, *Yukta*, *Surabhi*, *Sugantha moola*, *Suvatha*, *Surpagandha*, *Atirasa*, *Rasadhya*, *Nakuli*, *Gandhanakuli*, *Kulinjan*, *Aratha*, *Malay Vacha*, *sthulagranthi*, *etc.*[10] It is bitter (*Tikta rasa*), hot (*Ushna veerya* and *Katu Vipaka*), helps to pacify the *Kapha* and *Vata dosha*, and also works as an antidote for animal poisons, improves digestion, relieves flatulence and dyspepsia, acts as a carminative agent to relieve abdominal colic, and shows anti-inflammatory and analgesic effects when applied locally in the form of a paste or poultice. The official part of the plant used is the rhizome which forms the main ingredient of many Ayurvedic formulations.[10] Though *Rasna* is a plant of great therapeutic value in Ayurveda, it is one of the most controversial plants in Ayurveda. The official botanical name of *Rasna* is nominated as *Pluchea lanceolata* in the Ayurvedic Pharmacopoeia of India (API), an official publication of the Ministry of AYUSH, Government of India, New Delhi.[11] However, many other references are available which support the use of galangal [*Alpinia galanga* (L.) Willd.] as a *Rasna*. In different territories of India, about 34 plants have been identified yet as *Rasna*.[3,12] In South India, *Alpinia galanga* (L.) Willd. and *Alpinia calacarata* Roscoe; in North India, *Pluchea lanceolata*; in Bengal region, *Vanda tessellata*; in Kashmir and Nepal regions, *Viscum album* and *Inula racemosa*; in Punjab, *Withania coagulens*; in Bombay, *Aristolochia indica*, and throughout India, *Lochnera rosea*, *Enicostema littorale*, and *Rauwolfia lanceolata* have been used in the name of *Rasna*.[13] The most commonly used plant with the name *Rasna* in the Kerala region is shown in Figure 11.1.

11.2 Phytochemistry

11.2.1 Phytochemicals Present in the Galangal Species

The three galangals that are accepted under the term *Rasna*, namely *A. galanga* (L.) Willd., *A. officinarum* Hance, and *A. calcarata* (Roscoe) belong to the Zingiberaceae family,[14] and are rich in various organic chemicals and aromatic compounds. They have traditionally been used as a remedy for many diseases. Phytochemicals are heterogeneous compounds containing a variety of chemical compounds with characteristic medicinal activities.[15] The galangal plant species have revealed the presence of chemically rich and medicinally valuable phytochemicals like alkaloids, terpenoids, flavonoids, coumarins, saponins,

Figure 11.1 Rhizomes, leaves, and flowers of *Rasna*.

tannins, proteins, carbohydrates, amino acids, and essential oils, *etc.* The different plant parts of various galangal species have been studied and shown the presence of chemically enriched compounds with high medicinal value. The phytochemicals of galangal species and their medicinal properties are enlisted in Tables 11.1 and 11.2 and shown in Figures 11.2, 11.3 and 11.4.

11.3 *Alpinia galanga* (L.) Willd

Alpinia galanga (L.) Willd. is commonly used to treat different clinical conditions in Ayurveda and Siddha. It is the preferred botanical species used under the name of *Rasna* in most Ayurvedic formulations, especially in Keralian Ayurveda practice. It is commonly available with the name of *Kulanjan* in the local markets of Kerala, where it is often cultivated for its medicinal uses and commercial value.[30]

11.3.1 Synonyms

Alpinia alba (Retz.) Roscoe; Family: Zingiberaceae

11.3.2 English Names

It is also denominated as greater galangal in English.

11.3.3 Vernacular Names

Sanskrit: *Mahabaracach*, *Sugandha Vacha*, and *Rasna*; Hindi: *Kulanjan*; Gujrati and Marathi *Kulinjan*; Kannada: *Dhumasasmi*; Bengali: *Kulingjan*; Malayalam: *Arattha* and *Pararatta*; Tamil: *Pera-rattai*; and Telugu: *Pedda-dhumpa*.

11.3.4 Taxonomy

Kingdom – Plantae
Phylum – Tracheophyta
Class – Liliopsida
Order – Zingiberales
Family – Zingiberaceae
Genus – *Alpinia*
Species – *A. galanga*

11.3.5 Geographical Distribution

It is cultivated throughout South-East Asia and spread through the southern region of the Western Ghats and Himalayan regions. The Konkan area and various districts of Kerala in India also have rich cultivation of galangal (*Alpinia galanga* (L.) Willed) flora.[31,32]

Table 11.1 Phytochemicals and bioactive compounds present in different *Alpinia* species (*A. galanga*, *A. officinarum*, and *A. calcarata*).

Sr. no.	Galangal species	Plant part used	Phytochemicals	Name of bioactive Compound	Ref.
1.	*Alpinia galanga* (L.) Willd.	Leaves	Glycoside	Apigenin	(Shukla *et al.*, 2018)[16]
			Steroids	Chrysin	(Khairullah *et al.*, 2020)[14]
			Flavonoids	Galangin-3-methyl ether	(Das *et al.*, 2020)[17]
			Saponins	Galangin	(Kaushik, 2011)[18]
			Carbohydrates	Zingerone	(Beula *et al.*, 2016)[19]
			Alkaloids	Diterpene	
			Amino acids	Diterpene B	
			Chloride	1,8-Cineol	
			Phenols	α-Pinene	
			Terpenoids	Galangoisoflavonoid	
		Rhizome	Glycoside		
			Steroids		
			Flavonoids		
			Saponins		
			Carbohydrates		
			Tannins		
			Terpene		
		Flowers	Polyphenols	1′-Acetoxyeugenol acetate	(Tang *et al.*, 2018)[20]
			Essential Oils	α-Humulane, pentadecane	(Tang *et al.*, 2018)[20]
2.	*Alpinia officinarum* Hance	Rhizome	Tannins	Chrysin	(Bello *et al.*, 2018)[21]
			Phlobaphenes	Tectochrysin	(Ravindran and Balachandran, 2004)[22]
			Flavones	Apigenin	(Khairullah *et al.*, 2020)[14]
				Acacetin	(Das *et al.*, 2020)[17]
				Isorhmnetin	
				Kaempferide	
				Galangin	
				Alpinin	
				Quercetin	
				Kaempferol	
				Kaempferol-3-methyl ethers	

				Kaempferol-3-rhamnoside Pinocembrin 1,8-Cineol α-Pinene	
3.	*Alpinia calcarata* Roscoe	Leaves Flowers Rhizome	Tannins Polyphenols Terpenoids Alkaloids Flavonoids Glycosides Steroids Phytosterols Quinones Resins	2-Octanone camphene α-fenchyl acetate 2-hexanone vanillic acid protocatechuic acid syringic acid calacaratarins A–E shyobunone coumarins 1,8-cineol fenchyl acetate	(Rahman and Islam, 2015)[23] (Khandel *et al.*, 2018)[24] (Mohanasundari and Suja, 2015)[25] (Zhang *et al.*, 2016)[26]

Table 11.2 The bioactive compounds and their medicinal activities of *Alpinia* species (*A. galanga*, *A. officinarum*, and *A. calcarata*).

Sr. no.	Galangal species	Name of compound	Medicinal activity	References
1.	*Alpinia galanga* (L.) Willd.	1′-Acetoxyeugenol acetate	Antimicrobial	(Tang *et al.*, 2018)[20]
		4′-hydroxycinnamaldehyde	Antitumour	(Zhou *et al.*, 2018)[27]
		[di-(*p*-hydroxy-*cis*-styryl)] methane		(Ma *et al.*, 2017)[28]
		Phenolic compounds	Antioxidant	
		kaempferol		
		1′-Acetoxychavicol acetate	Anticancer	(Das *et al.*, 2020)[17]
		trans-p-Coumaryl alcohol		(Zhou *et al.*, 2018)[27]
		trans-p-Hydroxycinnamoyl acetate		
		trans-3,4-Dimethoxycinnamyl alcohol	Antiproliferative	
		Galangin	Anti-inflammatory	
		p-Hydroxycinnamaldehyde	Treatment of osteoarthritis	(Zhou *et al.*, 2018)[27]
		1,7-Bis(4-hydroxyphenyl)-1,4,6-heptatrien-3-one, bisdemethoxycurcumin	Inhibition of proliferation of melanoma cells and tyrosinase	(Ma *et al.*, 2017)[28]
		Galanganal, galanganols (A, B)	Cytotoxicity against cancer HeLa cells	(Ma *et al.*, 2017)[28]
		trans-p-Hydroxycinnamaldehyde, *trans-p*-hydroxycinnamoyl acetate	Antiallergic	(Ma *et al.*, 2017)[28]
		1*S*-1′-Acetoxychavicol acetate, 1-acetoxyeugenol acetate	Anti-ulcer, antileishmanial, and antitumour	(Ma *et al.*, 2017)[28]
		trans-p-Coumaryl diacetate	Anti-allergy, efflux pump inhibition, NO production inhibition, xanthine oxidase inhibition, antileishmanial, cytotoxicity, antibacterial	(Ma *et al.*, 2017)[28]
2.	*Alpinia officinarum* Hance	7-(4-Hydroxyphenyl)-1-phenyl-3-heptanone, Alpinoid-C	Anticancer	(Das *et al.*, 2020)[17]
		1-(4-Hydroxyphenyl)-7-(4″-hydroxy-3″-methoxyphenyl)-4*E*-en-3-heptanone		(Zhou *et al.*, 2018)[27] (Ma *et al.*, 2017)[28]

		1,7-Diphenyl-4-en-3-heptanone	Platelet-activating factor (PAF) receptor binding, antagonist activity	
		1,7-Diphenyl-3,5-heptanedione	Antibacterial, antimicrobial	
		5-hydroxy-7-(4″-hydroxy-3-methoxyphenyl)-1-phenyl-3-heptanone		
		4-phenethyl-1,7-diphenyl-1-heptene-3,5-dione		
		1,7-Diphenyl-5-ol-3-hepta-none	Anti-inflammatory, antigenotoxic	
		galangin		
		5-hydroxy-7-(4′-hydroxy-3′-methoxyphenyl)-1-phenyl-3-heptanone		
		Diarylheptanoid	Antiviral	
		7-(3,4-Dihydroxyphenyl)-1-(4-hydroxy-3 methoxyphenyl)-4-en-3-heptanone	Antitumour	(Ma *et al.*, 2017)[28]
		Dihydroxyashabushiketol	Anti-*helicobacter pylori* and hypolipidaemic activities	(Ma *et al.*, 2017)[28]
		7-(400-Hydroxy-300-methoxyphenyl)-1-phenyl-3,5-heptadione	Prostaglandin inhibitor	(Ma *et al.*, 2017)[28]
		Apigenin	Antioxidant	(Ma *et al.*, 2017)[28]
3.	*Alpinia calcarata* Roscoe	γ-Bicyclohomofarnesal, shyobunone	Anti-ACHE activity	(Ma *et al.*, 2017)[28]
		Isocoronarin D	Suppressor of proliferation in cancer cells	(Ma *et al.*, 2017)[28]
		bis-Labdanic diterpenoids	Cytotoxic activity against human KB cells	(Rahman and Islam, 2015)[23]
		Aalcaratarin (A–E)	Anticancer	(Rahman and Islam, 2015)[23]
		Camphene	Anti-inflammatory	(George, 2014)[29]
		1,8-Cineole, fenchol derivatives	Antiphlogistic, antibacterial, cooling activities	(George, 2014)[29]

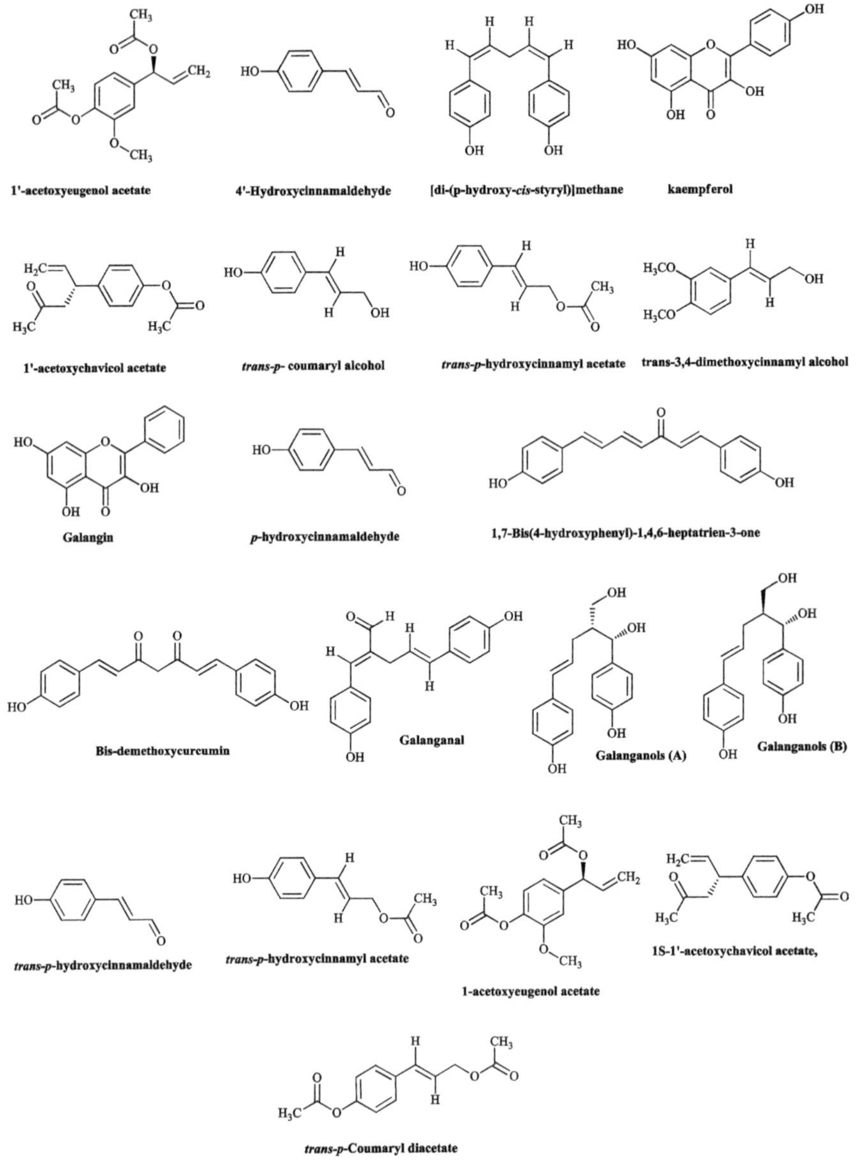

Figure 11.2 Chemical structures of phytochemicals present in *Alpinia galanga* (L.) Willd.

11.3.6 Traditional Medicinal Uses of *Alpinia galanga* (L.) Willd

The rhizome is the most commonly used botanical part of *Alpinia galanga* (L.) Willd. and possesses carminative and antiemetic properties, and acts as a digestive tonic.[33] It is also used for its antifungal, anthelmintic, antitumor,

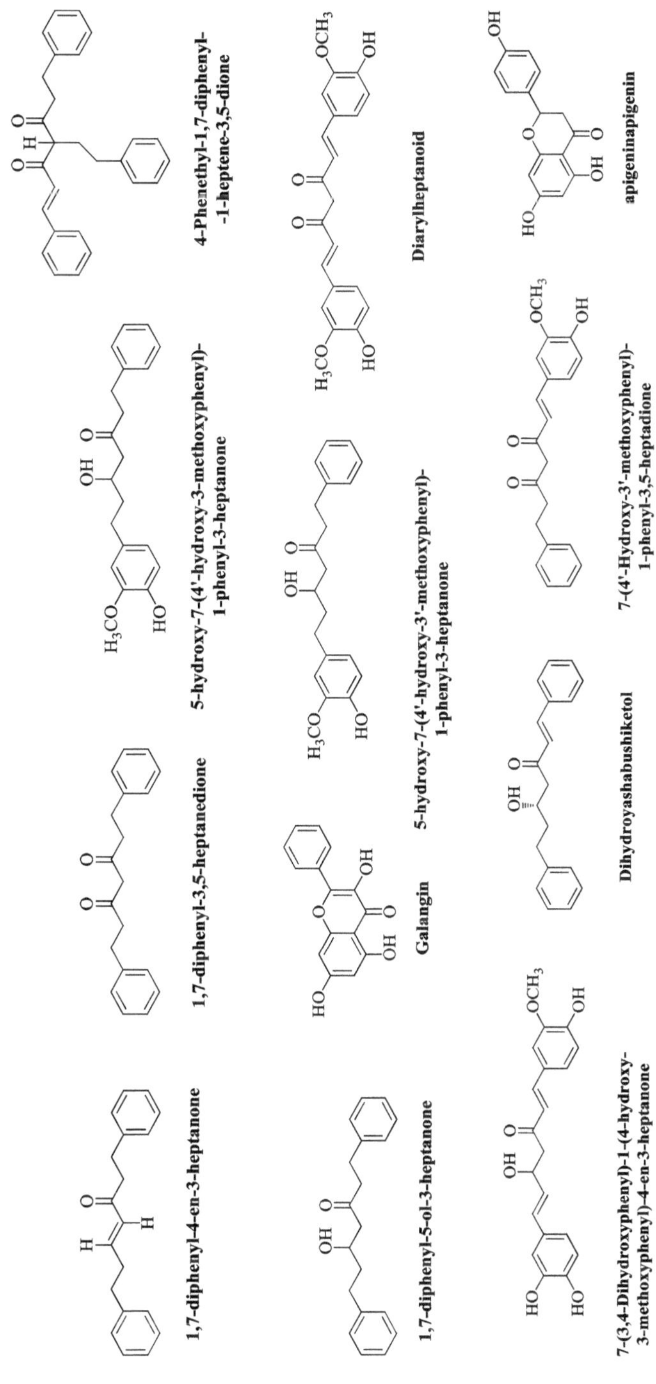

Figure 11.3 Chemical structures of phytochemicals present in *Alpinia officinarum* (Hance).

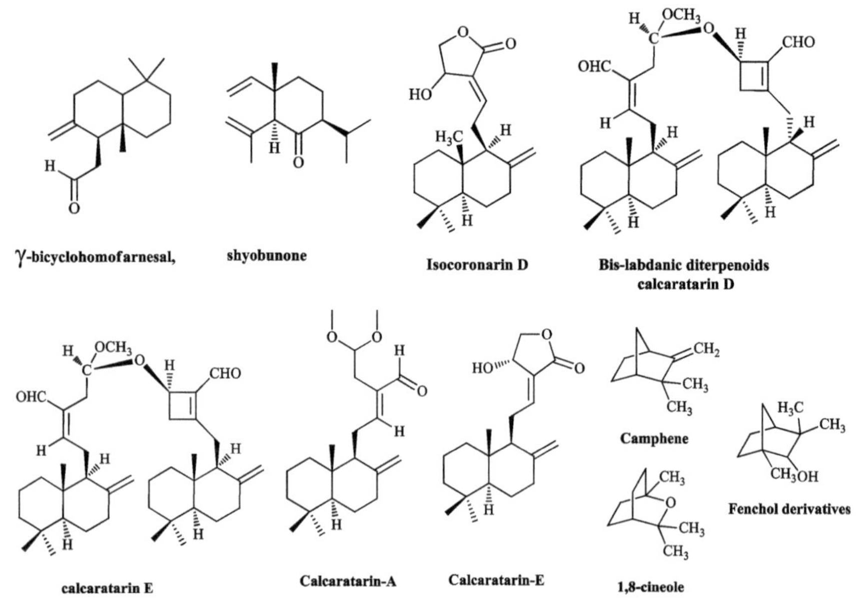

Figure 11.4 Chemical structures of phytochemicals present in *Alpinia calcarata* Roscoe.

antidiabetic, antidiuretic, antiulcerative, antidementia,[34] antitubercular,[35] antirheumatic,[36] antimicrobial,[32] antibacterial, anti-inflammatory,[37] cardioprotective, nephroprotective, neuroprotective, and aphrodisiac[32] actions. The seeds of *Alpinia galanga* (L.) Willd. are used for their cardiotonic, gastroprotective, diuretic, antifungal, antitumour, and antiplatelet activities.[38] Furthermore, the tubers of *Alpinia galanga* (L.) Willd. have been reported for their traditional uses in various conditions such as whooping cough in children and other respiratory diseases such as bronchitis and asthma, fever, and diabetes mellitus.[39]

11.4 Pharmacological Actions

11.4.1 Spermatogenesis Activity

Alpinia galanga (L.) Willd. exhibited spermatogenesis activity with improved sperm motility and serum testosterone levels *in vivo*.[40,41] A randomized and double-blinded clinical trial was conducted with the oral administration of four tablets of *Alpinia galanga* (L.) Willd. rhizome powder (764 mg per day) and four tablets of *P. granatum* fruit pomace extract (1000 mg per day) in divided doses of two tablets of each in the morning and two tablets in the evening. It was observed that for the 3 months of treatment, a 62% increase ($p = 0.026$) in the average total number of motile sperm was recorded. These results could have been due to the various active metabolites present in the standard extracts of both preparations. The exact mechanism through which the concurrent use of *Alpinia galanga* (L.) Willd. and *P. granatum* causes an improved count of ejaculated motile sperms should be investigated.[42]

11.4.2 Cytotoxic and Apoptotic Effects

Zeng *et al.*[43] evaluated the cytotoxic potential of isolated phytochemical constituents from *Alpinia galanga* (L.) Willd. seeds on cancer cell lines. Different fractions of ethanolic extracts (petroleum ether, ethyl acetate, and water) were evaluated for their cytotoxic activity on various cancer cell lines such as HepG-2, HeLa, A549, and SMMC-7721. Further, the protective effects of *Alpinia galanga* (L.) Willd. seeds against the damaging effects of H_2O_2 treatment have been evaluated in PC-12 cells. A significant inhibitory effect on HepG2 was observed with the total extracts. Both the ethyl acetate and petroleum ether fractions inhibited the growth of cancer cells more effectively than the total extracts. However, the ethyl acetate fraction (20 $\mu g\,mL^{-1}$) showed significant ($p<0.05$) protective effects against the damaging effects of H_2O_2-treated PC-12 cells. It also exhibited a more significant cytotoxic action ($p<0.05$) on cancer cell lines than petroleum and water fractions. The cytotoxic effects on HepG-2, HeLa, A549, and SMMC-7721 cell lines could be due to active metabolites such as 1′-*S*-1′-acetoxychavicol acetate, 1′-*S*-1′-acetoxyeugenol acetate, caryolane-1, 9 β-diol, and isocoronarin D present in the ethyl acetate fraction of ethanolic extract of *Alpinia galanga* (L.) Willd. seeds. These metabolites showed anticancer activity, with IC_{50} values ranging between 60 to 90 $\mu g\,mL^{-1}$.[43]

Samarghandian *et al.*[44] evaluated the cytotoxic and apoptotic effects of ethanolic extract of rhizome of *Alpinia galanga* (L.) Willd. in MCF-7 cells (cultured human breast carcinoma cells) and the effects were compared with MRC-5 cells (the non-malignant human fetal lung fibroblast cells). The extract reduced cell viability in MCF-7 cells and showed IC_{50} values at 400.0 ± 11.7 $\mu g\,mL^{-1}$ after 48 hours and at 170.0 ± 5.9 $\mu g\,mL^{-1}$ after 72 hours. Further *in vivo* and clinical studies are warranted into the use of *Alpinia galanga* (L.) Willd. as an anticancer agent in human breast cancer.[44]

Muangnoi *et al.*[45] reported the apoptotic potential of crude extract of rhizome of *Alpinia galanga* (L.) Willd. on normal epithelial cells derived from breast (MCF-12A cells) and normal fibroblast cells (CRL2522 cells). It was observed that the extract at a dose of 100 $\mu g\,mL^{-1}$ resulted in 15% apoptosis of MCF-12A cells, and at a dose of 1000 $\mu g\,mL^{-1}$ it showed 5% apoptosis of CRL2522 cells after 24 hours of treatment. However, the half-maximal inhibitory concentration of the crude extract was not studied in the normal standard fibroblast cell line previously.[45]

Sandra *et al.*[46] investigated the half-maximal inhibitory concentration (IC_{50}) of crude extract of rhizome of *Alpinia galanga* (L.) Willd. in NIH-3T3 cells (normal standard fibroblast cell line). The crude extract in various concentrations was applied in NIH-3T3 cells, and cells were counted by MTT assay. The IC_{50} of the crude extract of rhizome of *Alpinia galanga* (L.) Willd. was 620.5 $\mu g\,mL^{-1}$ after 24 hours and 666.6 $\mu g\,mL^{-1}$ after 48 hours of the treatment.[46] Further, the extract also did not show cytotoxic effects on

MRC-5 cells at a dose of 750 μg mL^{-1} for 48 hours.[44] These findings suggest that crude extract of rhizome of *Alpinia galanga* (L.) Willd. has a low cytotoxic impact on normal standard fibroblast cell lines, including MCF-12A, NIH-3T3, CRL2522 cells, and MRC-5 cells.

Furthermore, the IC_{50} of crude extract of rhizome of *Alpinia galanga* (L.) Willd. in COR-L23 cells (a lung cancer cell line) and MCF-7 cells was 5.4–13.3 μg mL^{-1} and 5.4–170 μg mL^{-1}, respectively.[47] Meanwhile, 100 μg mL^{-1} of the extract induced apoptosis (69%) in PC-3 cells (a prostate cancer cell line).[48] Overall, crude extract of the rhizome of *Alpinia galanga* (L.) Willd. exhibits potential cytotoxic activity in cancer cells.

Ahlina *et al.*[49] evaluated the potential of ethanolic extract of *Alpinia galanga* (L.) Willd. as a co-chemotherapeutic agent through its cytotoxic activity on 4T1 cells (metastatic breast cancer cells) and as an anti-ageing agent by examining its senescence inhibitory effect on normal fibroblast cells. The cytotoxicity test was performed with an MTT assay with doxorubicin (Dox) against 4T1 cells, whereas the senescence inhibitory effect was evaluated using an SA-β galactosidase-based assay. A flow cytometry-based assay was used to measure the level of ROS, apoptosis, and cell cycle. Also, the scratch wound healing assay was used to measure cell migration, and gelatin zymography assay was used to measure MMP-9 (matrix metalloproteinase) expression on 4T1 cells following treatment with ethanolic extract of *Alpinia galanga* (L.) Willd. It was observed that the IC_{50} value was 135 μg mL^{-1} in 4T1 cells, and at a concentration of 50 and 100 μg mL^{-1} the cytotoxic activity of doxorubicin was found to be increased. Without causing apoptosis, the extract increased the count of G2/M phase-arrested senescent cells featured with elevated intracellular ROS levels. However, the extract reduced senescence to normal in NIH 3T3 cells (a normal standard fibroblast cell line) with no change in ROS level under Dox-induced oxidative stress. In addition, the extract of *Alpinia galanga* (L.) Willd. inhibited the migration of 4T1 cells and suppressed Dox-induced MMP-9 expression.[49] Altogether, the ethanolic extract of *Alpinia galanga* (L.) Willd. is more selective for cancer cells. However, a greater number of studies should be conducted to illustrate its use as a co-chemotherapeutic and anti-ageing agent.

The metabolomic profile of *Alpinia galanga* (L.) Willd. extract revealed the presence of eugenol, sesquiterpenes, and coumarins. These phytoconstituents have been reported for their potential cytotoxic and antioxidant properties. Lakshmi *et al.*,[50] Kato *et al.*,[51] and Samarghandian *et al.*[44] have reported that the phenylpropanoid compounds present in *Alpinia galanga* (L.) Willd. extracts inhibited cell proliferation in lung cancer, hepatocarcinoma, and breast cancer by arresting cellular growth by inducing ROS-dependant premature senescence.[44,50,51]

Banjerdpongchai *et al.*[52] investigated the role and possible mechanistic pathways targeted by different metabolites present in *Alpinia galanga* (L.) Willd., which played a potential role in apoptotic cell death of cancer cells but not in normal cells. The 4′-HCA (4′-hydroxycinnamaldehyde) present in the *Alpinia galanga* (L.) Willd. was reported to have cytotoxic effects on HL60 and

U937 cells (human leukaemic cell lines) by Kaushik *et al.*[53] 4′-HCA (10 and 50 $\mu g\,mL^{-1}$) treated HL-60 and U937 cells for 4 hours resulted in apoptotic death by releasing cytochrome *c*, increasing the *Bax* level with declining the *Bcl-2* and *Bcl-xL* levels, elevating the Ca^{2+} contents of cytosol and mitochondria, and activating caspase-3. Overall, 4′-HCA from *Alpinia galanga* (L.) Willd. targeted the mitochondria and ER stress pathways to provoke apoptosis in HL-60 and U937 cells.[52,53] Further, galangin, a major phytoconstituent of *Alpinia galanga* (L.) Willd., exhibited antitumour activity (73.51% ± 4.742) in Ehrlich ascites carcinoma (EAC) cell-induced Swiss Albino mice without showing mortality. The results were compared with the standard drug, 5-fluorouracil.[54]

11.4.3 Antimicrobial Activity

Hamad *et al.*[55] reported the antimicrobial activities of *Alpinia galanga* (L.) Willd. flower extract and fractions against *Staphylococcus aureus* and *Listeria monocytogenes*, which are common food-borne pathogens. The activity was compared with control drugs: ampicillin and streptomycin. It was observed that both strains were susceptible to flower extract, whereas only *Listeria monocytogenes* was found to be susceptible to both control drugs. Further, amongst two strains, *Staphylococcus aureus* was found to be more sensitive to flower extract than *Listeria monocytogenes*. The MIC_{50} of flower phytochemical extracts against *S. aureus* and *L. monocytogenes* were 582 $\mu g\,mL^{-1}$ and 703 $\mu g\,mL^{-1}$, respectively. The antimicrobial action of flower essential oil extracts against *S. aureus* and *L. monocytogenes* was less than for the flower phytochemical extracts of *Alpinia galanga* (L.) Willd. 1′-Acetoxyeugenol acetate, an active metabolite of the galangal flower, presented potential antimicrobial activity at 34 $\mu g\,mL^{-1}$ and 68 $\mu g\,mL^{-1}$ (MIC_{50} values) against *S. aureus* and *L. monocytogenes*, respectively.[55]

Elsamma *et al.*[56] demonstrated the antibacterial activity of ether and ethyl acetate extracts of *Alpinia galanga* (L.) Willd. against multi-drug-resistant Gram-positive *Staphylococcus aureus* and Gram-negative *Klebsiella pneumonia* bacteria received from hospitalized patients. It was observed that *Staphylococcus aureus* and *Klebsiella pneumonia* were susceptible to both types of *Alpinia galanga* (L.) Willd. extracts. However, the ether extract showed more significant results.[56]

Quynh *et al.*[57] investigated the antimicrobial potential of essential oils derived from fresh and dried rhizomes and compared the results. It was reported that essential oils from fresh rhizomes of *Alpinia galanga* (L.) Willd. exhibited less antimicrobial activity than dried ones. *Staphylococcus aureus*, *Bacillus subtilis*, *Enterococcus faecalis*, *Escherichia coli* (*E. coli*), *Proteus vulgaris*, *Salmonella enteriditis*, *Saccharomyces cerevisiae*, and *Aspergillus niger* were found to be more susceptible to the essential oils from dried rhizomes with MIC values ranged from 1.25 to 12.5 $\mu L\,mL^{-1}$. It ranged from 2.5 to 20 $\mu L\,mL^{-1}$ for the essential oils from the fresh rhizome of *Alpinia galanga* (L.) Willd.[57]

Oonmetta-aree *et al.*[58] reported antimicrobial activity following the disc diffusion method for ethanol extracts of *Alpinia galanga* (L.) Willd. on

S. aureus and *E. coli*. The extract exhibited potent inhibitory action against *S. aureus*. The MIC of the ethanolic extract was 0.325 μg mL^{-1}. Further, it showed MBC (minimum bactericidal concentration) at 1.3 μg mL^{-1} using the broth dilution method. The ethanolic extract deranged the membrane function of *Staphylococcus aureus* cells by changing the internal pH and protein denaturation inside the cells. D,L-10-Acetoxychavicol acetate was the main phytoconstituent present in the extract, as confirmed by GC–MS and NMR.[58]

Latha *et al.*[59] investigated the antiplasmid activity of the crude acetone extract of the rhizomes of *Alpinia galanga* (L.) Willd, which showed 92% efficiency against *Salmonella typhi*, 82% against *E. coli*, and 8% against vancomycin-resistant *Enterococcus faecalis* at 400 μg mL^{-1} SIC. The acetone extract showed the presence of 1′-acetoxychavicol acetate, a phytoconstituent with antiplasmid activity against multi-drug-resistant bacteria.[59]

Yamsakul *et al.*[60] evaluated the antimicrobial properties of ethyl acetate, hexane, ethanol, and the essential oil fractions of *Alpinia galanga* (L.) Willd. extract against *E. coli*, *S. typhimurium*, *S. aureus*, *Salmonella enteritidis*, and *Pasteurella multocida*. Amongst all fractions, the essential oil of *Alpinia galanga* (L.) Willd. showed the most significant antibacterial and bactericidal properties, with MIC values of 8 mg mL^{-1} for *E. coli*, *S. typhimurium*, *S. aureus*, *Salmonella enteritidis*, and 16 mg mL^{-1} for *Pasteurella multocida*.[60]

Sunilson *et al.*[61] reported significant antibacterial and antifungal activities of different extracts of *Alpinia galanga* (L.) Willd. against common food-borne microbes such as *S. aureus*, *Bacillus cereus*, *Campylobacter jejuni*, *Salmonella enteriditis*, *Clostridium perfringens*, and *Escherichia coli* using an agar disc diffusion method. Further, the extracts also demonstrated antifungal properties against *Mucor mucedo*, *Candida albicans*, *Saccharomyces cerevisiae*, and *Hansenula anomala*, using the same method.[61]

11.4.4 Antifungal Activity

Khattak *et al.*[62] reported the antifungal activity of the ethanolic extract of *Alpinia galanga* (L.) Willd., against *Trichophyton longifusus*.[62] Further, n-hexane and DCM fractions exhibited potent antifungal and free radical scavenging activity due to their phenolic and flavonoid rich contents.[63]

11.4.5 Antioxidant Activity

Lu *et al.*[64] investigated the antioxidant potential of ethanolic extract of rhizome of *Alpinia galanga* (L.) Willd. using ABTS radical, DPPH radical, and FRAP (ferric-reducing antioxidant power) assays. The extract showed antioxidant ability at mean values of 593.90 μmol TE per g DW for the ABTS assay and 771.00 μmol TE per g DW for the FRAP assay. The ethanolic extract also exhibited a high level of DPPH inhibition (88.30%). These results were due to the presence of a high phenolic content in the extract. Ultra-high-performance liquid chromatography (UPLC) analysis also confirmed the presence of a large quantity of galangin (4333.5 μg g^{-1}) in the ethanolic extract.[64]

Tang *et al.*[65] assessed the antioxidant activity of different flower fractions by using a DPPH assay. The radical-scavenging ability of different flower fractions varied between the range of 2 to 4.45 mmol trolox per g phenolics. The flower fractions eluted by 20% and 60% methanol showed the highest antioxidant action. These radical scavenging effects of the 20% methanol fraction were due to the presence of flavonoids and gallic acid. The antioxidant activity of the flower was higher than for its rhizome due to the abundance of total phenolic content in galangal flowers.65 Further, the ethanolic extract of *Alpinia galanga* (L.) Willd. exhibited a higher antioxidant activity than a water extract and the essential oil when demonstrated with DPPH and ORAC methods. In the DPPH assay, the IC_{50} value for ethanolic extract was 10.66 mg mL^{-1}, whereas it was 55.48 and 455.43 mg mL^{-1} for water extract and essential oil, respectively. The ORAC value of the ethanolic extract was 8114.43 μmol TE per g, whereas it was 6845.40 and 1730.32 μmol TE per g extract for water extract and essential oil, respectively. The highest total phenolic content (31.49 $\pm$ 4.09 mg GAE per g extract) of ethanolic extract was responsible for its antioxidant activity. The HPLC analysis of water and ethanolic extract showed that catechin and myricetin were present in both extracts, whereas *p*-coumaric acid was found only in the ethanolic extract of *Alpinia galanga* (L.) Willd. These metabolites played an important role in the antioxidant activity of the ethanolic extract of *Alpinia galanga* (L.) Willd. Further, GC–MS analysis also revealed the presence of various volatile phenolic compounds in the essential oil and a phenylpropanoid, 1′-acetoxychavicol acetate (ACA), at a high concentration (10563.30 mg g^{-1} extract) in ethanolic extract of *Alpinia galanga* (L.) Willd.[66] Kubota *et al.*[67] reported the antioxidant properties of ACA using ferric thiocyanate and TBA assays. ACA exhibited a greater antioxidant action than α-tocopherol, and the result was comparable to butylated hydroxytoluene (BHT).[67] Overall, the ACA and the other phenolic compounds present in the ethanolic extract of *Alpinia galanga* (L.) Willd. could be responsible for its higher antioxidant activity than other extracts.

Wong *et al.*[68] also demonstrated the antioxidant activity of methanolic extracts of leaves and flowers of *Alpinia galanga* (L.) Willd. using DPPH, reducing power (RP), ferrous ion chelating, and β-carotene bleaching assays. Leaf and flower extracts showed the highest ferrous ion chelating ability (87.3 and 89.6%, respectively, at 7 mg mL^{-1}). The β-carotene bleaching ability of methanolic extract of leaves at 0.7 mg mL^{-1} was highest compared to quercetin at 3.3 mg mL^{-1}.[68]

11.4.6 Anti-inflammatory Activity

Matsuda *et al.*[69] demonstrated the antiallergic activity of *Alpinia galanga* (L.) Willd. rhizome *in vitro*. The 80% aqueous acetone extract of the rhizome inhibited the release of β-hexosaminidase as a marker of antigen-IgE-mediated degranulation in RBL-2H3 cells. Amongst the various isolated metabolites from rhizome extract, 1′*S*-1′-acetoxychavicol acetate and 1′*S*-1′-acetoxyeugenol acetate showed inhibitory action, with IC_{50} values of 15 and 19 μM, respectively.

Further, 1′*S*-1′-acetoxychavicol acetate and 1′*S*-1′-acetoxyeugenol acetate were found to inhibit the antigen-IgE-mediated TNF-α and IL-4 production.[69]

Pithak *et al.*[70] identified *p*-hydroxycinnamaldehyde from the purified acetone extract of *Alpinia galanga* (L.) Willd. with the help of NMR and mass spectrometry analyses. *p*-Hydroxycinnamaldehyde was found to suppress the expression levels of catabolic genes such as MMP-3 and MMP-13 (matrix metalloproteinases) and inflated the expression levels of anabolic genes like collagen II, SOX9, and aggrecan. These results illustrate the scope of therapeutic implementation of *Alpinia galanga* (L.) Willd. in the clinical management of osteoarthritis.[70]

Satish *et al.*[71] revealed the anti-inflammatory activity of the total alcoholic and aqueous extracts of rhizome of *Alpinia galanga* (L.) Willd. in acute and sub-acute rat models, using carrageenan-induced paw oedema and cotton pellet-induced granuloma methods, respectively. Both extracts at the dose level of 100 mg per kg body weight showed significant ($p<0.001$) anti-inflammatory activity. The results were comparable with the standard drug, phenylbutazone. *Alpinia galanga* (L.) Willd. rhizome extract may inhibit the cyclooxygenase pathway to reduce inflammation.[71]

Ghosh *et al.*[72] reported the anti-inflammatory and analgesic activities of a methanolic extract-prepared topical preparation from *Alpinia galanga* (L.) Willd. against carrageenan-induced oedema in rats. The results were compared with Piroxicam gel and methyl salicylate ointment (positive controls). The anti-inflammatory effect was comparable to the effect of Piroxicam gel at 3 hours after carrageenan injection.[72] The ethanolic extract of *Alpinia galanga* (L.) Willd. also manifested notable anti-inflammatory action in carrageenan-induced pleurisy rats.[73]

Unnisa and Parveen[74] investigated and reported the anti-inflammatory properties of methanolic, petroleum ether, chloroform, and aqueous methanolic (1 : 1) extracts of *Alpinia galanga* (L.) Willd. in carrageenan-induced paw oedema in Wistar rats. The results were compared with Ibuprofen (positive control). Amongst all the extracts, the methanolic extract (500 mg per kg body weight) exhibited maximum inhibition (79.51%) on carrageenan-induced paw oedema in rats.[74]

11.4.7 Antidiabetic Activity

Akhtar *et al.*[75] investigated the hypoglycaemic activity of powdered rhizome, methanolic, and aqueous extracts of *Alpinia galanga* (L.) Willd. rhizome in normal and alloxan-induced diabetic male New Zealand rabbits. The powdered rhizome at dose levels of 3 and 4 $g\,kg^{-1}$ exhibited a significant reduction in blood glucose level in normal rabbits. Similarly, the methanolic and aqueous extracts (4 $g\,kg^{-1}$) prepared from powdered rhizome showed a significant decline in the blood glucose level in normal rabbits. However, the powdered rhizome and its extracts did not exhibit any hypoglycaemic action in alloxan-induced diabetic rabbits. The results were comparable to gliclazide, a standard drug used in both normal and alloxan-induced diabetic rabbits.[75]

Kaushik *et al.*[76] also reported the antidiabetic potential of the alcoholic extract of rhizome of *Alpinia galanga* (L.) Willd. in STZ-induced diabetic Wistar rats. The study showed that oral administration of alcoholic extract of *Alpinia galanga* (L.) Willd. at doses of 50, 100, and 200 mg per kg body weight, once daily for 40 days, resulted in a significant ($p < 0.05$) reduction in blood glucose level. The extract also regulated dyslipidaemia and decreased the atherogenic index in diabetic rats. By reducing the blood urea nitrogen (BUN) level, albumin in urine, increased body weight, MDA, GSH, SOD, and CAT in diabetic-nephropathic rats, the alcoholic extract of *Alpinia galanga* (L.) Willd. showed a promising nephroprotective action in diabetes-induced kidney disease.[76] Jaju *et al.*[77] isolated galangogalloside, a new gallic acid glycoside present in the methanolic extract of rhizome of *Alpinia galanga* (L.) Willd. This metabolite was reported to have antidiabetic and anti-inflammatory properties.[77]

11.4.8 Antiulcer Property

The metabolites isolated from the seeds of *Alpinia galanga* (L.) Willd., namely 1′-acetoxychavicol acetate and 1′-acetoxyeugenol acetate, have been reported to have potent antiulcer properties.[78] Similarly, 1′*S*-1′-acetoxychavicol acetate and 1′*S*-1′-acetoxyeugenol acetate showed significant antiulcer activity on ethanol-induced gastric ulcer lesions in experimental rats.[79]

The antiulcer activity of the ethanolic extract of *Alpinia galanga* (L.) Willd. rhizome has been evaluated in experimental rats. The gastric ulcer was induced by pyloric ligation and hypothermic restraint stress in experimental rats. The ethanolic extract (500 mg per kg body weight) was found to reduce the gastric mucosal damage significantly in rats. The antisecretory and cytoprotective action of the extract of *Alpinia galanga* (L.) Willd. might be responsible for its antiulcer activity.[80]

11.4.9 Immunomodulatory Property

Bendjeddou *et al.*[81] reported the immuno-stimulating activity of the hot water-soluble polysaccharide extracts of *Alpinia galanga* (L.) Willd. in mice. Hot water polysaccharide extracts stimulated the reticuloendothelial system (RES) and increased the counts of peritoneal exudates cells (PECs) and spleen cells of mice.[81]

11.4.10 Neuroprotective Activity

Zhang *et al.*[82] investigated the neuroprotective role of *Alpinia galanga* (L.) Willd. against Alzheimer's disease in mice. *Alpinia galanga* (L.) Willd. exhibited the neuroprotective effect on Met-Leu-Gly-Ile-Ile-Ala-Gly-Lys-Asn-Ser-Gly amyloid-b (Ab)-induced amnesia in mice through Na^+, K^+-adenosine triphosphate (ATP)-ase increment and by decreasing the acetylcholinesterase level. It resulted in improved cognition by enhancing cholinergic transmission.[82]

11.4.11 Hepatoprotective Activity

The crude extract of *Alpinia galanga* (L.) Willd. (200 and 400 mg per kg body weight) exhibited a hepatoprotective role in paracetamol (3000 mg per kg body weight) induced hepatotoxicity in males Sprague-Dawley rats. The extract at a dose of 400 mg per kg body weight decreased liver MDA significantly ($p < 0.05$) against the group treated with 200 mg per kg body weight of *Alpinia galanga* (L.) Willd. crude extract.[83]

11.4.12 Anti-HIV Property

The researchers Ye and Li[84] reported the role of a metabolite, 1′*S*-1′-acetoxychavicol acetate (ACA), isolated from *Alpinia galanga* (L.) Willd. in the inhibition of HIV type 1 replication. 1′*S*-1′-acetoxychavicol acetate inhibited Rev transport by binding to chromosomal region maintenance. This resulted in accumulating the full-length HIV-1 RNA in the nucleus and blocking HIV-1 replication in peripheral blood mononuclear cells. Besides ACA, when used with didanosine, an anti-HIV drug, it inhibited HIV-1 replication by acting synergistically.[84]

11.5 Toxicity Profile of *Alpinia galanga* (L.) Willd

No significant mortality or no signs of acute or chronic toxicity were noted on the administration of ethanolic extracts of *Alpinia galanga* (L.) Willd. rhizome, up to the acute dose of 3.0 $g\,kg^{-1}$ and at a chronic dosage of 100 mg per kg body weight in Swiss albino mice, respectively.[85] The methanolic extract of *Alpinia galanga* (L.) Willd. also showed no behavioural changes and mortality with an LD_{50} value of more than 5000 mg per kg body weight in Wistar male rats.[74]

11.6 *Alpinia officinarum* Hance

11.6.1 Synonyms[86]

Languas officinarum (Hance) Farwell, *Languas officinarum* (Hance)

11.6.2 Family

Zingiberaceae

11.6.3 Common/English Names

Chinese ginger, China root, colic root, East India catarrh root, galangal, Indian root, lesser galanga, lesser galangal

11.6.4 Vernacular Names

Hindi – Kulinjan, Bengali – Punnagchampa, Sanskrit – Malayavaca, Tamil – Cirrarattai, Chitharathai.

11.6.5 Taxonomy[86]

Kingdom – Plantae
Phylum – Tracheophyta
Class – Liliopsida
Order – Zingiberales
Family – Zingiberaceae
Genus – *Alpinia*
Species – *Alpinia officinarum* Hance

11.6.6 Geographical Distribution[87]

Alpinia officinarum Hance is native to China and India. it is mainly found in the plains of West Bengal, Assam, and the Eastern Himalayas in India.

11.6.7 Traditional Medicinal Uses of *Alpinia officinarum* Hance

Alpinia officinarum Hance has a background in the Ayurvedic and Chinese systems of medicines.[88] *Alpinia officinarum* Hance rhizome is very effective in digestive disorders and acts as an antipyretic, analgesic, and antimicrobial (especially bacterial and fungal) agent. In Europe, it is very commonly used as a spice and medicament,[89] and effectively treats flatulence, dyspepsia, vomiting, and motion sickness.[90] *Alpinia officinarum* Hance has been used in indigestion, analgesic, and antiemetic treatments in Asian countries.[91,92] The dry root of *Alpinia officinarum* Hance has been used as a free radical scavenger, and for its antidiabetic, antiulcer, antidiarrheal, analgesic, anti-inflammatory, and anticoagulation effects in China.[93,94]

Alpinia officinarum Hance plant parts are shade-dried, ground, and used with food for the treatment of diseases, especially in gynaecological and musculoskeletal disorders.[95] The root is used by Moroccan tribes for the treatment of cough and to increase body weight.[96] The stem, or the whole plant, is prepared in powder form and added into a drink for the treatment of respiratory and digestive ailments.[96] The plant is effective in the treatment of rheumatism by Algerian tribes.[97] A combination of *Alpinia Officinarum* Hance, *Zingiber officinale*, honey, and lemon juice is administered orally as antitussive in Turkey[98,99] The *Alpinia officinarum* Hance rhizome decoction has proven efficacy in malaria, stomach pain, and ringworm infection in Vietnam.[100,101] The rhizome is used in common colds and coughs by Chinese.[102] In India, Mizoram tribes use the rhizome as a gargle to support speaking in children, in nocturnal enuresis, and in muscular spasticity.[103]

11.7 Pharmacological Properties

11.7.1 Antioxidant Activity

The methanolic extract of *Alpinia officinarum* Hance has been shown to have an antioxidative property using the ethyl linoleate system.[104] The methanolic extract of *Alpinia officinarum* Hance is enriched in total triterpenoid content and contributed to higher antioxidative property in a pyrogallol-luminol assay when compared to ascorbic acid and TROLOX-C.[105] The hydroalcoholic and methanolic isolates of *Alpinia officinarum* Hance rhizome revealed the concentration-based antioxidative activity by preventing the formation of diphenyl picrylhydrazyl (DPPH) free radicals.[106] The extract of *Alpinia officinarum* Hance augmented the viability against oxidative apoptosis injury in lung fibroblasts (V79-4 cells) in hamster.[107] The extract of *Alpinia officinarum* Hance also possesses potent activity against superoxide dismutase, catalase, and glutathione peroxidase in a dose-dependent manner.

11.7.2 Antiplatelet Activity

The diarylheptanoids from the rhizome of *Alpinia officinarum* Hance showed potent inhibitory activity towards PAF-receptor with an IC_{50} in range of 1.3–5.0 μM.[108]

11.7.3 Antiacetylcholinesterase Activity

Galangin isolated from *Alpinia officinarum* Hance rhizome exhibited inhibitory action on acetylcholinesterase with an inhibitory concentration (IC_{50}) of 120 μM and flavonoid inhibition constant of 74 μM.[109] This indicated that the galangin from *Alpinia officinarum* Hance possibly could a potential candidate for the development of new lead molecules for the treatment of Alzheimer's disease.

11.7.4 Cytotoxic/Antitumour Activity

The various studies suggest that *Alpinia officinarum* Hance possesses a potential role against several cancer cell lines. The methanolic extract of *Alpinia officinarum* Hance fortified the apoptosis of human oestrogen-positive breast cancer cells (MCF-7). The pathway responsible for this activity indicated the downregulation of *S*-phase-related regulatory proteins including cyclin-dependent protein kinase-2 (CDK2). The hexane and acetone extract of *Alpinia officinarum* Hance had a moderate antiproliferative effect on breast cancer cells with an IC_{50} of 50 $\mu g\,mL^{-1}$ and 5-α-reductase inhibition, respectively.[110] The exthanolic extract of *Alpinia officinarum* Hance and *Alpinia galanga* showed cytotoxic effects against PC-3 and MCF-7.[111,112]

The methanolic extract of *Alpinia officinarum* Hance inhibited MCF-7 cell proliferation in a concentration- and time-dependent manner and increased the cell apoptosis.[113,114] The methanolic, chloroform, dichloromethane extract of *Alpinia officinarum* Hance leaves and rhizomes exhibited anticancer activities on human AML cells over 48 h.[115] The diarylheptanoids from *Alpinia officinarum* Hance exhibited cytotoxic activity against human tumour cell lines (HepG2, MCF-7, and SF-268) and neuroblastoma cell lines, *viz.* IMR- 32, SK-N-SH, and NB-39, with the IC_{50} varying from 8.46 to 22.68 $\mu mol\,L^{-1}$.[116–118]

A study suggested that galangin induces apoptosis in hepatic cancerous cells by activating the caspase-8 and BH3 domain death mitochondrial cascade. These findings were supported by elevation of Ca^{2+} and GRP94, GRP78, and CHOP as well as increased free cytosolic Ca^{2+} level.[119,120] The molecular data showed that both FAK mRNA and protein level were reduced by galangin in a dose-dependent manner. Galangin inhibits the formation of tumour colonies in the lung tissue on C57BL/6J mouse lung (metastatic model) using B16F10 melanoma cells and significantly decreased the cell viability of B16F10 melanoma cells.[121] The apoptotic pathways were triggered by galangin and initiate the cleavage of procaspase-9, procaspase-3, and poly(ADP-ribose) polymerases in B16F10 cells. Galangin was also described as inhibiting the increase of HepG2 cells by activating AMPK.[122] Researchers discovered that galangin induced apoptosis and DNA condensation of the human colon cancer cells in a graded dose, and hence the probable mechanistic pathway is activation of caspase-3, caspase-9, and apoptosis-inducing factors from the mitochondria.[123]

11.7.5 Antihaemorrhagic Activity

A polyherbal formulation named Ankaferd blood stopper (ABS) consists of *Thymus vulgaris*, *Glycyrrhiza glabra*, *Vitis vinifera*, *Alpinia officinarum* Hance, and *Urtica dioica*. In Turkey, this formulation has been approved for the treatment of external haemorrhage and dental bleeding.[124] The traditional uses have been supported by experimental data where topical Ankaferd blood stopper decreased the period of bleeding and of the volume of blood lost by 68.4 and 54.6%, respectively. Purnak and co-workers studied the potential implications of Ankaferd blood stopper as an adjunctive medicine in a case of upper gastrointestinal bleeding in a patient suffering from defective haemostasis.[125] A detailed study by Beyazit illustrated that ABS was effective in controlling the cellular apoptotic responses to haemorrhagic stress.[126]

11.7.6 Antimicrobial Activity

The methanolic extract of *Alpinia officinarum* Hance showed antifungal property on clotrimazole-resistant *Candida albicans* at a MIC value of 2.5 $mg\,mL^{-1}$.[127] The acetone, ethanol, and hexane extract of *Alpinia officinarum*

Hance inhibited 50% of the growth of *Candida albicans*.[128,129] The methanolic extracts of *Alpinia officinarum* Hance possess moderate antibacterial activity against *Klebsiella pneumoniae* and *Staphylococcus aureus*, having a zone of inhibition at 12 mm and 12 mm, respectively.[130] Hot macerate (hydroalcoholic) of *Alpinia officinarum* Hance rhizome exhibited potent antibacterial activity against *Staphylococcus aureus* at MIC values of 31.25 $\mu g\,mL^{-1}$.[106,131]

A diarylheptanoid has been attributed as possessing antimicrobial property on multiple-drug-resistant *Escherichia coli* due to interaction with subunit A of DNA gyrase in *Escherichia coli*.[132] The three diarylheptanoids, *viz.* (4*E*)-7-(3,4-dihydroxy-5-methoxyphenyl)-1-phenylhept-4-en-3-one, (5*E*)-1,7- diphenylhept-5-en-3-one, and (1*E*)-1,7-diphenyl-4-(2-phenylethyl)hept-1-ene-3,5-dione were reported to exhibit antimicrobial activity against *Helicobacter pylori* (HP-Sydney strain 1 and HP-F44) and the MIC values were 9–12 $\mu g\,mL^{-1}$ and 25–30 $\mu g\,mL^{-1}$, respectively.[133] The diarylheptanoids stimulated a potent antitubercular effect against a dormant strain of *Mycobacterium tuberculosis* with IC_{50} values of 0.34–47.69 μM and 0.13–22.91 μM, respectively, when compared to rifampicin and isoniazid.[134] However, recent studies[123] on the extract of *Alpinia officinarum* Hance reported antimicrobial activities with a MIC value greater than 100 $\mu g\,mL^{-1}$.[135]

11.7.7 Antiviral Activity

The extract of *Alpinia officinarum* Hance and acyclovir has been revealed to prevent the growth of HSV-1 at an EC_{50} of 30.6 $\mu g\,mL^{-1}$ in an *in vitro* model.[136] The diarylheptanoid isolated from *Alpinia officinarum* Hance has been proven to display a potent anti-influenza activity on normal kidney MDCK cells infected with H_1N_1.[137] The mechanistic pathway studies revealed that (5*S*)-5-hydroxy-7-(4″-hydroxyphenyl)-1-phenyl-3-heptanone reduced the body weight and significantly extended the survival of mice infected with a virus, along with some minor side effects. The antiviral effect was attributed to suppression of antigen and viral messenger RNA.[138] The above-mentioned studies validate the potential use of *Alpinia officinarum* Hance extracts as an antiviral agent. Furthermore, *in vivo* and *in vitro* studies are required to establish the pharmacological mechanism of action and interaction of *Alpinia officinarum* Hance in viral diseases.

11.7.8 Hypolipidaemic Activity *In Vitro* Studies

The diarylheptanoid, *viz.* 5-hydroxy-7-(4′-hydroxy-3′-methoxyphenyl)-1-phenyl-3-heptanone (HPH), isolated from *Alpinia officinarum* Hance rhizome acts as an inhibitor of pancreatic lipase (IC_{50} value 1.5 $mg\,mL^{-1}$) which leads to antihyperlipidaemic activity.[139] It also lowered the serum TG level in triglyceridaemic mice (induced by feeding corn oil) and reduced serum triglyceride (TG) and cholesterol in Triton WR-1339-induced hyperlipidaemic mice. One study showed that *Alpinia officinarum* Hance extract (GE) could effectively inhibit fatty acid synthase (FAS) and the inhibition consisted of both reversible inhibitions with an IC_{50} value of 1.73 μg dried GE per mL and biphasic slow-binding inactivation.[140]

11.7.9 Hypolipidaemic Activity *In Vivo* studies

The aqueous and ethyl acetate extract of *Alpinia officinarum* Hance exhibited decreased level of serum triglyceride in triglyceridaemic mice and Triton WR-1339-induced hyperlipidaemic mice.[141] The galangin was isolated from *Alpinia officinarum* Hance as a potential inhibitor of pancreatic lipase with an $IC_{50} \sim 1.3$ mg mL^{-1}.[142] The extract decreases the total cholesterol, triglyceride, and LDL-cholesterol and improves the pathological changes in liver. The HDL-cholesterol concentration and the ratio of HDL-cholesterol/total cholesterol significantly increased compared with those of the HFD rats. Beattie *et al.*[143] found that administration of *Alpinia officinarum* Hance rhizome extract decreases the high-fat control group compared with the low-fat control group in mice.[143] The enzymes involved in β-oxidation of fatty acids (acetyl-coenzyme A, acyltransferase, and enoyl-CoA hydratase) were increased by consumption of phytochemical supplements. A study by Jung and co-workers described *Alpinia officinarum* Hance extract as having decreased lipid accumulation dose-dependently. This was attributed by downregulation of cytidine–cytidine–adenosine–adenosine–thymidine enhancer-binding protein-α, sterol regulatory element-binding protein-1 coded genes. The galangin also was reported to possess anti-adipogenic effects in 3T3-L1 cells.[144]

11.7.10 Melanogenesis Inhibitory Activity

The galangin isolated from *Alpinia officinarum* Hance reduces the melanin amount in B16 melanoma to 1.61 μg/10^5 cells.[145] The study results proposed that galangin acts as a lightening agent and is a favourable molecule for the prophylaxis of skin cancer. The water : acetone (80 : 20) extract of *A. officinarum* Hance exhibited an antimelanogenesis property in melanoma cells.[146] A study illustrated that four diarylheptanoids and two flavonoids inhibited melanogenesis with IC_{50} values in the range of 10–48 μM. Generally, these constituents inhibit the mRNA expression of tyrosinase-related proteins.

11.7.11 Anti-inflammatory Activity

The excessive production of nitric oxide (NO) in cells will result in an inflammatory pathway and cancer initiation. The *Alpinia officinarum* Hance possesses a good anti-inflammatory property in NO-induced inflammation. The water and ethyl acetate extract prevent prostaglandin (PG) synthetase and 5-lipoxygenase (5-LOX), which was required for leukotriene biosynthesis.[147] The diarylheptanoids have significant attenuation on nitric oxide and iNOS production in a mouse macrophage cell line (RAW 264.7) with variable IC_{50} values of 33–62 μM.[148–150] These metabolites prevent the release of interleukin-1β and tumour necrosis factor-α from peripheral blood mononuclear cells. The ethanolic extract of *Alpinia officinarum* Hance exhibited acute anti-inflammatory activity by decreasing the oedema volume in carrageenan-induced arthritis and NO production in rats.[150]

11.7.12 Antivitiligo Activity

A study illustrated the potential use of galangin in the treatment of vitiligo induced by hydroquinone in rodents and an increase in the concentration of tyrosinase protein.[151] There is elevation in melanin-containing epidermal cells and melanocytes when treated with a dose of 4.25 mg per kg of galangin isolated from *Alpinia officinarum* Hance.

11.7.13 Treatment of Erectile Dysfunction

The extract of *Alpinia officinarum* Hance presented as a potential aphrodisiac in Iran and has been used for the treatment of erectile dysfunction. The possible mechanistic pathway for this was the inhibition of cAMP and cGMP phosphodiesterase.[152]

11.8 Toxicity Profile of *Alpinia officinarum* Hance

Some studies suggest that there are no adverse effects of *Alpinia officinarum* Hance in various organs of animals after one week of treatment. The different parameters of adverse effects such as a change in body-weight, skin reaction, respiration rate, and locomotor activity of rabbits remained unchanged.[153–155] The LD_{50} of extracts and some metabolites (diarylheptanoids) were 1.3 $g\,kg^{-1}$ and 10.6 $g\,kg^{-1}$, respectively.[155] Galangin does not possess any significant toxicity in animals including alterations to biochemical parameters.[156]

11.9 *Alpinia calcarata* (Haw.) Roscoe

Alpinia calcarata (Haw.) Roscoe is an enduring herb that is commonly used in the traditional system of medicine in Sri Lanka. The mature branched rhizomes have dense structures with a light to dark brown colour. The leaf structure is simple, alternative, and of 25–32 cm length and 2.5–5 cm breadth.[157,158] *A. calcarata* Roscoe is cultivated in tropical countries like Sri Lanka, India, and Malaysia.159

11.9.1 Synonyms

Alpinia alata, A. Dietr., Alpinia bracteata Roscoe.

11.9.2 Family

Zingiberaceae.

11.9.3 Common/English Names

Snap ginger, lesser galangal.

11.9.4 Vernacular Names

Hindi – *Kulanjan*; Malayalam – *Chittaratha, Aratha, Kolinchi*; Tamil – *Amkolinji*; Sanskrit – *Rasna.*

11.9.5 Geographical Distribution[158]

The species is indigenous to Indo-Malaysia. In India, it is cultivated in Kerala.

11.9.6 Taxonomy[86]

Kingdom – Plantae
Phylum – Tracheophyta
Class – Liliopsida
Order – Zingiberales
Family – Zingiberaceae
Genus – *Alpinia*
Species – *Alpinia calcarata (Haw.)* Roscoe

11.9.7 Traditional Medicinal Uses of *Alpinia calcarata* Roscoe

The decoction of rhizomes is advised for internal administration to treat inflammatory conditions of joints. It is used also to treat snake poisoning. Pieces of rhizomes are used as a mouth freshener to remove unpleasant odours of the mouth. Small pieces of rhizomes are advised for stomachache, anorexia, and as an appetizer in chewable form. *Alpinia calcarata* Roscoe rhizomes, long pepper, and *Glycyrrhiza glabra* mixed with honey have been effectively used in the treatment of bronchitis.[159] Fresh crushed rhizomes mixed with lemon juice are used to treat fungal infections. Powder of rhizome is used to treat various skin diseases. For inflammation of joints, *Alpinia calcarata* Roscoe rhizomes are applied to the joints in the form of a paste or poultice.[159,160]

Rhizomes are frequently used as traditional remedies to treat respiratory ailments such as cough, asthma, and bronchitis in Sri Lanka.[161,162] It is widely used in treating common colds, swellings, stomachaches, and illnesses associated with the circulatory systems in the Indian region.[163] It is also grown in Bangladesh and is commonly used in the treatment of inflammatory disorders, cough, and other respiratory problems.[164] The beneficial health impacts of the plant broadly differ with the region of cultivation.

11.10 Pharmacological Activity of *Alpinia calcarata* Roscoe

11.10.1 Analgesic, Anxiolytic, and Seductive Activities

Ferdous *et al.*[165] evaluated the *in vivo* analgesic activity by using methanolic extract of leaves of *Alpinia calcarata* Roscoe with the help of an acetic acid-induced writhing test and formalin-induced paw licking test. The anxiolytic and sedative activities were investigated by an open-field, light–dark, elevated plus maze, thiopental sodium-induced sleeping time, and hole cross test behaviours in mice. The methanolic extract of leaves (200 and 400 mg per kg body weight) significantly ($p < 0.001$) decreased the number of abdominal writhes against an acetic acid-induced pain response. Similarly, the extract at a dose of 400 mg per kg body weight exhibited a significant reduction ($p < 0.001$) in paw-licking time. The results were compared with a standard drug, Aspirin. In the case of anxiolytic activity, the extract (400 mg per kg body weight) showed 35.83 ($p < 0.01$) and 16.33 ($p < 0.001$) square movements at 0 and 120 min, similar to a standard drug, diazepam. Further, the number of entries in the open arm of the apparatus for the methanolic extract of leaves in the elevated plus-maze test at doses of 200 and 400 mg per kg body weight were 11.6 and 12.0, and the durations observed were 128.2 and 145.5 s, respectively. The results were significant ($p < 0.001$) and comparable to diazepam. In a light–dark test, both extracts showed significant ($p < 0.001$) anxiolytic effects. The methanolic extract (400 mg per kg body weight) showed highly significant (8.33 min, $p < 0.001$) seductive effects and more than the standard drug, diazepam (11.5 min, $p < 0.001$).[165]

Altogether, the methanolic extract of *Alpinia calcarata* Roscoe leaves possesses potent analgesic, anxiolytic, and seductive activities. *Alpinia calcarata* Roscoe may block the related pathways involved in pain and anxiety. This is possibly due to the pharmacological actions of various metabolites present in the methanolic extract of leaves. Moreover, the findings support the anti-inflammatory and analgesic potential of *Alpinia calcarata* Roscoe and its traditional use in arthritis.[165]

11.10.2 Antinociceptive Activities

Arambewela *et al.*[166] investigated the antinociceptive action of hot water extract (HWE) and hot ethanol extract (HEE) derived from rhizomes of *Alpinia calcarata* Roscoe in rats. The activity was checked using three different methods, hot plate, tail flick, and formalin tests. In the hot plate test, HWE and HEE showed significant prolongation ($p < 0.05$) of the reaction time at 1-hour post-treatment. However, 5 hours post-treatment prolongation of reaction time was observed with HEE. In the tail-flick test, the reaction time was not significantly prolonged with HWE or HEE. Furthermore, a significant reduction in the licking time of the formalin-injected paw was noted in both early and late phases with HWE and HEE. The dose-dependent antinociceptive activity was highest at the 500 mg per kg dose for both extracts. The antinociceptive activity of both extracts was observed with induced opioid mechanisms. Further studies are warranted to uncover the role of the chemical compounds present in the *Alpinia calcarata* Roscoe rhizome.[166]

11.10.3 Cytotoxic Activity

Perveen *et al.*[167] reported the cytotoxic activity of ethanolic extract of rhizome of *Alpinia calcarata* Roscoe against Ehrlich ascites carcinoma (EAC) tumour in mice. The ethanolic extract (8 mg per kg body weight) significantly reduced the rate of tumour growth ($p<0.01$) and decreased tumour weight ($p<0.01$). The medium lethal concentration (LC_{50}) was 14.07 $\mu g\,mL^{-1}$ in the brine shrimp lethality bioassay *in vitro*, whereas the LD_{50} value of extract was 250 mg per kg body weight in mice. The various metabolites present in the ethanolic extract of rhizome of *Alpinia calcarata* Roscoe could be responsible for their potent antioxidant properties.[167]

11.10.4 Antiulcer Activity

The hot water extract of *Alpinia calcarata* Roscoe exhibited antiulcer activity against ethanol-induced gastric ulcers in rats. The oral administration of the hot water extract resulted in dose-dependent (500, 750, 1000 mg per kg body weight) and significant ($p<0.05$) protection against the ulcers, producing changes in the gastric mucosa by ethanol. The result was more promising than the standard drug, cimetidine. The flavonoids and alkaloids present in the extract could be responsible for the gastroprotective activity of *Alpinia calcarata* Roscoe.[168]

11.10.5 Aphrodisiac Activity

Ratnasooriya *et al.*[169] evaluated the aphrodisiac action of the hot water extract of *Alpinia calcarata* Roscoe rhizome (150, 250, and 500 mg per kg body weight) to check fertility and sexual competence in male rats. After 3 hours of administration of the extract, the sexual behaviour of the male rats was examined using receptive female rats. The hot water extract prolonged the latency for ejaculation. The extract also exhibited aphrodisiac action where the number of spontaneous penile erections was increased. The serum testosterone level was elevated in male rats. Further studies should be conducted on human participants to evaluate the aphrodisiac action and investigate the responsible phytoconstituents causing the aphrodisiac effects.[169]

11.10.6 Anti-inflammatory Activity

A study elaborated on the anti-inflammatory activity of aqueous and ethanol extract of *A. calcarata* Roscoe in carrageenan-induced inflammation and found that it reduces the inflammation significantly ($p < 0.05$) at different dose (250, 500, 750, and 1000 mg per kg).[170] The prominent effect was observed at 4 h after carrageenan administration and these effects were attributed to the inhibition of histamine and prostaglandin. The study established the ethnopharmacological uses of *Alpinia calcarata* Roscoe as an anti-inflammatory agent.[171,172]

11.10.7 Antioxidative Effects

The antioxidative properties of cold ethanolic, hot water, and hot ethanolic extract *Alpinia calcarata* Roscoe were studied using thiobarbituric acid reactive substances assay and DPPH free radical scavenging assay. The most potent antioxidative activity was shown by the cold ethanolic extract. In TBARS and DPPH assays, the antioxidant potential of cold ethanolic extract was as good and poor as that of butylated hydroxytoluene, respectively.[173]

11.10.8 Hypoglycaemic Effects

Researchers studied the impact of water and ethanol extract in STZ-induced diabetic rats and a normal control group.[174] In euglycaemic rats, the oral administration of ethanolic and water extract (250, 500, 750, 1000, and 1500 mg per kg) decreased the glucose levels significantly ($p < 0.05$). The ethanol extract possessed a higher hypoglycaemic effect when compared to the water extract. They also exhibited a reduction of absorption of glucose from the small intestine and the elevation of glycogen build-up in muscle and liver. One study investigated the ameliorative activity of ethanol extract of *Alpinia calcarata* Roscoe in alloxan-induced diabetes and observed a reduction in the blood glucose level at the total lipid profile at doses of 100, 200, and 300 mg per kg per day for 21 days.[175]

11.10.9 Antibacterial and Antifungal Effects

The hexane and methanolic extracts of *Alpinia calcarata* Roscoe exhibited noteworthy activity for *Escherichia coli*, *Staphylococcus aureus*, *S. pneumoniae*, *P. aeruginosa*, *S. paratyphi*, *V. cholerae*, and *B. subtilis*.[176] The essential oil from *Alpinia calcarata* Roscoe rhizome possesses antimicrobial property against various microbes and this activity is mainly attributed to the remarkable zone of inhibition of 15 mm each against *Escherichia coli* and *B. subtilis*, followed by *K. pneumoniae* (12 mm) and *Agrobacterium rhizogenes* (11 mm). The antifungal effect of the essential oil from *Alpinia calcarata* Roscoe against *A. aculeatus* and *A. awomori* exposed inhibitory zones of

10 mm and 7 mm, respectively. Therefore, the total antimicrobial activity of *Alpinia calcarata* Roscoe extract was better against bacteria than fungi.[177]

11.10.10 Toxicity Profile of *Alpinia calcarata* Roscoe

The toxicity profile of *Alpinia calcarata* Roscoe were investigated using aqueous and ethanolic extracts in rats at a dose of 1500 mg per kg. A long-term study (42 days) on animals did not show any signs of chronic toxicity or alterations in biochemical and haematological parameters.[178]

11.11 Conclusion

Although galangal is a controversial drug when considered with the name of *Rasna* in Ayurveda, all three species of galangal have high medicinal values and are substitutes for each other in many Ayurvedic formulations under the category of *Rasna*. The phytochemical profile of galangals reveals the possible scope for their therapeutic implementation in diverse disease conditions and the development of synthetic chemical moieties as a future adjunct therapy. The pharmacological actions of galangals also have provided valuable insights into their therapeutic potential in various diseases. This chapter has also explored the possible role of different phytoconstituents of galangals in targeting the pathways involved in the pathological progress of diseases. All three common species of galangal should be standardized with quality control methods to regulate their use in manufacturing various Ayurvedic formulations for safe and effective therapeutic outcomes.

Abbreviations

μM	Micromolar
ABTS	2,2′-Azinobis-(3-ethylbenzothiazoline-6-sulfonic acid)
API	Ayurvedic Pharmacopoeia of India
BC	Before Christ
BHT	Butylated hydroxy toluene
cAMP	Cyclic adenosine monophosphate
CAT	Catalase
cGMP	Cyclic guanosine monophosphate
cm	Centimetre
DPPH	2,2-Diphenyl-1-picrylhydrazyl
DW	Distilled water
FRAP	Ferric-reducing ability of plasma
g	Gram
GSH	Glutathione
i.p.	Intraperitoneal
IC_{50}	Half-maximal inhibitory concentration
kg	Kilogram

LD_{50}	Lethal dose
MBC	Minimum bactericidal concentration
MDA	Malondialdehyde
mg	Milligram
mL	Millilitre
mm	Millimetre
MMP	Matrix metalloproteinases
MTT	3-(4,5-Dimethylthiazol-2-yl)-2,5-diphenyltetrazolium bromide
nm	Nanometre
NO	Nitric oxide
PPAR-γ	Peroxisome proliferator-activated receptors-gamma
ROS	Reactive oxygen species
SOD	Superoxide dismutase
SREBP-1	Sterol regulatory element binding protein-1
TBARS	Thiobarbituric acid reactive substances

Acknowledgements

The authors are grateful to Maitreyi Nechiyil (B.A.M.S. final year student, Vaidyaratnam P S Varier Ayurveda College, Kottakkal, Kerala), for providing photographs of *Rasna* for this chapter.

References

1. G. Das, J. K. Patra, S. Gonçalves, A. Romano, E. P. Gutiérrez-Grijalva, J. B. Heredia, A. D. Talukdar, S. Shome and H. S. Shin, *Trends Food Sci. Technol.*, 2020, **101**, 50.
2. I. B. Abubakar, I. Malami, Y. Yahaya and S. M. Sule, *J. Ethnopharmacol.*, 2018, **224**, 45.
3. B. Vaidya, *Chaukhambha Orientalia*, Varanasi, 1982, p. 277.
4. V. V. Sivarajan and I. Balachandran, *Ayurvedic Drugs and Their Plant Sources*, Oxford & IBH Publishing Co. Pvt. Ltd, New Delhi, India, 1994, p. 119, 195, 398.
5. K. Raghunathan and R. Mitra, *Pharmacognosy of Indigenous Drugs Volume 2*, Central Council for Research in Ayurveda and Siddha, New Delhi, India, 1982, p. 834.
6. P. K. Warrier, V. P. K. Nambiar and C. Ramankutty, *Indian Medicinal Plants - A compendium of 500 species*, Orient Longman. Ltd., Madras, India, 1994, vol. 1, p. 106.
7. K. V. Nair, S. N. Yoganarasimhan, K. R. Kesavamurthy and Z. Mary, *Anc. Sci. Life*, 1982, **2**, 71.
8. Caraka, *Caraka Samhitha*, Vol. I Sutrasthana - English translation of text with Ayurveda Dipika Commentary of Cakrapani Datta,

P. V. Tewari (Ed.), Chaukhambha Visvabharati Oriental Publishers and Distributors, Varanasi, 1st edition, [ch.su 4/50], p. 88.

9. Caraka, *Caraka samhitha*, Vol. I Sutrasthana - English translation of text with Ayurveda Dipika commentary of Cakrapani Datta, P. V. Tewari (ed.), Chaukambha Visvabharati Oriental Publishers and Distributors, Varanasi, 1st edition, [ch. su. 25/40], p. 407.
10. P. V. Sharma, *Dravyaguna-Vijnana*, Chaukhambha Bharati Academy, Varanasi, 2015, vol. 2, p. 39.
11. Anonymous, *The Ayurvedic Pharmacopoeia of India*, Government of India, Ministry of Health and Family Welfare, Department of Indian System of Medicine & Homeopathy, New Delhi, India, 1st edn, 2001, part I, vol. III, p. 162.
12. S. Shukla, S. K. Ojha, G. K. Mishra, S. Gupta and T. V. G. Roy, *J. Pharmacogn. Phytochem.*, 2018, 7, 2410.
13. S. Mondal and M. Kundu, *Indian J. Ethnophytopharma*, 2018, **4**, 20.
14. L. Wilson, *Encyclopedia of Food and Health*, Elsevier Ltd., 2016.
15. A. R. Khairullah, T. I. Solikhah, A. Nur, M. Ansori, A. Fadholly and S. Cashyer, *Syst. Rev. Pharm.*, 2020, **11**, 387.
16. S. Shukla, S. K. Ojha, G. K. Mishra, S. Gupta and T. V. G. Roy, *J. Pharmacogn. Phytochem.*, 2018, 7, 2410.
17. G. Das, J. K. Patra, S. Gonçalves, A. Romano, E. P. Gutiérrez-Grijalva and J. B. Heredia, *Trends Food Sci. Technol.*, 2020, **101**, 50.
18. D. Kaushik, *J. Chin. Integr. Med.*, 2011, **9**, 1061.
19. R. K. R. Beula, S. K. Sundar and M. Murugan, *Asian J. Pharm. Clin. Res.*, 2016, **9**, 364.
20. X. Tang, C. Xu, Y. Yagiz, A. Simonne and M. R. Marshall, *Food Chem.*, 2018, **255**, 300.
21. O. A. Bello, O. I. Ayanda, O. S. Aworunse and B. I. Olukanmi, *Pharmacogn. Rev.*, 2018, **1**, 8.
22. P. N. Ravindran and I. Balachandran, *Galangal; Handbook of herbs and spices*, Elsevier Inc., 2004, vol. 21, p. 357.
23. A. Rahman and S. Islam, *Pharmacogn. Rev*, 2015, **9**, 55.
24. P. Khandel, S. K. Shahi, D. K. Soni, R. K. Yadaw and L. Kanwar, *Nano Convergence*, 2018, 1–17.
25. L. Mohanasundari and S. Suja, *J. Pharmacogn. Phytochem.*, 2015 **4**, 53.
26. W. Zhang, J. Luo and L. Kong, *World J. Tradit. Chin. Med.*, 2016 **2**, 26.
27. Y. Q. Zhou, H. Liu, M. X. He, R. Wang, Q. Q. Zeng, Y. Wang, W. C. Ye and Q. W. Zhang, *Natural and Artifiial Flavoring Agents and Food Dyes: Handbook of Food Bioengineering*, Elsevier Inc., 2018.
28. X. Ma, C. Xie, Z. Miao, Q. Yang and X. Yang, *RSC Adv.*, 2017, 7, 14114.
29. J. George, *Int. J. Pharma Sci. Res.*, 2014, **5**, 555.
30. P. P. Joy, J. Thomas, S. Mathew and B. P. Skaria, *J. Med. Aromat. Plant Sci.*, 2001, **23**, 343.

31. C. P. Khare, *A Dictionary of Indian Medicinal Plant*, Springer India Pvt. Ltd., 2007, p. 37.
32. R. P. Rastogi and B. N. Mahrotra, *Compendium of Indian Medicinal Plant*, CDRI & National Institute of Science Communication and Information, New Delhi, 2018, vol. IV, p. 6.
33. B.-E. Vanwyk and M. Wink, *Medicinal Plant of The World*, Briz Publication, South Africa, 1st edn, 2009, vol. 7, p. 43.
34. Manila, *Research Guidelines for Evaluating the Safety and Efficacy of Herbal Medicines*, WHO, Regional office for the Western Pacific, 1993, p. 1.
35. V. B. Rajpal and D. P. S. Kohli, *Herbal Drug Industry*, Business Horizons, New Delhi, 2nd edn, 2009, p. 100.
36. K. R. Kirtikar and B. D. Basu, *Indian Medicinal Plants*, International Book Distributors, Dehradun (Uttranchal), 2nd edn, 1935, p. 2444.
37. R. N. Chopra, S. L. Nayar and I. C. Chopra, *Glossary of Indian Medicinal Plant*, NISCAIR Press, New Delhi, 7th edn, 2006, p. 16.
38. A. DerMarderosian and J. A. Beutler, *The Review of Natural Product*, Fact and Comparison, 111 West Port Plaza, Missouri, 2nd edn, 2002, p. 29.
39. K. M. Nadkarni, *The Indian Materia Medica*, Bombay Popular Prakashan, 3rd edn, 2009, vol. I, p. 77.
40. S. Qureshi, A. H. Shah and A. M. Ageel, *Planta Med.*, 1992, **58**, 124.
41. M. W. Islam, M. N. M. Zakaria, R. Radhakrishnan, X. M. Liu, A. Ismail, K. Chan and A. Al-Attas, *J. Pharm. Pharmacol.*, 2000, **52**, 278.
42. M. D. K. Fedder, H. B. Jakobsen, I. Giversen, L. P. Christensen, E. T. Parner and J. Fedder, *PLoS One*, 2014, **9**, e108532.
43. Q. H. Zeng, C. L. Lu, X. W. Zhang and J. G. Jiang, *Food Funct.*, 2015, **6**, 431.
44. S. Samarghandian, M. A. Hadjzadeh, J. T. Afshari and M. Hosseini, *BMC Complementary Altern. Med.*, 2014, **14**, 192.
45. P. Muangnoi, M. Lu, J. Lee, A. Thepouyporn, R. Mirzayans, X. C. Le, M. Weinfeld and S. Changbumrung, *Planta Med.*, 2007, **73**, 748.
46. F. Sandra, J. Sudiono, P. Trisfilha and D. Pratiwi, *Indones. Biomed. J.*, 2017, **9**, 23.
47. C. C. Lee and P. Houghton, *J. Ethnopharmacol.*, 2005, **100**, 237.
48. S. Suja and P. Chinnaswamy, *Anc. Sci. Life*, 2008, **27**, 33.
49. F. N. Ahlina, N. Nugraheni, I. A. Salsabila, S. Haryanti, M. Da'i and E. Meiyanto, *Asian Pac. J. Cancer Prev.*, 2020, **21**, 107.
50. S. Lakshmi, S. Suresh and B. S. Rahul, *Med. Chem. Res.*, 2019, **28**, 39.
51. R. Kato, I. M. Yuasa, H. Azuma and A. K. Yuasa, *Chem. – Biol. Interact.*, 2014, **212**, 1.
52. R. Banjerdpongchai, P. Punyati, A. Nakrob, W. Pompimon and P. Kongtawelert, *Asian Pac. J. Cancer Prev.*, 2011, **12**, 593.
53. D. Kaushik, J. Yadav, P. Kaushik, D. Sacher and R. Rani, *J. Chin. Int. Med.*, 2011, **9**, 1061.
54. J. V. Jaiswal, P. A. Wadegaonkar and S. W. Hajare, *Appl. Biochem. Biotechnol.*, 2012, **167**, 1325.

55. A. Hamad, A. Alifah, A. Permadi and D. Hartanti, *Int. Food Res. J.*, 2016, **23**, 837.
56. T. Elsamma, J. Shanmugam and M. M. Rafi, *Biomedicine*, 1996, **16**, 15.
57. T. Quynh, T. Vuong and W. Duszkiewicz-Reinhard, *J. Essent. Oil-Bear. Plants*, 2004, 7, 165.
58. J. Oonmetta-aree, T. Suzuki, P. Gasaluck and G. Eumkeb, *LWT – Food Sci. Technol.*, 2006, **39**, 1214.
59. C. Latha, V. D. Shriram, S. S. Jahagirdar, P. K. Dhakephalkar and S. R. Rojatkar, *J. Ethnopharmacol.*, 2009, **123**, 522.
60. P. Yamsakul, S. Kongkhaew, T. Yano, V. Sukprasitch, W. Prakattagomol and S. Ogonoki, Proceedings of the 47th Kasetsart University Annual Conference, Kasetsart, 17–20 March, 2009, Subject: Veterinary Medicine, 2009, p. 208.
61. J. A. J. Sunilson, R. Suraj, G. Rejitha, K. Anandarajagopal, A. V. A. G. Kumari and P. Promwichit, *Am. J. Food Technol.*, 2009, **4**, 192.
62. S. Khattak, S. Rehman, H. Shah, W. Ahmad and M. Ahmad, *Fitoterapia*, 2005, **76**, 254.
63. A. Sharma, S. Jain, M. Bhatnagar and S. Ghosal, *Int. J. Phytomed.*, 2015, 7, 78.
64. M. Lu, B. Yuan, M. Zeng and J. Chen, *Food Res. Int.*, 2011, **44**, 530.
65. X. Tang, C. Xu, Y. Yagiz, A. Simonne and M. R. Marshall, *Food Chem.*, 2018, **255**, 300.
66. N. Mahae and S. Chaiseri, *Kasetsart J.: Nat. Sci.*, 2009, **43**, 358.
67. K. Kubota, Y. Ueda, M. Yasuda and A. Masuda, in *Food Flavors and Chemistry: Advances of the New Millennium*, ed. A. M. Spanier, F. Shahidi, T. H. Parliament and C.-T. Ho, Royal Society of Chemistry, Cambridge, UK, 2001, pp. 601–607.
68. L. F. Wong, Y. Y. Lim and M. Omar, *J. Food Biochem.*, 2009, **33**, 835.
69. H. Matsuda, T. Morikawa, H. Managi and M. Yoshikawa, *Bioorg. Med. Chem. Lett.*, 2003, **13**, 3197.
70. T. Phitak, K. Choocheep, P. Pothacharoen, W. Pompimon, B. Premanode and P. Kongtawelert, *Phytochemistry*, 2009, **70**, 237.
71. R. Satish and R. Dhananjayan, *Biomed*, 2003, **23**, 91.
72. A. K. Ghosh, M. Banerjee and N. K. Bhattacharyya, *Chron. Young Sci.*, 2011, **2**, 139.
73. K. R. Subash, G. B. Prakash, K. V. Reddy, K. Manjunath and K. U. Rao, *Nat. J. Physiol., Pharm. Pharmacol.*, 2016, **6**, 468.
74. A. Unnisa and T. D. Parveen, *Der. Pharm. Sin.*, 2011, **2**, 361.
75. M. S. Akhtar, M. A. Khan and M. T. Malik, *Fitoterapia*, 2002, **73**, 623.
76. P. Kaushik, D. Kaushik, J. Yadav and P. Pahwa, *Pak. J. Biol. Sci.*, 2013, **16**, 804.
77. S. Jaju, N. Indurwade, D. Sakarkar, N. Fuloria and M. Ali, *Int. J. Green Pharm.*, 2009, **3**, 144.
78. S. Mitsui, S. Kobayashi, H. Nagahori and A. Ogiso, *Chem. Pharm. Bull.*, 1976, **24**, 2377.
79. H. Matsuda, Y. Pongpiriyadacha, T. Morikawa, M. Ochi and M. Yoshikawa, *Eur. J. Pharm.*, 2003, **471**, 59.

80. M. A. Al-Yahya, S. Rafatullah, J. S. Mossa, A. M. Ageel, M. S. Al-Said and M. Tariq, *Phytother. Res.*, 1990, **4**, 114.
81. D. Bendjeddou, K. Lalaoui and D. Satta, *J. Ethnopharmacol.*, 2003, **88**, 155.
82. L. N. Zhang, Y. J. Sun, S. Pan, J. X. Li, Y. E. Qu and Y. Li, *Fundam. Clin. Pharmacol.*, 2013, **27**, 96.
83. B. Hemabarathy, S. B. Budin and V. Feizal, *J. Biol. Sci.*, 2009, **9**, 57.
84. Y. Ye and B. Li, *J. Gen. Virol.*, 2006, **87**, 2047.
85. S. Qureshi, A. Shah and A. Ageel, *Planta Med.*, 1992, **58**, 124.
86. The Plant List, 2013, http://www.theplantlist.org/ (accessed 25.6.2021).
87. Z. Tian, N. An, B. Zhou, P. Xiao, I. S. Kohane and E. Wu, *Cancer Chemother. Pharmacol.*, 2009, **63**, 1131.
88. D. Bown, *Encyclopaedia of herbs and their uses*, Dorling Kindersley, London, 1995, p. 424.
89. M. Grieve, *A modern herbal*, Vol. II, Dover Publications, New York, 1971, p. 919.
90. W. C. Tang and G. Eisenbrand, *Chemistry, Pharmacology, and Use in Traditional and Modern Medicine*, Springer, Heidelberg, 1992, p. 87.
91. L. Tao, Z. T. Wang, E. Y. Zhu, Y. H. Lu and D. Z. Wei, *South Afr. J. Bot.*, 2006, **72**, 163.
92. J. C. Luo, W. Rui, M. M. Jiang, Q. L. Tian, X. Ji and Y. F. Feng, *J. Chromatogr. Sci.*, 2010, **48**, 795.
93. Z. S. Xie, X. J. Xu, C. Y. Xie, J. Y. Huang, M. Yang and D. P. Yang, *J. Pharm. Anal.*, 2013, **3**, 215.
94. C. H. Jung, S. J. Jang, J. Ahn, S. Y. Gwon, T. I. Jeon, T. W. Kim and T. Y. Ha, *J. Med. Food*, 2012, **15**, 959.
95. I. Teixidor-Toneu, G. J. Martin, A. Ouhammou, R. K. Puri and J. A. Hawkins, *J. Ethnopharmacol.*, 2016, **188**, 96.
96. A. Ouarghidi, G. J. Martin, B. Powell, G. Esser and A. Abbad, *J. Ethnobiol. Ethnomed.*, 2013, **9**, 1.
97. B. Benarba, L. Belabid, K. Righi, A. A. Bekkar, M. Elouissi, A. Khaldi and A. Hamimed, *J. Ethnopharmacol.*, 2015, **175**, 626.
98. Y. Güzel, M. Güzelşemme and M. Miski, *J. Ethnopharmacol.*, 2015, **174**, 118.
99. E. Uzun, G. Sariyar, A. Adsersen, B. Karakoc, G. Ötük and E. S. Oktayoglu Pirildar, *J. Ethnopharmacol.*, 2004, **95**, 287.
100. N. P. Hanh, N. Q. Binh and B. S. Adhikari, *J. Biodiversity Endanger. Species*, 2014, 2, 1000121.
101. S. J. Pei, *J. Ethnopharmacol.*, 1985, **13**, 121.
102. X. Shang, C. Tao, X. Miao, D. Wang, D. Tangmuke, Y. Wang, Y. Yang and H. Pan, *J. Ethnopharmacol.*, 2012, **142**, 390.
103. H. K. Sharma, L. Chhangte and A. K. Dolui, *Fitoterapia*, 2001, **72**, 146.
104. B. J. Kim, J. H. Kim, H. P. Kim and M. Y. Heo, *Int. J. Cosmet. Sci.*, 1997, **19**, 299.

105. C. L. Chang, C. S. Lin and G. H. Lai, *Evidence-Based Complementary Altern. Med.*, 2012, **2012**, 984295.
106. A. R. Srividya, S. P. Dhanabal, V. K. Misra and G. Suja, *Indian J. Pharm. Sci.*, 2010, **72**, 145.
107. S. E. Lee, H. J. Hwang, J. S. Ha, H. S. Jeong and J. H. Kim, *Life Sci.*, 2003, **73**, 167.
108. G. J. Fan, Y. H. Kang, Y. N. Han and B. H. Han, *Bioorg. Med. Chem. Lett.*, 2007, **17**, 6720.
109. A. J. Y. Guo, H. Q. Xie, R. C. Y. Choi, K. Y. Z. Zheng, C. W. C. Bi, S. L. Xu, T. T. X. Dong and K. W. K. Tsim, *Chem. – Biol. Interact.*, 2010, **187**, 246.
110. Y. U. Kim, H. K. Son, H. K. Song, M. J. Ahn, S. S. Lee and S. K. Lee, *Planta Med.*, 2003, **69**, 72.
111. S. Suja and P. Chinnaswamy, *Anc. Sci. Life*, 2008, **27**, 33.
112. K. Yasukawa, Y. Sun, S. Kitanaka, N. Tomizawa, M. Miura and S. Motohashi, *J. Nat. Med.*, 2008, **62**, 374.
113. C. C. Lee and P. Houghton, *J. Ethnopharmacol.*, 2005, **100**, 237.
114. P. Houghton, R. Fang, I. Techatanawat, G. Steventon, P. J. Hylands and C. C. Lee, *Methods*, 2007, **42**, 377.
115. S. N. Omoregie, F. O. Omoruyi, V. F. Wright, L. Jones and P. V. Zimba, *J. Med. Food*, 2013, **16**, 647.
116. C. L. Lu, H. Y. Zhao and J. G. Jiang, *Int. J. Food Sci. Nutr.*, 2013, **64**, 28.
117. N. An, Z. M. Zou, Z. Tian, X. Z. Luo, S. L. Yang and L. Z. Xu, *Fitoterapia*, 2008, **79**, 27.
118. D. Liu, Y. W. Liu, F. Q. Guan and J. Y. Liang, *Fitoterapia*, 2014, **96**, 76.
119. H. T. Zhang, J. Wu, M. Wen, L. J. Su and H. Luo, *J. Asian Nat. Prod. Res.*, 2012, **14**, 626.
120. L. Su, X. Chen, J. Wu, B. Lin, H. Zhang, L. Lan and H. Luo, *Food Chem. Toxicol.*, 2013, **62**, 810.
121. W. J. Zhang, Y. Lan, Q. L. Huang and Z. C. Hua, *Cytotechnology*, 2013, **65**, 447.
122. H. T. Zhang, N. Li, J. Wu, L. J. Su, X. Y. Chen, B. Y. Lin and H. Luo, *Eur. J. Pharmacol.*, 2013, **718**, 235.
123. T. K. Ha, M. E. Kim, J. H. Yoon, S. J. Bae, J. Yeom and J. S. Lee, *Exp. Biol. Med.*, 2013, **238**, 1047.
124. A. Kosar, H. S. Cipil, A. Kaya, B. Uz, I. C. Haznedaroglu, H. Goker, O. Ozdemir, S. Ercetin, S. Kirazli and H. C. Firat, *Blood Coagulation Fibrinolysis*, 2009, **20**, 185.
125. T. Purnak, E. Ozaslan, Y. Beyazit and I. C. Haznedaroglu, *Phytother. Res.*, 2011, **25**, 312.
126. Y. Beyazit, M. Kekilli, I. C. Haznedaroglu, E. Kayacetin and M. Basaranoglu, *World J. Gastroenterol.*, 2011, **17**, 3962.
127. G. H. Shahidi Bonjar, *Fitoterapia*, 2004, **75**, 74.
128. Q. Liu, W. Luyten, K. Pellens, Y. Wang, W. Wang, K. Thevissen, Q. Liang, B. P. A. Cammue, L. Schoofs and G. Luo, *J. Ethnopharmacol.*, 2012, **143**, 772.

129. Z. Zhang, A. S. Ravipati, S. R. Koyyalamudi, S. C. Jeong, N. Reddy, J. Bartlett, P. T. Smith, M. de la Cruz, M. C. Monteiro, A. Melguizo, E. Jiménez and F. Vicente, *Asian Pac. J. Trop. Med.*, 2013, **6**, 673.
130. G. H. Shahidi Bonjar, *Fitoterapia*, 2004, **75**, 231.
131. D. Lakshmanan, J. Nanda and K. Jeevaratnam, *Nat. Prod. Res.*, 2017, **6419**, 1.
132. K. Subramanian, C. Selvakkumar, K. S. Vinaykumar, N. Goswami, S. Meenakshisundaram, A. Balakrishnan and B. S. Lakshmi, *Int. J. Antimicrob. Agents*, 2009, **33**, 244.
133. B. Zhang, Y. Dai, Z. Liao and L. Ding, *Fitoterapia*, 2010, **81**, 948.
134. V. S. Honmore, S. R. Rojatkar, L. U. Nawale, M. A. Arkile, V. M. Khedkar, A. D. Natu and D. Sarkar, *Nat. Prod. Res.*, 2016, **30**, 2825.
135. P. Cos, A. J. Vlietinck, D. Berghe and L. Maes, *J. Ethnopharmacol.*, 2006, **106**, 290.
136. M. Kurokawa, K. Nagasaka, T. Hirabayashi, S. Uyama, H. Sato, T. Kageyama, S. Kadota, H. Ohyama, T. Hozumi, T. Namba and K. Shiraki, *Antiviral Res.*, 1995, **27**, 19.
137. R. Sawamura, Y. Sun, K. Yasukawa, T. Shimizu, W. Watanabe and M. Kurokawa, *J. Nat. Med.*, 2010, **64**, 117.
138. R. Sawamura, T. Shimizu, Y. Sun, K. Yasukawa, M. Miura, M. Toriyama, S. Motohashi, W. Watanabe, K. Konno and M. Kurokawa, *Antiviral Chem. Chemother.*, 2010, **21**, 33.
139. J. E. Shin, M. J. Han, M. C. Song, N. I. Baek and D. H. Kim, *Biol. Pharm. Bull.*, 2004, **27**, 138.
140. B. H. Li and W. X. Tian, *J. Enzyme Inhib. Med. Chem.*, 2003, **18**, 349.
141. J. E. Shin, H. M. Joo and D. H. Kim, *Biol. Pharm. Bull.*, 2003, **26**, 854.
142. D. Z. Xia, X. F. Yu, H. M. Wang, Q. Y. Ren and B. M. Chen, *J. Med. Food*, 2010, **13**, 785.
143. J. H. Beattie, F. Nicol, M. J. Gordon, M. D. Reid, L. Cantlay, G. W. Horgan, I. S. Kwun, J. Y. Ahn and T. Y. Ha, *Mol. Nutr. Food Res.*, 2011, **55**, S203.
144. C. H. Jung, S. J. Jang, J. Ahn, S. Y. Gwon, T. I. Jeon, T. W. Kim and T. Y. Ha, *J. Med. Food*, 2012, **15**, 959.
145. Y. H. Lu, W. Z. T. Lin-Tao, D. Z. Wei and H. B. Xiang, *J. Enzyme Inhib. Med. Chem.*, 2007, **22**, 433.
146. H. Matsuda, S. Ando, T. Kato, T. Morikawa and M. Yoshikawa, *Bioorg. Med. Chem.*, 2006, **14**, 138.
147. F. Kiuchi, S. Iwakami, M. Shibuya, F. Hanaoka and U. Sankawa, *Chem. Pharm. Bull.*, 1992, **40**, 387.
148. P. N. Yadav, Z. Liu and M. M. Rafi, *J. Pharmacol. Exp. Ther.*, 2003, **305**, 925.
149. H. J. Lee, J. S. Kim and J. H. Ryu, *Planta Med.*, 2006, **72**, 68.
150. J. S. Lee, K. A. Kim, S. Jeong, S. Lee, H. J. Park, N. J. Kim and S. Lim, *J. Ethnopharmacol.*, 2009, **126**, 258.
151. S. X. Huo, X. M. Liu, C. H. Ge, L. Gao, X. M. Peng, P. P. Zhao and M. Yan, *Phytother. Res.*, 2014, **28**, 1533.

152. M. Khanavi, H. Azimi, S. Ghiasi, S. Hassani, R. Rahimi, S. Nikfar, Y. Ajani, M. R. Shams-Ardekani and M. Abdollahi, *Int. J. Pharmacol.*, 2012, **8**, 161.
153. J. X. Liu, Y. H. Sun and C. P. Li, *Exp. Ther. Med.*, 2015, **9**, 1304.
154. O. S. Balcik, M. Koroglu, H. Cipil, O. Kaftan, S. Maral and A. Gurel, *J. Hematol. Oncol.*, 2012, **22**, 267.
155. W. L. Cui, Unpublished Master thesis, Guangdong Pharmaceutical University, 2008.
156. K. W. Ren, W. Z. Zhang, G. Wu, J. Z. Ren, H. B. Lu and Z. M. Li, *Biomed. Pharmacother.*, 2016, **84**, 1748.
157. M. D. Dassanayake and F. R. Fosberg, *A Revised Hand Book to the flora of Ceylon*, Amreind, New Delhi, 1981, vol. 4, p. 517.
158. D. M. A. Jayaweera, *Medicinal Plants Used in Ceylon*, National Science Council of Sri Lanka, 1982, vol. 5, p. 213.
159. L. Ramanayake, *Osu Visithuru*, Department of Ayurveda, Colombo, Sri Lanka, 1994, vol. 1.
160. L. S. R. Arambewela, C. S. Basnayake, P. Serasinghe, M. S. A. Tissera, S. Dias and D. R. Weerasekara, *Traditional treatment in Sri Lanka for chronic Arthritis*, NARESA Printing Unit, Colombo, Sri Lanka, 1995.
161. D. M. Jayaweera, *Medicinal Plants Used in Ceylon*, National Science Council of Sri Lanka, Colombo, 1982, p. 213.
162. L. Ramanayake, *Osu Visithuru*, Department of Ayurveda, Colombo, Sri Lanka, 1994, p. 68.
163. A. D. Ahemed, G. Medine, S. Meryem, O. Hatice, S. Fikerettin and K. Isa, *Turk. J. Biol.*, 2005, **29**, 155.
164. J. U. Chowdhury, M. Yusuf and M. M. Husain, *Indian Perfum.*, 2003, **47**, 355.
165. M. Ferdous, A. F. M. S. Ud Daula, S. Naznin, F. Yeasmin and M. A. Basher, *J. Med. Plants Res.*, 2020, **14**, 155.
166. L. S. R. Arambewela, L. D. A. M. Arawwawala and W. D. Ratnasooriya, *J. Ethnopharmcol.*, 2004, **95**, 311.
167. R. Perveen, F. Islam, J. Khanum and T. Yeasmin, *Asian Pac. J. Trop. Med.*, 2012, **5**, 121.
168. L. S. R. Arambewela, L. D. A. M. Arawwawala and W. D. Ratnasooriya, *Pharmacogn. Mag.*, 2009, **5**, 226.
169. W. D. Ratnasooriya and J. R. Jayakody, *Acta Biol. Hung.*, 2006, **57**, 23.
170. D. A. Arawwawala, S. R. Arambewela and D. W. Ratnasooriya, *Nat. Prod. J.*, 2012, **2**, 263.
171. A. K. Sharma and R. H. Singh, *Bull. Med. Res.*, 1980, **11**, 262.
172. M. A. Rahman, M. M. Rahman and N. U. Ahmed, *J. Biol. Act. Prod. Nat.*, 2012, **2**, 365.
173. L. S. Arambewela, L. D. Arawwawala and N. Athauda, *J. Ayurveda Integr. Med.*, 2010, **1**, 199.
174. L. S. Arambewela, L. D. Arawwawala and W. D. Ratnasooriya, *Phcog. Mag.*, 2009, **5**, 412–418.

175. N. Raj, S. Nadeem, S. Jain, C. Raj and C. K. Nandi, *Dig. J. Nanomat. Biost.*, 2011, **6**, 991.
176. J. P. Robinson, V. Balakrishnan, S. Raj and S. J. Britto, *Adv. Biol. Res.*, 2009, **3**, 185.
177. B. Joseph, G. George and M. Helen, *Asian Pac. J. Trop. Biomed.*, 2012, **5**, 1.
178. L. Arambewela, L. D. A. M. Arawwawala and W. D. Ratnasooriya, *Bol. Latinoam. Caribe Plant. Med. Aromat.*, 2011, **10**, 435.

CHAPTER 12

Phytochemical and Pharmacological Aspects of Cissus quadrangularis *in Ayurveda*

DHANALEKSHMI UM, JAWAID AKHTAR AND SHAH ALAM KHAN*

College of Pharmacy, National University of Science and Technology, PO Box 620, PC 130, Muscat, Sultanate of Oman
*Email: shahalam@nu.edu.om

12.1 Introduction

Ayurveda is an Indian traditional system of medicine that means science of life in Sanskrit (Ayur = longevity; Veda = knowledge or science). It is one of the ancient systems of holistic medicine that emphasizes treating the root cause of the diseases rather than symptoms and helps in the restoration of good health. Ayurvedic medicines are mainly prepared from different parts of medicinal plants and are used to treat both acute and chronic ailments. One of the important indigenous medicinal plants used to prepare herbal Ayurvedic formulations for the healing of fractured bones is *Cissus quadrangularis* Linn. (CQ). CQ belongs to the family Vitaceae and is commonly known as Veldt Grape, or Devil's Backbone (English), Hadjod or hadjora (Hindi), Pindia or pirandai (Tamil), Changlam parranda (Malayalam), and Mogarolli (in Kannada). It is a perennial climber that prefers a hot climate. Geographically, it is distributed in the tropical regions of India and other

Chemistry, Biological Activities and Therapeutic Applications of Medicinal Plants in Ayurveda
Edited by Augustine Amalraj, Sasikumar Kuttappan and Karthik Varma A. C.

Published by the Royal Society of Chemistry, www.rsc.org

Southeast Asian and African countries. CQ herb grows up to 2.5 m in height and has quadrangular branches (internodes of 1.5–1.8 cm and 10–15 cm long).[1] Its flowers are calyx cup shaped. The stems are fleshy and have four edges (quadrangular) that look like the bones and joints in the body. Its leaves are broad, ovate (2–5 cm) having 3–7 lobes (Figure 12.1). The other name of the plant Asthisamharaka or Asthi Samhara in Sanskrit is rightly suited to the plant which means relief from the bone disease.[2] CQ is an edible plant and curries, chutneys, pickles, and soups made from its stems are very popular in South India. The plant stem and root parts have been used for the preparation of herbal formulations for treating different diseases.[3] In the Unani system of medicines, CQ is applied externally to heal the fractured bones while Siddha practitioners prescribe the stem part for treating digestive disorders.

In Ayurveda, CQ is known as a bonesetter, and therefore it is one of the important constituents in marketed Ayurvedic formulations namely; Bonoheal™ (ZocVida), Bonoxtil® Herbal (Visham Lifecare), Bonzoc™ (Zoic Pharmaceuticals), C3 plus® (BSA Pharma Inc), Ciss Q™ (Weldew remedies), C-Quad® (Kaplin Healthcare Pvt Ltd), Fevijoint Forte™ (BSA Pharma), Hadjod (Himalaya Pvt Ltd), Hudcura® (Vee Remedies), Laksha Guggulu® (Dabur India), Lakshadi Guggulu® (Baidyanath), ONE-SG™ (Akshar molecules), Ordex™ (BSA Pharma), Orthociss® (Stenroz Biotech), Osta-Heal (Ravenbhel Pharmaceuticals Pvt Ltd), Osteofed-CQ (Fedley Healthcare Pvt Ltd), Patanjali Divya Lakshadi Guggul® (Patanjali Ayurved Limited), Reco BLT® (Solace Biotech Ltd), Sitzod (Sitnez Biocare Pvt Ltd), and Zeotone Plus soft gel capsule® (AVN Ayurveda formulations Pvt Ltd)[4] available in the form of capsules, tablets, and syrup dosage forms. These drugs are commonly used for weight loss, to treat bone fractures and

Figure 12.1 *Cissus quadrangularis* plant: (A) fresh young stem of an adamant creeper sprout with tendrils and ivy-like leaves. Image credit: JP Sweety, Tamil Nadu. (B) Climbing vine over trees with quadrangular sectioned branches. Image credit: KM Badsha, Kerala. (C) Devils backbone hanging over the trees with a highly branched stem. Image credit: KM Badsha, Kerala.

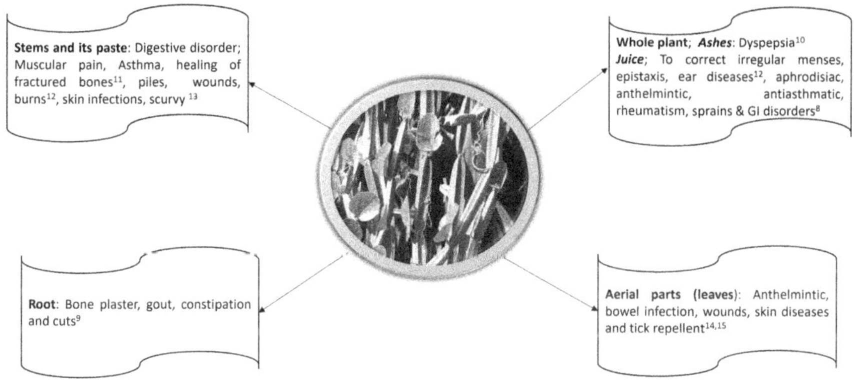

Figure 12.2 Folkloric/traditional uses of CQ and its various parts.[8–15]

dislocation of joints, and its average daily dose for adults is 1000–1500 mg. Other parts of CQ have been used in stomach disorders such as dyspepsia, flatulence, and other abnormalities, including cancer, epilepsy, inflammation, and obesity. The root part is used for the treatment of fractured bones, whereas dyspepsia and indigestion are treated using the young shoot. The stem paste is used for the management of asthma.[5] The plant extracts have been reported to exhibit important biological activities including antibacterial, antifungal, bone fracture healing, anthelmintic, analgesic, and anti-hemorrhoidal activities. The pharmacological activities of the herbs are related to the presence of triterpenoids, carotenoids, and ascorbic acid.[6,7] It also exhibits gastroprotective activities when used in combination with NSAID therapy. The juice of CQ is traditionally used to control irregular menses and to stop nose bleeding. The decoction of the whole plant is claimed to possess anthelmintic activity.[8] The traditional/folkloric uses of CQ and its various parts are presented in Figure 12.2.

A study on humans showed that nutraceutical agents from the CQ plant induce weight loss without any adverse effects.[16] The CQ plant is considered as a panacea for the treatment of bone disorders such as healing of fractured bones, bone regeneration in dentistry, dislocation of hip, osteoporosis, osteoarthritis, *etc.* Owing to its numerous beneficial therapeutic uses, this plant has received considerable attention worldwide and therefore several preclinical and clinical studies have been conducted to validate its traditional uses. This chapter aims to provide an overview of its traditional uses with emphasis on Ayurveda, phytochemistry, pharmacological uses, toxicology, clinical studies, patents granted, and application of CQ in nanotechnology and biomedical fields.

12.2 Phytochemical Analysis

CQ has been reported to be a rich source of minerals, vitamins, amino acids, and other nutrients. Calcium (1.76, 1.68, and 1.51 g per 100 g) and phosphorus (1.82, 1.42, and 1.26 g per 100 g) are the main elements present in stem, leaves, and roots, respectively, confirming that the mineral content is higher in stems as compared to roots and aerial parts. Other elements present include potassium, iron, zinc, copper, *etc.* Surprisingly, root (11.1 g per 100 g) contains the highest amount of protein followed by the stem (10.4 g per 100 g). Carbohydrate content was found in the following order root > stem > leaf (37.5, 34.3, and 33.7 g per 100 g) while the fat content was highest in the stem (1.33 g per 100 g) and lowest in the root (0.66 g per 100 g).[17] It is also a good source of antioxidants such as ascorbic acid (480 mg per 100 g) and β-carotene (267 units per 100 g).[18] Qualitative fingerprinting of organic and aqueous extracts of different parts of CQ plant have been shown to contain carbohydrates, flavonoids, steroids, terpenoids, phytosterols, phenolic acids, fatty acids, lignans, iridoids, stilbene derivatives, volatile oils, *etc.*[19] Phytochemical screening of the ethanolic extract of CQ was found to be rich in phytoestrogen.[20,21] This fraction acted as anti-osteoporotic in ovariectomized rats. Phytochemical investigation of the stems of CQ have resulted is characterization of many phytoconstituents (Table 12.1). The chemical structures of some selected phytoconstituents isolated from the stems of CQ are presented in Figure 12.3. The hepatoprotective effect of the methanol extract of CQ in rats was attributed to the presence of β-carotene, which has key antioxidant properties.[22,23] Experimental studies indicated that iridoid picroside I possesses significant osteogenic activity and is capable of increasing the alkaline phosphatase (ALP) activity and can stimulate the formation of mineralized nodules.[24] Muthu *et al.*, extracted antibacterial cysteine protease (Cp) from a buffered solution of CQ. The optimum pH and temperature for the protease activity were 6 and 50 °C, respectively. The Cp showed potent antibacterial action and showed 90% efficiency in the pH range of 3 to 10. The extracted Cp degraded the peptidoglycan layer of the bacteria *B. cereus* (64.49%) and was confirmed by transmission electron microscopy (TEM) analysis.[25]

Ethanolic extract of CQ roots has yielded stilbene derivatives, phenolic as well flavonoid glycosides. Major chemicals isolated from the roots are rosmarinic acid-*O*-hexoside, sinapic acid-3-*O*-glucoside, apigenin coumaroylhexoside, naringinin-7-*O*-rutinoside, resveratrol, luteolin, and quercetin.[26] Most of these phytochemicals also occur in other parts of the plant.

Aerial parts of CQ have been reported to contain an unsymmetrical tetracyclic triterpenoid (7-oxoonocer-8-ene-3β-21-α-diol) and was shown to neutralize the antianabolic effects of cortisone in the treatment of bone fractures.[5] Gupta *et al.*, in 1991 isolated some alicyclic lipids and stilbene derivatives (resveratrol, piceatanon, pallidol, and parthenocissus).[37] Major fatty acids present in the leaves are *n*-hexadecanoic acid, ethan-1,1-diethoxy, and 9,12,15-octadecatrienoic acid-methyl ester (*Z*,*Z*,*Z*).[38] Few other studies also showed the occurrence of flavonoids in leaves including quercetin, kaempferol 3-*O*-rhamnoside, quercetin 3-*O*-rhamnoside (quercitrin), vitexin (apigenin 8-*C*-glucoside), apigenin-7-glucoside, and isovitexin (apigenin 6-*C*-glucoside).[39] Flavonoids, especially quercetin, present in the plant were

Table 12.1 Various chemical constituents isolated from the stems of *Cissus quadrangularis (CQ)*.

S. no.	Chemical class	Compound isolated[ref]
1	Flavonoid and derivatives	Kaempferol acetyl-hexoside[26]
		Isoquercitrin[26]
		Apigenin-7-*O*-glucoside[26]
		Vitexin (apigenin-8-*C*-glucoside)[26]
		Quercetin-3-*O*-glucoside
		Ononin[26]
		Kampferol-3-*O*-glucoside[26]
		Kaempferol-3-*O*-pentoside[26]
		Kaempferol-*O*-deoxyhexoside[26]
		Pentahydroxyflavone[26]
		Quercetin[27]
		Quercetin glucoside[27]
		Kaempferol[28]
		Petunidin[28]
		Luteolin[29]
		Lupeol[30]
		Myricetin[31]
		Genistein and diadzein[27,32]
2	Steroids and derivatives	β-Sitosterol[27]
		β-Sitosterol glycoside[27]
		Friedelin-3-one,[3,20,30]
		Sitosterol oleate[32]
		Sitosterol-3-*O*-β-glucopyranoside[32]
		60-Palmitoyl-3-*O*-β-glucopyranosyl-sitosterol[32]
		60-Stearoyl-3-*O*-β-glucopyranosyl-sitosterol[32]
		60-Linoleoyl-3-*O*-β-glucopyranosyl-sitosterol[32]
		60-Linolenoyl-3-*O*-β-glucopyranosyl-sitosterol[32]
		β-Sitosterol-3,5,6-triol[32]
		(6*R*)-9,10-Dihydroxy-4,7-megastigmadiene-3-one-9-*O*-β-D-glucopyranoside[33]
		(6*R*)-6,9-Dihydroxy-4,7-megastigmadiene-3-one-9-*O*-β-D-glucopyranoside[33]

Table 12.1 (*Continued*)

S. no.	Chemical class	Compound isolated[ref]
		(6*R*)-6,9,10-Trihydroxy-4,7-megastigmadiene-3-one-9-*O*-β-D-glucopyranoside[33]
		Cholest-4-ene-3-one[33]
		5β-Cholestane-3α,7α,12α,25,26-pentol[34]
		3β,7β-Dihydroxy-12-oxo-5β-cholan-24-oic acid[34]
		5α-Cholestane-3β,7α,26-triol[34]
		3β,12β-Dihydroxy-5β-cholanoic acid[34]
		5α-Cholestane-3α,7α,12α,25,26,27-hexol[34]
3	Fatty acids	Trihydroxy-octadecenoic acid[26]
		Iso-pentadecanoic acid[32,35]
		9-Octadecenoic acid methylester[32,35]
		9,12-Octadecadienoic acidmethylester[32,35]
		9,12,15-Octadecatrienoicacidmethylester[32,35]
		Hexadecanoic acid[32,35]
		13,17-Octadecadienoic acid[34]
		2-Propyl-tridecanoic acid[34]
		5-Octadecylenic acid[34]
		2-Bromo-dodecanoic acid[34]
		Hexadecanedioic acid[34]
4	Iridoids	6-*O*-[2,3-Dimethoxy]-*trans*-cinnamoyl catalpol[27]
		6-*O*-*meta*-Methoxybenzoyl catalpol[27]
		Iridoid picroside 1[27]
5	Stilbene derivatives	Quadrangularins A, B, and C[27,37]
		Resveratrol[27,37]
		Piceatannol[27,37]
		Pallidol[27,37]
		Parthenocissine A[27,37]
		Trans-resveratrol-3-*O*-glucoside[35]
6	Glycerolipids	3-*O*-Linoleoylglycerol (4), 3-*O*-linolenoylglycerol[32]
		1-*O*-Stearoyl-2-*O*-palmitoyl-3-*O*-β-galactopyranosylglycerol[32]
		1-*O*-Oleoyl-2-*O*-palmitoyl-3-*O*-β-galactopyranosylglycerol[32]
		1-*O*-Linoleoyl-2-*O*-palmitoyl-3-*O*-β-galactopyranosylglycerol[32]

		1-*O*-Linolenoyl-2-*O*-palmitoyl-3-*O*-β-galactopyranosylglycerol[32]
7	Triterpenoids	δ-Amyrone (8.4%)[38]
		δ-Amyrin (12%)[38]
		Unsymmetric tetracyclic triterpenoids:
		Onocer-7-ene-3α,21β-diol[38]
		Onocer-7-ene-3β,21α-diol[38]
		7-Oxoonocer-8-ene-3β-21-α-diol[5]
		β-Amyrone[33]
		24-Methyl-dammara-20,25-diene-3β-yl-acetate[33]
		24-Methyl-dammara-20,25-diene-3β-yl-palmitate[33]
		24-Methyl-dammara-20,25-diene-3β-yl-stearate[33]
		24-Methyl-dammara-2,20,25-triene-1-one[33]
		24-Methyl-dammara-20,25-diene-3-one[33]
		δ-Amyrin acetate[35]
8	Phenolic acid and derivatives	Caffeic acid-*O*-hexoside[26]
		Quinic acid[26]
		Cissusic acid[36]
9	Lignan glycoside	Cissuside[33]
		Cissusol[33]
10	Phytols	1-*O*-β-Glucopyranosyl-2-[(2-hydroxy-octadecanoyl)amido]-4,8-octadecadiene-1,3-diol[32]
11	Miscellaneous	*N*-Acetylphenylalanine[26]
		Ascorbic acid[36]
		β-Carotene[36]
		Cimidahurine[33]
		Cuminaldehyde
		Vanilline[33]
		Eugenol[33]
		(−)-β-Caryophyllene oxide[33]

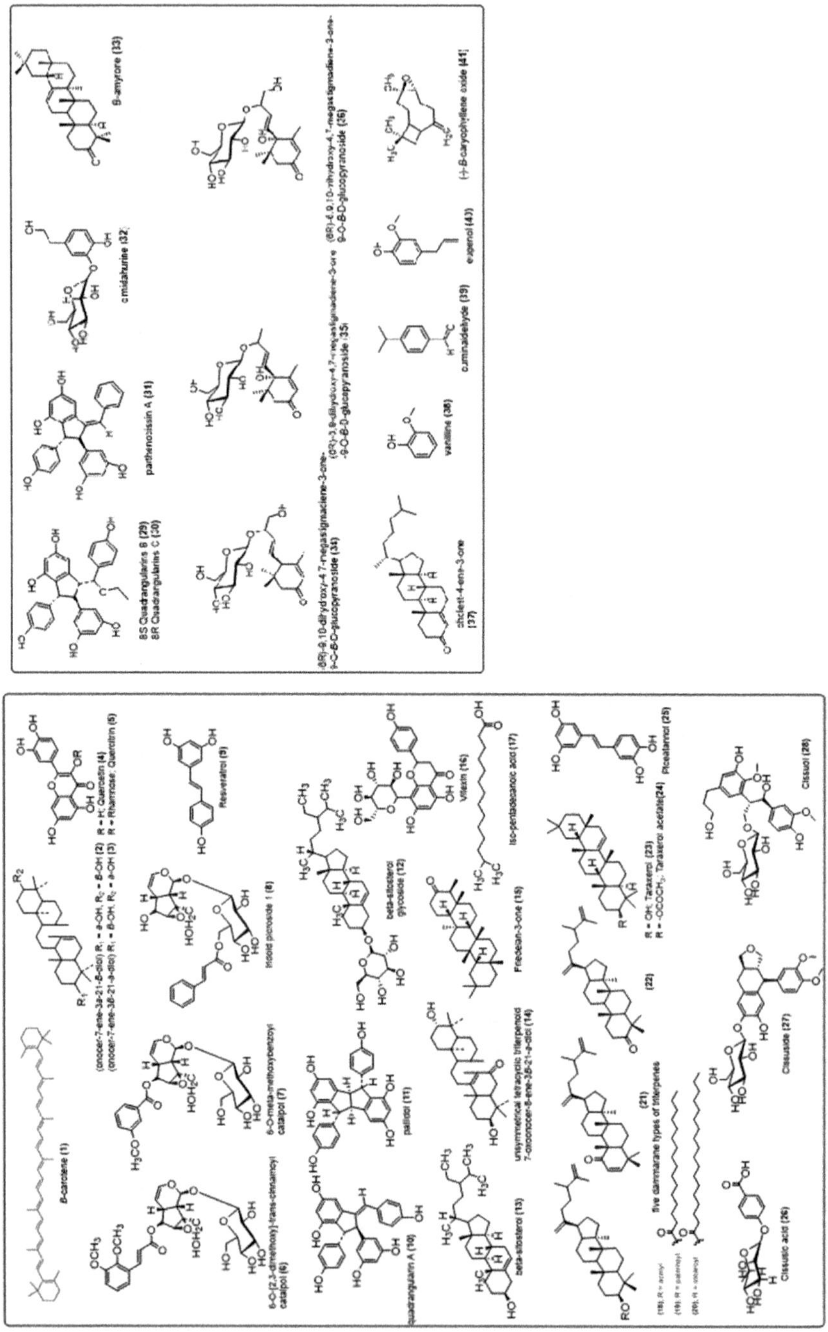

Figure 12.3 Chemical structures of some selected phytoconstituents isolated from the CQ stem.

found to be responsible for its cytoprotective action against oxidative stress and inhibition of tumorigenesis.[40] A total of 12 lipid constituents in the hexane fraction from the aerial part of the plant were isolated and characterized with the help of mass, infra-red (IR), and nuclear magnetic resonance (NMR) spectral data. The lipid compounds isolated from the aerial hexane fractions included esters (taraxeryl acetate), heptadecyl octadecenoate, Icosanyl icosanoate, ketone (4-hydroxy-2-methyl-tricos-2-en-22-one, friedelan-3-one, 7-hydroxy-20-oxodocosanyl cyclohexane), alcohols (taraxerol, 31-methyltritriacontanol, β-sitosterol), fatty acids (31-methyltritriacontanoic acid, iso-pentacosanoic acid), and alkenes (9-methyl-octadec-9-ene).[37]

12.3 Preclinical (*In Vitro/In Vivo)* and Clinical Pharmacology of CQ

Medicinal plants have been good resources for therapeutic treatment in local communities around the world for millions of years of tradition and are the primary source of treatment. Pharmacological evaluation methods and modern technological advancements are useful when analyzing and categorizing medicinal plants effectively and aid in using them in a better therapeutic and valuable manner. Many of the traditionally useful plants lack reports about scientific evidence and hence the preclinical and clinical studies in assessing and evaluating the medicinal property of the plant remain as important today and in the future also. CQ is easily available in the southern part of India and is considered as one of the best traditional medicines in remote rural communities. It is one of the edible medicinal plants and has an imperative role in retentive human health and longevity. This plant has multiple nutritional and pharmacological values to promote health, especially of the skeletal system. It offers an affordable health care regimen in traditional systems. To date, traditional medicine represents a well-established therapeutic modality to other forms of drugs. In order to update and conserve the knowledge and therapeutic benefits of CQ, the following section consolidated the reported scientific evidence, plausible mechanisms, and several therapeutic activities of CQ by *in vitro* and *in vivo* studies as well as by clinical reports.

12.3.1 Osteoporotic and Osteoblast-related Activity/Bone-related Activity

CQ is one of the well-recognized plants in the conventional system of medicine for treating bone fracture and bone-related complications (Figure 12.4).

Ethanolic extract of aerial parts of CQ was found to be rich in phytoestrogen and reported to be effective in the management of osteoporosis in ovariectomized rats. The study also showed that CQ could increase bone thickness, prevent bone loss and decrease alkaline phosphatase (ALKP)

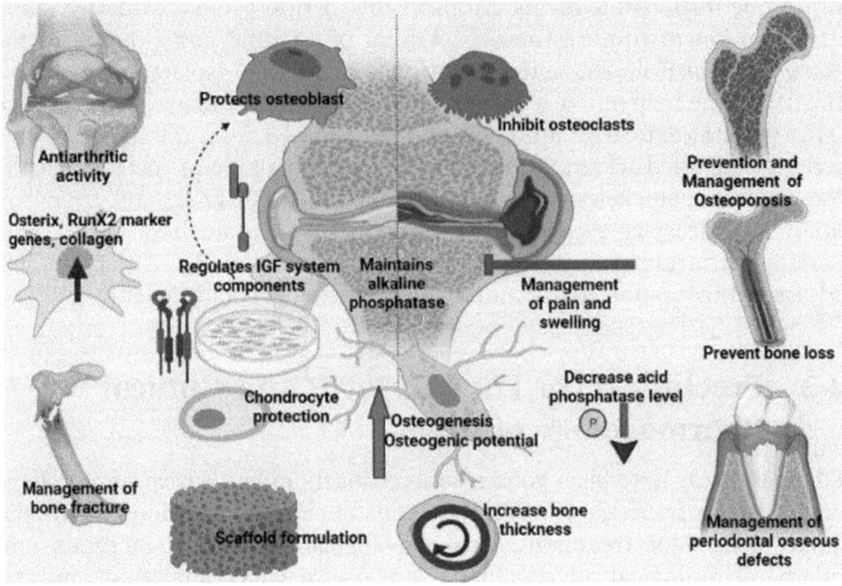

Figure 12.4 Bone related activity of CQ.

levels.[20] Potu *et al.*, in 2008 evaluated the efficacy of petroleum ether extract of CQ plant towards the development of long bones of the rat fetus during the intrauterine developmental stage. CQ extract in a dose of 500 mg per kg of body weight was given to the pregnant rats from gestation day 9 until the delivery period and the ossified cartilage (bone) in the newborn litter was analyzed to confirm its efficacy.[41] Their group further demonstrated that petroleum ether extract of CQ can effectively increase the thickness of cortical and trabecular bone in ovariectomized rats.[32,42] Another study reported the anti-osteoporotic effect of CQ hexane extract in an ovariectomized mice model. CQ successfully produced its osteoporotic effect by inhibiting bone loss in the cancellous and cortical bones of the femur and proximal tibia of C57BL/6 mice.[43] The chloroform fraction of the ethanolic extract of CQ showed a remarkable reduction in tartrate-resistant acid phosphatase levels and showed anti-osteoclastic activity without affecting osteoblastic activity. Shirwaikar and co-workers studied the anti-osteoporotic effect of the ethanol extract of CQ in ovariectomized mice at the two dose levels of 500 and 700 mg per kg. The treatment led to an increase in biomechanical strength, higher osteoblastic and lower osteoclastic activity.[44] The anti-osteoporotic activity of the extract could be attributed to the presence of steroids that acts as phytoestrogens to prevent or reduce bone loss.[45] In continuation with the investigation of the osteogenic potential of CQ, it has been reported that hexane (CQH) and dichloromethane (CQD) fractions of CQ extract delayed the mineralization of MC3T3-E1 cells whereas CQH enhanced the mineralization and expression profile of osteoblast marker genes such as osterix, RunX2, ALKP, and collagen. CQH extract showed the greatest osteogenic potential in comparison to the ethanol and ethyl acetate extracts. CQH because of its non-cytotoxicity or metabolic side effects can be used as a medicine for osteoporosis.[46] Ethanolic extract of CQ plays a prominent role in the regulation of insulin-like growth factor (IGF) system components in human osteoblast cells like SaOS-2 cells. CQ increased the mRNA expression of IGF-I, II, and interleukin (IL) in cultured cells.[47] IGF II is the most abundant in human bone and after the CQ treatment, its level increased two-fold as compared to IGF-I. The rise in mRNA expression and IGFs revealed the stimulatory effects of the CQ extract which further aids in fracture healing. CQ extract being rich is flavonoids, increases the cyclic adenosine monophosphate (cAMP), mRNA expression, and protein levels. Phytoestrogens present in the extract mimic the action of estradiol and could increase the expression of IGF proteins. The compound β-sitosterol is a weak estrogen receptor agonist, binds to estrogen receptor-β (ER-β), and exerts estrogen-like action. Iridoid picroside I isolated from CQ showed an effective increase in the ALKP activity and stimulated the formation of mineralized nodules.[48]

The effect of CQ in the long bone of ovariectomized rats was reported by Guerra *et al.*, (2019). They suggested that physiological changes can trigger the ability of CQ on bone turnover. CQ showed its maximum effect on bone protection in the distal femur of the long bone followed by the proximal tibia with the protection of trabecular thickness, and connectivity density. It was hypothesized that CQ accelerated bone formation by influencing more than one metabolic pathway probably by stimulating the protein matrix and protecting collagen breakdown.[21] Ruangsuriya *et al.*, in 2020 carried out molecular docking studies followed by ethanolic, hexane fraction estimation to narrow down the detection of the active ingredients for osteogenic activity in CQ. They found that ethanolic extract enriched with rutin and quercetin showed the highest therapeutic potential. It activated osteoprotegrin (OPG) and inhibited Receptor Activator of Nuclear Factor K Ligand (RANKL) expression.[49] Another *in silico* study reported that filtered molecules like cosmosiin, apiin, and calcitriol identified in CQ extract possess greater binding interactions with the targets of osteoarthritis and thus can promote and support bone growth.[34]

12.3.1.1 CQ in Bone-related Clinical Studies

Several clinical studies have been conducted in India and in other parts of the world to scientifically prove the traditional use and the worth of this miraculous herb of the Indian traditional system of medicine. A study highlighted the importance of CQ in the treatment and management of bone fracture by survey analysis and documentation of different traditionally used medicinal plants and their formulations. Out of 38 plant species belonging to 24 families, twenty-six formulations of 30 plant species were used in the treatment and management of bone fractures and CQ had the highest use in these formulations when compared with all other plants.[50] Oral administration of the CQ pulp powder in capsules with a dose of two capsules three times daily for 3 months with milk produced encouraging results in a clinical trial report for bone strengthening.[51] CQ is routinely used in the Indian subcontinent to accelerate the process of healing in bone fractures. A clinical report constituting of 60 patients in the age group of 20–35 years (divided into treatment and control groups) assessed fracture healing after the patients received CQ extract in capsule form. Clinical and radiological analyses demonstrated better healing of fractures in the treatment group. All the samples were examined for osteopontin expression. The obtained data showed that CQ accelerated fracture healing and caused early remodeling of fracture callus.[52] Another clinical study carried out in menopausal patients evaluated the therapeutic efficacy of CQ against osteoporosis as the bone resorption and reformation are slow in post-menopausal patients due to less estrogen. The results showed a percent decrease in pain (41.15%) and stooped posture (17.34%).[53]

Rajagopalan *et al.*, in 2018 documented the effect of CQ in bone fracture reunion and its ability to reduce pain and swelling. They carried out a clinical case study on a woman aged 78 years admitted with an accidentally fractured distal radius of the left hand with depression and evaluated the effect of CQ in the process of bone healing. She was prescribed 750 mg per kg of CQ along with calcium supplements for a period of 30 days. Remarkably, the post-treatment radiographic reports were amazing and displayed completely healed bone. This preliminary report was from their orthopedic center in Tamil Nadu, India and they are further working on it to assess the efficacy of CQ in a large number of patients.[54] Similarly, another clinical case study reported a 26 year-old male with Bennet's fracture was managed successfully within 6 weeks of treatment with CQ. Radiographic reports revealed a united fracture with a good amount of callus formation after treatment with CQ. The authors recommended the use of CQ as a supplement in various types of bone fractures for quicker healing with the early remodeling of fracture callus.[55] The role of CQ in the management of periodontal osseous defects, periodontitis, and bone growth, adjunct in the stability of orthodontic implants, pain management, and bone healing after implant placement in dentistry has been reviewed.[11] CQ showed better osseous healing than the control group in a clinical trial study involving 20 patients with periodontitis.[56]

12.3.2 Hepatoprotective Activity

CQ showed a hepatoprotective effect against a rifampicin-induced liver toxicity model. A potent antioxidant β-carotene is considered as the main chemical constituent responsible for the hepatoprotective action of the extract. The extract also decreased lipid peroxidation and augmented the antioxidant activities.[57] CQ also showed hepatoprotective activity against lead acetate-induced liver injury. It was able to restore the tissue level of superoxide dismutase (SOD), reduced glutathione (GSH), and catalase (CAT) enzymes.[58] Ethanolic extract of CQ prevented the symptoms of steatohepatitis (elevated serum lipid content, aminotransferase, inflammation in lobes, ballooning in liver section, and presence of steatosis). Pro-inflammatory markers NADPH oxidase 4 (NOX4) with levels of reactive oxygen species (ROS), oxidative stress markers, and inflammatory cytokines were also decreased after the treatment with ethanolic extract.[31]

12.3.3 Antimicrobial Activity

CQ has been reported to contain a number of phytoconstituents that are primarily responsible for their broad spectrum of biological activity. CQ has been shown to exhibit antimicrobial activity against bacteria, fungi, and virus. CQ biosynthesized nanoparticles play a major role in antimicrobial activity. The details of the reported scientific evidence on the antibacterial, antifungal, and antiviral properties of various parts of CQ are given in Table 12.2.

12.3.4 Antioxidants and Their Related Activity

Phytochemicals such as phenolics and flavonoids are associated with antioxidant properties. The presence of flavonoids in ethanolic and methanolic extract of the aerial parts of the CQ scavenge nitric oxide (NO) and hydroxyl (OH^-) radicals, whereas both phenolics and flavonoids found in the plant scavenged superoxide anions. The methanolic extract of CQ showed higher activity than the ethanolic extract. Among the flavonoids the activity for antioxidant activity decreased in the order; apigenin > fisetin > kaempferol. Quercetin present in the plant was found to be responsible for its cytoprotective action against oxidative stress and inhibition of tumorigenesis.[40] The hydroxyl (mono or dihydroxyl part) in the flavonoid ring and its hydrogen donating ability confer these compounds with protection against oxidative damage induced by free radicals. The antioxidant effects of CQ also help to reduce cardiovascular risk factors.[65]

Sapsrithong *et al.*, in 2012 evaluated the protective role of the ethanolic extract of CQ constituent's quercetin and resveratrol, which were found to inhibit ROS in hydrogen peroxide (H_2O_2) injured human umbilical vein endothelial ECV304 cells. The quercetin and resveratrol in CQ ethanolic extract led to an increase in antioxidant protein enzyme expression such as superoxide dismutase (SOD), glutathione peroxidase (GPx), and endothelial nitric oxide synthase as compared to the control.[66] It was also observed that quercetin and resveratrol along with other natural antioxidant ascorbic acid and kaempferol showed synergistic antioxidant activity to restore the ROS levels in endothelial cells. The CQ methanol extract showed antioxidative activity with an IC_{50} value of 522.0 $\mu g\,mL^{-1}$, and inhibitory effects on tyrosinase activity and melanin production in B16 melanocytes in a dose-dependent manner (100–2000 $\mu g\,mL^{-1}$).[67] The CQ aqueous extract also actively inhibited oxidative stress led steroidogenesis and male infertility

Table 12.2 Antimicrobial activity of various parts of CQ.

Plant part	Extract	Method	Organism	Standard	Observation	Ref.
Stem	Ethyl acetate, methanol, aqueous	Agar well diffusion	*B. subtilis* *B. cereus* *S. aureus* *P. aeruginosa* *Streptococcus* sp. *E. coli*	Tetracycline	Ethyl acetate extract showed better activity Gram-positive bacteria are highly susceptible	59
Whole plant	Aqueous/protease effect	Agar well diffusion	*B. cereus* *B. subtilis* *B. megaterium* *B. licheniformis*	None reported	Effective protease action	25
Stems	Aqueous Biosynthesized CaO nanoparticles using the extract	Agar well diffusion	*S. aureus* *E. coli* *P. aeruginosa* *K. pneumoniae* *S. typhi* *Shigella dysenteriae* *V. cholera*	None reported	Effective against Gram-positive and Gram-negative bacteria	60
Not specified	*Cissus quadrangularis* (CQ–silver nanoparticles)	Agar well diffusion	*B. licheniformis* *B. pumilus* *P. aeruginosa, Vibrio parahaemolyticus*	None reported	Effective against Gram-negative bacteria	61

Stem	Methanol Ethanol Petroleum ether Aqueous	Disc diffusion	*Escherichia species* *Klebsiella species* *Salmonella* sp. *Pasteurella* sp. *Staphylococcus*	Ciprofloxacin	Methanol extract showed potential activity against all the tested organisms	62
Stem	Methanol Ethanol Petroleum ether Aqueous	Clinical and Laboratory Standard Institute (CLSI) method using potato dextrose	*Aspergillus*	Ketoconazole	Methanol extract showed potential antifungal activity	62
Leaf	Aqueous Biosynthesized CuO nanoparticles using the extract	CLSI method using potato dextrose	*niger* *A. flavus*	Carbendazim	Better fungicidal activity than standard	63
Stem	Methanol	Dye uptake assay	*Herpes simplex virus (HSV) Type 1 and 2* *Vero cells*	None	Active against HSV1 and 2 at 1 : 400 dilution	64

induced by quinalphos. The extract further brought the testosterone levels and sperm count to normal through their androgenic activity. Phytochemical analysis of the extract showed the presence of high levels of steroids. The high content of ascorbic acid plays a significant role in spermatogenesis. The plant extract was also involved in scavenging ROS generated by quinalphos and elevating the levels of the antioxidant enzymes such as GPx, catalase (CAT), SOD, and ascorbic acid. The plant extract increased the levels of testicular cholesterol and activated the cholesteryl ester hydrolase and steroidogenic acute regulatory protein (StAR) expression and steroidogenic enzymes.[68] The ethyl acetate fraction of fresh and dry CQ stems showed 61.6% antioxidant activity in a 2,2 diphenyl picrylhydrazyl (DPPH) assay and 64.8% antioxidant activity in a β-carotene–linoleic acid model. The antioxidant activity of methanol, *n*-hexane, and aqueous extracts were comparatively less than ethyl acetate extract.[59] An ethanol extract of the aerial part of CQ showed prominent dose-dependent free radical scavenging activity with IC_{50} values of 98, 125, and 96 $\mu g\,mL^{-1}$ and the isolated flavonoid fraction showed 10, 12, and 14 $\mu g\,mL^{-1}$ against NO, DPPH, and OH radicals.[69] Ethanolic extract of CQ was able to reverse the decrease in the activities of antioxidant enzymes such as SOD, GPx, GST, ALKP, and collagen in ovariectomy rats. The extract also decreased the lipid peroxidation and tartrate resistant acid phosphatase (TRAP) (a marker of the osteoclast in bone resorption). The osteoclast bone resorption generates high levels of H_2O_2 and superoxide anions. These free radicals modulate both inter- and intra-cellular signaling for bone loss.[70]

12.3.5 Antidiabetic Activity

Methanolic extract of CQ stems produced protective action against insulin resistance and maintained an oxidant–antioxidant balance in rats fed with a high fat–high fructose diet. The protective action such as improvement in insulin sensitivity, reduction in liver damage, and restoring the levels of antioxidant and lipids to normal could be due to the high content of polyphenols, vitamins C and E. The total polyphenolic contents estimated were 585.40 ± 0.16 mg gallic acid equivalent (GAE) per 100 g and the levels of vitamin C and E were 327 mg per 100 g and 697 mg per 100 g of the powdered extract.[71] The ethyl acetate fraction of the CQ stem is shown to be effective against hyperglycemia-mediated oxidative stress and inflammation in the nicotinamide/streptozotocin diabetes mellitus rat model. The extract reversed the increase in blood glucose, liver toxicity markers, and lipid peroxidation enzymes. The extract led to an increase in the activities of antioxidant enzymes (SOD, CAT, GPx, and GSH). The histopathological studies of the liver and pancreas revealed no alteration after treatment with the CQ extracts.[72]

12.3.6 Antiarthritic Activity

Kumar *et al.*, investigated the ability of CQ in the down-regulation of pro-inflammatory cytokines and inhibition of angiogenesis in an adjuvant-induced arthritis model.[73] CQ hydroalcoholic extract produced a dose-dependent inhibition of joint swelling in both formaldehyde-induced and adjuvant-induced arthritis. The treatment also reduced the serum tumor necrosis factor-α (TNF-α) level, oxidative stress, and the synovial expression of inflammatory and angiogenesis markers. This immunomodulatory activity on pro-inflammatory cytokines could be attributed to the presence of kaempferol, quercetin, and β-sitosterol.[73] The anti-arthritic effect of the active fraction of CQ (AFCQ) has been reported. Complete Freunds Adjuvant (CFA) induced rat arthritis model was used and analyzed for hematological and biochemical parameters. AFCQ at the dose of 100 mg per kg body weight demonstrated an efficient anti-arthritic effect when compared with celecoxib and methotrexate. The results were further analyzed through histopathology and radiography to confirm its efficacy and they were found to be effective in the management and treatment of arthritis.[4] CQ stem and root extracts were tested against standard drug naproxen in a mono sodium iodoacetate induced experimental *in vivo* osteoarthritis (OA) model. CQ stem extract showed better therapeutic activity than root extracts. The difference in the activity could be due to the chemical nature and quantity of bioactive constituents present in the stem that facilitated the recovery from joint tissue damage and reduced the risk factors for the OA incidence and progression.[26] An ethanolic extract of CQ was subjected to immunomodulatory evaluation through a carbon clearance test, the effect on serum immunoglobulins, determination of delayed type hypersensitivity, and *Candida albicans* phagocytosis test. The study results confirmed its immunoprophylactic potential which might be due to the presence of active phytoconstituents with immunomodulatory action.[74]

12.3.6.1 CQ Clinical Studies for an Antiarthritic Effect

A clinical study investigated the effect of specialized topical phytotherapeutic formulation considered during muscle atrophy, weakness inflammatory disorders, and pain with chronic arthropathy in humans. This phytotherapeutic formulation contained *Cissus quadrangularis, Calotropis gigantea, Zingiber officinalis, Rosemarinus officinale, Boswellia serrata, Curcuma longa*, and *Withania somnifera* mixed with sesame oil and beehives wax. Baseline data were collected from 153 patients in the age group of 59.89 ± 11.37 years suffering from muscular dystrophy during osteoarthritic disorders (MD-OADs) for 7.89 ± 1.9 years. After six weeks of treatment with this topical phytotherapeutic herbal composition, serum creatine kinase-MM and aldolase-A levels were measured and compared with baseline levels. All the participants underwent standard physical and radiographic examinations and the results were also promising.[75] Kumar *et al.*, 2020[34] analyzed 23 active compounds identified from the ethanol extract of the stem part of CQ by selecting the targets which have a major role in the degeneration and destabilization of the knee joints. The study was carried out to identify small molecules for the antiarthritic activity.

A clinical study was conducted in South India using CQ and *Zingiber officinalis* (ZO) to support the recommendations of clinical guidelines to use non-pharmacological strategies as the first-line management of OA. A total of 60 patients enrolled in the study were divided into three groups and were treated with CQ-5g, ZO-5g, and CQ + ZO-5g, respectively. Patients experienced a significant reduction in cracking pain in joints after treatment with CQ as well as ZO individually. A prominent reduction in joint pain was noticed after treatment with the combination of CQ + ZO which confirmed the therapeutic efficacy.[76]

12.3.7 Antiulcer Activity

Methanolic extract of CQ was shown to exert a protective action on an aspirin-induced gastric ulcer. It increased the mucus production and decreased lipid peroxides levels and myeloperoxidase (MPO) activity. The antioxidant phytoconstituents such as β-carotene, ascorbic acid, ketosteroids, triterpenoids, calcium, β-amyrin, and α-amyrin may be involved in protecting the aspirin-induced lesions.[77] In another mechanistic study, CQ was found to increase in levels of GSH, SOD, CAT, GPX, and glutathione S-transferase (GST). It decreased the lipid peroxidation and reduced oxidative damage to DNA, thus blocking cell death.[78] Different selected doses of CQ extract of 250, 500, and 750 mg per kg showed 40, 71.2, and 72.6% protection compared to standard ranitidine (30 mg per kg of 71.9% protection). It has also been stated that ascorbic acid, the major constituent isolated from the extract of CQ was involved in the increase of interleukin (IL)-1β, IL-6, TNF-α, and interferon-gamma (IFN-γ) and the decrease in IL-10, IL-4, and prostaglandin E_2 (PGE_2) levels. Thus, it could be inferred that the extract prevented the aspirin-induced gastric ulcer due to its immunomodulatory effect.[79]

12.3.8 Analgesic, Anti-inflammatory, and Antihemorrhoidal Activity

CQ root extract has been shown to inhibit the acetic acid-induced writhing's and increased tail-flick withdrawal responses in mice.[76] An aqueous extract of the aerial parts of CQ was evaluated for anti-arthritis activity against formaldehyde-induced arthritis. The extract was found to be superior to indomethacin in inhibiting joint inflammation. It also showed a dose-dependent increase in antioxidant markers and a decrease in levels of pro-oxidant markers.[73] The use of methanolic extract of CQ in the treatment of hemorrhoids is due to its analgesic, anti-inflammatory, and venotonic effects. It also reduced ear edema formation in rats induced by phenyl propiolate and paw edema induced by carrageenan and arachidonic acid. The extract exerted a venotonic effect in isolated human umbilical veins like the mixture of bioflavonoids (90% diosmin and 10% hesperidin).[80]

The active acetone fraction of CQ was found to inhibit cyclooxygenase-1 (COX-1), cyclooxygenase-2 (COX-2), and 5-lipooxygenase (5-LOX) enzyme activities. In addition to that, it also inhibits pro-inflammatory mediators like inducible nitric oxide synthase (*i*NOS) and TNF-α. It was involved in the translocation of nuclear transcription factor nuclear transcription factor-2 (Nrf2) important for cellular protection and cell survival. The anti-inflammatory action of CQ could be due to the presence of luteolin and β-sitosterol.[81] The ethyl acetate fraction of CQ inhibited the lipopolysaccharide (LPS) induced NO. It also induced heme oxygenase-1 (HO-1) gene expression and suppressed the *i*NOS and nuclear translocation of p65 NF-κBF.[82] The exact chemical responsible for the induction of HO-1 and anti-inflammatory action remains unclear, however quercetin is known to inhibit the LPS-stimulated NO generation. Other chemicals such as resveratrol and piceatannol are inducers of HO-1 expression both *in vitro* and *in vivo*.

An ethanol extract of the whole plant of CQ at the dose of 500 mg per kg showed 50% inhibition of carrageenan induced rat paw edema and was nearly equivalent to the standard diclofenac at 10 mg per kg. It also showed 78% venoconstrictive effect (at the dose of 1 $\mu g\,mL^{-1}$) on endothelium intact and denuded rat veins with EC_{50} values of 0.01 and 0.03 $\mu g\,mL^{-1}$.[83] Following the preclinical study, a double-blind randomized controlled clinical trial was carried out in hemorrhoid patients. A total of 160 patients were involved in the trial and they were administered with enteric film-coated tablets of CQ extract (100 mg CQ extract standardized to contain 10–11 mg quercetin) at a dose of 200 mg twice per day for seven days. The findings revealed CQ to be effective in the management of bleeding, pain, and prolapse of hemorrhoid tissue.[83] They also concluded that CQ extract is safe and effective in the management and treatment of hemorrhoids along with other traditional methods of treatment.

12.3.9 Metabolic Syndromes: Weight Loss and Obesity-related Activity

CQ extract in hot water showed anti-obesity action against 3T3-L1 adipocytes. The extract inhibited lipid accumulation with an absence of any cytotoxic effects against 3T3-L1 adipocytes. There are other studies that report the presence of compounds such as friedelin, α-amyrone, resveratrol, piceatannol, and pallidol.[3] Some evidence also supported that CQ inhibits enzyme lipase and amylase enzyme, thus blocking the dietary uptake of the carbohydrate and fat uptake. This helps in reducing fats, body weight, and markers associated with metabolic disorders.[65]

12.3.9.1 CQ Clinical Studies for Weight Loss and Obesity

It has been well documented that a change in homeostasis in food intake and energy expenditure is related to obesity. Several plant-based products are available in the market for weight loss and the comparative effect of each product is still in question. CQ formulation (CQR-300) was tested for its effect on weight loss and central obesity in a randomized, double-blind, placebo controlled clinical trial. The study involved 123 overweight and obese subjects including males (47.2%) and females (52.8%) in the age group of 19–50 years. They were divided into two groups based on body mass index parameter (BMI) into CQ formulation and placebo group. All the study participants were treated with two daily doses of formulation or placebo and remained on a normal or calorie-controlled diet for 8 weeks. At the end of the clinical trial, net reductions in weight and central obesity, as well as in fasting blood glucose, total cholesterol, low density lipoprotein (LDL)-cholesterol, triglycerides, and C-reactive protein (CRP) were observed in participants. This study also confirmed the effect of CQ in the management of weight loss and metabolic syndrome.[84] Similar results were also reported by another clinical trial using CQ aqueous leaf and stem extracts at the dose range of 300 mg per kg body weight in humans and they concluded that CQ was effective in the management of metabolic syndromes in overweight individuals.[85] Encouraged by the results, an additional pilot clinical study for eight weeks was conducted to confirm the effect of CQ in reducing the percentage of body fat measured using two different methods-bioelectrical impedance assays *versus* dual-energy X-ray absorptiometry (DEXA). It was a double blinded, placebo-controlled study that enrolled 67 participants. The participants were counseled to maintain normal exercise and diet. They were further divided into two groups placebo (32 participants) and the CQR-300 group (35 participants) and were given 300 mg of corn starch or CQR-300 daily. After 8 weeks of the trial, the placebo group showed only a 1.05% decrease in body fat, whereas the CQR-300 treated group showed 8.9 and 12.8% decrease in the body fat as measured by impedance and DEXA, respectively. The CQR treated group also showed a significant ($p<0.05$) decrease in waist and hip circumferences, systolic and diastolic blood pressures, total cholesterol, triglycerides, fasting blood glucose, as well as leptin levels but an increase in high density lipoprotein (HDL) cholesterol level and adiponectin levels.[86]

A preclinical *in vivo* study evaluated the appetite suppression efficacy of well-known products including CQ. They tested pure caffeine, potato skin extract, CQ extract, *Garcinia cambogia* extract, *Crocus sativus* extract, raspberry ketone isolated from *Rubus idaeus*, one commercial product (Appetrex), and one novel composition (UP601). The CQ extract showed a significant reduction in food consumption at the 2 h time point with 44.3% after food provision at an equivalent human dosage of 10 g.[87] CQ showed significant weight loss activity in another randomized clinical trial conducted using seven herbal medicines. Three (*C. quadrangularis, S. indicus* combined with *G. mangostana,* and *I. gabonensis*) of the seven herbal medicines showed prominent beneficial effects and CQ was one of them. As per the consolidated report of Maunder *et al.*, this trial was conducted only for a duration of 8–10 weeks and there existed some reporting and confirmation bias.

Therefore, more evidence is needed to take a decision about its effect on weight loss. They also recommended further investigation in a large population to confirm the dose, dosage, long term safety, and efficacy because one single clinical study is not sufficient to recommend a herbal drug for weight loss/obesity management.[88]

Chatree *et al.*, in a clinical study, elucidated the effect of CQ extract on obesity, lipolysis, and browning of human white adipocytes (hWA). It was observed that treatment of obese subjects with CQ significantly reduced waist and hip circumferences at weeks 4 and 8 when compared with the baseline value. The effective reduction in the serum leptin level, hWA associated glycerol level were also noted. They also reported that CQ extract can enhance the UCP1mRNA expression in hWA in a dose dependent manner suggestive of its browning effect on hWA.[89]

12.3.10 Wound Healing Activity

Aqueous and methanolic extract of the whole aerial part of CQ showed an effective wound healing property in BALB/c female mice with respect to the standard oxytetracycline ointment. It was noted that CQ extract had a superior wound healing property as the wound area completely returned to the normal structure by day 15 of the experiment with complete epithelization and the presence of mature collagen fibers in the skin.[90] The role of herbal medicines in stem cell therapy is reviewed by Udalamaththa and group. They suggested that CQ might play a prominent role in stem cell therapy in the future based on its effect in scaffold and bone regeneration therapy.[91]

12.3.11 Neurological Activity

CQ is traditionally used for its anticonvulsant effect. Neuropharmacological effects of methanolic extract of CQ roots were evaluated in a mice model. CQ extract at the dose of 50, 100, and 200 mg per kg showed a reduction in exploratory behavior, motor coordination, spontaneous motor activity, and prolonged pentobarbitone sleeping time. This study confirmed that the methanolic root extract of CQ contains some phytoconstituents that act centrally and produce CNS related action especially the sedative effect.[92] Anxiolytic and antiepileptic effects of the aqueous extract of CQ were evaluated in mice using an elevated plus maze, open field paradigms model and convulsion induced model. Four different doses of CQ extract were used for evaluation (37.22, 93.05, 186.11, and 372.21 mg per kg, per oral) and the effects were compared with standard drug sodium valproate (300 mg per kg). The involvement of GABAergic neurotransmission was determined by measuring the levels of malondialdehyde, GSH, gamma amino benzoic acid (GABA), and GABA-transaminase in the hippocampus of sacrificed epileptic mice. CQ extract significantly decreased the number and duration of seizures and also increased the number and time spent in the open arm and reduced the number and time spent in the closed arm. This evaluation supported the anticonvulsant and anxiolytic activities of CQ and the plausible mechanism through GABA transmission.[93]

12.3.12 Anticancer Activity

The ethanol extract and isolated flavonoid fraction of the aerial part of CQ were subjected to anticancer evaluation in breast cancer cell lines (MCF-7) by MTT assay. The results depicted that the flavonoid fraction exhibits potent anticancer activity with an IC_{50} value of 40 $\mu g\,mL^{-1}$.[69] Another study explored the genotoxic and apoptotic effects of ethanolic extract of CQ in A431 cells by 3-(4,5-dimethylthiazolyl-2)-2,5-diphenyltetrazolium bromide (MTT) assay. CQ extract showed dose dependent cytotoxic effects and also displayed significant production of intracellular ROS in the cells. Genotoxicity analysis by acridine orange staining depicted significant induction of chromosomal condensation. DNA fragmentation analysis showed that CQ extract caused DNA damage in A431 cells in a concentration-dependent manner. The results of this study correlate the effect of CQ toward apoptosis and genotoxicity.[94]

12.3.13 Role in Sarcopenia: Muscle Wasting Diseases

In recent years, there has been an increase in the incidence rate of sarcopenia and muscle wasting diseases with the increase in global life expectancy. In an attempt to identify a herbal formulation, a preclinical study was conducted using a combination of *Boswellia serrata*, *Cissus quadrangularis*, and *Withania somnifera* (HIM-CHX) for the management of sarcopenia with the dose range of 120, 250, and 500 mg per kg body weight. The effect of this formulation on muscle mass, grip strength, motor coordination, gait, and locomotor activity in rats was evaluated. Molecular level analysis like the measurement of inflammatory cytokines, myokines, and growth hormone has also been evaluated. The results showed that HIM-CHX displayed a significant improvement in gastrocnemius muscle weight, carcass weight, locomotor, gait, and endurance. It also reduced TNF-α, IL-6, and myostatin with an increase in IGF-1 level. This study confirmed the potentiality of the formulation in correcting the pathophysiological changes associated with sarcopenia.[95]

12.4 CQ in Nanotechnology

CQ plants owing to their rich content of bioactive chemical constituents have been used extensively for the green synthesis of nanoparticles including gold, silver, and copper oxide. The use of CQ plant extract offers a simple, eco-friendly, and cheaper alternative method to synthesize stable, uniform, spherical, and crystalline nanoparticles with a broad spectrum of biological activities.

12.4.1 Silver Nanoparticles (AgNPs)

In an eco-friendly and green synthetic approach, aqueous extract of CQ stem was used to synthesize AgNPs as a potential antiparasitic agent. FTIR suggested the presence of phenolic compounds on the surface of AgNPs. CQ synthesized AgNPs exhibited potent adulticidal and larvicidal activity against *Hippobosca maculate* (LC_{50} 6.30 mg L^{-1}) and *Rhipicephalus microplus* (LC_{50} 7.61 mg L^{-1}).[96] Acute toxicity of the AgNPs of CQ against *Poecilia reticulata* larvae and antibiofilm activity against Gram-positive and Gram-negative bacteria was investigated by Sathappan *et al.* Treatment with the CQ nanoparticles reduced the average thickness in both bacteria. The toxicity study against *P. reticulata* larvae showed 100% mortality in 40% CQ AgNPs (20 μg mL^{-1}) whereas no mortality was observed with CQ extract up to 500 μg mL^{-1}.[61]

12.4.2 Gold Nanoparticles (AuNPs)

Microwave irradiation of hydrogen tetrachloroaurate with CQ aqueous extract (for just 40 s at pH 9) have been used to synthesize gold nanoparticles. The AuNPs were of small size, stable, spherical, and crystalline in nature which were observed to be biocompatible in a hemolysis assay. It was proposed that the presence of flavonoids, stilbenes, and ascorbic acid in CQ extract are responsible for the reduction of gold particles. Furthermore, Fourier transform-IR and thermogravimetric analysis revealed that CQ extract components were also capping the synthesized AuNPs.[97]

12.4.3 Copper Oxide Nanoparticles (CuONPs)

Devipriya and Roopan in 2017 used CQ aqueous extract for the green synthesis of CuONPs. The synthesized NPs were spherical and crystalline with 65.85 and 18.98% of Cu and O, respectively exhibiting good fungicidal activity against *Aspergillus niger* and *A. flavus*.[63]

12.5 Biomedical Applications of CQ

Parvathi *et al.*, fabricated poly-L-lactic acid (PLLA) nanofibers loaded with CQ extract (20 and 40 w/w%) using an electrospinning technique for bone tissue engineering and drug delivery applications. The herbal hydrophilic fibrous sheet containing high percentages of CQ was designed to heal critical sized segmental bone defects. The hydrophobicity of the PLLA fibers was reduced and the power of adhesion, proliferation, and differentiation of the mesenchymal cells was significantly increased. The fiber was also found to be capable of promoting mineralization in one of the simulated body fluids. It showed enhanced protein absorption and an osteoinductive property making it an ideal biomaterial candidate for bone grafting. The improved osteoinductivity of nanofibers was attributed to the presence of ascorbic acid, β-sitosterol, δ-amyrin, δ-amyrone, and quercetin phytochemicals in CQ. The presence of ascorbic acid has been known to induce osteoblast specific gene expression and proline and lysine hydroxylation in collagen.[98] Nanofibers scaffolds of CQ prepared with hydroxyapatite and polycaprolactone have also been reported to show beneficial effects in bone tissue regeneration.[99] The 3D printed β-tricalcium phosphate (TCP) loaded polydopamine loaded CQ implants with improved *in vivo* new bone formation was developed by Robertson and Bose.[100] Polydopamine showed increased proliferation and alkaline phosphatase activity in osteoblast cell culture which was further enhanced with the addition of CQ extract. The implants showed more new bone growth compared to β-tricalcium phosphate. Figure 12.5 shows the polydopamine and CQ coating to β-TCP scaffolds.

A novel herbal scaffold of 2% natural biopolymer alginate, *O*-carboxymethyl chitosan, and herbal extract of CQ was developed by Soumya *et al.*[101] Initial cell attachment and proliferation of herbal extract was well spread across the scaffold surface. The scaffold enhanced the activity of ALKP and differentiation of human mesenchymal stem cells (hMSCs) to osteoblasts with increased calcified mineral deposition. The stimulation of osteoblastogenesis is because of the presence of phytogenic steroids in CQ extract and its action on estrogen receptors. The differentiation, proliferation of osteoblast and increased mineralization of hMSCs led to bone growth.

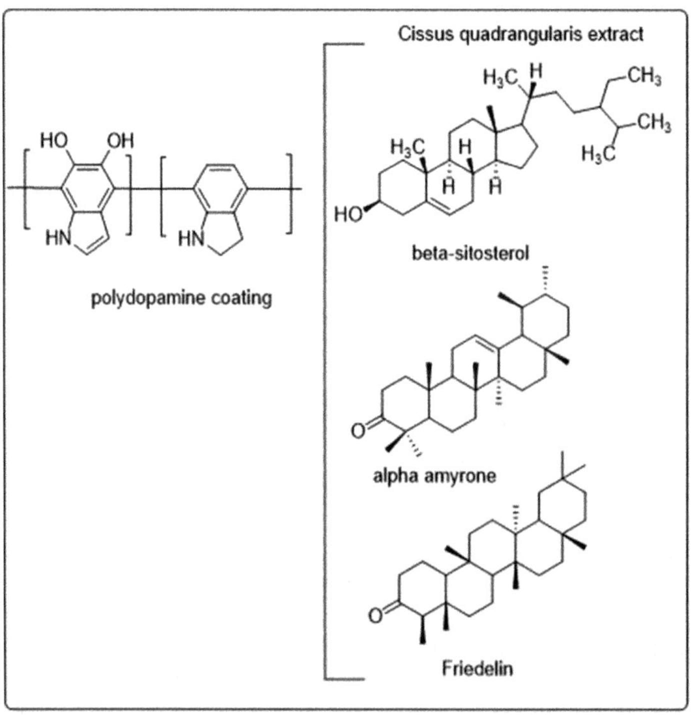

Figure 12.5 Coating and doping of the beta-TCP scaffolds.

Samarasinghe *et al.*, 2014 formulated nano-formulation C-CP-NCs to deliver CQ extracts. They used mice as the animal model and demonstrated the effect of the extract to ameliorate OA disease progression by inducing chondral protective and disease-modifying activity. A signaling mechanism modulated by alginate enclosed C-CP-NCs in chondrocyte was proposed. This study also established the non-toxic and biodegradable nature of the formulation with superior therapeutic activity.[102]

A phytochemical scaffold was fabricated by mixing CQ extract with a chitosan/Na-carboxymethyl cellulose blend *via* a lyophilization technique for bone regeneration. It was found that the addition of CQ increased the compression modulus of the composites and were suitable for bone regeneration. The fabricated scaffolds exhibited an osteoinductive property suggesting their potential application in bone tissue engineering.[103] Thongtham *et al.*, in 2020 fabricated the bone tissue engineering composite scaffolds using chitosan/collagen/hydroxyapatite biopolymers to encapsulate CQ loaded polycaprolactone nanoparticles prepared by the double emulsion method. The prepared biopolymer scaffold was shown to prevent the sudden release of CQ and showed an improved release profile along with a prolonged release rate due to a double sustained mechanism. The fabricated scaffolds were also found to help in osteoblast proliferation and were non-toxic to MC3T3-E1 murine osteoblasts.[104]

12.6 CQ Safety Assessment in Animal Models and in Clinical Studies

CQ stems, extracts, and powders are commonly used in traditional systems of medicine and also consumed by the local community as a general health tonic, to promote bone and tissue healing, to treat infections, and to promote weight loss and management. Scientific reports established its toxicity level by using animal models and they support the use of CQ in the health system. Some studies concluded that CQ appears to be very safe at the doses commonly used in toxicity studies. Some other studies recommended further standardization and clinical toxicity data.[105] The sub-chronic toxicity of CQ stem powder was evaluated in Wistar rats by Attawish *et al.*, in 2002. They found that oral treatment with a CQ powder suspension in water at four different concentrations (0.03, 0.3, 3.0, and 3.0 g per kg body weight per day) for three months did not reveal any significant changes in hematological and biochemical parameters, body weight before and after the study, as well as in histopathological studies indicating it to be possibly safe at a 1000-fold therapeutic dose in humans.[106]

Kothari *et al.*, in 2011 investigated the sub-chronic toxicity of CQR-300 in Sprague Dawley rats for 90 days. CQR extract at doses (0–2500 mg per kg body weight) did not produce any clinical toxicity, including mortality, change in body weight, microscopic or macroscopic abnormalities, and changes in hematological, biochemical, urine analysis, and ophthalmological parameters.

Furthermore, results of an Ames assay of mutagenicity indicated that CQR-300 is not genotoxic. Based on the results of this study, it was suggested that CQ extract up to a dose of 150 g per person per day won't produce any adverse effect and hence is very safe for human consumption.[107] Similarly, administration of a standardized phytoestrogen-rich fraction of CQ to the rats in a eight week-long study also showed no signs of toxicity.[20] In a sub-acute toxicity study of a CQ hexane extract, chronic administration of it did not produce any physiological and pathological changes as compared to normal rats.[73] In another study, the sub-chronic toxicity of enteric film coated tablets containing CQ extract was evaluated in Wistar rats. Animals were given tablets containing 7.14–714 mg per kg of CQ extract for 90 days. None of the animals showed any sign of dose related toxicity on hematological parameters, serum chemistry, urine analysis, and macroscopic abnormalities.[83] Kavitha *et al.*, in 2018 demonstrated that an aqueous extract of the whole plant of CQ at a dose of 3000 mg per kg bodyweight upon administration to Swiss albino mice for 14 days did not produce any toxicity.[108] Contrary to preclinical trials, a clinical case report stated that herbal capsules containing CQ extract induced thrombocytopenia in a renal allograft patient. This patient was under an immunosuppressive regimen for his ailment and suddenly reported severe thrombocytopenia after consuming CQ herbal capsules. A prompt improvement in the condition of the patient has been observed after the withdrawal of the CQ herbal capsules. Hence the physician related this effect to the CQ extract for thrombocytopenia and advised avoiding CQ herbal capsules for patients under immunosuppressive therapy as there is no supporting evidence about the drug interactions and also a lack of available data about the adverse drug reaction (ADR).

The data from the acute and sub-chronic animal studies[20,106,107] indicate that daily use of CQ stems powder or extract for 90 days is not associated with any unwanted side effects confirming its safety. However, there is a scarcity of data on their long-term clinical use in humans.

12.7 Intellectual Property Rights Related to CQ

Granted patents and applications related to the use of CQ for various disorders are presented in Table 12.3.

12.8 Conclusion and Future Perspectives on CQ Clinical Use

Existing systematic and meta-analyses of reviews related to CQ suggest that CQ is possibly safe and more effective in the management of bone fractures, bone related disorders, and obesity/weight loss management, but multiple clinical evidence is lacking. The effect of CQ in obesity/weight loss management is inconclusive because of the lack of multiple evidence. Meta-analysis reviews suggest that CQ is effective in combination with other ingredients for the management of obesity, but the effectiveness is inconclusive. Many studies reported the effect of CQ in bone fractures, but the

Table 12.3 Summary of patents/patent applications related to CQ.

S. no.	Patent details	Inventors	Title of invention	Description
1.	US20050048141A1 2007-01-19	Brian Cortes Lal Hingorani Vijay Thawani	Use of the plant CQ or its extracts for osteo-health including prevention or mitigation of degenerative disease, fracture healing, and anabolic increase in osteo tissues	This invention describes the composition and preparation of a nutritional product or a dietary supplement containing CQ or its extracts to restore bone health and in the treatment of osteo-health related conditions.
2.	WO2008081233A3 2009-11-19	Villoo M Patell	CQ plant extracts for treating osteoporosis and the extraction process thereof	It is related to the preparation of CQ extracts and its efficacy in the treatment of osteoporosis.
3.	US8318213B2 2012-11-27	Julius E. Oben	Plant extract mixtures and their uses	A composition comprising an effective amount of CQ leaves and stems extract claims to be beneficial in weight loss in obesity.
4.	WO2020089944A1 2020-05-07	Nilima Mohanty; Somashekara Nirvanashetty; Sanjib K. Panda Vivek A Parachur	Process for preparing a herbal extract and compositions thereof	It describes a novel process for the isolation of selective compounds and resins in specific concentrations for preparation of a herbal extract and compositions comprising the same.
5.	US2008166419A1 10-07-2008	Scott F. Sones	Krill oil compositions	This invention claims a composition containing CQ standardized to 5% ketosteroids which is beneficial in reducing one or more risk factors associated with cardiovascular diseases.
6.	WO2020236023A1 2020-11-26	Anton L. Prozumentov Georgiy V. Sarychev Vera K. Vedmina Ivan S. Martynenko	Dietary supplement formulation for weight loss	It is related to composition of a dietary supplement containing various plant extracts including CQ stalks for weight loss.
7.	EP2941261B1 24-07-2019	Ganga R. Gokaraju Rama R. Gokaraju Venkata K. R. R. Gokaraju	Synergistic curcuma and cissus compositions for enhancing physical performance and energy levels	It claims a synergistic dietary supplement compositions having extracts (20% to 80%) of CQ and *Curcuma longa* for improving physical performance and mental alertness.

Table 12.3 (*Continued*)

S. no.	Patent details	Inventors	Title of invention	Description
		Kiran Bhupathiraju Trimurtulu Golakoti Krishanu Sengupta Venkata K. R. Alluri		
8.	US7582316B1 01-09-2009	Jacob Geissler	Ketone-steroid extracts from the CQ plant and methods thereof	This invention discloses composition and preparation of ketone steroid extract of the roots of CQ which can be used to promote lean body mass, treat damaged or diseased connective tissue, and provide an analgesic effect while reducing cortisol.
9.	US10542989B2 28-01-2020	Ganga R. Gokaraju Rama R. Gokaraju Venkata K. R. R. Gokaraju Kiran Bhupathiraju Trimurtulu Golakoti Krishanu Sengupta Venkata K. R. Alluri	Synergistic dietary supplement compositions for enhancing physical performance and energy levels	It discloses the composition of a novel synergistic dietary supplement containing at least two plant extracts to enhance the body energy levels that can augment physical and mental health.
10.	US7935367B2 03-05-2011	Marvin Heuer Ken Clement Shan Chaudhuri	Compositions and methods for increasing adipose metabolism, lipolysis, or lipolytic metabolism *via* thermogenesis	This invention discloses a composition of a dietary supplement containing extracts of CQ (0.05 g to about 0.3 g) for enhancing lipid metabolism.
11.	WO2007101066A3 2008-10-30	Shil C Kothari	Methods and compositions for improving cardiovascular risk factors and metabolic risk factors that cause syndrome *x*	This invention discloses the beneficial use of 300 mg CQ extract per day for six weeks in obese patients such as weight loss, appetite suppression, decrease in LDL, and blood sugar and triglyceride levels.

12.	WO02085394A1 2002-10-31	Bakulesh M. Khamar, Vandana D. Mody Kalpesh K. Pandya	The process of preparing the topical anti-inflammatory/ analgesic preparation	This invention provides a method of preparation of a synergistic herbal preparation containing CQ and Costus species as main ingredients along with other volatile oils. The preparation is claimed to be effective in relieving painful and inflammatory conditions.
13.	WO2008028268A1 2008-03-13	Marvin Heuer Ken Clement Shan Chaudhuri	Diet supplement for causing weight loss comprising CQ, green tea, and caffeine	It claims a dietary supplement containing CQ stem and leaf extract (0.01 g to 1 g), green tea leaf extract, and anhydrous caffeine (max 0.3 g) for causing weight loss.
14.	WO2010054469A9 2010-07-08	Marvin A. Heuer Ken Clement Megan Thomas Jason Peters	Appetite suppressing weight management composition comprising *Caralluma fimbriata*, *Hoodia gordonii*, and CQ extracts	It claims a method of preparation of a composition containing five ingredients to reduce body weight and suppress appetite.
15.	US7659313B2 2010-02-09	Julius E. Oben	Methods and related compositions using specific flavonoids and indanes to reduce weight and inhibit lipase, alpha-amylase, and alpha-glucosidase activity in mammals	It claims a composition (50–200 mg per day) prepared by using flavonoids and indanes extracted from the stem and leaves of CQ for weight loss. Selected flavonoids or indanes include 3-*O*-rhamnopyranosyl kaempferol; 3-(4-hydroxybenzylidene)-2-(2,5-dihydroxyphenyl)-1-(4-hydroxyphenyl)indane-4,6-diol; quercitrin; rhamnitrin; rhamnocitrin; quercitrin-3-*O*″-acetate; parthenocissin A, *etc.*

depth of the study and the plausible mechanism is still not clear. Several clinical studies reported the pharmacological activity of CQ but they have their own limitations like small sample size, poor design and methodological incompliances, inadequate reporting, and medical interventions. Sawangjit *et al.*, in 2017 pooled the effects of CQ in different dosage forms, dose and clinical conditions especially for bone related ailments. They concluded that future studies are warranted to determine whether the dose, dosage form, and the clinical condition correlates to the pharmacological effects of CQ and also the combinational therapies for better therapeutic management.[109] They also mentioned that most of the included studies in meta-analyses (six of nine) indicated that CQ is safe for human consumption for a period of 1 to 10 weeks. However, there is limited data on the long-term safety and clinical use in different ailments and future research reports with strong evidence are recommended.[109] Extra randomized clinical trials are needed to establish the safety and efficacy of CQ for the management and treatment of various ailments.

This chapter consolidated the significance of CQ phytochemical constituents and pharmacological actions with the hope that comprehensive information about CQ is essential for future scientific investigations in order to generate novel therapeutically effective lead molecules. The next level of phytochemical investigations should be carried out to isolate potent molecules from different fractions of the CQ extract and expedite the development of medicine. *In vitro* and *in vivo* preclinical studies are extensively reported, but still a large gap exists between the relationship of active phytoconstituents – targets – molecular mechanisms – therapeutic effects. Substantial clinical study to support the pharmacological activity and human safety of CQ is imperative and inevitable. With the advancement of modern analytical and molecular techniques coupled with the conventional methods, future research should make a significant contribution towards the development of newer therapeutic agents from CQ. Several daunting challenges need to be addressed effectively for the development of lead molecules from medicinal plants through collaborative efforts of anthropologists, ethnobotanists, phytochemists, pharmacists, and physicians. Future studies should focus on bridging the gap between *in vitro*/*in vivo* and clinical effects to assure the therapeutic efficacy of this incredible herbal plant.

Abbreviations

AgNPs	Silver nanoparticles
ALKP	Alkaline phosphatase
AuNPs	Gold nanoparticles
BMI	Body mass index
cAMP	Cyclic adenosine monophosphate
CAT	Catalase
CFA	Complete Freunds Adjuvant

COX	Cyclooxygenase
Cp	Cysteine protease
CQ	*Cissus quadrangularis*
CuONPs	Copper oxide nanoparticles
DEXA	Dual-energy X-ray absorptiometry
DNA	Deoxy ribose nucleic acid
DPPH	2,2-Diphenyl picrylhydrazyl
ER	Estrogen receptor
GABA	Gamma amino butyric acid
GAE	Gallic acid equivalent
GPx	Glutathione peroxidase
GSH	Reduced glutathione
GST	Glutathione *S*-transferase
H_2O_2	Hydrogen peroxide
HDL	High density lipoprotein
hMSCs	Human mesenchymal stem cells
HO-1	Heme oxygenase-1
hWA	Human white adipocytes
IGF	Insulin-like growth factor
IL	Interleukin
*i*NOS	Inducible nitric oxide synthase
IR	Infra-red
LDL	Low density lipoprotein
LOX	Lipooxygenase
LPS	Lipopolysaccharide
MD-OADs	Muscular dystrophy during osteoarthritic disorders
MTT	(3-(4,5-Dimethylthiazolyl-2)-2,5-diphenyl-tetrazolium bromide)
NMR	Nuclear magnetic resonance
NO	Nitric oxide
Nrf	Nuclear transcription factor
NSAID	Non-steroidal anti-inflammatory drugs
OA	Osteoarthritis
OPG	Osteoprotegrin
PGE2	Prostaglandin E2
PLLA	Poly-L-lactic acid
RANKL	Receptor Activator of Nuclear Factor K Ligand
ROS	Reactive oxygen species
SOD	Superoxide dismutase
StAR	Steroidogenic acute regulatory protein
TCP	β-Tricalcium phosphate
TEM	Transmission electron microscopy
TNF-α	Tumor necrosis factor-α
TRAP	Tartrate resistant acid phosphatase

References

1. S. Indran, R. E. Raj and V. Sreenivasan, *Carbohydr. Polym.*, 2014, **110**, 423.
2. R. S. Dipti and J. Sarmah, *J. Ayurveda Integr. Med.*, 2018, **9**, S16.
3. H. J. Lee, B. Le, D.-R. Lee, B.-K. Choi and S. H. Yang, *Toxicol. Rep.*, 2018, **5**, 608.
4. A. M. Bhujade, S. Talmale, N. Kumar, G. Gupta, P. Reddanna, S. K. Das and M. Patil, *J. Ethnopharmacol.*, 2012, **141**, 989.
5. M. M. Gupta and R. K. Verma, *Phytochemistry*, 1990, **29**, 336.
6. V. Gopinath, S. Priyadarshini, N. M. Priyadharsshini, K. Pandian and P. Velusamy, *Mater. Lett.*, 2013, **91**, 224.
7. G. Mishra, S. Srivastava and B. Nagori, *Int. J. Pharmtech Res.*, 2010, **2**, 1298.
8. E. Williamson, *Major Herbs of Ayurveda*, Churchill Livingstone, New York, 2002.
9. K. K. Singh and J. K. Maheshwari, *J. Econ. Taxon. Bot.*, 1983, **4**, 829.
10. M. Kumbhojkar, D. Kulkarni and A. Upadhye, *Ethnobotany*, 1991, **3**, 21.
11. M. Jain, M. Nivedhitha, S. Deepak and K. S. Rajesh, *J. Evol. Med. Dent. Sci.*, 2020, **9**, 2833.
12. Anonymous, *The wealth of India: A Dictionary of Indian Raw materials and Industrial Products Raw materials*, Publication and Information Directorate CSIR, New Delhi, 1992, 593–594.
13. L. Asolkar, K. Kakkar and O. Chakre, *Glossary of Indian medicinal plants with active principles (part 1)*, National Institute of Science Communication And Information Resources (CSIR), New Delhi, 1997.
14. R. P. Rastogi and B. Mehrotra, *Compendium of Indian medicinal plants*, CSIR, New Delhi, India, 1993, vol. 1.
15. D. Luseba and D. Van der Merwe, *Onderstepoort J. Vet. Res.*, 2006, **73**, 115.
16. M. Kazemipoor, G. A. Cordell, M. M. R. Sarker, C. W. J. B. W. M. Radzi, M. Hajifaraji and P. En Kiat, *Int. J. Food Prop.*, 2015, **18**, 1942.
17. R. Udayakumar, M. Sundaran and R. Krishna, *Anc. Sci. Life*, 2004, **24**, 79.
18. K. C. Murthy, A. Vanitha, M. M. Swamy and G. Ravishankar, *J. Med. Food*, 2003, **6**, 99.
19. M. Jainu and C. S. Devi, *Chem. - Biol. Interact.*, 2006, **161**, 262.
20. U. M. Aswar, V. Mohan and S. L. Bodhankar, *Indian J. Pharmacol.*, 2012, **44**, 345.
21. J. M. Guerra, M. A. Hanes, C. Rasa, N. Loganathan, W. Innis-Whitehouse, E. Gutierrez, S. Nair and J. Banu, *J. Bone Miner. Metab.*, 2019, **37**, 780.
22. K. Ganesan, M. Jayachandran and B. Xu, *Crit. Rev. Food Sci. Nutr.*, 2018, **58**, 1165.
23. A. H. Swamy, R. V. Kulkarni, B. C. Koti, P. C. Gadad, A. H. Thippeswamy and A. Gore, *Indian J. Pharm. Sci.*, 2012, **74**, 183.
24. M. Kumar, P. Rawat, P. Dixit, D. Mishra, A. K. Gautam, R. Pandey, D. Singh, N. Chattopadhyay and R. Maurya, *Phytomedicine*, 2010, **17**, 993.

25. S. Muthu, V. B. Gopal, S. N. Karthik, P. Sivaji, S. Malairaj, M. Lakshmikanthan, N. Subramani and P. Perumal, *Int. J. Biol. Macromol.*, 2017, **103**, 878.
26. D. K. Lakshmanan, G. Ravichandran, A. Elangovan, P. Jeyapaul, S. Murugesan and S. Thilagar, *Food Funct.*, 2020, **11**, 7842.
27. G. Singh, P. Rawat and R. Maurya, *Nat. Prod. Res.*, 2007, **21**, 522.
28. R. Kumar, Y. K. Gupta, S. Singh and S. Arunraja, *J. Ethnopharmacol.*, 2015, **175**, 346.
29. A. M. Bhujade, S. Talmale, N. Kumar, G. Gupta, P. Reddanna, S. K. Das and M. B. Patil, *J. Ethnopharmacol.*, 2012, **141**, 989.
30. G. Rao, T. Annamalai, T. Mukhopadhyay and M. Madhavi, *Res. J. Chem. Sci.*, 2011, **1**, 25.
31. A. A. Syed, M. I. Reza, A. Husain, P. Singh and J. R. Gayen, *Phytomed. Plus*, 2021, **1**, 100021.
32. T. Pathomwichaiwat, P. Ochareon, N. Soonthornchareonnon, Z. Ali, I. A. Khan and S. Prathanturarug, *J. Ethnopharmacol.*, 2015, **160**, 52.
33. P. Kumar, K. Dev, K. Sharma, M. Sahai and R. Maurya, *Nat. Prod. Res.*, 2019, **33**, 233.
34. L. D. Kumar, R. Prathiviraj, M. Selvakumar, R. Guna, E. Abbirami and T. Sivasudha, *J. Mol. Struct.*, 2020, **1199**, 127048.
35. A. Thakur, V. Jain, L. Hingorani and K. Laddha, *Pharmacogn. Res.*, 2009, **1**, 213.
36. M. Mehta, N. Kaur and K. Bhutani, *Phytochem. Anal.*, 2001, **12**, 91.
37. M. M. Gupta and R. K. Verma, *Phytochemistry*, 1991, **30**, 875.
38. T. S. Kumar, A. Anandan, M. Jegadeesan, A. Appl and S. Res, *Arch. Appl. Sci. Res.*, 2012, **4**, 1782.
39. S. S. Vandana Jain and K. S. Laddha, *Eur. J. Biomed. Pharm. Sci.*, 2017, **4**, 614.
40. S. Dhanasekaran, *Saudi J. Biol. Sci.*, 2020, **27**, 1302.
41. B. K. Potu, M. S. Rao, N. G. Kutty, K. M. Bhat, M. R. Chamallamudi and S. R. Nayak, *Clinics*, 2008, **63**, 815.
42. B. K. Potu, M. S. Rao, G. K. Nampurath, M. R. Chamallamudi, K. Prasad, S. R. Nayak, P. K. Dharmavarapu, V. Kedage and K. M. Bhat, *Upsala J. Med. Sci.*, 2009, **114**, 140.
43. J. Banu, E. Varela, A. N. Bahadur, R. Soomro, N. Kazi and G. Fernandes, *J. Osteoporosis*, 2012, **2012**, 101206.
44. A. Shirwaikar, S. Khan and S. Malini, *J. Ethnopharmacol.*, 2003, **89**, 245.
45. A. C. Kaveri Kuttappa and R. Mahinder, *J. Complementary Med. Altern. Healthcare*, 2020, **11**, 555808.
46. R. H. Toor, R. Tasadduq, A. Adhikari, M. I. Chaudhary, J. B. Lian, J. L. Stein, G. S. Stein and A. R. Shakoori, *J. Cell. Physiol.*, 2019, **234**, 10300.
47. S. Muthusami, I. Ramachandran, S. Krishnamoorthy, R. Govindan and S. Narasimhan, *Growth Horm. IGF Res.*, 2011, **21**, 343.
48. M. Kumar, P. Rawat, P. Dixit, D. Mishra, A. K. Gautam, R. Pandey, D. Singh, N. Chattopadhyay and R. Maurya, *Phytomedicine*, 2010, **17**, 993.

49. J. Ruangsuriya, S. Charumanee, S. Jiranusornkul, P. Sirisa-Ard, B. Sirithunyalug, J. Sirithunyalug, T. Pattananandecha and C. Saenjum, *BMC Complementary Med. Ther.*, 2020, **20**, 1.
50. V. Upadhya, H. V. Hegde, S. Bhat, P. J. Hurkadale, S. Kholkute and G. Hegde, *J. Ethnopharmacol.*, 2012, **142**, 557.
51. A. K. Gupta, N. Shah and A. Thakar, *Ayurveda*, 2012, **33**, 110.
52. N. Singh, V. Singh, R. Singh, A. Pant, U. Pal, L. R. Malkunje and G. Mehta, *Natl. J. Maxillofac. Surg.*, 2013, **4**, 52.
53. S. Zaki, R. Malathi, V. Latha and G. Sibi, *Int. J. Clin. Microbiol. Biochem. Technol.*, 2020, **3**, 49.
54. A. Rajagopalan, C. S. Sasikumar and V. K. Ramachandran, *J. Complementary Med. Alt. Healthcare*, 2018, **5**, 555663.
55. N. Kumar and D. Joshi, *Int. J. Res. Orthop.*, 2019, **6**, 224.
56. N. Lal and J. Dixit, *J. Oral Biol. Craniofacial Res.*, 2012, **2**, 36.
57. A. V. Swamy, R. V. Kulkarni, B. Koti, P. Gadad, A. Thippeswamy and A. Gore, *Indian J. Pharm. Sci.*, 2012, **74**, 183.
58. N. Jayan, S. Das, S. Kumar, S. E. Prince and A. Devi, *Asian J. Pharm. Clin. Res.*, 2017, **10**, 182.
59. K. Chidambara Murthy, A. Vanitha, M. Mahadeva Swamy and G. Ravishankar, *J. Med. Food*, 2003, **6**, 99.
60. G. Marquis, B. Ramasamy, S. Banwarilal and A. P. Munusamy, *J. Photochem. Photobiol., B*, 2016, **155**, 28.
61. S. Sathappan and V. Baskaralingam, *Fish Shellfish Immunol.*, 2016, **53**, 99.
62. J. Chenniappan, A. Sankaranarayanan and S. Arjunan, *J. Sci. Res.*, 2020, **64**, 90.
63. D. Devipriya and S. M. Roopan, *Mater. Sci. Eng., C*, 2017, **80**, 38.
64. P. Balasubramanian, K. Jayalakshmi, N. Vidhya, R. Prasad, A. K. Sheriff, G. Kathiravan, K. Rajagopal and S. M. Sureban, *J. Basic Clin. Pharm.*, 2009, **1**, 37.
65. K. Poddar, S. Kolge, L. Bezman, G. E. Mullin and L. J. Cheskin, *Nutr. Clin. Pract.*, 2011, **26**, 539.
66. T. Sapsrithong, W. Kaewprem, S. Tongumpai, P. Nusuetrong and D. Meksuriyen, *J. Ethnopharmacol.*, 2012, **143**, 664.
67. I. Suzu, H. Goto, N. Hiwatashi, S. Hattori, K. Rotjanapan, W. Leeanansaksiri and S. Okada, *Nat. Prod. Commun.*, 2013, **8**, 629.
68. P. Kokilavani, U. Suriyakalaa, P. Elumalai, B. Abirami, R. Ramachandran, A. Sankarganesh and S. Achiraman, *Pestic. Biochem. Physiol.*, 2014, **109**, 18.
69. A. Vijayalakshmi, P. Kumar, S. Sakthi Priyadarsini and C. Meenaxshi, *J. Chem.*, 2013, **2013**, 150675.
70. S. Muthusami, V. Gopalakrishnan, J. A. Stanley, S. Krishnamoorthy, R. Ilangovan, V. K. Gopalakrishnan and N. Srinivasan, *Biomed. Pharmacother.*, 2016, **81**, 416.
71. J. Chidambaram and A. C. Venkatraman, *Food Chem. Toxicol.*, 2010, **48**, 2021.

72. R. Lekshmi, R. Rajesh and S. Mini, *Phytomedicine*, 2015, **22**, 952.
73. R. Kumar, Y. K. Gupta, S. Singh and S. Arunraja, *J. Ethnopharmacol.*, 2015, **175**, 346.
74. P. Yadav, A. Ganeshpurkar, N. Sonkar, D. Bansal and N. Dubey, *Niger J. Exp. Clin. Biosci.*, 2014, **2**, 49.
75. A. Ganguly, *Caspian J. Intern. Med.*, 2019, **10**, 183.
76. J. Viswanath, S. Sankaraiah, C. Cheekavolu and R. Dixit, *J. Med. Sci. Clin. Res.*, 2017, **5**, 23841.
77. M. Jainu, *J. Clin. Biochem. Nutr.*, 2004, **34**, 43.
78. M. Jainu and C. S. S. Devi, *Chem. – Biol. Interact.*, 2006, **161**, 262.
79. M. Jainu and K. V. Mohan, *Int. Immunopharmacol.*, 2008, **8**, 1721.
80. A. Panthong, W. Supraditaporn, D. Kanjanapothi, T. Taesotikul and V. Reutrakul, *J. Ethnopharmacol.*, 2007, **110**, 264.
81. A. Bhujade, S. Talmale and M. Patil, *J. Clin. Cell. Immunol.*, 2015, **6**, 2.
82. K. Srisook, M. Palachot, N. Mongkol, E. Srisook and S. Sarapusit, *J. Ethnopharmacol.*, 2011, **133**, 1008.
83. P. A. Pirshahid, C. Thisayakorn, T. Hemthanon, T. Suntorntanasat, C. Banchonglikitkul, J. Eiamwat and S. Sookwhan, *Acta Hortic.*, 2016, **1125**, 161.
84. J. Oben, D. Kuate, G. Agbor, C. Momo and X. Talla, *Lipids Health Dis.*, 2006, **5**, 1.
85. D. Kuate, R. J. Nash, B. Bartholomew and Y. Penkova, *Nat. Prod. Commun.*, 2015, **10**, 1281.
86. R. Nash, B. Azantsa, D. Kuate, H. Singh and J. Oben, *J. Altern. Complement. Med.*, 2019, **25**, 98.
87. M. Yimam, P. Jiao, M. Hong, L. Brownell, Y.-C. Lee, E.-J. Hyun, H.-J. Kim, T.-W. Kim, J.-B. Nam and M.-R. Kim, *J. Diet. Suppl.*, 2019, **16**, 86.
88. A. Maunder, E. Bessell, R. Lauche, J. Adams, A. Sainsbury and N. R. Fuller, *Diabetes, Obes. Metab.*, 2020, **22**, 891.
89. S. Chatree, C. Sitticharoon, P. Maikaew, K. Pongwattanapakin, I. Keadkraichaiwat, M. Churintaraphan, C. Sripong, R. Sririwichitchai and S. Tapechum, *Sci. Rep.*, 2021, **11**, 1.
90. A. Marume, G. Matope, S. Katsande, S. Khoza, I. Mutingwende, T. Mduluza, T. Munodawafa-Taderera and A. R. Ndhlala, *Front. Pharmacol.*, 2017, **8**, 544.
91. V. L. Udalamaththa, C. D. Jayasinghe and P. V. Udagama, *Stem Cell Res. Ther.*, 2016, **7**, 1.
92. A. H. M. V. Swamy, A. H. M. Thippeswamy, D. V. Manjula and C. B. M. Kumar, *Afr. J. Biomed. Res.*, 2006, **9**, 69.
93. F. C. Moto, A. Arsa'a, G. T. Ngoupaye, G. S. Taiwe, J. S. Njapdounke, A. K. Kandeda, G. C. Nkantchoua, J. P. Omam Omam, S. Pale and N. E. Kouemou, *Front. Pharmacol.*, 2018, **9**, 751.
94. M. Arshad, S. Siddiqui and D. Ali, *Caryologia*, 2016, **69**, 128.
95. M. M. Azeemuddin, C. M. Rao, M. Rafiq, U. V. Babu and P. Rangesh, *Exp. Gerontol.*, 2019, **125**, 110663.

96. T. Santhoshkumar, A. A. Rahuman, A. Bagavan, S. Marimuthu, C. Jayaseelan, A. V. Kirthi, C. Kamaraj, G. Rajakumar, A. A. Zahir and G. Elango, *Exp. Parasitol.*, 2012, **132**, 156.
97. S. Bhuvanasree, D. Harini, A. Rajaram and R. Rajaram, *Spectrochim. Acta, Part A*, 2013, **106**, 190.
98. K. Parvathi, A. G. Krishnan, A. Anitha, R. Jayakumar and M. B. Nair, *Int. J. Biol. Macromol.*, 2018, **110**, 514.
99. S. Suganya, J. Venugopal, S. Ramakrishna, B. Lakshmi and V. Giri Dev, *J. Appl. Polym. Sci.*, 2014, **131**, 39835.
100. S. F. Robertson and S. Bose, *J. Mech. Behav. Biomed. Mater.*, 2020, **111**, 103945.
101. S. Soumya, K. Sajesh, R. Jayakumar, S. Nair and K. Chennazhi, *Carbohydr. Polym.*, 2012, **87**, 1787.
102. R. M. Samarasinghe, R. K. Kanwar, K. Kumar and J. R. Kanwar, *Nanomedicine*, 2014, **9**, 819.
103. S. Tamburaci, C. Kimna and F. Tihminlioglu, *J. Bioact. Compat. Polym.*, 2018, **33**, 629.
104. N. Thongtham, P. Chai-in, O. Unger, S. Boonrungsiman and O. Suwantong, *Polym. Adv. Technol.*, 2020, **31**, 1496.
105. S. J. Stohs and S. D. Ray, *Phytother. Res.*, 2013, **27**, 1107.
106. A. Attawish, P. Chavalittumrong, S. Chivapat, A. Chuthaputti, S. Rattanajarasroj and S. Punyamong, *Songklanakarin J. Sci. Technol.*, 2002, **24**, 39.
107. S. C. Kothari, P. Shivarudraiah, S. B. Venkataramaiah, K. P. Koppolu, S. Gavara, R. Jairam, S. Krishna, R. K. Chandrappa and M. G. Soni, *Food Chem. Toxicol.*, 2011, **49**, 3343.
108. A. Kavitha, A. N. Babu and R. R. Nadendla, *Panacea J. Pharm. Pharm. Sci.*, 2018, 7, 748.
109. R. Sawangjit, P. Puttarak, S. Saokaew and N. Chaiyakunapruk, *Phytother. Res.*, 2017, **31**, 555.

CHAPTER 13

Ayurvedic Knowledge Inspired Approach to Modern Drug Discovery†

PREMALATHA BALACHANDRAN

National Center for Natural Products Research, Research Institute of Pharmaceutical Sciences, School of Pharmacy, University of Mississippi, MS 38677, USA
Email: prembala@olemiss.edu

13.1 Introduction

Since the beginning of humanity, our goal has been to alleviate diseases or their symptoms and we have focused our attention on nature to discover medicines for these diseases. From a manufacturing point of view, for a drug to be successful it should be stable, less complicated, easily producible in large quantities, and deliverable at affordable costs. From efficacy's standpoint, a drug should exhibit good ADMET characteristics like absorption, distribution, metabolism, excretion, and toxicity, and should be able treat the targeted disease with specific efficacy.[1]

In the pharmaceutical field, drug discovery (DD) is an innovative multidisciplinary process of finding new medications for any disease or health

†Disclaimer: Opinions, views, and recommendations expressed in this chapter are solely those of the author and do not express the views or opinions of the National Center for Natural Products Research, the School of Pharmacy, or the University of Mississippi.

Chemistry, Biological Activities and Therapeutic Applications of Medicinal Plants in Ayurveda
Edited by Augustine Amalraj, Sasikumar Kuttappan and Karthik Varma A. C.

Published by the Royal Society of Chemistry, www.rsc.org

problems. Historically, this DD process was started from the active ingredients of traditionally-used medicines or through accidental surprise discoveries.[2] But in recent days, it is done through a well-planned process involving bioactivity guided fractionation and isolation of thousands of pure compounds. These compound libraries are screened through high throughput methods against several molecular targets, and interesting leads from this initial screening process are subjected to pre-clinical trials in animals for efficacy. The modern drug discovery process involves the optimization/refinement of those lead compounds to augment their binding affinity, specificity, efficacy, stability, and favorable bioavailability. Potential compounds that pass through this rigorous process enter the drug development stage and further proceed in clinical trials; the average development time is about 12 years.[3]

Ayurveda, a traditional Indian system of medicine not only includes medical practices and beliefs related to the prevention and the treatment of diseases but also overall holistic health, well-being, and spiritual healing. Ayurveda dates to 3000 BC and its classic texts *Atharvaveda, Charak Samhita,* and *Sushrut Samhita* describe more than 700 medicinal plants. In India, more than 3000 plants possess established medicinal values. Additionally, over 6000 plants are used as traditional and folk medicines.[4] Ayurvedic drugs have been time tested over centuries and developed based on extensive knowledge on pharmacognosy, medicinal chemistry, pharmacology, and clinical case studies that are continuing to be proven useful for modern drug discovery. Ayurvedic medicines are composed of a unique combination of natural products (herbs, minerals, metals in oxidized non-toxic forms, and rarely animal products) and these components provide a strong basis for the discovery of novel modern drugs which could either be a single molecule drug or a combo medicine.

Millions of people in developed countries use some form or other alterative medicines for therapeutic, food-additive, and cosmetic benefits. This has led to a new revolution in the drug industry and leading pharmaceutical companies are moving their attention to natural products-based drug discovery. New strategies with advanced biological applications are being established to facilitate the process.[5] Currently, modern medicine uses natural products in scientific ways and these products have contributed to a fair share of approved drugs in the 21st century and the approach is also shifting towards combining existing agents based on traditional knowledge.[6]

13.2 Overview of the Modern Drug Development Process

Jean *et al.*[2] described this process in three major steps *viz.*, target identification, *i.e.*, identifying the direct molecular target and its validation, screening for desired lead compounds, and their optimization.

13.2.1 Target Identification

Initial stages of drug discovery start from this step and various sectors such as academic researchers, scientists, physicians, and commercial pharmaceutical companies, all play a significant role in the identification of these disease targets. For a drug to be considered safe enough, it should fulfill both clinical and commercial requirements, in other words, "druggable". The first step in this drug discovery process begins with the identification of possible therapeutic targets (a gene or protein), not only with a unique role in the disease but is also safe and effective. This is followed by the characterization of their mechanism of action and ability to attack or modify the target gene, responsible for the disease. The ideas for drug targets are generated from basic science research, literature searches, functional screening, pathway and phenotypic analysis, or data mining using bioinformatics. Among these techniques, bioinformatics data mining has an advantage as it can identify, select, and prioritize potential disease targets and this technique could advance with the development of artificial intelligence methods.

After this critical step, the identified targets are tested for their ability to demonstrate their therapeutic effect on the disease onset or its progression. Functional analysis or validation processes are performed through the analysis of signaling pathways, cell-based *in vitro* assays, protein interactions, and genetic approaches to evaluate the relationship between the pharmacology and the phenotype of interest, and would determine the safety profile of the target.[2,7,8]

13.2.2 Lead Compound Screening

This stage could be performed *via* two different methods of screening. For knowledge-based screening, large libraries of test compounds, selected based on prior knowledge like literature evidence, traditional use, patent precedents, *etc.* are screened using automated robots through high throughput screening (HTS). As HTS is time and cost efficient, millions of assays could be performed with great accuracy in a short period of time. HTS coupled with molecular biology tools are widely adopted to identify hit/lead compound(s) with a desired biological or pharmacological activity, and the compounds emerging from these screening processes exhibit high potential to be developed as a drug. Another method called "focused screening" involves screening of a smaller pool of compounds narrowed on the basis of an existing proven activity with the drug target.[2,7,8]

13.2.3 Lead Compound Optimization

Once identified, these lead compounds are optimized for potency, selectivity, toxicity, safety, mechanism of action, and biodistribution. At this stage, the compounds which have strong toxicity, low bioavailability, and improper pharmacokinetics are modified with desired characteristics before moving on to the developmental stage which includes pre-clinical studies in animals and several phases of clinical trials. The drugs that have emerged successfully after phase 3 clinical trials can be filed for NDA (New Drug Application) with the FDA for review and approval. NDA approval opens the door for commercial production and marketing and the drug eventually becomes available for use. The FDA would still continue to monitor drug safety and other issues even after the drug has been on the market for several years.[2]

13.3 Challenges in the Modern Drug Discovery Approach

According to a 2016 study by the Tufts center for the study of Drug Development, the approximate cost for pharmaceutical companies to develop a new drug (from screening to US FDA approval) is estimated to be around $1.395 billion US dollars just for out-of-pocket costs. It costs $2.558 billion US dollars when we calculate complete capitalization costs.[9] Based on the availability of resources and methodological differences this cost could be different. Another study shows the median cost of developing a cancer drug is $648 million in out-of-pocket costs and $757 million in capitalized costs.[10] Furthermore, data shows that developing non-orphan drugs is expensive compared with orphan (rare disease) drugs.[11] These costs are largely due to the various phases of clinical trials required to generate a safety profile for the drugs.

13.4 Ayurvedic Drug Discovery

13.4.1 Basic Principles of Ayurveda

Ayurveda means "the science of life" and it is known as the "mother of all healing". It promotes health maintenance by paying close attention to the balance in one's life, right thinking, healthy diet, lifestyle, and the use of medicinal herbs at the appropriate time and quantity. All human beings are unique in their own fingerprint or constitution, which is a particular form of energy – a combination of physical, mental and emotional characteristics, and behavior. This constitution is determined at the time of birth by several factors, including external and internal. These factors influence balanced health and any disturbance to this balanced state is reflected as a change in one's constitution leading to a diseased condition. Ayurvedic physicians have the expertise to understand these factors and take appropriate action to nullify or minimize the effects or eliminate the causes of imbalance and re-establish one's original constitution. According to Ayurveda, "balance is the natural order while imbalance is disorder"; "health is order while disease is disorder", and there is a constant oscillation between order and disorder.[12] Ayurveda identifies 3 basic functional principles *Vata*, the subtle energy associated with movement, *Pitta*, the energy of digestion based on the body's metabolism, and *Kapha*, the energy that forms the body's structure

(Tridoshas). According to Ayurvedic philosophy, the entire cosmos is an interaction of the energies of the five great elements – space, air, fire, water, and earth. Vata, pitta, and kapha are combinations and variations of these elements that demonstrate as patterns present in all human beings.[12]

"Panchakarma therapy" in Ayurveda focuses on body rejuvenation, and cleansing leading to longevity of life. As the name implies, the *Pancha karma* is composed of five karmas (actions) that are used for the removal of toxic substances from our body tissues and fluids. They are the *Virechan* (purgation using powders, pastes, or decoction), *Vaman* (forced therapeutic emesis using medicines), *Basti* (use of enemas prepared from medicated oils), *Rakta moksha* (detoxification of blood and body fluids), and *Nasya* (nasal administration of medicines like decoctions, oils, and fumes).[13]

It is always surprising to realize the wealth of knowledge possessed by our ancestors. This oldest system of medicine was subdivided into eight disciplines and they are called "*Äshtanga Ayurveda*". They are *Kayachikitsa* (internal medicine treatment), *Bhootavidya* (treatment of psychological disorders), *Kaumar Bhritya* (pediatric treatment), *Rasayana* (study of geriatrics), *Vajikarana* (treatment through aphrodisiacs and eugenics), *Shalya* (surgical treatment), *Shalakya* (otorhinolaryngological and ophthalmological treatment), and *Agada Tantra* (toxicological studies).[13] This subdivision enables Ayurvedic physicians to earn expertise in a particular field and focus and adopt various treatment modalities. Also, this classification helps modern scientists to narrow the target screening process purely based on this knowledge.

13.4.2 Integrating Ayurvedic Principles into Modern Drug Discovery

Although use of natural products for health improvement or for the treatment of diseases, has been described throughout history in the form of traditional medicines and various remedies, their bioactive natural ingredients (active constituents) are largely unidentified.[14] The ultimate goal of any drug discovery is to achieve safe, effective, personalized medicine and the depth of ayurvedic knowledge offers immense solutions for many diseases and remains an appropriate approach, but unfortunately this ancient wisdom and depth of knowledge are still not extensively explored through modern scientific methodologies. The medicinal plants that have been used in Ayurveda for centuries could provide a platform for modern drugs and advanced technologies could be developed from ayurvedic principles. Pharma companies from various countries are pivoting their strategies, paving a way for new herbal-based therapeutic channels from what has long been known as traditional remedies.[15] As the safety profile of these medicines is well established, this approach has added the benefit of bringing patients closer to nature while cutting down the cost and time of drug development.

Current trends in the drug discovery process, although being revolutionized by the developments of advanced technologies, have their roots in natural resources (plant- or mineral-based). Numerous examples are available among modern drugs that were developed from herbal-based resources. One such classic example is opium-based morphine, which was produced from the latex of the poppy plant, *Papaver somniferum*, 200 years ago. In these footsteps, several drugs developed from natural/plant sources have certainly revolutionized the entire world. These drugs, such as antibiotics (*e.g.*, erythromycin, penicillin, tetracycline), antiparasitics (*e.g.*, avermectin), antimalarials (*e.g.*, quinine, artemisinin), hypolipidemic (*e.g.*, lovastatin and analogs), immunosuppressants for organ transplants (*e.g.*, rapamycin, cyclosporine), anticancer drugs (*e.g.*, irinotecan, paclitaxel), and antidiabetic drugs (*e.g.*, metformin)[16] have saved millions of lives.

High throughput screening methods, molecular docking *in silico* methods, understanding of signal transduction pathways and their interconnection with various disorders, and advanced tools in molecular biological techniques have paved a new way in drug discovery methodologies.

Despite the availability of these, only 410 drugs were approved by the US FDA over the past decade. Although an average of 40 drugs per year, the highest number of novel drugs approved any given year was 59 (in 2018).[17] These low numbers are largely due to the strict regulatory processes, especially towards the safety profile aspect of new drugs as it cannot be evaluated with certainty for several years after its market launch.[18] Due to these challenges in modern drug discovery, Ayurveda's basic principles, experiential wisdom, holistic approach and systemic database could offer efficient drugs[19] and there has been a massive reemergence of interest in the Ayurvedic system of medicine.[20] The therapeutic principles of Ayurveda focus on *Prakriti* and tridoshas. *Prakriti* or body constitution determines how an individual could respond or react to medications, environmental conditions, and dietary factors. 'Ayurgenomics', recent research in exploration, bridges this gap between genomics and Ayurveda and serves as a new tool to understand the differences between individuals in responses to therapies in various diseases.[21] As it could evaluate inter-individual genetic variances in patients from identical ethnic backgrounds,[13] it offers the basis for gene targeted therapy and facilitates the development of customized therapeutic approaches.[22,23] Ayusoft, a database developed for various Ayurvedic drugs, includes the network of diet patterns, lifestyles, diseases, and treatment principles.[24] This advanced, handy software is very helpful as a quick index tool for practitioners to refer to and access information for their diagnosis and treatment. Herboprint uses three-dimensional High-Performance Liquid Chromatography (HPLC) and it is in the process of developing techniques for bioactivity-based botanical standardization.[25]

Rotti *et al.*,[26] have done extensive research on finding a correlation between the concept of *prakriti* in Ayurveda and modern-day science. One of the reports indicate the correlation of *prakriti* with the Body Mass Index (BMI) and birthplace of individuals. In another study, the individuals with various *prakriti* types, *viz.* vata, pitta, and kapha constitutions were evaluated and compared based on their molecular differences, as these mechanisms play a significant role in the response to the treatment of every individual. In addition, these authors have also classified human population with respect to their DNA methylation signatures, and compared them with the traditional Ayurveda concept of *prakriti*.[27] In a study involving a genome-wide SNP (single nucleotide polymorphism) consisting of 262 male volunteers, belonging to three different *prakritis*, they found a classical association between the *PGM1* gene and energy production and this gene is more homogeneous in the pitta *prakriti*, than the kapha or vata *prakriti*.[28] These scientific approaches to ayurvedic concepts serve as a catalyst and motivate modern scientists to further explore the scientific basis of ayurveda.

13.4.3 Suggested Sequential Steps in Ayurvedic Drug Discovery

Although traditional medicine like Ayurveda serves as a rich source of therapeutic leads and provides fertile ground for modern drug development,[1] not all the potential lead molecules from plant sources can enter the drug development process. Success in identifying a new biologically active plant-based natural product can be influenced by several factors, including random or non-random plant selection, virtual screening through computational methods, *in vitro* screening through biological methods (high throughput screening), screening from ethnomedical knowledge and use of other approaches like ecological approaches, *in vivo* toxicity, bioavailability, and SAR relationships. A review of natural products estimates the presence of about 300 000 to 500 000 species of plants, but less than 10% of these plants have been studied for 1 or more bioactivities.[14] Using traditional Ayurvedic knowledge to explore these medicinal plants offers a new innovative strategy in the drug discovery process and facilitates not only the development of new single drug candidates but also synergistic botanical formulations as well.[5] The flow chart depicted in Figure 13.1. proposes a graphical summary of the sequential steps for the drug discovery from the plants used in Ayurveda. This concept was partially adapted from Farnsworth *et al.* on the study of plants from traditional medicine.[29] The problems and possible solutions in this process are provided in Table 13.1.[30]

13.4.4 Reverse Pharmacology: Promising Road to Ayurvedic Drug Discovery

Reverse pharmacology is an accomplished scientific approach to develop new drugs utilizing available knowledge, wisdom, and guidance from traditional medicines through sound clinical and pre-clinical research studies. As the name implies, it is a reverse process to modern drug discovery. As we have seen earlier in this chapter, while modern drug discovery begins with

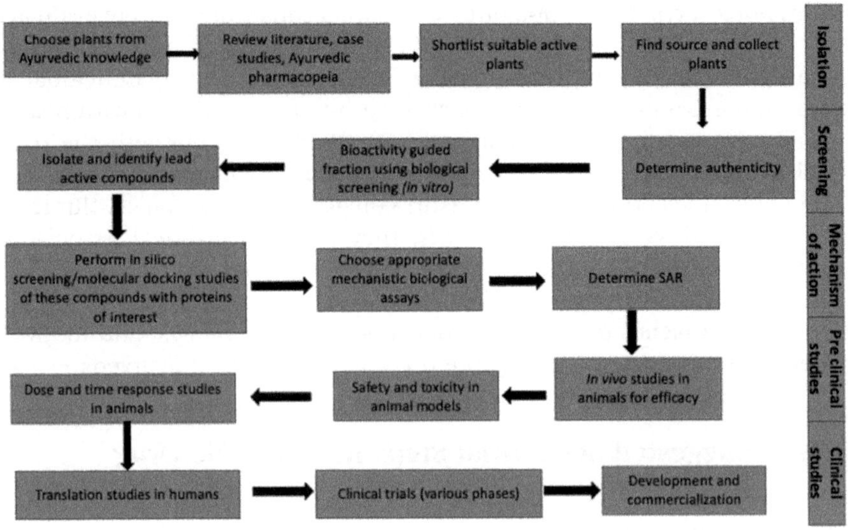

Figure 13.1 The Ayurvedic drug discovery-suggested sequence of steps.

Table 13.1 The problems and solutions in the pure compound isolation process.[30]

Steps in bioactivity guided fractionation	Problems	Solutions
Plant collection	Not possible to grow them out of the natural habitat	New tissue culture methods to grow these plants
	Desired active compounds are not available when taken out of natural habitat	New methods for synthesis induction
Crude extract preparation	Presence of compounds that do have drug activities or bioactive present in insufficient quantities	New methods for extraction and pre-fractionation
Isolated compounds	Unknown molecular targets identified with phenotypic assays	New methods for molecular mode of action elucidation

lead identification and optimization followed by *in vivo* preclinical studies (phase-I) and human clinical trials (phase-II and phase-III), this reverse direction could work best for the Ayurvedic drug discovery process and a schematic diagram for this process was depicted in our earlier publication.[31] This concept of reverse pharmacology in Ayurvedic Sciences was founded by eminent *vaidyas* Mahamahopadhyaya Gananath Sen[32] and evolved as a powerful scientific approach for newer drug discovery from ayurvedic medicine. Ashok DB Vaidya[33] described reverse pharmacology as a newer discipline in science that integrates well established clinical and experiential evidences, providing leads to *trans*-disciplinary exploratory studies. These leads could be further developed into potential drugs of interest triggered by experimental and clinical research. Thus, Ayurvedic drug discovery follows a retrogressive path based on clinical observations noted with herbal products. Several research institutions in India with the financial support from the CSIR (Council of Scientific and Industrial Research) and ICMR (The Indian Council for Medical Research), have successfully conducted many clinical trials based on 'reverse-pharmacology' and the Director Generals from these Institutes have detailed the prospects of this approach in national and international forums for future drug discoveries. Some of the plants that have been successfully analyzed by this approach include *Mucuna pruriens* for Parkinson's disease, *Zingiber officinale* for nausea/vomiting, *Picrorhiza kurroa* for hepatitis, *Curcuma longa* for oral cancer, *Panchavalkal species* for burns and wounds, and *Azadirachta indica* for malaria,[31] and the list continues.

Since the clinical implication of the herbs or their formulations have already been described in ayurvedic medicine, there could be a significant reduction of failure rates during the drug development process. This is one of the biggest advantages of this reverse pharmacology approach. In addition, this approach gives an opportunity to seek more effective synergistic formulations with increased bioavailability and these innovative strategies that could play a significant role in drug development. For *e.g.*, piperine studies reported earlier revealed that it improves the bioavailability of synthetic drugs such as propranolol, theophylline, and rifampicin, when used in combination.[34]

13.4.5 Essential Considerations in the Ayurveda-based Drug Development Process

Fundamental basic research in physiology, pathology, pharmacology, and pharmacokinetics is required to replace the faith and suppositions of Ayurvedic medicines with scientific reasoning and this data should be complemented with facts and figures. The objectives of the investigation in the fundamental research are categorized into the human body (*Purusha*), the disease (*Vyadhi*), the medicine (*Aushadha*), and the right time for action (*Kriyakala*).[35]

To develop drugs based on Ayurvedic knowledge and principles, it is necessary to consider several factors. First, it demands the incorporation of alternate and additional steps to the current drug development process to perform a more versatile Ayurvedic library screen not just a chemical compound library screen. The prospects of this elaborate, extensive-screening approach could be numerous, but essentially allows chemical and biological characterization of chemical compounds belonging to different classes to successfully hit 'druggable' Ayurvedic leads with defined biochemical specificities.[36] Also, it is well-known that natural products-based extract-libraries as opposed to existing chemical libraries harbor predominantly (>90%) compound structures that follow Lipinski's 'Rule-Of-Five'[37] properties that are ideal for oral administrations. Such a cohort of compounds are those drug molecules already developed and commercialized.[36] In addition, Ayurvedic drug discovery mandates the inclusion of Ayurvedic medicines based on existing knowledge and blends its concept and principles with modern drug development technologies for more successful pipeline drugs. To fulfill these requirements and face challenges, Ayurvedic practitioners and researchers are currently focusing on many effective, yet feasible methods that could document detailed experimentation on Ayurvedic herbs and products for the eventual implementation of standardized Ayurvedic drug discovery methodologies.[33,38,39] We reviewed the scientific basis and resource availability of Ayurvedic medicines and discussed its relevance to current and potential future drug development strategies, along with the major scientific challenges to be faced.[31]

Most of the research methodology of Ayurveda is based on examination/diagnostic tools called *Pareeksha*, that are inspired from the philosophical term *Pramaana*, which means evidence.[40] These diagnostic tools include (i) the direct observation (*Pratyaksha*), (ii) the inference (*Anumana*), and (iii) the authoritative testimonies or literature (*Aptopadesha*). Modern-day disease diagnosis also depends on these three basic tools to some extent, but efficiency and accuracy have been largely boosted by the utility of the innovative technologies and the use of modern devices. These advanced diagnostic methods are essential to modernize Ayurveda practices from principle-based to scientific-based medicine. If high quality research merges with modern scientific techniques, it would expand Ayurvedic knowledge worldwide and Ayurveda could show its presence in the path of current medical science.[41]

13.5 Development of New Drugs

13.5.1 From Natural Resources

Despite several challenges and hindrances, natural products continue to play a highly significant role in the drug discovery and development process. During the last four decades (1981 to 2019), 71 drugs were developed from natural products in an unaltered form, 14 were botanical drugs (including defined mixtures), 356 drugs were derivatives of natural products, 207 drugs were natural product pharmacophores, and 217 drugs were the mimics of natural products. These 865 drugs are listed in Table 13.2 along with their target diseases and year of approval.[42]

13.5.2 From Ayurvedic Medicinal Plants

Since most Ayurvedic drugs have been in practice since ancient times, the drug development process from Ayurveda is unique and based on either reverse pharmacology (starts with clinical and goes to mechanistic studies) or "drug rediscovery" (research to understand their mechanism or identifying new molecular target). These processes involve several stages and

Table 13.2 New drugs developed from natural sources. Reproduced from ref. 42 with permission from the American Chemical Society, Copyright 2020.

No.	Generic name	Trade name	Disease	Year	Source
1	Aclarubicin	Aclacin	Anticancer	1981	Natural product Unaltered
2	Netilimicin sulfate	Netromicine	Antibacterial	1981	Natural product Unaltered
3	Peplomycin	Pepleo	Anticancer	1981	Natural product Unaltered
4	Isotrentoin	Accutane	Antiacne	1982	Natural product Unaltered
5	Micronomicin sulfate	Sagamicin	Antibacterial	1982	Natural product Unaltered
6	Chenodiol	Chenix	Hypocholesterolemic	1983	Natural product Unaltered
7	Cyclosporine	Sandimmun	Immunosuppressant	1983	Natural product Unaltered
8	Meglutol	Lipoglutaren	Hypolipidemic	1983	Natural product Unaltered
9	Sodium cellulose PO4	Calcibind	Hypocalciuric	1983	Natural product Unaltered
10	Mizoribine	Bredinin	Immunosuppressant	1984	Natural product Unaltered
11	Miokamycin	Miocamycin	Antibacterial	1985	Natural product Unaltered
12	Mupirocin	Bactroban	Antibacterial	1985	Natural product Unaltered
13	Podophyllotoxin	Condyline	Genital warts	1985	Natural product Unaltered
14	Dronabinol	Marinol	Antinauseant	1986	Natural product Unaltered
15	Lentinan	Lentinan	Immunostimulant	1986	Natural product Unaltered
16	Schizophyllan	Sonifilan	Immunostimulant	1986	Natural product Unaltered
17	Artemisinin	Artemisin	Antiparasitic	1987	Natural product Unaltered
18	Bucillamine	Rimatil	Immunomodulator	1987	Natural product Unaltered
19	Ganglioside GM1	Sygen	Antithrombotic	1987	Natural product Unaltered
20	Guglipid	Guglip	Lipoprotein disorders	1987	Natural product Unaltered
21	Ivermectin	Mectizan	Antiparasitic	1987	Natural product Unaltered
22	Lovastatin	Mevacor	Hypocholesterolemic	1987	Natural product Unaltered
23	Plaunotol	Kelnac	Antiulcer	1987	Natural product Unaltered
24	Ubenimex	Bestatin	Immunostimulant	1987	Natural product Unaltered
25	Carumonam	Amasulin	Antibacterial	1988	Natural product Unaltered
26	Fosfomycin trometamol	Monuril	Antibacterial	1988	Natural product Unaltered
27	Gamolenic acid	Epogam	Antiallergic	1988	Natural product Unaltered
28	Isepamicin	Isepacin	Antibacterial	1988	Natural product Unaltered
29	Simvastatin	Zocor	Hypocholesterolemic	1988	Natural product Unaltered

Table 13.2 *(Continued)*

No.	Generic name	Trade name	Disease	Year	Source
30	Teicoplanin	Targocid	Antibacterial	1988	Natural product Unaltered
31	Adenosine	Adenocord	Antiarrhythmic	1989	Natural product Unaltered
32	Azelaic acid	Skinoren	Antiacne	1989	Natural product Unaltered
33	Pravastatin	Mevalotin	Hypocholesterolemic	1989	Natural product Unaltered
34	RV-11	Zalig	Antibacterial	1989	Natural product Unaltered
35	Pumactant	Alec	Respiratory distress syndrome	1990	Natural product Unaltered
36	Masoprocol	Actinex	Anticancer	1992	Natural product Unaltered
37	Pentostatin	Nipent	Anticancer	1992	Natural product Unaltered
38	Paclitaxel	Taxol	Anticancer	1993	Natural product Unaltered
39	Tacrolimus	Prograf	Immunosuppressant	1993	Natural product Unaltered
40	Angiotensin II	Delivert	Anticancer	1994	Natural product Unaltered
41	Voglibose	Basen	Antidiabetic	1994	Natural product Unaltered
42	Carperitide	Hanp	Congestive heart failure	1995	Natural product Unaltered
43	Galanthamine HBr	Nivalin	Alzheimer's disease	1995	Natural product Unaltered
44	Betain anhydrous	Cystadane	Homocystinuria	1996	Natural product Unaltered
45	Arglabin	?	Anticancer	1999	Natural product Unaltered
46	Sirolimus	Rapamune	Immunosuppressant	1999	Natural product Unaltered
47	Aminolevulinic acid HCl	Levulan	Anticancer	2000	Natural product Unaltered
48	Daptomycin	Cubicin	Antibacterial	2003	Natural product Unaltered
49	Mycophenolate sodium	Myfortic	Immunosuppressant	2003	Natural product Unaltered
50	Paclitaxel liposomal	Lipusu	Anticancer	2003	Natural product Unaltered
51	Ach-02	Acheflan	Analgesic	2005	Natural product Unaltered
52	Paclitaxel nanoparticles	Abraxane	Anticancer	2005	Natural product Unaltered
53	Ziconotide	Prialt	Neuropathic pain	2005	Natural product Unaltered
54	Citrofen		Antiarthritic	2007	Natural product Unaltered
55	Paclitaxel nanoparticles	Nanoxel	Anticancer	2007	Natural product Unaltered
56	Paclitaxel nanoparticles	Genexol-PM	Anticancer	2007	Natural product Unaltered
57	Trabectedin	Yondelis	Anticancer	2007	Natural product Unaltered
58	Romidepsin	Istodax	Anticancer	2010	Natural product Unaltered
59	Fidaxomicin	Dificid	Antibacterial	2011	Natural product Unaltered
60	Spinosad	Natroba	Insecticide	2011	Natural product Unaltered

61	Cysteamine hydrochloride	Cystaran	Cystinosis	2012	Natural product Unaltered
62	Homoharringtonine	Ceflatonin	Anticancer	2012	Natural product Unaltered
63	Ingenol mebutate	Picato	Anticancer	2012	Natural product Unaltered
64	Cholic acid	Orphacol	Inborn errors of bile synthesis	2013	Natural product Unaltered
65	PICN		Anticancer	2014	Natural product Unaltered
66	Omega-3-carboxylic acids	Epanova	Hypertriglyceridemia	2014	Natural product Unaltered
67	Deoxycholic acid sodium salt	Kybella	Dermatological disorders	2015	Natural product Unaltered
68	Prasterone	Intrarosa	Dyspareunia	2016	Natural product Unaltered
69	Angiotensin II acetate	Giapreza	Antihypertensive	2017	Natural product Unaltered
70	Aplidine	Aplidin	Anticancer	2018	Natural product Unaltered
71	Cannabidiol	Epidiolex	Antiepileptic	2018	Natural product Unaltered
72	Extract of serenoa repens	Permixon	Benign prostatic hypertrophy	1982	Botanical drug
73	Solamargines	Curaderm	Anticancer	1989	Botanical drug
74	Cannabis sativa L extract	Sativex	Neuropathic pain	2005	Botanical drug
75	Kunecatechins	Veregen	Genital warts	2006	Botanical drug
76	NIX-0699	Hemoxin	Anemia, sickle cell	2006	Botanical drug
77	Curemisinin		Antiallergic	2007	Botanical drug
78	XP-828L	Dermylex	Antipsoriatic	2007	Botanical drug
79	Oralair Grasses		Allergic rhinitis	2008	Botanical drug
80	Crofelemer	Fulyzaq	Antidiarrheal	2012	Botanical drug
81	PG-201	Layla	Antiarthritic	2012	Botanical drug
82	Jap cedar pollen	Cedartolen	Immediate allergy	2014	Botanical drug
83	Ragweed pollen extract	Ragwitek	Immediate allergy	2014	Botanical drug
84	Extract from birch bark	Episalvan	Vulnerary	2016	Botanical drug
85	IZN-6D4	Curasite	Topical ulcers	2017	Botanical drug
86	Budesonide	Pulmicort	Antiasthmatic	1981	Natural product derivative
87	Cefoperazone sodium	Cefobis	Antibacterial	1981	Natural product derivative
88	Cefotiam HCl	Pansporin	Antibacterial	1981	Natural product derivative
89	Cefsoludin sodium	Takesulin	Antibacterial	1981	Natural product derivative
90	Desogestrel/ethinylestradiol	Marvelon	Contraception	1981	Natural product derivative
91	Sulprostone	Nalador	Abortifacient	1981	Natural product derivative
92	Testosterone undecanoate	Restandol	Hypogonadism	1981	Natural product derivative
93	Apalcillin sodium	Lumota	Antibacterial	1982	Natural product derivative

Table 13.2 (*Continued*)

No.	Generic name	Trade name	Disease	Year	Source
94	Atracurium besilate	Nimbex	Muscle relaxant	1982	Natural product derivative
95	Ceftizoxime sodium	Epocelin	Antibacterial	1982	Natural product derivative
96	Ceftriaxone sodium	Rocephin	Antibacterial	1982	Natural product derivative
97	Lisuride maleate	Dopergin	Antihyperprolactinemia	1982	Natural product derivative
98	Moxalactam disodium	Shiomarin	Antibacterial	1982	Natural product derivative
99	Vecuronium bromide	Norcuron	Muscle relaxant	1982	Natural product derivative
100	Cefmenoxime HCl	Tacef	Antibacterial	1983	Natural product derivative
101	Ceftazidime	Fortam	Antibacterial	1983	Natural product derivative
102	Elliptinium acetate	Celiptium	Anticancer	1983	Natural product derivative
103	Epoprostenol sodium	Flolan	Platelet aggreg. inhib.	1983	Natural product derivative
104	Gemeprost	Cervagem	Abortifacient	1983	Natural product derivative
105	Halometasone	Sicorten	Antiinflammatory	1983	Natural product derivative
106	Hydrocortisone butyrate	Pandel	Antiinflammatory	1983	Natural product derivative
107	Oxitropium bromide	Oxivent	Bronchodilator	1983	Natural product derivative
108	Promegestrone	Surgestone	Progestogen	1983	Natural product derivative
109	Aztreonam	Azactam	Antibacterial	1984	Natural product derivative
110	Buserelin acetate	Suprefact	Hormone	1984	Natural product derivative
111	Cefonicid sodium	Monocid	Antibacterial	1984	Natural product derivative
112	Ceforanide	Precef	Antibacterial	1984	Natural product derivative
113	Cefotetan disodium	Yamatetan	Antibacterial	1984	Natural product derivative
114	Epirubicin HCI	Farmorubicin	Anticancer	1984	Natural product derivative
115	Ibopamine HCI	Inopamil /Scandine	Cardiotonic	1984	Natural product derivative
116	Leuprolide acetate	Enantone	Hormone	1984	Natural product derivative
117	Melinamide	Artes Oil	Hypocholesterolemic	1984	Natural product derivative
118	Naltrexone HCI	ReVia	Antinarcotic	1984	Natural product derivative
119	Nicorandil	Sigmart	Vasodilator; coronary	1984	Natural product derivative
120	Sofalcone	Solon	Antiulcer	1984	Natural product derivative
121	Temocillin disodium	Temopen	Antibacterial	1984	Natural product derivative
122	Teprenone	Selbex	Antiulcer	1984	Natural product derivative
123	Alclometasone dipropionate	Vaderm	Antiinflammatory	1985	Natural product derivative
124	Astromycin sulfate	Fortimicin	Antibacterial	1985	Natural product derivative

125	Cefbuperazone sodium	Tomiporan	Antibacterial	1985	Natural product derivative
126	Cefpiramide sodium	Sepatren	Antibacterial	1985	Natural product derivative
127	Cimetropium bromide	Alginor	Antispasmodic	1985	Natural product derivative
128	Dalteparin sodium	Fragmin	Anticoagulant	1985	Natural product derivative
129	Doxofylline	Ansimar	Bronchodilator	1985	Natural product derivative
130	Elcatonin	Diatin	Calcium metabolism	1985	Natural product derivative
131	Enprostil	Gardrin	Antiulcer	1985	Natural product derivative
132	Heme arginate	Normosang	Porphyria	1985	Natural product derivative
133	Imipenem/cilastatin	Zienam	Antibacterial	1985	Natural product derivative
134	Mefloquine HCI	Fansimef	Antiparasitic	1985	Natural product derivative
135	Misoprostol	Cytotec	Antiulcer	1985	Natural product derivative
136	Nabumetone	Relifex	Antiinflammatory	1985	Natural product derivative
137	Nadroparin calcium	Fraxiparine	Anticoagulant	1985	Natural product derivative
138	Rifaximin	Rifacol	Antibacterial	1985	Natural product derivative
139	Rosaprostol	Rosal	Antiulcer	1985	Natural product derivative
140	Thymopentin	Sintomodulina	Immunomodulator	1985	Natural product derivative
141	Brovincamine fumarate	Sabromin	Vasodilator; cerebral	1986	Natural product derivative
142	Deflazacort	Deflan	Antiinflammatory	1986	Natural product derivative
143	Gestrinone	Dimetrose	Antiprogestogenic	1986	Natural product derivative
144	Gonadoreline-6-D-trp acetate	Decapeptyl	Hormone	1986	Natural product derivative
145	Idebenone	Avan	Nootropic	1986	Natural product derivative
146	Levacecarnine HCI	Nicetile	Nootropic	1986	Natural product derivative
147	Nedocromil sodium	Tildade	Antiallergic	1986	Natural product derivative
148	Nomegestrol acetate	Lutenyl	Progestogen	1986	Natural product derivative
149	Norgestimate	Cilest	Progestogen	1986	Natural product derivative
150	Prednicarbate	Dermatop	Antiinflammatory	1986	Natural product derivative
151	Rokitamycin	Ricamycin	Antibacterial	1986	Natural product derivative
152	Sulbactam sodium	Sulperazone	Beta-lactamase inhibitor	1986	Natural product derivative
153	Triptorelin	Decapeptyl	Anticancer	1986	Natural product derivative
154	Amlexanox	Solfa	Antiasthmatic	1987	Natural product derivative
155	Artemether	Artemetheri	Antiparasitic	1987	Natural product derivative
156	Artenusate	Arinate	Antiparasitic	1987	Natural product derivative
157	Aspoxicillin	Doyle	Antibacterial	1987	Natural product derivative

Table 13.2 (*Continued*)

No.	Generic name	Trade name	Disease	Year	Source
158	Cefixime	Cefspan	Antibacterial	1987	Natural product derivative
159	Cefminox sodium	Meicelin	Antibacterial	1987	Natural product derivative
160	Cefpimizole	Ajicef	Antibacterial	1987	Natural product derivative
161	Cefteram pivoxil	Tomiron	Antibacterial	1987	Natural product derivative
162	Cefuroxime axetil	Zinnat	Antibacterial	1987	Natural product derivative
163	Cefuzonam sodium	Cosmosin	Antibacterial	1987	Natural product derivative
164	Enoxaparin	Lovenox	Anticoagulant	1987	Natural product derivative
165	Gestodene/ethinylestradiol	Femodene	Contraception	1987	Natural product derivative
166	Goserelin	Zoladex	Hormone	1987	Natural product derivative
167	Hidrosmin	Venosmil	Cardiovascular disease	1987	Natural product derivative
168	Lenampicillin HCI	Varacillin	Antibacterial	1987	Natural product derivative
169	Metergolin	Liserdol	Antihyperprolactinemia	1987	Natural product derivative
170	Micinicate	Micivas	Cardiovascular disease	1987	Natural product derivative
171	Midoriamin	Midoriamine	Antiulcer	1987	Natural product derivative
172	Mometasone furoate	Elocon	Antiinflammatory	1987	Natural product derivative
173	Ornoprostil	Alloca	Antiulcer	1987	Natural product derivative
174	Rifamixin	Normix	Antibacterial	1987	Natural product derivative
175	Roxithromycin	Rulid	Antibacterial	1987	Natural product derivative
176	Sultamycillin tosylate	Unasyn	Antibacterial	1987	Natural product derivative
177	Teriparatide acetate	Parathar	Calcium metabolism	1987	Natural product derivative
178	Azithromycin	Sunamed	Antibacterial	1988	Natural product derivative
179	Choline alfoscerate	Delecit	Nootropic	1988	Natural product derivative
180	Erythromycin acistrate	Erasis	Antibacterial	1988	Natural product derivative
181	Flomoxef sodium	Flumarin	Antibacterial	1988	Natural product derivative
182	Flutropium bromide	Flubron	Antitussive	1988	Natural product derivative
183	Halopredone acetate	Haloart	Antiarthritic	1988	Natural product derivative
184	Hydrocortisone aceponate	Retef	Antiinflammatory	1988	Natural product derivative
185	Limaprost	Opalmon	Antithrombotic	1988	Natural product derivative
186	Mifepristone	Mifegyne	Abortifacient	1988	Natural product derivative
187	Octreotide	Sandostatin	Antisecretory	1988	Natural product derivative
188	Pergolide mesylate	Permax	Alzheimer's disease	1988	Natural product derivative

189	Pirarubicin	Pinorubicin	Anticancer	1988	Natural product derivative
190	Propentofylline propionate	Hextol	Vasodilator; cerebral	1988	Natural product derivative
191	Rifapentine	Rifampin	Antibacterial	1988	Natural product derivative
192	Secalciferol	Osteo D	Calcium metabolism	1988	Natural product derivative
193	Tibolone	Livial	Anabolic metabolism	1988	Natural product derivative
194	Acitretin	Neotigason	Antipsoriatic	1989	Natural product derivative
195	Cefpodoxime proxetil	Banan	Antibacterial	1989	Natural product derivative
196	Deslorelin	Somagard	Hormone	1989	Natural product derivative
197	Ipriflavone	Osten	Calcium metabolism	1989	Natural product derivative
198	LMW heparin	Sandoparin	Anticoagulant	1989	Natural product derivative
199	Tiopronin	Thiota	Urolithiasis	1989	Natural product derivative
200	Vigabatrin	Sabril	Antiepileptic	1989	Natural product derivative
201	Vinorelbine	Navelbine	Anticancer	1989	Natural product derivative
202	Acarbose	Glucobay	Antidiabetic	1990	Natural product derivative
203	Arbekacin	Habekacin	Antibacterial	1990	Natural product derivative
204	Cefodizime sodium	Neucef	Antibacterial	1990	Natural product derivative
205	Clarithromycin	Klaricid	Antibacterial	1990	Natural product derivative
206	Eflornithine HCl	Ornidyl	Antiparasitic	1990	Natural product derivative
207	Ethyl icosapentate	Epadel	Antithrombotic	1990	Natural product derivative
208	Fluticasone propionate	Flixonase	Antiinflammatory	1990	Natural product derivative
209	Idarubicin hydrochloride	Zavedos	Anticancer	1990	Natural product derivative
210	Nafarelin acetate	Synarel	Hormone	1990	Natural product derivative
211	Rebamipide	Mucosta	Antiulcer	1990	Natural product derivative
212	Calcipotriol	Daivonex	Antipsoriatic	1991	Natural product derivative
213	Cefdinir	Cefzon	Antibacterial	1991	Natural product derivative
214	Cloricromen	Proendotal	Antithrombotic	1991	Natural product derivative
215	Halobetasol propionate	Ultravate	Antiinflammatory	1991	Natural product derivative
216	LMW heparin	Logiparin	Anticoagulant	1991	Natural product derivative
217	Romurtide	Nopla	Immunostimulant	1991	Natural product derivative
218	Beraprost sodium	Dorner	Platelet aggreg. inhib.	1992	Natural product derivative
219	Cefetamet pivoxil HCl	Globocef	Antibacterial	1992	Natural product derivative
220	Cefpirome sulfate	Cefrom	Antibacterial	1992	Natural product derivative
221	Cefprozil	Cefzil	Antibacterial	1992	Natural product derivative

Table 13.2 (*Continued*)

No.	Generic name	Trade name	Disease	Year	Source
222	Ceftibuten	Seftem	Antibacterial	1992	Natural product derivative
223	Danaparoid sodium	Orgaran	Anticoagulant	1992	Natural product derivative
224	Deprodone propionate	Eclar	Antiinflammatory	1992	Natural product derivative
225	Finasteride	Proscar	5 Alpha-reductase inhibitor	1992	Natural product derivative
226	Iloprost	Ilomedin	Platelet aggreg. inhib.	1992	Natural product derivative
227	Loracarbef	Lorabid	Antibacterial	1992	Natural product derivative
228	Rifabutin	Mycobutin	Antibacterial	1992	Natural product derivative
229	Sapropterin hydrochloride	Biopten	Hyperphenylalaninemia	1992	Natural product derivative
230	Tazobactam sodium	Tazocillin	Antibacterial	1992	Natural product derivative
231	Alendronate sodium	Alendros	Calcium metabolism	1993	Natural product derivative
232	Cabergoline	Dostinex	Antihyperprolactinemia	1993	Natural product derivative
233	Cefepime	Maxipime	Antibacterial	1993	Natural product derivative
234	Cladribine	Leustatin	Anticancer	1993	Natural product derivative
235	Cytarabine ocfosfate	Starsaid	Anticancer	1993	Natural product derivative
236	Dirithromycin	Nortron	Antibacterial	1993	Natural product derivative
237	Ecabet sodium	Gastrom	Antiulcer	1993	Natural product derivative
238	Formestane	Lentaron	Anticancer	1993	Natural product derivative
239	Histrelin	Supprelin	Hormone	1993	Natural product derivative
240	Miltefosine	Miltex	Anticancer	1993	Natural product derivative
241	Parnaparin sodium	Fluxum	Anticoagulant	1993	Natural product derivative
242	Reviparin sodium	Clivarin	Anticoagulant	1993	Natural product derivative
243	Tacalcitol	Bonalfa	Antipsoriatic	1993	Natural product derivative
244	Tretinoin tocoferil	Olcenon	Antiulcer	1993	Natural product derivative
245	Betamethasone butyrate propion	Antebate	Antiinflammatory	1994	Natural product derivative
246	Cefditoren pivoxil	Meiact	Antibacterial	1994	Natural product derivative
247	Docarpamine	Tanadopa	Cardiotonic	1994	Natural product derivative
248	Gusperimus trihydrochloride	Spanidin	Immunosuppressant	1994	Natural product derivative
249	Irinotecan hydrochloride	Campto	Anticancer	1994	Natural product derivative
250	Meropenem	Merrem	Antibacterial	1994	Natural product derivative
251	Panipenem/betamipron	Carbenin	Antibacterial	1994	Natural product derivative
252	Polaprezinc	Promac	Antiulcer	1994	Natural product derivative

253	Quinagolide hydrochloride	Norprolac	Antihyperprolactinemia	1994	Natural product derivative
254	Rocuronium bromide	Zemuron	Muscle relaxant	1994	Natural product derivative
255	Trimetrexate glucuronate	Neutrexin	PCP/Toxoplasmosis	1994	Natural product derivative
256	Unoprostone isopropyl ester	Rescula	Antiglaucoma	1994	Natural product derivative
257	Zinostatin stimalamer	Smancs	Anticancer	1994	Natural product derivative
258	Cefozopran HCl	Firstcin	Antibacterial	1995	Natural product derivative
259	Docetaxel	Taxotere	Anticancer	1995	Natural product derivative
260	Estrogen/medroxyproesterone	Prempro	Hormone replacement therapy	1995	Natural product derivative
261	Estrogen/medroxyproesterone	Premphase	Hormone replacement therapy	1995	Natural product derivative
262	Lanreotide acetate	Somatuline	Hormone	1995	Natural product derivative
263	Mycophenolate mofetil	CellCept	Immunosuppressant	1995	Natural product derivative
264	Nalmefene HCl	Revex	Antinarcotic	1995	Natural product derivative
265	Rimexolone	Vexol	Antiinflammatory	1995	Natural product derivative
266	Tirilazad mesylate	Freedox	Subarachnoid hemorrhage	1995	Natural product derivative
267	Certoparin sodium	Alphaparin	Anticoagulant	1996	Natural product derivative
268	Etoposide phosphate	Etopophos	Anticancer	1996	Natural product derivative
269	Glycopin	Likopid	Immunostimulant	1996	Natural product derivative
270	Latanoprost	Xalatan	Antiglaucoma	1996	Natural product derivative
271	Prezatide copper acetate	Lamin gel	Vulnerary	1996	Natural product derivative
272	Topotecan HCl	Hycamptin	Anticancer	1996	Natural product derivative
273	Zinc hyaluronate	Curioscii	Vulnerary	1996	Natural product derivative
274	Ardeparin sodium	Normiflo	Anticoagulant	1997	Natural product derivative
275	Cefcapene pivoxil	Flomox	Antibacterial	1997	Natural product derivative
276	Flurithromycin ethylsuccinate	Ritro	Antibacterial	1997	Natural product derivative
277	Fropenam	Farom	Antibacterial	1997	Natural product derivative
278	Lepirudin	Refludan	Antithrombotic	1997	Natural product derivative
279	Bemiparin sodium	Hibor	Anticoagulant	1998	Natural product derivative
280	Cefoselis	Wincef	Antibacterial	1998	Natural product derivative
281	Desirudin	Revasc	Anticoagulant	1998	Natural product derivative
282	Levonorgestrel/ethinyl estradi	Preven	Contraception	1998	Natural product derivative
283	Loteprednol etabonate	Lotemax	Antiallergic	1998	Natural product derivative
284	Miglitol	Diastabol	Antidiabetic	1998	Natural product derivative
285	Orlistat	Xenical	anti-obesity	1998	Natural product derivative

Table 13.2 (*Continued*)

No.	Generic name	Trade name	Disease	Year	Source
286	Paricalcitol	Zemplar	Calcium metabolism	1998	Natural product derivative
287	Sermorelin acetate	Geref	Hormone	1998	Natural product derivative
288	Alitretinoin	Panretin	Anticancer	1999	Natural product derivative
289	Cetrorelix	Cetrotide	Hormone	1999	Natural product derivative
290	Colforsin daropate HCl	Adele	Cardiotonic	1999	Natural product derivative
291	Conjugated estrogens a	Conestin	Hormone replacement therapy	1999	Natural product derivative
292	Dalfopristin	Synercid	Antibacterial	1999	Natural product derivative
293	Doxercalciferol	Hectorol	Calcium metabolism	1999	Natural product derivative
294	Exemestane	Aromasin	Anticancer	1999	Natural product derivative
295	Melevodopa (CHF-1301)	Levomet	Parkinson's disease	1999	Natural product derivative
296	Oseltamivir	Tamiflu	Antiviral	1999	Natural product derivative
297	Quinupristin	Synercid	Antibacterial	1999	Natural product derivative
298	Rapacuronium bromide	Raplon	Muscle relaxant	1999	Natural product derivative
299	Valrubicin	Valstar	Anticancer	1999	Natural product derivative
300	Zanamivir	Relenza	Antiviral	1999	Natural product derivative
301	Arteether	Artemotil	Antiparasitic	2000	Natural product derivative
302	Atosiban	Tractocile	Premature birth	2000	Natural product derivative
303	Bivalrudin	Angiomax	Antithrombotic	2000	Natural product derivative
304	Dosmalfate	Diotul	Gastroprotectant	2000	Natural product derivative
305	Drospirenone	Yasmin	Contraception	2000	Natural product derivative
306	Egualen sodium	Azuloxa	Antiulcer	2000	Natural product derivative
307	Ganirelix acetate	Orgalutran	Ovulation	2000	Natural product derivative
308	Gemtuzumab ozogamicin	Mylotarg	Anticancer	2000	Natural product derivative
309	Maxacalcitol	Prezios	Calcium metabolism	2000	Natural product derivative
310	Medroxyprogesterone/estradiol	Lunelle	Contraception	2000	Natural product derivative
311	Taltirelin	Ceredist	CNS stimulant	2000	Natural product derivative
312	Trimegestrone/estradiol	Totelle Sekvens	Hormone replacement therapy	2000	Natural product derivative
313	Aminolevulinic acid me ester	Metvix	Anticancer	2001	Natural product derivative
314	Bimatoprost	Lumigan	Antiglaucoma	2001	Natural product derivative
315	Caspofungin acetate	Cancidas	Antifungal	2001	Natural product derivative
316	Dutaseride	Duagen	Benign prostatic hypertrophy	2001	Natural product derivative

317	Falecalcitrol	Horner / Fulstan	Sec. hyperthyroidism	2001	Natural product derivative
318	Fudosteine	Cleanal/Spelear	Expectorant	2001	Natural product derivative
319	Nesiritide	Natrecor	Congestive heart failure	2001	Natural product derivative
320	Norelgestromin/ethinylestradio	Ortho Evra	Contraception	2001	Natural product derivative
321	Telithromycin	Ketek	Antibacterial	2001	Natural product derivative
322	Tenofovir disoproxil fumarate	Viread	Antiviral	2001	Natural product derivative
323	Travoprost	Travatan	Antiglaucoma	2001	Natural product derivative
324	Amrubicin HCl	Calsed	Anticancer	2002	Natural product derivative
325	Biapenem	Omegacin	Antibacterial	2002	Natural product derivative
326	Eplerenone	Inspra	Antihypertensive	2002	Natural product derivative
327	Ertapenem sodium	Invanz	Antibacterial	2002	Natural product derivative
328	Etonogestrel/ethinylestradiol	NuvaRing	Contraception	2002	Natural product derivative
329	Fondaparinux sodium	Arixtra	Anticoagulant	2002	Natural product derivative
330	Fulvestrant	Faslodex	Anticancer	2002	Natural product derivative
331	Micafungin sodium	Fungard	Antifungal	2002	Natural product derivative
332	Pimecrolimus (SDZ-ASM-981)	Elidel	Antiallergic	2002	Natural product derivative
333	Sodium oxybate	Xyrem	Antinarcolepsy	2002	Natural product derivative
334	Teriparatide	Forteo	Osteoporosis	2002	Natural product derivative
335	Treprostinil sodium	Remodulin	Antihypertensive	2002	Natural product derivative
336	Carglumic acid	Carbaglu	Hyperammonemia	2003	Natural product derivative
337	Drospirenone/estradiol	Angeliq	Hormone replacement therapy	2003	Natural product derivative
338	Enfuvirtide	Fuzeon	Antiviral	2003	Natural product derivative
339	Estradiol/nomegestrol acetate	Naemis	Hormone replacement therapy	2003	Natural product derivative
340	Miglustat	Zavesca	GAUCHER'S disease	2003	Natural product derivative
341	Belotecan hydrochloride	Camtobell	Anticancer	2004	Natural product derivative
342	Estradiol/levonorgestrel	Climara Pro	Hormone replacement therapy	2004	Natural product derivative
343	Everolimus	Certican	Immunosuppressant	2004	Natural product derivative
344	Hexyl aminolevulinate	Hexvix	Anticancer	2004	Natural product derivative
345	Pregabalin	Lyrica	Antiepileptic	2004	Natural product derivative
346	Talaporfin sodium	Laserphyrin	Anticancer	2004	Natural product derivative
347	Vapreotide acetate	Docrised	Anticancer	2004	Natural product derivative
348	Ciclesonide	Alvesco	Antiasthmatic	2005	Natural product derivative
349	Conjugated estrogens b	Enjuvia	Hormone replacement therapy	2005	Natural product derivative

Table 13.2 (*Continued*)

No.	Generic name	Trade name	Disease	Year	Source
350	Doripenem	Finibax	Antibacterial	2005	Natural product derivative
351	Extenatide	Byetta	Antidiabetic	2005	Natural product derivative
352	L-alanyl-L-glutamine	Dipeptiven	Mucositis	2005	Natural product derivative
353	Pralmorelin	GHRP Kaken 100	Treatment of GH deficiency	2005	Natural product derivative
354	Tigecycline	Tygacil	Antibacterial	2005	Natural product derivative
355	Triproamylin acetate	Normylin	Antidiabetic	2005	Natural product derivative
356	Zotarolimus	Endeavor	Restenosis	2005	Natural product derivative
357	Anecortave acetate	Retaane	Macular degeneration	2006	Natural product derivative
358	Anidulafungin	Eraxis	Antifungal	2006	Natural product derivative
359	Lubiprostone	Amitiza	Constipation	2006	Natural product derivative
360	Fluticasone furoate	Allermist	Antiallergic	2007	Natural product derivative
361	Ixabepilone	Ixempra	Anticancer	2007	Natural product derivative
362	Lisdexamfetamine mesilate	Vyvanse	ADHD	2007	Natural product derivative
363	Retapamulin	Altabax	Antibacterial	2007	Natural product derivative
364	Temsirolimus	Toricel	Anticancer	2007	Natural product derivative
365	Biolimus a9	BioMatrix	Restenosis	2008	Natural product derivative
366	Ceftobiprole medocaril	Zeftera	Antibacterial	2008	Natural product derivative
367	Dienogest	Dinagest	Endometriosis	2008	Natural product derivative
368	Difluprednate	Durezol	Ocular inflammation	2008	Natural product derivative
369	Icatibant acetate	Firazyr	Hereditary angioedema	2008	Natural product derivative
370	Methylnaltrexone bromide	Relistor	Constipation	2008	Natural product derivative
371	Sugammadex sodium	Bridion	Antidote	2008	Natural product derivative
372	Tafluprost	Taflotan	Antiglaucoma	2008	Natural product derivative
373	Liraglutide	Victoza	Antidiabetic	2009	Natural product derivative
374	Nalfurafine HCl	Remitch	Uremic pruritis	2009	Natural product derivative
375	Pralatrexate	Folotyn	Anticancer	2009	Natural product derivative
376	Telavancin HCl	Vibativ	Antibacterial	2009	Natural product derivative
377	Ulliprisnil acetate	ellaOne	Contraception	2009	Natural product derivative
378	Cabazitaxel	Jevtana	Anticancer	2010	Natural product derivative
379	Diquafosol tetrasodium	Diquas	Dry eye syndrome	2010	Natural product derivative
380	Eribulin	Halaven	Anticancer	2010	Natural product derivative

381	Fingolimod HCl	Gilenya	Multiple sclerosis	2010	Natural product derivative
382	Laninamivir octanoate	Inavir	Antiviral	2010	Natural product derivative
383	Mifamurtide	Junovan	Anticancer	2010	Natural product derivative
384	Vinflunine	Javlor	Anticancer	2010	Natural product derivative
385	Zucapsacin	Civanex	Osteoarthritis	2010	Natural product derivative
386	Abiraterone acetate	Zytiga	Anticancer	2011	Natural product derivative
387	Brentuximab vedotin	Adcetris	Anticancer	2011	Natural product derivative
388	Ceftaroline fosamil acetate	Teflaro	Antibacterial	2011	Natural product derivative
389	Eldecalcitol	Edirol	Osteoporosis	2011	Natural product derivative
390	Tesamorelin	Egrifta	Lipodystrophy syndrome [HIV]	2011	Natural product derivative
391	Zucapsaicin	Zuacta	Antiarthritic	2011	Natural product derivative
392	bf-200 ala	Ameluz	Anticancer	2012	Natural product derivative
393	Carfilzomib	Kyprolis	Anticancer	2012	Natural product derivative
394	Hyaluronic acid crosslinked	Gel-One	Antiarthritic	2012	Natural product derivative
395	Linaclotide	Constella	Irritable bowel syndrome	2012	Natural product derivative
396	Pasireotide	Signifor	CUSHINGS syndrome	2012	Natural product derivative
397	Teduglutide	Gattex	Short bowel syndrome	2012	Natural product derivative
398	Dimethyl fumarate	Tecfidera	Multiple sclerosis	2013	Natural product derivative
399	Lixisenatide	Lyxumia	Antidiabetic	2013	Natural product derivative
400	Trastuzumab emtansine	Kadcyla	Anticancer	2013	Natural product derivative
401	Aafamelanotide	Scenesse	Phototoxicity in adults	2014	Natural product derivative
402	Cetolozane/tazobactam	Zerbaxa	Antibacterial	2014	Natural product derivative
403	Dalbavancin	Dalvance	Antibacterial	2014	Natural product derivative
404	Naloxegol	Movantik	Opioid induced constipation	2014	Natural product derivative
405	Oritavancin	Orbactiv	Antibacterial	2014	Natural product derivative
406	Vorapaxar	Zontivity	Coronary artery disease	2014	Natural product derivative
407	Cangrelor tetrasodium	Kangreal	Coronary artery disease	2015	Natural product derivative
408	Hyaluronate / triamcinolone	Cingal	Antiarthritic	2015	Natural product derivative
409	Padeliporfin potassium	Stakel	Anticancer	2015	Natural product derivative
410	Uridine triacetate	Xuriden	Orotic aciduria	2015	Natural product derivative
411	Dronabinol hemiacetate	Syndros	Anorexia	2016	Natural product derivative
412	Migalastat hydrochloride	Galafold	FABRY'S disease	2016	Natural product derivative
413	Obeticholic acid	Ocaliva	Anticirrhotic	2016	Natural product derivative

Table 13.2 (*Continued*)

No.	Generic name	Trade name	Disease	Year	Source
414	Abaloparatide	Tymlos	Osteoporosis	2017	Natural product derivative
415	Dinalbuphine sebacate	Naldebain	Analgesic	2017	Natural product derivative
416	Forodesine HCL	Mundesine	Anticancer	2017	Natural product derivative
417	Inotuzumab ozogamicin	Besponsa	Anticancer	2017	Natural product derivative
418	Latanoprostene bunod	Vyzulta	Antiglaucoma	2017	Natural product derivative
419	Meropenem/vaborbactam	Vabomere	Antibacterial	2017	Natural product derivative
420	Midostaurin	Rydapt	Anticancer	2017	Natural product derivative
421	Naldemedine tosylate	Symproic	Opioid-induced constipation	2017	Natural product derivative
422	Plecanatide	trulance	Irritable bowel syndrome	2017	Natural product derivative
423	Semaglutide	Ozempic	Antidiabetic	2017	Natural product derivative
424	Tenofovir disoproxil orotate	virreal	Antiviral	2017	Natural product derivative
425	Benzhydrocodone/ acetoaminophen	Apadaz	Analgesic	2018	Natural product derivative
426	Eravacycline	Xerava	Antibacterial	2018	Natural product derivative
427	Moxidectin		Antiparasitic	2018	Natural product derivative
428	Omadacycline	Nuzyra	Antibacterial	2018	Natural product derivative
429	Plazomicin	Zemdri	Antibacterial	2018	Natural product derivative
430	Sarecycline	Seysara	Antibacterial	2018	Natural product derivative
431	Segesterone acetate/ethinylest	Annovera	Contraception	2018	Natural product derivative
432	Tafenoquine succinate	Etaquine	Antiparasitic	2018	Natural product derivative
433	Brexanolone	Zulresso	Antidepressant	2019	Natural product derivative
434	Dihydrospirorenone	Slinda	Contraception	2019	Natural product derivative
435	Imi-cilast_relebactam	Recarbrio	Antibacterial	2019	Natural product derivative
436	Lefamulin	Xenlita	Antibacterial	2019	Natural product derivative
437	Mirogabalin besylate	Tarlige	Neuropathic pain	2019	Natural product derivative
438	Monomethyl fumarate	Bafiertam	Multiple sclerosis	2019	Natural product derivative
439	PEG-loxenatide	Fulaimei	Antidiabetic	2019	Natural product derivative
440	Polatuzumab vedotin	Polivy	Anticancer	2019	Natural product derivative
441	Siponimod fumarate	Mayzent	Multiple sclerosis	2019	Natural product derivative
442	Cloranolol HCl	Tobanum	Antihypertensive	1981	Natural product pharmacophore
443	Mepindolol sulfate	Corindolan	Antihypertensive	1981	Natural product pharmacophore

444	Penbutanol sulfate	Betapressin	Antihypertensive	1981	Natural product pharmacophore
445	Tulobuterol HCl	Berachin	Bronchodilator	1981	Natural product pharmacophore
446	Carteolol HCl	Teoptic	Antiglaucoma	1982	Natural product pharmacophore
447	Methypranolol	Minims metipranolol	Antiglaucoma	1982	Natural product pharmacophore
448	Betaxolol HCI	Kerlone	Antihypertensive	1983	Natural product pharmacophore
449	Pirbuterol HCL	Exirel	Bronchodilator	1983	Natural product pharmacophore
450	Bucladesine sodium	Actosin	Cardiotonic	1984	Natural product pharmacophore
451	Enalapril maleate	Reniten	Antihypertensive	1984	Natural product pharmacophore
452	Bopindolol	Sandonorm	Antihypertensive	1985	Natural product pharmacophore
453	Levobunolol HCI	Betagan	Antiglaucoma	1985	Natural product pharmacophore
454	Arotinolol HCI	Almarl	Antihypertensive	1986	Natural product pharmacophore
455	Bisoprolol fumarate	Concor	Antihypertensive	1986	Natural product pharmacophore
456	Formoterol fumarate	Atock	Bronchodilator	1986	Natural product pharmacophore
457	Mabuteroi HCl	Broncholin	Bronchodilator	1986	Natural product pharmacophore
458	Bevantolol HCl	Ranestol	Antihypertensive	1987	Natural product pharmacophore
459	Enalaprilat	Renitec	Antihypertensive	1987	Natural product pharmacophore
460	Esmolol HCI	Brevibloc	Antiarrhythmic	1987	Natural product pharmacophore
461	Lisinopril	Prinivil	Antihypertensive	1987	Natural product pharmacophore
462	Tertatolol HCl	Artex	Antihypertensive	1987	Natural product pharmacophore
463	Alacepril	Cetapril	Antihypertensive	1988	Natural product pharmacophore
464	Amosulalol	Lowgan	Antihypertensive	1988	Natural product pharmacophore
465	Denopamine	Kalgut	Cardiotonic	1988	Natural product pharmacophore
466	Nipradilol	Hypadil	Antihypertensive	1988	Natural product pharmacophore
467	Perindopril	Coversyl	Antihypertensive	1988	Natural product pharmacophore
468	Xamoterol fumarate	Corwin	Cardiotonic	1988	Natural product pharmacophore
469	Citalopram	Cipramil	Antidepressant	1989	Natural product pharmacophore
470	Delapril	Adecut	Antihypertensive	1989	Natural product pharmacophore
471	Dilevalol	Levadil	Antihypertensive	1989	Natural product pharmacophore
472	Dopexamine	Dopacard	Cardiotonic	1989	Natural product pharmacophore
473	Droxidopa	DOPS	Parkinson's disease	1989	Natural product pharmacophore
474	Quinapril	Accupro	Antihypertensive	1989	Natural product pharmacophore
475	Ramipril	Triatec	Antihypertensive	1989	Natural product pharmacophore

Table 13.2 (*Continued*)

No.	Generic name	Trade name	Disease	Year	Source
476	Argatroban	Slonnon	Antithrombotic	1990	Natural product pharmacophore
477	Bambuterol	Bambec	Bronchodilator	1990	Natural product pharmacophore
478	Benazepril hydrochloride	Cibacen	Antihypertensive	1990	Natural product pharmacophore
479	Cilazapril	Inhibace	Antihypertensive	1990	Natural product pharmacophore
480	Ondansetron hydrochloride	Zophran	Antiemetic	1990	Natural product pharmacophore
481	Remoxipride hydrochloride	Roxiam	Antipsychotic	1990	Natural product pharmacophore
482	Salmeterol xinafoate	Serevent	Bronchodilator	1990	Natural product pharmacophore
483	Carvedilol	Dilatrend	Antihypertensive	1991	Natural product pharmacophore
484	Fosinopril sodium	Staril	Antihypertensive	1991	Natural product pharmacophore
485	Granisetron hydrochloride	Kytril	Antiemetic	1991	Natural product pharmacophore
486	Moxonidine	Cynt	Antihypertensive	1991	Natural product pharmacophore
487	Sumatriptan succinate	Imigran	Antimigraine	1991	Natural product pharmacophore
488	Tilisolol hydrochloride	Daim	Antihypertensive	1992	Natural product pharmacophore
489	Tropisetron	Novaban	Antiemetic	1992	Natural product pharmacophore
490	Acetorphan	Tiorfan	Antidiarrheal	1993	Natural product pharmacophore
491	Brodimoprin	Hyprim	Antibacterial	1993	Natural product pharmacophore
492	imidaprii HCI	Tanatril	Antihypertensive	1993	Natural product pharmacophore
493	Trandolapril	Odrik	Antihypertensive	1993	Natural product pharmacophore
494	Fluvastatin sodium	Lescol	Hypocholesterolemic	1994	Natural product pharmacophore
495	Nazasetron hydrochloride	Serotone	Antiemetic	1994	Natural product pharmacophore
496	Temocapril hydrochloride	Acecol	Antihypertensive	1994	Natural product pharmacophore
497	Itopride HCl	Ganaton	Gastroprokinetic	1995	Natural product pharmacophore
498	Moexipril HCl	Univasc	Antihypertensive	1995	Natural product pharmacophore
499	Pranlukast	Onon	Antiasthmatic	1995	Natural product pharmacophore
500	Saquinavir mesylate	Invirase	Antiviral	1995	Natural product pharmacophore
501	Spirapril HCl	Setrilan	Antihypertensive	1995	Natural product pharmacophore
502	Indinavir sulfate	Crixivan	Antiviral	1996	Natural product pharmacophore
503	Loprinone HCl	Cortec	Cardiotonic	1996	Natural product pharmacophore
504	Raltitrexed	Tomudex	Anticancer	1996	Natural product pharmacophore
505	Ramosetron	Nasea	Antiemetic	1996	Natural product pharmacophore
506	Ritonavir	Norvir	Antiviral	1996	Natural product pharmacophore

507	Ropinirole HCl	ReQuip	Parkinson's disease	1996	Natural product pharmacophore
508	Sertindole	Serdolect	Neuroleptic	1996	Natural product pharmacophore
509	Tiagabine	Gabitril	Antiepileptic	1996	Natural product pharmacophore
510	Anagrelide HCl	Agrylin	Hematological	1997	Natural product pharmacophore
511	Atorvastin calcium	Lipitor	Hypocholesterolemic	1997	Natural product pharmacophore
512	Cerivastatin	Lipobay	Hypocholesterolemic	1997	Natural product pharmacophore
513	Naratriptan HCl	Naramig	Antimigraine	1997	Natural product pharmacophore
514	Neflinavir mesylate	Viracept	Antiviral	1997	Natural product pharmacophore
515	Tazarotene	Zorac	Antipsoriatic	1997	Natural product pharmacophore
516	Tolcapone	Tasmar	Parkinson's disease	1997	Natural product pharmacophore
517	Zomitriptan	Zomig	Antimigraine	1997	Natural product pharmacophore
518	Dolasetron mesylate	Anzemet	Antiemetic	1998	Natural product pharmacophore
519	Entacapone	Comtess	Parkinson's disease	1998	Natural product pharmacophore
520	Fomivirsen sodium	Vitravene	Antiviral	1998	Natural product pharmacophore
521	Rizatriptan benzoate	Maxalt	Antimigraine	1998	Natural product pharmacophore
522	Sildenafil citrate	Viagra	Male sexual dysfunction	1998	Natural product pharmacophore
523	Amprenavir	Agenerase	Antiviral	1999	Natural product pharmacophore
524	Eptilfibatide	Integrilin	Antithrombotic	1999	Natural product pharmacophore
525	Levalbuterol HCl	Xopenex	Antiasthmatic	1999	Natural product pharmacophore
526	Temozolomide	Temodal	Anticancer	1999	Natural product pharmacophore
527	Almotriptan	Almogran	Antimigraine	2000	Natural product pharmacophore
528	Alosetron HCL	Lotronex	Irritable bowel syndrome	2000	Natural product pharmacophore
529	Bexarotene	Targretine	Anticancer	2000	Natural product pharmacophore
530	Cevimeline HCl	Evoxac	Antixerostomia	2000	Natural product pharmacophore
531	Lopinavir	Kaletra	Antiviral	2000	Natural product pharmacophore
532	Ziprasidone HCl	Zeldox	Neuroleptic	2000	Natural product pharmacophore
533	Zofenapril calcium	Zantipres	Antihypertensive	2000	Natural product pharmacophore
534	Eletriptan	Relpax	Antimigraine	2001	Natural product pharmacophore
535	Perspirone hydrochloride	Lullan	Antipsychotic	2001	Natural product pharmacophore
536	Tegaserod maleate	Zelmac	Irritable bowel syndrome	2001	Natural product pharmacophore
537	Escitalopram	Cipralex	Antidepressant	2002	Natural product pharmacophore
538	Frovatriptan	Frova	Antimigraine	2002	Natural product pharmacophore
539	Landiolol	Onoact	Antiarrhythmic	2002	Natural product pharmacophore

Table 13.2 (*Continued*)

No.	Generic name	Trade name	Disease	Year	Source
540	Tiotropium bromide	Spiriva	COPD	2002	Natural product pharmacophore
541	Atazanavir	Reyataz	Antiviral	2003	Natural product pharmacophore
542	Atomoxetine	Strattera	ADHD	2003	Natural product pharmacophore
543	Fosamprenevir	Lexiva	Antiviral	2003	Natural product pharmacophore
544	Palonosetron	Aloxi	Antiemetic	2003	Natural product pharmacophore
545	Pitavastatin	Livalo	Hypocholesterolemic	2003	Natural product pharmacophore
546	Rosuvastatin	Crestor	Hypocholesterolemic	2003	Natural product pharmacophore
547	Tadalafil	Cialis	Male sexual dysfunction	2003	Natural product pharmacophore
548	Vardenafil	Levitra	Male sexual dysfunction	2003	Natural product pharmacophore
549	Abarelix	Plenaxis	Anticancer	2004	Natural product pharmacophore
550	Indisetron hydrochloride	Sinseron	Antiemetic	2004	Natural product pharmacophore
551	Pemetrexed disodium	Alimta	Anticancer	2004	Natural product pharmacophore
552	Tamibarotene	Amnoid	Anticancer	2005	Natural product pharmacophore
553	Tipranavir	Aptivus	Antiviral	2005	Natural product pharmacophore
554	Udenafil	Zydena	Male sexual dysfunction	2005	Natural product pharmacophore
555	Favipiravir	Avigan	Antiviral	2006	Natural product pharmacophore
556	Vorinostat	Zolinza	Anticancer	2006	Natural product pharmacophore
557	Aliskiren fumarate	Tekturna	Antihypertensive	2007	Natural product pharmacophore
558	Arformoterol tartrate	Brovana	COPD	2007	Natural product pharmacophore
559	Alvimopan hydrate	Entereg	Irritable bowel syndrome	2008	Natural product pharmacophore
560	Dabigatran etexilate	Pradaxa	Anticoagulant	2008	Natural product pharmacophore
561	Lacosamide	Vimpat	Antiepileptic	2008	Natural product pharmacophore
562	Rivaroxaban	Xarelto	Venous thromboembolism	2008	Natural product pharmacophore
563	Degarelix	Firmagon	Anticancer	2009	Natural product pharmacophore
564	Indacaterol maleate	Onbrez	COPD	2009	Natural product pharmacophore
565	Apixaban	Eliquis	Anticoagulant	2011	Natural product pharmacophore
566	Avanafil	Zepeed	Male sexual dysfunction	2011	Natural product pharmacophore
567	Boceprevir	Victrelis	Antiviral	2011	Natural product pharmacophore
568	Crizotinib	Xalkori	Anticancer	2011	Natural product pharmacophore
569	Edoxaban tosilate	Lixiana	Anticoagulant	2011	Natural product pharmacophore
570	Mirabegron	Betanis	Urinary incontinence	2011	Natural product pharmacophore

571	Ruxolitinib phosphate	Jakafi	Anticancer	2011	Natural product pharmacophore
572	Telaprevir	Incivek	Antiviral	2011	Natural product pharmacophore
573	Ticagrelor	Britique	Antithrombotic	2011	Natural product pharmacophore
574	Vandetanib	Caprelsa	Anticancer	2011	Natural product pharmacophore
575	Vemurafenib	Zeboraf	Anticancer	2011	Natural product pharmacophore
576	Axitinib	Inlyta	Anticancer	2012	Natural product pharmacophore
577	Bosutiniib	Bosulif	Anticancer	2012	Natural product pharmacophore
578	Cabozantinib *S*-malate	Cometriq	Anticancer	2012	Natural product pharmacophore
579	Dapagliflozin	Forxiga	Antidiabetic	2012	Natural product pharmacophore
580	Peginesatide	Omontys	Anemia	2012	Natural product pharmacophore
581	Radotinib	Supect	Anticancer	2012	Natural product pharmacophore
582	Regorafenib	Stivarga	Anticancer	2012	Natural product pharmacophore
583	Teriflunomide	Aubagio	Multiple sclerosis	2012	Natural product pharmacophore
584	Tofacitinib	Xeljanz	Antiarthritic	2012	Natural product pharmacophore
585	Afatinib	Gilotrif	Anticancer	2013	Natural product pharmacophore
586	Canagliflozin	Invokana	Antidiabetic	2013	Natural product pharmacophore
587	Dabrafenib mesilate	Tafinlar	Anticancer	2013	Natural product pharmacophore
588	FF/GW-642444	Breo	COPD	2013	Natural product pharmacophore
589	Ibrutinib	Imbruvica	Anticancer	2013	Natural product pharmacophore
590	Olodaterol hydrochloride	Striverdi	COPD	2013	Natural product pharmacophore
591	Ponatinib	Iclusig	Anticancer	2013	Natural product pharmacophore
592	Simeprevir	Sovriad	Antiviral	2013	Natural product pharmacophore
593	Trametinib DMSO	Mekinist	Anticancer	2013	Natural product pharmacophore
594	Belinostat	Beleodaq	Anticancer	2014	Natural product pharmacophore
595	Eliglustat tartrate	Cerdelga	Gaucher's disease	2014	Natural product pharmacophore
596	Empagliflozin	Jardiance	Antidiabetic	2014	Natural product pharmacophore
597	Idelalisib	Zydelig	Anticancer	2014	Natural product pharmacophore
598	Ipragliflozin proline	Suglat	Antidiabetic	2014	Natural product pharmacophore
599	Luseogliflozin	Lusefi	Antidiabetic	2014	Natural product pharmacophore
600	Ombitasvir	Viekira Pak	Antiviral	2014	Natural product pharmacophore
601	Paritaprevir	Viekira Pak	Antiviral	2014	Natural product pharmacophore
602	Tofogliflozin	Apleway	Antidiabetic	2014	Natural product pharmacophore
603	Vaniprevir	Vanihep	Antiviral	2014	Natural product pharmacophore

Table 13.2 (*Continued*)

No.	Generic name	Trade name	Disease	Year	Source
604	Eluxadoline	Viberzi	Irritable bowel syndrome	2015	Natural product pharmacophore
605	Lenvatinib mesylate	Lenvima	Anticancer	2015	Natural product pharmacophore
606	Osimertinib mesylate	Tagrisso	Anticancer	2015	Natural product pharmacophore
607	Palbociclib	Ibrance	Anticancer	2015	Natural product pharmacophore
608	Safinamide mesilate	Xadago	Antiepileptic	2015	Natural product pharmacophore
609	Selexipag	Uptravi	PAH	2015	Natural product pharmacophore
610	Narlaprevir	arlansa	Antiviral	2016	Natural product pharmacophore
611	Olmutinib	Olita	Anticancer	2016	Natural product pharmacophore
612	Opicapone	Ongentys	Parkinson's disease	2016	Natural product pharmacophore
613	Abemaciclib mesylate	Verzenio	Anticancer	2017	Natural product pharmacophore
614	Acalabrutinib	Calquence	Anticancer	2017	Natural product pharmacophore
615	Baricitinib	Olumiant	Antiarthritic	2017	Natural product pharmacophore
616	Brigatinib	Alunbrig	Anticancer	2017	Natural product pharmacophore
617	Copanlisib	Aliqopa	Anticancer	2017	Natural product pharmacophore
618	Deutetrabenazine	Austedo	Tardive dyskinesia	2017	Natural product pharmacophore
619	Ertugliflozin	Steglatro	Antidiabetic	2017	Natural product pharmacophore
620	Glecaprevir/pibrentasvir	Maviret	Antiviral	2017	Natural product pharmacophore
621	Macimoralin acetate	Macrilen	GH deficit	2017	Natural product pharmacophore
622	Neratinib	Nerlynx	Anticancer	2017	Natural product pharmacophore
623	Pitavastatin magnesium	Zypitamag	Hypocholesterolemic	2017	Natural product pharmacophore
624	Ribociclib	Kisqali	Anticancer	2017	Natural product pharmacophore
625	Telotristat etiprate	Xermelo	Ulcerative colitis	2017	Natural product pharmacophore
626	Tivozanib	Fotivda	Anticancer	2017	Natural product pharmacophore
627	Valbenazine	Ingrezza	Tardive dyskinesia	2017	Natural product pharmacophore
628	Albuvirtide	Aikening	Antiviral	2018	Natural product pharmacophore
629	Dacomitinib	Vizimpro	Anticancer	2018	Natural product pharmacophore
630	Danoprevir	Ganovo	Antiviral	2018	Natural product pharmacophore
631	Duvelisib	Copiktra	Anticancer	2018	Natural product pharmacophore
632	Fostamatinib disodium	Tavalisse	Idiopathic thrombocytopenia	2018	Natural product pharmacophore
633	Fruquintinib	Elunate	Anticancer	2018	Natural product pharmacophore
634	Gilteritinib	Xospata	Anticancer	2018	Natural product pharmacophore

635	Ivosidenib	Tibsovo	Anticancer	2018	Natural product pharmacophore
636	Larotrectinib sulfate	Vitrakvi	Anticancer	2018	Natural product pharmacophore
637	Pyrotinib	Airuini	Anticancer	2018	Natural product pharmacophore
638	Revefenacin	Yupelri	COPD	2018	Natural product pharmacophore
639	Vibegron	Beova	Overactive bladder	2018	Natural product pharmacophore
640	Bremelanotide	Vyleesi	Female sexual dysfunction	2019	Natural product pharmacophore
641	Entrectinib	Rozlytrek	Anticancer	2019	Natural product pharmacophore
642	Erdafitinib	Balversa	Anticancer	2019	Natural product pharmacophore
643	Fedratinib HCl	Inrebic	Anticancer	2019	Natural product pharmacophore
644	Peficitinib hydrobromide	Smyraf	Antiarthritic	2019	Natural product pharmacophore
645	Remogliflozin etaborate	Remo	Antidiabetic	2019	Natural product pharmacophore
646	Solriamfetol hydrochloride	Sunosi	Antinarcolepsy	2019	Natural product pharmacophore
647	Sotagliflozin	Zynquista	Antidiabetic	2019	Natural product pharmacophore
648	Upadacitinib tartrate	Rinvoq	Antiarthritic	2019	Natural product pharmacophore
649	Selegiline HCl	Jumex	Parkinson's disease	1981	Natural product mimic
650	Amrinone	Inocor	Cardiotonic	1983	Natural product mimic
651	Fluvoxamine maleate	Floxyfral	Antidepressant	1983	Natural product mimic
652	Tizanidine	Sirdalud	Muscle relaxant	1984	Natural product mimic
653	Toloxatone	Humoryl	Antidepressant	1984	Natural product mimic
654	Bunazosin HCl	Detandol	Antihypertensive	1985	Natural product mimic
655	Camostat mesylate	Foipan	Anticancer	1985	Natural product mimic
656	Ketanserin	Serefrex	Antihypertensive	1985	Natural product mimic
657	Trimazosin HCl	Supres	Antihypertensive	1985	Natural product mimic
658	Amisulpride	Socian	Antipsychotic	1986	Natural product mimic
659	Fluoxetine HCI	Prozac	Antidepressant	1986	Natural product mimic
660	Nafamostat mesylate	Futhan	Pancreatitis	1986	Natural product mimic
661	Dapiprazole HCl	Glamidolo	Antiglaucoma	1987	Natural product mimic
662	Flumazenil	Anexate	Benzodiazepine antagonist	1987	Natural product mimic
663	Levosulpiride	Levopraid	Antipsychotic	1987	Natural product mimic
664	Picotamide	Plactidil	Antithrombotic	1987	Natural product mimic
665	Alpiropride	Rivestel	Antimigraine	1988	Natural product mimic
666	Apraclonidine HCl	Lopidine	Antiglaucoma	1988	Natural product mimic
667	Cilostazol	Retal	Antithrombotic	1988	Natural product mimic
668	Doxazosin mesylate	Carduran	Antihypertensive	1988	Natural product mimic

Table 13.2 (*Continued*)

No.	Generic name	Trade name	Disease	Year	Source
669	Enoximone	Perfan	Cardiotonic	1988	Natural product mimic
670	Ozagrel sodium	Xanbon	Antithrombotic	1988	Natural product mimic
671	Ibudilast	Ketas	Antiasthmatic	1989	Natural product mimic
672	Milrinone	Corotrope	Cardiotonic	1989	Natural product mimic
673	Setiptiline	Tecipul	Antidepressant	1989	Natural product mimic
674	Tolrestat	Alredase	Antidiabetic	1989	Natural product mimic
675	Toremifene	Fareston	Anticancer	1989	Natural product mimic
676	Cinitapride	Cidine	Gastroprokinetic	1990	Natural product mimic
677	Moclobemide	Aurorix	Antidepressant	1990	Natural product mimic
678	Setraline hydrochloride	Lustral	Antidepressant	1990	Natural product mimic
679	Alpidem	Ananxyl	Anxiolytic	1991	Natural product mimic
680	Centchroman	Choice 7	Contraception	1991	Natural product mimic
681	Clospipramine hydrochloride	Cremin	Neuroleptic	1991	Natural product mimic
682	Dezocine	Dalgan	Analgesic	1991	Natural product mimic
683	Nemonapride	Emilace	Neuroleptic	1991	Natural product mimic
684	Paroxetine	Seroxat	Antidepressant	1991	Natural product mimic
685	Terbinafine hydrochloride	Lamisil	Antifungal	1991	Natural product mimic
686	Butenafine hydrochloride	Mentax	Antifungal	1992	Natural product mimic
687	Epalrestat	Kinedak	Antidiabetic	1992	Natural product mimic
688	Risperidone	Risperdal	Neuroleptic	1993	Natural product mimic
689	Sarpogrelate HCI	Anplag	Platelet aggreg. inhib.	1993	Natural product mimic
690	Tacrine HCI	Cognex	Alzheimer's disease	1993	Natural product mimic
691	Tamsulosin HCI	Harnal	Benign prostatic hypertrophy	1993	Natural product mimic
692	Broxaterol	Summair	Antiasthmatic	1994	Natural product mimic
693	Losartan potassium	Cozaar	Antihypertensive	1994	Natural product mimic
694	Mirtazapine	Remeron	Antidepressant	1994	Natural product mimic
695	Nefazodone hydrochloride	Serzone	Antidepressant	1994	Natural product mimic
696	Venlafaxine	Effexor	Antidepressant	1994	Natural product mimic
697	Anastrozole	Arimidex	Anticancer	1995	Natural product mimic
698	Bicalutamide	Casodex	Anticancer	1995	Natural product mimic
699	Dorzolamide HCl	Trusopt	Antiglaucoma	1995	Natural product mimic

700	Fadrozole HCl	Afema	Anticancer	1995	Natural product mimic
701	Seratrodast	Bronica	Antiasthmatic	1995	Natural product mimic
702	Brimonidine	Alphagan	Antiglaucoma	1996	Natural product mimic
703	Letrazole	Femara	Anticancer	1996	Natural product mimic
704	Meloxicam	Mobec	Antiarthritic	1996	Natural product mimic
705	Olanzapine	Zyprexa	Neuroleptic	1996	Natural product mimic
706	Talipexole	Domin	Parkinson's disease	1996	Natural product mimic
707	Tandospirone	Sediel	Anxiolytic	1996	Natural product mimic
708	Valsartan	Diovan	Antihypertensive	1996	Natural product mimic
709	Zafirlukast	Accolate	Antiasthmatic	1996	Natural product mimic
710	Budipine	Parkinsan	Parkinson's disease	1997	Natural product mimic
711	Candesartan cilexetil	Atacand	Antihypertensive	1997	Natural product mimic
712	Dexfenfluramine	Redux	Antiobesity	1997	Natural product mimic
713	Donepezil HCl	Aricept	Alzheimer's disease	1997	Natural product mimic
714	Eprosartan	Teveten	Antihypertensive	1997	Natural product mimic
715	Irbesartan	Avapro	Antihypertensive	1997	Natural product mimic
716	Milnacipran	Ixel	Antidepressant	1997	Natural product mimic
717	Nebivolol	Nebilet	Antihypertensive	1997	Natural product mimic
718	Pramipexole HCl	Mirapex	Parkinson's disease	1997	Natural product mimic
719	Quetiapine fumarate	Seroquel	Neuroleptic	1997	Natural product mimic
720	Reboxetine	Edronax	Antidepressant	1997	Natural product mimic
721	Rivastigmin	Exelon	Alzheimer's disease	1997	Natural product mimic
722	Tolterodine	Detrusitol	Urinary incontinence	1997	Natural product mimic
723	Troglitazone	Rezulin	Antidiabetic	1997	Natural product mimic
724	Zileuton	Zyflo	Antiasthmatic	1997	Natural product mimic
725	Brinzolamide	Azopt	Antiglaucoma	1998	Natural product mimic
726	Clopidogrel hydrogensulfate	Plavix	Antithrombotic	1998	Natural product mimic
727	Fenoldopam mesylate	Corlopam	Antihypertensive	1998	Natural product mimic
728	Fomepizole	Antizol	Antidote	1998	Natural product mimic
729	Leflunomide	Arava	Antiarthritic	1998	Natural product mimic
730	Montelukast sodium	Singulair	Antiasthmatic	1998	Natural product mimic
731	Mosapride citrate	Gasmotin	Gastroprokinetic	1998	Natural product mimic
732	Raloxifene HCl	Evista	Calcium metabolism	1998	Natural product mimic
733	Sibutramine	Meridia	Antiobesity	1998	Natural product mimic
734	Tirofiban HCl	Aggrastat	Antithrombotic	1998	Natural product mimic

Table 13.2 (*Continued*)

No.	Generic name	Trade name	Disease	Year	Source
735	Celecoxib	Celebrex	Antiarthritic	1999	Natural product mimic
736	Naftopidil	Flivas	Dysuria	1999	Natural product mimic
737	Rofecoxib	Vioxx	Antiarthritic	1999	Natural product mimic
738	Rosiglitazone maleate	Avandia	Antidiabetic	1999	Natural product mimic
739	Telmisartan	Micardis	Antihypertensive	1999	Natural product mimic
740	Dexmedetomidine HCl	Precedex	Sedative	2000	Natural product mimic
741	Levosimendan	Simdax	MI, acute	2000	Natural product mimic
742	Liranaftate	Zefnart	Antifungal	2000	Natural product mimic
743	Bosentan	Tra-cleer	Antihypertensive	2001	Natural product mimic
744	Imatinib mesilate	Gleevec	Anticancer	2001	Natural product mimic
745	Aripiprazole	Abilify	Neuroleptic	2002	Natural product mimic
746	Etoricoxib	Arcoxia	Antiarthritic	2002	Natural product mimic
747	Gefitinib	Iressa	Anticancer	2002	Natural product mimic
748	Olmesartan medoxil	Benicar	Antihypertensive	2002	Natural product mimic
749	Parecoxib sodium	Dynastat	Analgesic	2002	Natural product mimic
750	Sivelestat sodium hydrate	Elaspol	Respiratory distress syndrome	2002	Natural product mimic
751	Temoporfin	Foscan	Anticancer	2002	Natural product mimic
752	Valdecoxib	Bextra	Antiarthritic	2002	Natural product mimic
753	Aprepitant	Emend	Antiemetic	2003	Natural product mimic
754	Bortezomib	Velcade	Anticancer	2003	Natural product mimic
755	Duloxetine hydrochloride	Cymbalta	Antidepressant	2004	Natural product mimic
756	Erlotinib hydrochloride	Tarceva	Anticancer	2004	Natural product mimic
757	Darifenacin hydrobromide	Emselex	Urinary incontinence	2005	Natural product mimic
758	Lumiracoxib	Prexige	Osteoarthritis	2005	Natural product mimic
759	Ramelteon	Rozerem	Insomnia	2005	Natural product mimic
760	Conivaptan HCl	Vaprisol	Diuretic	2006	Natural product mimic
761	Darunavir	Prezista	Antiviral	2006	Natural product mimic
762	Dasatinib	Sprycel	Anticancer	2006	Natural product mimic
763	Mozavaptan	Physuline	Diuretic	2006	Natural product mimic
764	Rimonabant	Acomplia	Antiobesity	2006	Natural product mimic
765	Rotigotine	Neupro	Parkinson's disease	2006	Natural product mimic

766	Sitagliptin	Januvia	Antidiabetic	2006	Natural product mimic
767	Sitaxsentan sodium	Thelin	PAH	2006	Natural product mimic
768	Sunitinib malate	Sutent	Anticancer	2006	Natural product mimic
769	Varenicline tartrate	Chantix	Smoking cessation	2006	Natural product mimic
770	Ambrisentan	Letairis	PAH	2007	Natural product mimic
771	Lapatinib ditosylate	Tykerb	Anticancer	2007	Natural product mimic
772	Nilotinib hydrochloride	Tasigna	Anticancer	2007	Natural product mimic
773	Paliperidone	Invega	Schizophrenia	2007	Natural product mimic
774	Vildagliptin	Galvus	Antidiabetic	2007	Natural product mimic
775	Blonanserin	Lonasen	Antipsychotic	2008	Natural product mimic
776	Desvenlafaxine succinate	Pristiq	Antidepressant	2008	Natural product mimic
777	Fesoterodine fumarate	Toviaz	Urinary incontinence	2008	Natural product mimic
778	Fosaprepitant dimeglumine	Ivemend	Antiemetic	2008	Natural product mimic
779	Asenapine maleate	Saphris	Antipsychotic	2009	Natural product mimic
780	Dapoxetine HCl	Priligy	Premature ejaculation	2009	Natural product mimic
781	Febuxostat	Adenuric	Hyperuricemia	2009	Natural product mimic
782	Lasofoxifene tartrate	Fablyn	Calcium metabolism	2009	Natural product mimic
783	Niacin/laropiprant	Tredaptive	Hypocholesterolemic	2009	Natural product mimic
784	Pazopanib	Votrient	Anticancer	2009	Natural product mimic
785	Prasugrel	Efient	Antithrombotic	2009	Natural product mimic
786	Saxagliptin	Onglyza	Antidiabetic	2009	Natural product mimic
787	Tapentadol HCL	Nucynta	Analgesic	2009	Natural product mimic
788	Tolvaptan	Samsca	Hyponatremia	2009	Natural product mimic
789	Alogliptin benzoate	Nesina	Antidiabetic	2010	Natural product mimic
790	Bazedoxifene acetate	Conbriza	Calcium metabolism	2010	Natural product mimic
791	Iloperidone	Fanapt	Antipsychotic	2010	Natural product mimic
792	Peramivir	PeramiFlu	Antiviral	2010	Natural product mimic
793	Prucalopride	Resolor	Constipation	2010	Natural product mimic
794	Roflumilast	Daxas	COPD	2010	Natural product mimic
795	Linagliptin	Tradjenta	Antidiabetic	2011	Natural product mimic
796	Lurasidone hydrochloride	Latuda	Antipsychotic	2011	Natural product mimic
797	Vilazodone hydrochloride	Vibryd	Antidepressant	2011	Natural product mimic
798	Aclidinium bromide	Tudorza	COPD	2012	Natural product mimic
799	Anagliptin	Suiny	Antidiabetic	2012	Natural product mimic
800	Azilsartan medoxomil	Azilva	Antihypertensive	2012	Natural product mimic

Table 13.2 (*Continued*)

No.	Generic name	Trade name	Disease	Year	Source
801	Enzalutamide	Xtandi	Anticancer	2012	Natural product mimic
802	Gemiglptin	Zemiglo	Antidiabetic	2012	Natural product mimic
803	Lorcaserin hydrochloride	Belviq	Anti-obesity	2012	Natural product mimic
804	Teneligliptin hydrobromide	Tenelia	Antidiabetic	2012	Natural product mimic
805	Cobicistat	Tybost	Antiviral	2013	Natural product mimic
806	Levomilnacipran	Fetzima	Antidepressant	2013	Natural product mimic
807	Macitentan	Opsumit	PAH	2013	Natural product mimic
808	Ospemifene	Osphena	Dyspareunia	2013	Natural product mimic
809	Topiroxostat	Topiloric	Gout	2013	Natural product mimic
810	Vortioxetine	Brintellix	Antidepressant	2013	Natural product mimic
811	Alectinib hydrochloride	Alecensa	Anticancer	2014	Natural product mimic
812	Apatinib mesylate		Anticancer	2014	Natural product mimic
813	Apremilast	Otezla	Antipsoriatic	2014	Natural product mimic
814	Asunaprevir	Sunvepra	Antiviral	2014	Natural product mimic
815	Ceritinib	Zykadia	Anticancer	2014	Natural product mimic
816	Ledipasvir	Harvoni	Antiviral	2014	Natural product mimic
817	Netupitent	Akynzeo	Antinauseant	2014	Natural product mimic
818	Nintedanib	Ofev	Idiopathic pulmonary fibrosis	2014	Natural product mimic
819	Suvorexant	Belsomra	Insomnia	2014	Natural product mimic
820	Tasimelteon	Hetlioz	Free-running circadian disorder	2014	Natural product mimic
821	Umeclidinium bromide	Incruse	COPD	2014	Natural product mimic
822	Aripiprazole laurate	Aristada	Schizophrenia	2015	Natural product mimic
823	Brexpiprazole	Rexulti	Schizophrenia	2015	Natural product mimic
824	Cariprazine hydrochloride	Vraylar	Schizophrenia	2015	Natural product mimic
825	Chidamide	Epidaza	Anticancer	2015	Natural product mimic
826	Evogliptin hydrochloride	Suganon	Antidiabetic	2015	Natural product mimic
827	Flibanserin	Addyi	Female sexual dysfunction	2015	Natural product mimic
828	Ixazomib citrate	Ninlaro	Anticancer	2015	Natural product mimic
829	Lusutrombopag	Mulpleta	Idiopathic thrombocytopenia	2015	Natural product mimic
830	Omarigliptin	Marizev	Antidiabetic	2015	Natural product mimic
831	Panobinostat lactate	Farydak	Anticancer	2015	Natural product mimic
832	Polmacoxib	Acelex	Antiinflammatory	2015	Natural product mimic

833	Rolapitant hydrochloride	Varubi	Antinauseant	2015	Natural product mimic
834	Trelagliptin succinate	Zafatek	Antidiabetic	2015	Natural product mimic
835	Crisaborole	Eucrisa	Antipsoriatic	2016	Natural product mimic
836	Elbasvir	Erelsa	Antiviral	2016	Natural product mimic
837	Gosogliptin hydrochloride	SatRx	Antidiabetic	2016	Natural product mimic
838	Grazoprevir	Grazyna	Antiviral	2016	Natural product mimic
839	Lifitegrast	Xiidra	Dry eye syndrome	2016	Natural product mimic
840	Pimavanserin tartrate	Nuplazid	Parkinson's disease	2016	Natural product mimic
841	Pitolisant hydrochloride	Wakix	Antinarcolepsy	2016	Natural product mimic
842	Betrixaban	Bevyxxa	Anticoagulant	2017	Natural product mimic
843	Enasidenib mesylate	Idhifa	Anticancer	2017	Natural product mimic
844	Niraparib	Zejula	Anticancer	2017	Natural product mimic
845	Solifenacin tartrate	Vesigeum	Urinary incontinence	2017	Natural product mimic
846	Apalutamide	Erleada	Anticancer	2018	Natural product mimic
847	Avatrombopag maleate	Doptelet	Idiopathic thrombocytopenia	2018	Natural product mimic
848	Baloxavir marboxil	Xofluza	Antiviral	2018	Natural product mimic
849	Binimetinib	Mektovi	Anticancer	2018	Natural product mimic
850	Binimetinib	Mektovi	Anticancer	2018	Natural product mimic
851	Elagolix sodium	Orilissa	Endometriosis	2018	Natural product mimic
852	Elobixibat	Goofice	Chronic idiopathic constipation	2018	Natural product mimic
853	Encorafenib	Braftovi	Anticancer	2018	Natural product mimic
854	Glasdegib maleate	Daurismo	Anticancer	2018	Natural product mimic
855	Glycopyrrolate tosylate	Qbrexza	Hyperhydrosis	2018	Natural product mimic
856	Ivosidenib	Tibsovo	Anticancer	2018	Natural product mimic
857	Lorlatinib	Lorbrena	Anticancer	2018	Natural product mimic
858	Omidenepag isopropyl	Eybelis	Antiglaucoma	2018	Natural product mimic
859	Tecovirimat	Arestvyr	Antiviral	2018	Natural product mimic
860	Alpelisib	Piqray	Anticancer	2019	Natural product mimic
861	Darolutamide	Nubeqa	Anticancer	2019	Natural product mimic
862	Esaxerenone	Minnebro	Antihypertensive	2019	Natural product mimic
863	Pexidartinib	Turalio	Anticancer	2019	Natural product mimic
864	Qizartinib	Vanflyta	Anticancer	2019	Natural product mimic
865	Relugolix	Relumina	Uterine fibroids	2019	Natural product mimic

approaches for conducting preclinical and clinical research. One of the effective approaches could be to adopt retrospective studies in hospitals and clinicians' practice of a disease or herb. Since uniform practice methods are not adopted in documentation and record maintenance, this approach is challenging.[43] Analytical and chemical fingerprinting studies on ayurvedic phytochemicals could be a field of interest and investigation as they offer a direct path for the emergence of newer therapeutic agents. Furthermore, when chemical structures of the active molecules are identified, it provides a foundation platform for synthetic and medicinal chemists to enhance their efficacy or nullifies toxicity properties or even creates synthetic analogs with varied pharmacological actions targeting different diseases. For the convenience of Ayurvedic drug discovery, the plants widely used in Ayurveda could be classified into 2 groups.

(a) The plants with known active constituents for which sufficient leads are available and their biological targets have been identified. Some well-known Ayurvedic leads and their molecular targets in cancer therapeutics have been discussed in detail by Aggarwal *et al.*[44] A comprehensive summary of Ayurvedic plants and their biological targets are listed in Table 13.3.[31]
(b) Botanicals with known treatment targets and proven medicinal value based on scientific evidence are listed in Table 13.4.[31] The linkage between "the past" and "the future" of medicine could give us new directions for better understanding health, disease, and possible solutions.

13.6 Ayurveda-based Approach in Drug Discovery

13.6.1 Advantages

The plants used in Ayurveda have a long history of human use enriched with bioactive compounds belonging to wide chemical classes when compared to synthetic small molecules and the active molecules isolated from these plants are more likely to be safe and active. In addition, this minimizes the challenges in synthesizing new compounds and evaluating their efficacy and safety. Quinine from *Cinchona officinalis*, reserpine from *Rauwolfia serpentina*, and digitalis from *Digitalis purpurea* are some examples of the drugs identified from traditional knowledge.[45] In some instances, natural compounds with less potency and/or more toxicity are chemically modified to enhance their efficacy. Analogs of these compounds can be developed through chemical compounds followed by chemical derivatization through semi-synthesis using the botanical product as a starting point. One of the classic examples is Salicin from the Salix tree. This compound was hydrolyzed into salicylic acid and its semi-synthetic derivative is acetyl salicylic acid which has been developed as aspirin, a pain reliever.[46] There are numerous examples of natural product-based phytoconstituents that have been modified chemically, *viz.*, morphine (*Papaver somniferum*), colchicine (*Colchicum autumnale*), penicillin G (*Penicillium citrinum*), paclitaxel (*Taxus brevifolia*), metformin (*Galega officinalis*), *etc.*[20] The proven success of these examples set the stage for ongoing drug discovery efforts from natural resources.

13.6.2 Concerns

As medicinal plants gain popularity with increased scientific and commercial attention, there is a downside which is causing a burden on the wild, naturally grown plant populations from which these plants are harvested. Overharvesting leads to the risk of extinction of many valuable medicinal species. When these plants are exploited for commercial production, it results in the unavailability of indigenous medicine to native people who live in these areas and have relied on them for many generations. Unfortunately, about 15 000 medicinal plant species may be threatened with extinction worldwide and experts estimate that one potential drug is lost from the Earth every two years.[47] In India, the Government has banned the use of about 29 plants and their worthy products as these species are considered endangered.[48]

Discovering drugs from natural resources does come with some risks, in particular to the environment. To avoid the excessive harvesting of medicinal plants, trees, and shrubs, the production could be achieved through cultivation. If only the leaves and flowers are required, it is necessary to leave the root, and whole-plant harvesting results in the extinction of that species. Even though naturally occurring botanical compounds exhibit desirable relevant bioactivities, obtaining intellectual property (IP) rights for these nonmodified compounds is a hurdle as a new patent law doesn't permit patenting of naturally-occurring compounds in their original form, but legal frameworks vary between countries.[49]

13.7 Future of Botanical Drug Development

13.7.1 Pharmaceutical Industry – Future Trends

As per IQVIA's recent report, the global pharma market is expected to grow at the rate of 3–6% all the way through to 2025, and the total market size could reach about $1.6 trillion US dollars. This excludes spending on covid-19 vaccines, which is projected to be $157 billion. Immunology, oncology, and neurology are the therapeutic areas of focus and they represent the largest aggregate contributors and a fair share to the growth during the next 5 years from the continued flow of new medicines. Among these, oncology and immunology are forecast to grow at a rate of 9–12% until 2025. Oncology alone is projected to include about 100 new treatment drugs to the total spending of $260 billion US dollars. Neurology is expected to see novel migraine therapies and potential therapies for memory disorders, and

Table 13.3 Successful lead compounds isolated from Ayurvedic plant products, their biological targets, and therapeutic effects. Adapted from ref. 31 with permission from Taylor & Francis, Copyright 2007.

Plant name	Ayurvedic name	Lead compound (structures are given in the figure)	Biological targets/mechanism of action	Therapeutic effects
Adhatoda Vasika	*Vasaka*	Vasicine	Inhibition of histamine release, clearance of airway passages. Uterine stimulant	Bronchodilator and expectorant
Allium sativum	*Lasuna*	Allicin	Inhibition of cysteine protease and alcohol dehydrogenase enzymes	Antimicrobial
Andrographis paniculata	*Kalamegha*	Andrographolide	Inhibits NO synthesis in macrophages by reducing iNOS protein expression	Antiinflammatory
Bacopa monnieri	*Brahmi*	Bacosides	Repair of damaged neurons by enhancing kinase activity, neuronal synthesis, and restoration of synaptic activity, and ultimately nerve impulse transmission	Cognitive function enhancer
Bergenia ligulata	*Pashanbhedi*	Gallic acid	NFkB inhibitor, TNF-α induced p65 nuclear translocation	Anti-inflammatory, anti-allergic
Boswellia serrata	*Shallaki*	Boswellic acid	By non-redox inhibition of pro-inflammatory enzyme, 5-lipoxygenase, inhibition of NFkB	Anti-inflammatory, anti-arthritic
Commiphora mukul	*Guggulu*	Guggulsterone	Bile acid receptor antagonist, cox-2 inhibition, NFkB inhibition	Hypolipidemic agents, anti-inflammatory, promotes cardiovascular health
Curcuma longa	*Haridra*	Curcumin	Regulation of aryl hydrocarbon receptors, cytochrome P450 enzymes, GST kinases, COX, iNOS, MMP, p53 signaling, NFkB inhibition, apoptosis inducer	Cancer, inflammation, diabetes, and chemoprevention
Eugenia caryophyllata	*Lavangaha*	Eugenol	Antibacterial activity through membrane disruption	Toothache
Glychyrrhiza glabra	*Madhu yasti*	Glycyrrhizin	P21CIP1/WAF1 inhibitor	Constipation, body ache, *etc.*
Mucuna Pruriens	*Kapikachhu*	L-dopa	Production of dopamine	Parkinson's disease, nerve tonic

Picrorhiza Kurroa	*Kutki*	Kutkoside	Stimulate the action of nucleolar polymerase A, resulting in ribosomal protein synthesis and thus stimulates the regenerative ability of the liver and the formation of new hepatocytes	Hepatoprotective against jaundice, hepatitis
Piper longum	*Pipalli*	Piperin	Unknown	Bioenhancer
Punica granatum	*Dadima*	Ellagic acid	NFkB inhibition, Cox-2 inhibition	Anticancer, anti-inflammatory
Rauwolfia serpentina	*Sarpagandha*	Reserpine	Disrupts norepinephrine vesicular storage, adrenergic blocking, inhibition of the ATP/Mg^{2+} pump located in the presynaptic neuron	Antihypertensive, adrenergic uptake inhibitor, used for snake bites
Semecarpus anacardium	*Bhallataka*	Biflavonoids	Inhibition of NFkB	Anticancer, anti-inflammatory
Tinospora cordifolia	*Guduchi*	Tinocordifolin	Cox-2 inhibition, cytokine modulator, NFkB translocation	Asthma, rheumatism, immune stimulant
Vitis vinefera	*Draksa*	Resveratrol	NF-kB Inhibition, inhibition of the CYP1A1, COX-2 enzymes	Anticancer
Withania somnifera	*Aswagantha*	*Withaferin A*	Inhibition of IκB kinase β, NFκB	Anti-inflammatory
Zingiber officinale	*Sunthi*	Gingerols	Capsaicin-activated VR1 (Vannilloid) receptors agonists	Antipyretic, analgesic, antitussive, sedative

Table 13.4 List of Ayurvedic botanicals with potential for drug discovery. Adapted from ref. 31 with permission from Taylor & Francis, Copyright 2007.

Botanical name	Ayurvedic/ common name	Disease conditions
Acorus calamus	Vacha	Obesity, hypertension, speech disorders, sedative, transquilizer, mental distress
Adhatoda zeylanica	Vasa	Bronchiolar constriction
Anethum Sowa	Soya	Appetizer, carminative
Amorphophallus campanulatus	Soorana	Piles, amenorrhea, indigestion, and anorexia
Asparagus racemosus	Shatavari	Aphrodisiac, dysuria, galactagogue, leucorrohea, peptie ulcer
Aloe vera	Kumari	Promotes health, heals burns, skin diseases
Azadirachta indica	Nimba	Gastritis, bleeding piles, wounds, fumigative, disinfectant against bacterial and viral infections.
Berberis aristata	Daruharidra	Liver disorders, conjunctivitis, dysentery, anti-pyretic, antiperiodic, diaphoretic
Berginia ligulate	Pashanabheda	Stomach, urinary and urinogenital disorders
Boerhavia diffusa	Punarnava	Eye diseases, diabetes, anti-diuretic, hepatoprotective, inflammation
Carum Bulbocastanum	Krishna jiraka	Digestion disorders and urinary tract infections
Crataeva nurvala	Varun	Erysipelas, lymphadenopathy, urolithiasis
Cassia augustifolia	Rechani	Severe constipation
Cassia tora	Chakramard	Leprosy, ringworm infection, skin diseases
Centella asiatica	Mandukarparni	Neurological problems
Cleordendrum serratum	Bharangi	Common cold, sinusitis, chronic respiratory problems
Commiphora wightii	Balsamodendron mukul	Arthritis
Cuminium cyminum	Sweta jiraka	Chronic diarrhea and dyspepsia, diabetes, and cancer
Emblica officinalis	Amalaki (Amla)	Anemia, jaundice, hemorrhage, acidity, immune stimulant
Eugenia jambolana	Jamboo	Diabetes
Ferula foetida	Hingu (Heeng)	Septic conditions, digestive stimulant, expectorant, carminative
Garinia pedunculata	Amlavetasa	Cardiac stimulant, digestive, and laxative
Juniperous communis	Hapusha	Diuretic, anti-arthritis
Mallotus philippinensis	Kampillak (Kamila)	Anthelminthic, aphrodisiac
Moringa oleifera	Shigru (Sehanjana)	Abscess, septic conditions, wound healing, piles, inflammation, neuritis, joint diseases
Melia azedarach	Mahanimba	Acne, burns, measles, and prickly heat
Myristica fragrans	Jaiphal	nausea and vomiting, diarrhea, leucorrhea.
Nigella sativa	Kalajaji	Respiratory and digestive tract
Nelumbo nucifera	Kamal beej/ Kamal Gatta	Hypoglycemic
Piper longum (root)	Pippalimool	Malaria, insomnia, worms, piles

Table 13.4 (*Continued*)

Botanical name	Ayurvedic/ common name	Disease conditions
Phyllanthus emblica	Amla	Anti-aging
Pongamia pinnata	Karanja	Tumors, piles, skin diseases, and ulcers
Psoralea caryllifotia	Bakuchi	Vitiligo, bronchial asthma
Randia spinosa	Madanaphala	Gastrointestinal and hepatic disorders
Ricinus communis	Eranda Patra	Arthritis, muscle aches, constipation
Saraca asoca	Ashok	Meno-metrorrhagia, neurological disorders, spasmogenic, uterine infections
Terminalia arjuna	Arjun	Cardiac disorders
Terminalia chebula	Haritaki (harad)	Laxative
Trachyspermum ammi	Ajwain	Anthelminthic, spasmogenic carminative
Tribulus terrestris	Gokshur (Gokkru)	Diuretic, aphrodisiac, lithontriptic

Alzheimer's and Parkinson's diseases.[50] This forecast predicts a huge opportunity and an urgent need to discover new drugs from Ayurveda.

13.7.2 Current Challenges and Recommendations

In 2004, the FDA published draft guidelines for the development and approval of botanical-based medicines, and it provided the road map for researchers and manufacturers to follow. After the publication of these guidelines, although more than 400 investigational new drug (IND) applications were granted for botanical medicines, surprisingly only three medicines have since reached the new drug application (NDA) stage of development and received approval.

1. A green tea ointment derived from *Camellia Sinensis* (green tea), used for the treatment of genital warts (sinecatechins).
2. An extract of the latex of the "Devil's blood" tree (Crofelemer) used for the treatment of HIV/AIDS.
3. Cannabidiol (CBD), extracted from the cannabis plant and approved for the treatment of three rare epilepsy conditions in children.

Regulatory approval is problematic and almost impossible if the active principle is an extract, a mixture, or a composition rather than a pure isolated compound with a known structure. US FDA is understandably very reluctant to approve these drug combinations consisting of multiple molecules.[51] Polyphenon E, a topical antiviral ointment prepared from green tea catechins was the first botanical-based drug approved in 2006, after the publication of the FDA guidelines.[1] Cannabidiol was most recently approved as a botanical medicine but it presents as a classic example of various issues and challenges encountered during the development of herbal medicine. Despite Cannabidiol's successful emergence as a drug, another closely related major cannabinoid, tetrahydro cannabidiol (THC), couldn't move up the development ladder as it provokes seizures in toxicology studies. For these safety reasons, the FDA requires strong evidences for herbal medicines which is the same as the level of evidence required for synthetic compounds. This includes an understanding of drug metabolism, interactions, and mechanism of action. Another big challenging part of botanical medicine is to present the evidence that compounds from one batch and another are identical in composition and do not affect efficacy and safety.[52]

In addition, there are several problems arising in clinical trials of botanical drugs. In addition to authentication and adulteration issues, there is no specific protocol on how to conduct clinical trials for botanical drugs to accommodate their complex nature where the mechanism of action may differ from that of a single compound. Synergistic and antagonistic effects of these combo-mixtures could favor or be unfavorable for the desired outcome. Among a total of 195 botanical drug clinical trials registered with clinical trails.gov (between 2016 to 2019) only 81 trials are in advanced stages *viz.*, phase II or phase II/III. Among these, 95% of all phase II and phase II/III studies have only very few (~100) participants per arm 100, from which efficacy outcome is hard to determine with accuracy.[53] For ayurvedic drugs, a phase I trial is not necessary as they have an extensive history of human use and their efficacy is already proven. This is the biggest advantage of translating botanicals used in Ayurveda into modern drugs. It could save time and costs in drug development. Although ayurvedic research is carried out in various parts of India, due to a lack of appropriate communication, the outcome is so fragmented and often not deep in its own foundational theories or in its interaction with modern science.[54] Modern scientists often use Ayurvedic formulations as the starting point (similar to reverse pharmacology) and try to isolate the single-molecule and determine their molecular mechanisms and mode of action.[55]

The history of botanical drug usage differs greatly between countries. In China, like India, plant-based medicines have been used for thousands of years and the knowledge has been passed from one generation to the next in the form of case studies and experiences from physicians. The difficulty lies in transforming this wealth of experience into transparent, auditable, reproducible, high-quality data, known as real-world data (RWD) which is essential to analyze and generate safety and efficacy evidence in order to attain regulatory approval.[54] China, which conducts 50% of registered botanical drug clinical trials, is implementing an Electronic Medical Record (EMR) system to standardize their results and data.[56] Similarly, if India also could translate all Ayurvedic clinical-use data into an Electronic Medical record, this would speed up the process of discovering modern drugs from our traditional knowledge. Although translating Ayurveda to comply with modern drug standards is difficult, it is doable with dedication and efforts.

For botanical drugs with known active ingredients and molecular structure, Randomized Controlled Trials (RCT) are the best choice as regulatory agencies like the FDA are familiar with this pattern, trial design, and the regulators could understand and review the entire process. In addition, it can be done with a small number of patients in minimal time. Inclusion criteria require information about target binding sites and the possible biological interactions and endpoint should be measurable.[54] For botanical drugs with unknown active ingredients, it is a challenge to determine appropriate inclusion/exclusion criteria and trial size. Sun and Qian[54] suggested conducting a controlled observational phase II/III study with a large sample size that could broaden the eligibility criteria and increase enrollments to include different types of evidence, which may increase the understanding of the botanicals benefit-risk profile. Since active ingredients are well known for our Ayurvedic medicines, biological tests could be performed (as shown in Figure 13.1) to determine their target binding sites and interactions. This would help to choose RCTs and generate EMR for further regulatory approval.

13.8 Conclusion – Could West Meet East?

The clinical efficacy of Ayurvedic medicines is well established and they have been proven to be safe and could help to alleviate the burden of mortality and morbidity caused by modern drugs accompanied with numerous side effects. Furthermore, they are effective against several infectious diseases where these harmful microbes developed drug resistance. Botanical drugs hold the best options for finding novel drugs when worked from the basis of ayurvedic knowledge using advanced technological tools and in conjunction with drug discovery scientists and it could offer the potential to discover targeted drugs effective against a variety of human diseases. These futuristic goals can be achieved when biomedical scientists dive deep into the system, principles, and knowledge of Ayurveda and continue their research upon exploring the scientific basis of their mechanism of action. Interdisciplinary research is the pathway to the success of integrating this old value system of medicine into modern drug discovery for wider acceptance.

References

1. T. W. Corson and C. M. Crews, Molecular understanding and modern application of traditional medicines: triumphs and trials, *Cell*, 2007, **130**(5), 769–774.
2. L. U. Jean, M. Kerri, I. Smith and G. Warren, Traditional Drug Development Process in Actuarial Perspectives on Prescription drug financing, May 2020, https://theactuarymagazine.org/traditional-drug-development-process/.
3. The drug development process: Step 3: Clinical research US Food and Drug Administration. 4 January 2018. Retrieved 23 December 2021.
4. P. E. Rajasekaran, Herbal medicine, in *World of Science*, Employment News, 2002, p. 3.

5. S. V. Chennupati, Traditional medicine (Ayurveda): Inspired approaches to drug discovery, International Conference on Pharmacognosy, Phytochemistry and Natural Products, *Nat. Prod. Chem. Res.*, 2015, **3**, 6.
6. B. Patwardhan and R. A. Mashelkar, Traditional medicine-inspired approaches to drug discovery: can Ayurveda show the way forward?, *Drug Discovery Today*, 2009, **14**(15–16), 804–811.
7. J. Chen, X. Luo, H. Qiu, V. Mackey, L. Sun and X. Ouyang, Drug discovery and drug marketing with the critical roles of modern administration, *Am. J. Transl. Res.*, 2018, **10**(12), 4302.
8. J. P. Hughes, S. Rees, S. B. Kalindjian and K. L. Philpott, Principles of early drug discovery, *Br. J. Pharmacol.*, 2011, **162**(6), 1239–1249.
9. J. A. DiMasi, H. G. Grabowski and R. W. Hansen, Innovation in the pharmaceutical industry: new estimates of R&D costs, *J. Health Econ.*, 2016, **47**, 20–33.
10. V. Prasad and S. Mailankody, Research and development spending to bring a single cancer drug to market and revenues after approval, *JAMA Intern. Med.*, 2017, **177**(11), 1569–1575.
11. K. Jayasundara, A. Hollis, M. Krahn, M. Mamdani, J. S. Hoch and P. Grootendorst, Estimating the clinical cost of drug development for orphan versus non-orphan drugs, *Orphanet J. Rare Dis.*, 2019, **14**(1), 1–10.
12. V. Lad, Ayurveda: A brief introduction and guide, The Ayurvedic Institute, https://www.ayurveda.com/ayurveda-a-brief-introduction-and-guide/.
13. Y. S. Jaiswal and L. L. Williams, A glimpse of Ayurveda–The forgotten history and principles of Indian traditional medicine, *J. Tradit. Complementary Altern. Med.*, 2017, 7(1), 50–53.
14. D. A. Dias, S. Urban and U. Roessner, A historical overview of natural products in drug discovery, *Metabolites*, 2012, **2**(2), 303–336.
15. A. M. Rouhi, Rediscovering natural products, *Chem. Eng. News*, 2003, **81**(41), 77–91.
16. A. N. M. Alamgir, Drugs: their natural, synthetic, and biosynthetic sources, in *Therapeutic use of medicinal plants and their extracts:* Springer, Cham, 2017, Volume 1, 105–123.
17. S. Pal, Recent trends in approvals of novel drugs, *US Pharm.*, 2021, **46**, 14.
18. K. E. Lasser, P. D. Allen, S. J. Woolhandler, D. U. Himmelstein, S. M. Wolfe and D. H. Bor, Timing of new black box warnings and withdrawals for prescription medications, *Jama*, 2002, **287**(17), 2215–2220.
19. B. Patwardhan and R. A. Mashelkar, Traditional medicine-inspired approaches to drug discovery: can Ayurveda show the way forward?, *Drug Discovery Today*, 2009, **14**(15–16), 804–811.
20. S. B. Mishra, A. Mukerjee and S. Singh, Global Approach for Drug Discovery and Development from Indian Traditional Medicine, in *In-Evidence Based Validation of Traditional Medicines*, Springer, Singapore, 2021, pp. 3–27.

21. P. D. Gupta, Pharmacogenetics, pharmacogenomics and ayurgenomics for personalized medicine: a paradigm shift, *Indian J. Pharm. Sci.*, 2015, **77**(2), 135.
22. B. Patwardhan, Ayugenomic: integration of customized medicine, *Indian J. Nat. Prod.*, 2003, **19**, 16–23.
23. B. Patwardhan and G. Bodeker, Ayurvedic genomics: establishing a genetic basis for mind–body typologies, *J. Altern. Complementary Med.*, 2008, **14**(5), 571–576.
24. H. Bhargav, N. Jasti, P. More, V. Kumar, U. Chikkanna, R. K. Kumar and S. Varambally, Correlation of prakriti diagnosis using AyuSoft prakriti diagnostic tool with clinician rating in patients with psychiatric disorders, *J. Ayurveda Integr. Med.*, 2021, **12**(2), 365–368.
25. N. Verma, Current regulatory challenges and approaches in the registration of herbal drugs in Europe, *Clin. Res. Regul. Aff.*, 2016, **33**(1), 9–24.
26. H. Rotti, R. Raval, S. Anchan, R. Bellampalli, S. Bhale, R. Bharadwaj, B. K. Bhat, A. P. Dedge, V. R. Dhumal, G. G. Gangadharan and T. K. Girijakumari, Determinants of prakriti, the human constitution types of Indian traditional medicine and its correlation with contemporary science, *J. Ayurveda Integr. Med.*, 2014, **5**(3), 167.
27. H. Rotti, S. Mallya, S. P. Kabekkodu, S. Chakrabarty, S. Bhale, R. Bharadwaj, B. K. Bhat, A. P. Dedge, V. R. Dhumal, G. G. Gangadharan and P. M. Gopinath, DNA methylation analysis of phenotype specific stratified Indian population, *J. Transl. Med.*, 2015, **13**(1), 1–2.
28. P. Govindaraj, S. Nizamuddin, A. Sharath, V. Jyothi, H. Rotti, R. Raval, J. Nayak, B. K. Bhat, B. V. Prasanna, P. Shintre and M. Sule, Genome-wide analysis correlates Ayurveda Prakriti, *Sci. Rep.*, 2015, **5**(1), 1–2.
29. N. R. Farnsworth, O. Akerele, A. S. Bingel, D. D. Soejarto and Z. Guo, Medicinal plants in therapy, *Bull. W. H. O.*, 1985, **63**(6), 965.
30. A. G. Atanasov, S. B. Zotchev, V. M. Dirsch and C. T. Supuran, Natural products in drug discovery: Advances and opportunities, *Nat. Rev. Drug Discovery*, 2021, **20**(3), 200–216.
31. P. Balachandran and R. Govindarajan, Ayurvedic drug discovery, *Expert Opin. Drug Discovery*, 2007, **2**(12), 1631–1652.
32. A. Vaidya, *Reverse Pharmacology approach, CSIR NMITLI Herbal Drug Development Programme*, 2002.
33. A. D. Vaidya, Reverse pharmacological correlates of ayurvedic drug actions, *Indian J. Pharmacol.*, 2006, **38**(5), 311.
34. I. Kola and J. Landis, Can the pharmaceutical industry reduce attrition rates?, *Nat. Rev. Drug Discovery*, 2004, **3**(8), 711–716.
35. *Sushruta Samhita, Sutra Sthana*, ed. P. V. Sharma, Choukhamba Visvabharati, Varanasi, 2013, ch. 1, ver. 22, vol. 1, p. 16.
36. T. V. Padma, Ayurveda, *Nature*, 2005, **436**(7050), 486.
37. C. A. Lipinski, F. Lombardo, B. W. Dominy and P. J. Feeney, Experimental and computational approaches to estimate solubility and

permeability in drug discovery and development settings, *Adv. Drug Delivery Rev.*, 1997, **23**(1–3), 3–25.

38. B. Patwardhan, A. D. Vaidya and M. Chorghade, Ayurveda and natural products drug discovery, *Curr. sci.*, 2004, 789–799.
39. R. A. Mashelkar, *Chitrakoot declaration*, Convention of National Botanical Research Institute, NBRI, Lucknow, 2003.
40. *Charaka Samhita of Agnivesha, Sutra Sthana*, ed. P. V. Sharma, Choukhamba Orientalia, Varanasi, 1995, ch. 11, ver. 2, p. 114.
41. A. Chauhan, D. K. Semwal, S. P. Mishra and R. B. Semwal, Ayurvedic research and methodology: Present status and future strategies, *Ayu*, 2015, **36**(4), 364.
42. D. J. Newman and G. M. Cragg, Natural products as sources of new drugs over the nearly four decades from 01/1981 to 09/2019, *J. Nat. Prod.*, 2020, **83**(3), 770–803.
43. B. Patwardhan and K. Khambholja, Drug Discovery and Ayurveda: Win-Win Relationship Between Contemporary and Ancient Sciences, in *Drug Discovery and Development - Present and Future*, 2011, ed. I. Kapetanović, ISBN: 978-953-307-615-7.
44. B. B. Aggarwal, H. Ichikawa, P. Garodia, P. Weerasinghe, G. Sethi, I. D. Bhatt, M. K. Pandey, S. Shishodia and M. G. Nair, From traditional Ayurvedic medicine to modern medicine: identification of therapeutic targets for suppression of inflammation and cancer, *Expert Opin. Ther. Targets*, 2006, **10**(1), 87–118.
45. A. G. Atanasov, B. Waltenberger, E. M. Pferschy-Wenzig, T. Linder, C. Wawrosch, P. Uhrin, V. Temml, L. Wang, S. Schwaiger, E. H. Heiss and J. M. Rollinger, Discovery and resupply of pharmacologically active plant-derived natural products: A review, *Biotechnol. Adv.*, 2015, **33**(8), 1582–1614.
46. F. Jamshidi-Kia, Z. Lorigooini and H. Amini-Khoei, Medicinal plants: Past history and future perspective, *J. HerbMed Pharmacol.*, 2018, **7**(1), 1–7.
47. E. Roberson, Medicinal plants at risk, Nature's pharmacy, our treasure chest, Why we must conserve our natural heritage, A native plant conservation campaign report, 2008, https://www.biologicaldiversity.org/publications/papers/Medicinal_Plants_042008_lores.pdf.
48. S. Sen and R. Chakraborty, Revival, modernization and integration of Indian traditional herbal medicine in clinical practice: Importance, challenges and future, *J. Tradit. Complementary Med.*, 2017, 7(2), 234–244.
49. C. Harrison, Patenting natural products just got harder, *Nat. Biotechnol.*, 2014, **32**(5), 403–405.
50. https://www.iqvia.com/insights/the-iqvia-institute/reports/global-medicine-spending-and-usage-trends-outlook-to-2025.
51. B. M. Schmidt, D. M. Ribnicky, P. E. Lipsky and I. Raskin, Revisiting the ancient concept of botanical therapeutics, *Nat. Chem. Biol.*, 2007, **3**(7), 360–366.
52. S. Wright, The development of plant-based medicines for prescription. in Technology networks drug discovery, Published March 2021,

https://www.technologynetworks.com/drug-discovery/blog/the-development-of-plant-based-medicines-for-prescription-346230.
53. Y. Sun and J. Qian, Botanical drug clinical trial: Common issues and future options, *Acta Pharm. Sin. B*, 2021, **11**(1), 300–303.
54. B. Patwardhan, Good publications need good research, *J. Ayurveda Integr. Med.*, 2015, **6**(2), 73.
55. S. Ramaswamy, Reflections on current Ayurveda research, *J. Ayurveda Integr. Med.*, 2018, **9**(4), 250–251.
56. J. Liang, Y. Li, Z. Zhang, D. Shen, J. Xu, X. Zheng, T. Wang, B. Tang, J. Lei and J. Zhang, Adoption of Electronic Health Records (EHRs) in China During the Past 10 Years: Consecutive Survey Data Analysis and Comparison of Sino-American Challenges and Experiences, *J. Med. Int. Res.*, 2021, **23**(2), e24813.

CHAPTER 14

Ayurvedic Formulations and Their Clinical Uses

KARTHIK VARMA A. C.,* SASIKUMAR KUTTAPPAN, GEORGE JOSEPH, SHINTU JUDE AND AUGUSTINE AMALRAJ

R&D Centre, Aurea Biolabs (P) Ltd, Kolenchery, Cochin-682 311, Kerala, India
*Email: ackvarmaapm@gmail.com

14.1 Introduction

Traditional systems of medicine have always gained the accolade from ancient days onwards because of the wide range of pharmacological activities. Herbs and spices have special socio-economic importance and are a traditional system of medicine.[1] The diverse range of phytochemicals is at the root of the modern nutraceutical industry for their multi varied uses.[2] Plant-derived classes of phytochemicals include polyphenols, flavonoids, tannins, minerals, vitamins, *etc.*[3]

Ayurveda has marked its potential as a traditional Indian system of medicine, which is mainly composed of polyherbal formulations for treating various disorders. Literature suggests that Ayurveda uses almost 1200 species of plant followed by 900 for Siddha and 700 for Unani.[4] The majority of the Asian population relies on the use of traditional systems of medicine.[5] Ayurveda is described as the "science of life" owing to its effects on the mental, physical, spiritual, and social wellbeing of human beings.[6]

Ayurvedic systems of medicine concentrate of the development of a product covering the three doshas and the majority of them are not clinically evaluated for their indented use. A modern system of medicine concentrates on the use of specific bioactive components, which are attributed to the desired pharmacological activities.

Chemistry, Biological Activities and Therapeutic Applications of Medicinal Plants in Ayurveda
Edited by Augustine Amalraj, Sasikumar Kuttappan and Karthik Varma A. C.

Published by the Royal Society of Chemistry, www.rsc.org

This chapter tries to present an overview on the need for clinical trials, the phytochemicals used in ayurvedic medicines, the hurdles in the clinical trials, and the future prospective for the need of clinical trials for product development.[7]

14.2 The Need for Randomized Clinical Trials (RCT) and Compatibility With Ayurveda

A randomized clinical trial (RCT) is one of the predominant study designs that are commonly employed to evaluate the pharmacokinetics and pharmacodynamics profile of any single molecule or for a polyherbal formulation. A typical RCT undergoes various steps like drafting of the study design, which includes protocol development, enrolment of subjects, and drafting of the primary and secondary outcomes.[7] A study conducted by Manohar has identified two major concerns in designing the RCT for Ayurveda products: (1) complex diagnostic assessment in ayurveda that significantly differs from allopathic diagnosis and (2) multiple complexity of treatments offered in Ayurveda.[8] Charaka Samhita, one of the classical texts in ayurveda describes that ten important patient assessments are required in addition to clinical assessment, which include prakriti (physico-mental constitution), agni (metabolic profile), samhanana (anthropometry), *etc.*[9] Similarly, Patwardhan *et al.*[10] pointed out that a clinical trial in allopathy is restricted to the evaluation of the effect of a single herb, but in ayurvedic practices of medicine the drug interventions are complex and often involve evaluation of multiple medications, physical therapies, and diet and lifestyle.[10] Results from a study also showed that the study designs in Ayurveda are further complicated by the dynamic assessment of progression and regression of disease in real-time clinical practice that requires a periodic assessment and modification of therapeutic interventions even during the course of the trial.[11] Fonnebo *et al.*[12] pointed out that studies claiming little/no efficacy of alternative medicine therapies further supported this argument, and reports of substantial benefits in real-time practice show there is a mismatch between current RCT evaluation of Ayurveda therapies and drugs.[12] Some of the other struggles in designing Ayurvedic trials also include:

i. One of the major hurdles in designing the study design involves standardization of bioactive components in polyherbal formulation. In Ayurveda practices the system involves the mixing of the herbs in a fixed ratio and is not based on the percentage of bioactives.
ii. Studies have also established hurdles in the estimation of "active" molecules in body fluids in the pharmacokinetic studies in phase I of drug development in ayurvedic practices.
iii. Dose fixation is also one of the biggest hurdles in Ayurveda as many of the formulations use crude herbs instead of using concentrated extracts – the use of crude raw herb can have a high degree of variation based on the geographic and seasonal variation.

iv. Ayurvedic practices involve supplementation of the drug based on the body nature of particular individuals and not on a group of the population.
v. Another drawback in the traditional systems of medicines is the method of treatment as many traditional systems of medicine concentrate on a holistic approach which involves treating the mind, body, and soul at the same time.

14.3 Ayurvedic Medicinal Plants in Preclinical and Clinical Trials

The boom and efficacy of traditional plant-based systems of medicine, including ayurvedic drugs, is gaining much importance nowadays because of their efficacy. Traditional ayurvedic medicines are evaluated for their efficacy in various health disorders like bronchial asthma, rheumatoid arthritis, ischemic heart disease, *etc.*[13–16] The most common herbs which are evaluated in ayurvedic systems of medicine include turmeric, ginger, boswellia, pepper, ashwagandha, shatavari, *etc.* for their biological activities. The majority of the ayurvedic drugs are polyherbal formulations whose combined synergistic action contributes to the bioefficacy of the product. The US National Centre for Complementary and Alternative Medicine has given funding for several research studies based on the ayurvedic system of medicine, out of which the most evaluated include turmeric, ginger, black pepper, and boswellia. Most of the ayurvedic preparations are based on the concept of polyherbal formulations.

14.3.1 Polyherbal Formulations

Triphala, one of the most well-known formulations in ayurvedic preparations made using the fruits of *Terminalia chebula*, *Terminalia bellirica*, and *Emblica officinalis* was clinically evaluated for its positive role in controlling constipation and gastric problems. The study showed promising effects on the role of the formulation in controlling the symptoms associated with constipation and gastric problems.[17] Human studies involving a triphala water decoction found it to be as effective as chlorhexidine gluconate germicidal mouthwash in the prevention of dental cavities. Studies revealed that Triphala at a dosage of 6% mouthwash promoted a significant reduction in oral streptococcus colonies.[18] Triphala mouthwash was also found to reverse precancerous oral lesions associated with tobacco usage in the young Indian population.[19] The most promising effect of Triphala is in weight management. In a 12 week, double blind, randomized, placebo-controlled trial, human subjects treated with triphala was found to decrease their body weight by approximately 5 kg compared to the placebo group.[20] Triphala also exhibits a promising role in reducing the blood glucose levels. A human clinical study on non-insulin dependent diabetes mellitus subjects showed that supplementation of 5 g of Triphala for a duration of 45 days significantly lowered blood glucose levels.[21]

Trikatu, another promising herbal formulation comprising black pepper, long pepper, and ginger, is also used for many disorders in ayurvedic formulations. A human study conducted by Saxena *et al.* evaluated that Trikatu Churna was found to be effective in the management of sthaulya. Twenty-four subjects who were supplemented with Trikatu churna had effectively lowered signs and symptoms of sthaulya.[22]

Guduchyadi ghrita medhya rasayana, an ayurvedic formulation containing *Tinospora cordifolia*, *Achyranthes aspera*, *Embelia ribes, Convolvulus prostratus*, *Acorus calamus*, *Terminalia chebula*, *Saussurea lappa*, and *Asparagus racemosa* showed positive effects in clinical evaluations by improving the mental health, memory, stress, and depression conditions and can further lead to slowing down of the aging process.[23]

A clinical study of Suvarna Bindu Prashan evaluated its use in twenty healthy male and female children within the age group of three to four years for a duration of six months. The group, which received the test product, showed significant improvement in the scores of eating habits, behavior, mood, temperament, and in the scores of events of illness.[24] Chyawanprash, one of the most widely used ayurvedic formulations with the ingredients *Withania somnifera*, *Emblica officinalis*, *Piper longum*, *Tribulus terrestris*, *Asparagus racemosa, Bacopa monnieri* and *Tinospora cordifolia* is regarded as an omnipotent formulation for boosting the immunity and many other disease ailments. There are many studies indicating the various health benefits of Chyawanprash. In a double-blind placebo-controlled study, the subjects showed significant improvement in Hamilton-D (HAM-D) scores compared to the placebo.[25] Similarly, the nootropic effect of Chyawanprash significantly improved the cognitive functions in college students. Parameters like alertness, attention, and concentration were improved in the study population.[26] The immune boosting studies of Chyawanprash were confirmed by reducing disease symptoms of seasonal influences, modulated IgE and immunity markers levels, improved pulmonary functions, reduction in cortisol levels, and improvement in quality of life (QoL).[27] Chyawanprash was evaluated in children of 5–6 years in age, in a 6-month long randomized, open labeled, prospective, multi-center, clinical study in children and it was shown to lead to significant improvement in immunity, energy levels, physical strength, vigor, and quality of life assessed through KIDSCREEN QOL-27 questionnaires in children.[28] Studies have also confirmed various other biological activities like adaptogenic,[29] diabetes management,[30] health promotion, *etc.*[31]

HFPM-01 a novel ayurvedic formulation containing herbs like *Boswellia serrata*, *Zingiber officinale, Withania somnifera*, *Curcuma longa,* and *Pluchea lanceolate* as the key ingredients was evaluated for its efficacy in controlling osteoarthritis (OA) of the knee. The study showed that the herbal formulation was found to be effective in controlling the symptoms associated with knee pain and thereby controlling the OA[32] (Table 14.1).

Table 14.1 Clinical evaluation of important ayurvedic formulations/herbs.

Herb/formulation	Target of study	Ref.
Triphala	Gastric problems	17
	Anti-microbial mouthwash	18 and 19
	Weight management	20
	Blood glucose level management	21
Triphala	Sthaulya	22
Guduchyadi ghrita medhya rasayana	Mental health	23
Suvarna Bindu Prashan	Overall well-being	24
Chyawanprash	Cognitive function	25
	Nootropic effect	26
	Immune boosting	27
	Adaptogenic	29
	Diabetes management	30
	Health promotion	31
HFPM-01	Osteoarthritis	32
Bacopa Monnieri	Improving memory	33
	Increasing brain function	34
	Promoting longevity	35
	Protective effects on β-amyloid toxicity	36
	Cognitive performance	37
	Alzheimer's disease[46]	46
Asparagus racemose	Rejuvenating tonic	54
	Digestive disorders	55
	Female sexual health	56
Curcuma longa	Rheumatoid arthritis	57
	Chronic anterior uveitis	58
	Anti-inflammatory	59–65
	Skin disorders	66–69
	Eye health	70–73
	Nervous system	74–79
	Cardiovascular system	80–84
	Gastro-intestinal disorders	85–92
	Urogenital system disorders	93–95
	Metabolic disorders	96–100
	Intoxication	101
	Malignant diseases	102–106
Withania somnifera	Stress and anxiety	107–110
	Sexual health	111–116
	Cognitive performance	122–124
	Pain	125
	Remove fatigue	126–127
	Schizophrenia	128
	Diabetes management	129
	Obsessive compulsive disorder	130
	Insomnia	131
	General well-being	132
Piper Nigrum	Gastric problems	133
Boswellia Serrata	Sandhigata vata	135
	Brain tumors	136
Zingiber officinale	Nausea and vomiting	137
	Migraine	138
	Knee osteoarthritis	139

Table 14.1 (*Continued*)

Herb/formulation	Target of study	Ref.
	Active rheumatoid arthritis	140
	Diabetes mellitus	141
	Heavy menstrual bleeding	142
	Primary dysmenorrhea	143
	Muscle pain	144
	Irritable bowel syndrome	145
Ocimum sanctum	Diabetes mellitus	146
	Gouty arthritis	147
	Hypertension	148
	Immuno modulation	149
	Cognitive function	150
Emblica officinalis	Diabetes mellitus	153–155
	Hyperacidity	156
	Pulmonary tuberculosis	157
	Cardiac health	158
	Idiopathic oligoasthenoteratospermia	159
	Vitiligo	160
	Dental disorders	161–162

14.3.2 Most Studied Herbs in Ayurveda

There are many herbs which are used in traditional systems of medicine including ayurvedic practices for the treatment of various disorders as they possess potential pharmacological activities when used individually and when used in combination (Table 14.1). *Bacopa monnieri* (Brahmi) is one of the most promising and widely evaluated herbs for its promising potential in neuroprotective and cognitive effects. It has been used for over 3000 years in ayurvedic practices for improving memory,[33] increasing brain function,[34] promoting longevity,[35] its protective effects on β-amyloid toxicity,[36] and cognitive performance,[37] *etc.* The bioactive components responsible for the pharmacological activities include steroidal saponins (bacoside A includes bacosides A2 and A3; bacoside B includes bacopasides N1, N2, IV and V, bacopasides I–VIII, bacopaside X, and bacopasaponins A–G and triterpenoidal saponins. Several human studies involving different extracts, different study dosages, subjects' criteria, and study duration confirmed the efficacy of the bioactive components in various studies.[38–45] Multiple clinical studies have also highlighted the potential of Brahmi in Alzheimer's disease.[46–49]

Asparagus racemosa (*A. racemosa*) is one of such herbs, which was used in traditional ayurvedic practices even in the ayurvedic system of medicine. *A. racemosa* commonly referred to as shatavari is referred to in many traditional systems of medicine. In traditional systems of medicine, shatavari is referred to as the "Queen of Herbs" owing to the miraculous benefit of the herb to females.[50] The plant is well known in ayurvedic practices in terms of anti-aging, female sexual health, dyspepsia, tumors, *etc.*[51] In ayurvedic practices, shatavari is regarded as "Rasayana" which implies that the drug is used for promoting wellness and vitality.[52] The use of asparagus species was mentioned in the ancient ayurvedic literature Charaka Samhita.[53] Charaka Samhita written by Charak and Ashtang Hridyam by Vagbhata, the two main classical texts on Ayurvedic practices, list *A. racemosa* as part of the formulas to treat women's health disorders. In many traditional systems of medicine, shatavari is used in epilepsy, vata disorders, and digestive disorders,[54] as a rejuvenating tonic,[55] and in female sexual health,[56] *etc.*

Another important herb, which is most commonly used in ayurvedic practices include turmeric, which is commonly referred to as Haridra in ayurvedic practices. Turmeric is used in many ayurvedic formulations from vedic dates onwards. The bioactive components responsible for the pharmacological activities include the so-called phenolic curcuminoids. The role of curcuminoids is clinically validated for many disorders including rheumatoid arthritis,[57] chronic anterior uveitis,[58] in anti-inflammation,[59–65] skin disorders,[66–69] eye disorders,[70–73] the nervous system,[74–79] cardiovascular system,[80–84] gastro intestinal disorders,[85–92] urogenital system disorders,[93–95] metabolic disorders,[96–100] intoxication,[101] and malignant diseases.[102–106]

Withania somnifera is very revered herb of the Indian Ayurvedic system of medicine, which is referred to as Rasyana in ayurvedic literatures. In traditional languages like Sanskrit (a language of ancient India), ashwagandha means "horse's smell" (ashwa = horse and gandha = smell), and when consumed it is believed to provide horse-like power.[107] The bioactive components that contribute to the pharmacological activities include the so-called withanolides. Many clinical studies have highlighted the different biological activities as well. The majority of the clinical studies have focused on the positive effect of ashwagandha in symptoms related to stress, anxiety, and having aphrodisiac activities. Different randomized studies have studied the positive effect of different dosages and extracts of ashwagandha and their role in clinical biomarkers associated with stress and anxiety.[108–110] Another clinical area in which extensive research was carried out on ashwagandha include the sexual function and fertility in both men and women. Significant improvement in parameters like sperm count, motility, and morphology was found in males and a positive impact in the female population was also observed.[111–116] The use of ashwagandha extract was evaluated for its potential to be used in improving athletic performance, which was modulated by multiple mechanisms of action.[117–121] The use of ashwagandha was also evaluated in improving the cognitive performance,[122–124] pain,[125] to remove fatiguness,[126,127] schizophrenia,[128] diabetes management,[129] obsessive compulsive disorder,[130] insomnia,[131] and general well-being,[132] *etc.*

Black pepper/long pepper belonging to the piper family are most common among the spices which are widely used in the spice and food industry. The piperine is one of the key bioactive components, which is most commonly used as a bioavailability enhancer in traditional ayurvedic practices and most modern nutraceutical industries. The majority of the clinical studies of the piperine is concentrated on enhancing the bioavailability of active components like curcuminoids, resveratrol, *etc.* In a single dose case indicative study, the effects of black pepper on the gastric mucosa were assessed using double-blind intragastric administration of the spice (1.5 g) to healthy human subjects, with aspirin (655 mg) as the positive control. The study proved that administration of piperine increases in parietal secretion, pepsin secretion, and potassium loss and black pepper administration and mucosal micro bleeding was also seen.[133]

Boswellia serrata is a common herb used in ayurvedic practices for treating inflammatory disorders. The herb is used as a combination and/or as a single herb. The bioactive components responsible for the pharmacological activities include the so-called boswellic acids.[134] Sandhigata vata commonly referred to as osteo arthritis in modern medicine is a sort of inflammatory condition particularly affecting the joints. In a clinical study 56 patients fulfilling the diagnostic criteria of Sandhigata vata were divided into two groups, in which one group was supplemented with a 500 mg capsule of Shallaki, 6 g per day (in three divided doses) with lukewarm water ($n = 29$), and the second group was given a capsule Shallaki as above, along with local application of Shallaki ointment on the affected joints ($n = 23$). After the two months supplementation, symptomatic improvement was observed in both study groups.[135] In another case indicative study 12 patients with brain tumors and progressive edema were treated with B. *serrata* (H15) extract. The MRI showed a significant reduction of edema in two of the seven patients with glioblastoma.[136]

Ginger (*Zingiber officinale*) belonging to the Zingiberaceae family is a well-known ingredient in traditional systems of medicine, which is most commonly used for digestive disorders. The pungency of ginger and the pharmacological activities are mainly attributed to the active components gingerols and its dehydrated form shogaols. There are a number of clinical studies which used either raw ginger powder or ginger in the form of a concentrated extract. Studies have concentrated on the use of ginger for various disorders like nausea and vomiting,[137] migraines,[138] knee osteoarthritis,[139] active rheumatoid arthritis,[140] Type 2 diabetes mellitus,[141] heavy menstrual bleeding,[142] primary dysmenorrhea,[143] muscle pain,[144] irritable bowel syndrome,[145] *etc.*

Ocimum sanctum an aromatic shrub in the family Lamiaceae is believed to be cultivated throughout the Asian countries. In ayurvedic texts, tulsi is known as "The Incomparable One," "Mother Medicine of Nature" and is revered as an "elixir of life" that is without an equal in terms of both its medicinal and spiritual properties.[146] Multiple clinical studies have highlighted its diverse range of biological activities. The biological activities of tulsi and its different extracts have been widely evaluated in various sorts of metabolic disorders like diabetes mellitus,[146] gouty arthritis,[147] hypertension,[148] *etc.* The immunomodulatory potential of tulsi extract was evaluated in healthy patients, subjects with asthma, and those with viral infections.[149] The phenolic compounds in tulsi leaves have a high degree of anti-oxidant potential which accounts for the antioxidant potential of the herb. The leaf extract was clinically evaluated for its potential to be used in neuroprotective disorders.[150]

Indian gooseberry (*Emblica officinalis*), commonly referred to as Amla belonging to the Phyllanthaceae family is one of the important herbs in most of the traditional systems of medicine including the ayurvedic practices. Phytochemical evaluations have either quantified or identified the constituents like ellagic acid, chebulinic acid, apigenin, gallic acid, quercetin, chebulagic acid, isostrictiniin, corilagin, methyl gallate, luteolin, *etc.*[151] Emblicanin A, emblicanin B, phyllaemblicin B, punigluconin, and pedunculagin are the tannins which are also present in the fruit extract of *E. officinalis.*[152] Supplementation of raw amla powder or the concentrated extract has a positive impact on controlling type 2 diabetes mellitus which was proved clinically in subjects with diabetes mellitus.[153–155] Dried fruit choorna of amla called amlaki given at a dose of 5 g for 45 days was evaluated to check the possibility of controlling the hyperacidity or Amalpitta. The study showed that above 40% of the study subjects showed improvement in the symptoms associated with hyperacidity.[156] Six grams of Jawarish amla, a Unani product of amla, was evaluated for its role in pulmonary tuberculosis. Significant improvements in the clinical indications were observed.[157] The capsules of amla extract were evaluated for their potential to be used in cold pressor induced cardiovascular changes. Amla extract showed a significant decrease in the cold pressor stress test induced changes on aortic wave reflections and a significant decrease was reported in the indices of arterial stiffness.[158] Addyzoa, a herbo-mineral agent that contains *Withania somnifera* and *Emblica officinalis* as its principal constituent was evaluated for its role in idiopathic oligoasthenoteratospermia in a randomized study. The study showed significant improvement in the total and progressive motility of the semen in men with idiopathic oligoasthenoteratospermia after 3 months of therapy.[159] An oral supplement containing amla (100 mg), vitamin E (10 mg), and carotenoids (4.7 mg) for a duration of six months showed an increase in the effectiveness of other vitiligo treatments.[160] Clinical studies have also focused on the role of amla in various sorts of dental disorders.[161,162]

14.4 Future Prospective

The clinical efficacy of the ayurvedic formulations if properly evaluated in a suitable study design can provide more insights into the efficacy of the herbs. Another area which needs to be focused on is the study design as ayurvedic drug administration varies from person to person and not a particular group of subjects. Standardization of the bioactive in the formulation and the extract can prevent a too high dosage for human consumption.

14.5 Conclusion

The Ayurvedic system of medicine if appropriately evaluated using suitable clinical studies can open a door to curing many diseases with natural extracts. Most modern nutraceutical industries can utilize the principles of this traditional system of medicine to create new formulations.

An appropriate standardization method is also needed in fingerprinting the formulations and herbs used in ayurvedic principles.

References

1. K. Ramakrishnappa, Impact of cultivation and gathering of medicinal plants on biodiversity: Case studies from India, in Food and Agriculture Organization of the United Nations (FAO), Biodiversity and the Ecosystem Approach in Agriculture, Forestry and Fisheries, Satellite event on the occasion of the Ninth Regular Session of the Commission on Genetic Resources for Food and Agriculture, Rome, Italy, FAO, Rome, October 12–13, 2002.
2. S. Sen, R. Chakraborty and B. De, *J. Herb. Med.*, 2011, **1**, 67.
3. A. Altemimi, N. Lakhssassi, A. Baharlouei, D. G. Watson and D. A. Lightfoot, *Plants*, 2017, **6**, 42.
4. S. Kannaiyan, Diversity, sustainable use and conservation of medicinal plants, Inaugural address to the International Seminar on Medicinal Plants and Herbal Products, Tirupati, India, March 7–9, 2008.
5. M. Borins, *Can. Fam. Physician*, 1987, **33**, 1061.
6. R. K. Singla, V. Yadav and S. Jayalakshmi, *Webmedcentral*, 2012, **3**(4), WMC003299.
7. J. M. Kendall, *Emerg. Med.*, 2003, **20**, 164.
8. P. R. Manohar, *Anc. Sci. Life*, 2013, **33**, 1.
9. P. V. Sharma, *Charaka samhita of agnivesha, vimana sthana, rogabhishagjiteeyam vimanam*, Chaukhambha Orientalia, 2nd edn, 1981, ch. 8, ver. 94.
10. B. Patwardhan, *J. Ayurveda. Integr. Med.*, 2011, **2**, 1.
11. P. A. Gautama, *J. Ayurveda. Integr. Med.*, 2021, **12**, 556.
12. V. Fønnebø, S. Grimsgaard, H. Walach, C. Ritenbaugh, A. J. Norheim, H. MacPherson, G. Lewith, L. Launsø, M. Koithan, T. Falkenberg, H. Boon and M. Aickin, *BMC. Med. Res. Methodol.*, 2007, 7, 2.
13. B. Patwardhan, D. Warude, P. Pushpangadan and N. Bhatt, *Evidence-Based Complementary Altern. Med.*, 2005, **2**, 465.
14. A. Chopra, P. Lavin, B. Patwardhan and D. Chitre, *J. Rheumatol.*, 2000, **27**, 1365.
15. R. Gupta, S. Singhal, A. Goyle and V. N. Sharma, *J. Assoc. Physicians. India.*, 2000, **49**, 231.
16. P. U. Kumar, P. Adhikari, P. Pereira and P. Bhat, *J. Assoc. Physicians. India.*, 1999, **47**, 685.
17. T. Balganesh, T. K. Kundu, T. K. Chakraborty and S. Roy, *ACS. Med. Chem. Lett.*, 2014, **5**, 724.
18. J. Srinagesh, P. Krishnappa and S. N. Somanna, *Indian. J. Dent. Res.*, 2012, **23**, 696.
19. A. Deshpande, S. Tandon and N. Deshpande, *Ayu*, 2014, **35**, 160.
20. S. H. Kamali, A. R. Khalaj, S. Hasani-Ranjbar, M. M. Esfehani, M. Kamalinejad, O. Soheil and S. A. Kamali, *Daru*, 2012, **20**, 33.

21. S. S. Rajan and S. Antony, *Ancient. Sci. Life*, 2008, **27**, 45.
22. A. Saxena, S. K. Gupta, J. Saxena and S. Niranjan, *int. Ayurvedic Med. J.*, 2020, 7, 2382.
23. D. N. Shukla, H. M. Chandola and B. Ravishankar, *J. Res. Ayurveda Siddha*, 2008, **29**, 1.
24. P. Aniket, D. Pallavi, A. Aziz, K. Avinash and S. Vikas, *J. Ayurveda Integr. Med. Sci.*, 2017, **2**, 11.
25. A. Ghosh, D. Laloo and N. K. Singh, *Int. J. Pharm. Pharm. Sci.*, 2013, **5**, 801.
26. K. S. Sailesh, R. Archana, S. Mishra and J. K. Symphoria Mukkadan, *Int. J. Res. Ayurveda Pharm*, 2014, **5**, 435.
27. J. L. N. Sastry, A. Gupta, N. B. Brindavanam, S. Kanjilal, S. Kumar, M. Setia, S. Vedula and R. Srivastava, *Ind. J. Appl. Res.*, 2014, **4**, 205.
28. A. Gupta, S. Kumar, S. Dole, S. Deshpande, V. Deshpande, S. Singh and V. Sasibhushan, *Ancient Sci. Life*, 2017, **36**, 141.
29. J. S. Yadav, S. Thakur and P. Chadha, *Int. J. Hum. Genet.*, 2003, **3**, 33.
30. S. Kumar, A. Gupta, R. Ghungralekar and V. S. Deshpande, *Altern. Integr. Med.*, 2017, **6**, 230.
31. S. K. Dalai, L. Dwivedi, G. Dadheech, S. Mishra and P. Sharma, *Anc. Sci. Life*, 2005, **24**, 183.
32. D. Veena, S. Bharatbhushan, G. Gayatrhin and G. Umesh, *Int. J. Res. Rev.*, 2021, **8**, 2349.
33. K. Abascal and E. Yarnell, *Altern. Complementary Ther.*, 2011, **17**, 21.
34. C. Calabrese, W. L. Gregory, M. Leo, D. Kraemer, K. Bone and B. Oken, *J. Altern. Complementary Med.*, 2008, **14**, 707.
35. A. Morgan and J. Stevens, *J. Altern. Complementary Med.*, 2010, **16**, 735.
36. N. L. Impeanchob, S. Jaipan, S. Rattanakaruna, W. Phrompittayarat and K. Ingkaninan, *J. Ethnopharmacol.*, 2008, **120**, 112.
37. N. Uabundit, J. Wattanathorn, S. Mucimapura and K. Ingkaninan, *J. Ethnopharmacol.*, 2010, **127**, 26.
38. C. Calabrese, W. L. Gregory, M. Leo, D. Kraemer, K. Bone and B. Oken, *J. Altern. Complementary Med.*, 2008, **14**, 707.
39. J. Morgan and J. Stevens, *J. Altern. Complementary Med.*, 2010, **16**, 735.
40. S. Raghav, H. Singh, P. K. Dalai, J. S. Srivastava and O. P. Asthana, *Indian J. Psychiatry*, 2006, **48**, 238.
41. S. Roodenrys, D. Booth, S. Bulzomi, A. Phipps, C. Micallef and J. Smoker, *Neuropsychopharmacol.*, 2006, **27**, 279.
42. C. Stough, L. A. Downey, J. Lloyd, B. Silber, S. Redman, C. W. Hutchison, K. Wesnes and P. J. Nathan, *Phytother. Res.*, 2008, **22**, 1629.
43. C. Stough, J. Lloyd, L. A. Downey, C. W. Hutchison, T. Rodgers and P. J. Nathan, *Psychopharmacol.*, 2001, **156**, 481.
44. T. Peth-Nui, J. Wattanathorn, S. Muchimapura, T. Tong-Un, N. Piyavhatkul, P. Rangseekajee, K. Ingkaninan and S. Vittaya-areekul, *Evidence-Based Complementary Altern. Med.*, 2012, **2012**, 606424.
45. V. Sathyanarayanan, T. Thomas, S. J. L. Einöther, R. Dobriyal, M. K. Joshi and S. Krishnamachari, *Psychopharmacol.*, 2013, **227**, 299.

46. T. Mitra-Ganguli, S. Kalita, S. Bhushan, C. Stough, J. Kean, N. Wang, V. Sethi and A. Khadilkar, *Front. Pharmacol.*, 2017, **8**, 678.
47. K. S. Chaudhari, N. R. Tiwari, R. R. Tiwari and R. S. Sharma, *Ann. Neurosci.*, 2017, **24**, 111.
48. P. J. Nathan, J. Clarke, J. Lloyd, C. W. Hutchison, L. Downey and C. Stough, *Hum. Psychopharmacol. Clin. Exp.*, 2001, **16**, 345.
49. S. Raghav, H. Singh, P. K. Dalal, J. S. Srivastawa and O. P. Asthana, *Indian J. Psychiatry*, 2006, **48**, 238.
50. V. M. Gogte, *Ayurvedic pharmacology and therapeutic uses of medicinal plants*, SPARC, Mumbai, 2000.
51. Atreya, *Ayurvedic healing for women*, Samuel Weiser Inc, York, 1999.
52. H. S. Puri, *'Rasayana' – Ayurvedic herbs for longevity and rejuvenation*, Taylor and Francis, London, 2003.
53. R. K. Sharma and B. Dash, *Charaka samhita-text with English translation and critical exposition based on Chakrapani Datta's Ayurveda dipika*, Chowkhamba Varanasi, India, 2003.
54. W. K. B. D. S. Fernando and K. C. Perera, *Eur. J. Biomed. Pharm. Sci.*, 2019, **3**, 26.
55. K. Sharma and M. Bhatnagar, *Int. J. Pharm. Bio.l Arch.*, 2011, **2**, 855.
56. S. A. Dayani Siriwardene, L. Karunathilaka, N. D. Kodituwakku and Y. Karunarathne, *Ayu*, 2010, **31**, 24.
57. A. Amalraj, K. Varma, J. Jacob, C. Divya, A. B. Kunnumakkara, S. J. Stohs and S. Gopi, *J. Med. Food*, 2017, **20**, 1022.
58. B. Lal, A. Kapoor, O. Asthana, P. Agrawal, R. Prasad, P. Kumar and R. Srimal, *Phytother. Res.*, 1999, **13**, 318.
59. P. Usharani, A. Mateen, M. Naidu, Y. Raju and N. Chandra, *Drugs R D*, 2008, **9**, 243.
60. S. Dcodhar, R. Sethi and R. Srimal, *Indian J. Med. Res.*, 1980, **71**, 632.
61. R. Kulkarni, P. Patki, V. Jog, S. Gandage and B. Patwardhan, *J. Ethnopharmacol.*, 1991, **33**, 91.
62. G. Belcaro, M. R. Cesarone, M. Dugall, L. Pellegrini, A. Ledda, M. G. Grossi, S. Togni and G. Appendino, *Altern. Med. Rev.*, 2010, **15**, 337.
63. N. Kertia, A. Asdie and W. Rochmah, *Acta Med Indones.*, 2012, **44**, 105.
64. V. Kuptniratsaikul, S. Thanakhumtorn, P. Chinswangwatanakul, L. Wattanamongkonsil and V. Thamlikitkul, *J. Altern. Complementary Med.*, 2009, **15**, 891.
65. R. Satoskar, S. Shah and S. Shenoy, *Int. J. Clin. Pharmacol. Ther.*, 1986, **24**, 651.
66. K. Madhu, K. Chanda and M. Saji, *Inflammopharmacology*, 2013, **21**, 129.
67. M. Heng, M. Song, J. Harker and M. Heng, *Br. J. Dermatol. Suppl.*, 2000, **143**, 937.
68. S. K. Kurd, N. Smith, A. VanVoorhees, A. B. Troxel, V. Badmaev, J. T. Seykora and J. M. Gelfand, *J. Am. Acad. Dermatol.*, 2008, **58**, 625.

69. Y. Panahi, A. Sahebkar, M. Amiri, S. M. Davoudi, F. Beiraghdar, S. L. Hoseininejad and M. Kolivand, *Br. J. Nutr.*, 2012, **108**, 1272.
70. J. M. Ringman, S. A. Frautschy, E. Teng, A. N. Begum, J. Bardens, M. Beigi, K. H. Gylys, V. Badmaev, D. D. Heath and L. G. Apostolova, *Alzheimer's Res. Ther.*, 2012, **4**, 43.
71. S. R. Rainey-Smith, B. M. Brown, H. R. Sohrabi, T. Shah, K. G. Goozee, V. B. Gupta and R. N. Martins, *Br. J. Nutr.*, 2016, **115**, 2106.
72. N. Hishikawa, Y. Takahashi, Y. Amakusa, Y. Tanno, Y. Tuji, H. Niwa, N. Murakami and U. Krishna, *Ayu*, 2012, **33**, 499.
73. K. H. Cox, A. Pipingas and A. B. Scholey, *J. Psychopharmacol.*, 2015, **29**, 642.
74. L. Baum, C. W. K. Lam, S. K.-K. Cheung, T. Kwok, V. Lui, J. Tsoh, L. Lam, V. Leung, E. Hui and C. Ng, *J. Clin. Psychopharmacol*, 2008, **28**, 110.
75. J. M. Ringman, S. A. Frautschy, E. Teng, A. N. Begum, J. Bardens, M. Beigi, K. H. Gylys, V. Badmaev, D. D. Heath and L. G. Apostolova, *Alzheimer's. Res. Ther.*, 2012, **4**, 43.
76. N. Hishikawa, Y. Takahashi, Y. Amakusa, Y. Tanno, Y. Tuji, H. Niwa, N. Murakami and U. Krishna, *Ayu*, 2012, **33**, 499.
77. K. H. Cox, A. Pipingas and A. B. Scholey, *J. Psychopharmacol.*, 2015, **29**, 642.
78. K. Soni and R. Kuttan, *Indian J. Physiol. Pharmacol.*, 1992, **36**, 273.
79. A. Ramırez-Boscá, A. Soler, M. A. Carrion, J. Dıaz-Alperi, A. Bernd, C. Quintanilla, E. Q. Almagro and J. Miquel, *Mech. Ageing. Dev.*, 2000, **119**, 41.
80. I. Alwi, T. Santoso, S. Suyono, B. Sutrisna, F. D. Suyatna, S. B. Kresno and S. Ernie, *Acta Med. Indones.*, 2008, **40**, 201.
81. R. A. DiSilvestro, E. Joseph, S. Zhao and J. Bomser, *Nutr. J.*, 2012, **11**, 79.
82. S. Durgaprasad, C. G. Pai and J. F. Alvres, *Indian J. Med. Res.*, 2005, **122**, 315.
83. A. Rasyid and A. Lelo, *Aliment. Pharmacol. Ther*, 1999, **13**, 245.
84. Y. S. Yang, Y. F. Su, H. W. Yang, Y. H. Lee, J. I. Chou and K. C. Ueng, *Phytother. Res.*, 2014, **28**, 1770.
85. M. R. Adhvaryu, N. M. Reddy and B. C. Vakharia, *World J. Gastroenterol.*, 2008, **14**, 4753.
86. S. W. Kim, K. C. Ha, E. K. Choi, S. Y. Jung, M. G. Kim, D. Y. Kwon, H. J. Yang, M. J. Kim, H. J. Kang and H. I. Back, *BMC Complementary Altern. Med.*, 2013, **13**, 58.
87. S. Rahmani, S. Asgary, G. Askari, M. Keshvari, M. Hatamipour, A. Feizi and A. Sahebkar, *Phytother. Res.*, 2016, **30**, 1540.
88. R. Bundy, A. F. Walker, R. W. Middleton and J. Booth, *J. Altern. Complementary Med.*, 2004, **10**, 1015.
89. S. Durgaprasad, C. G. Pai and J. F. Alvres, *Indian J. Med. Res.*, 2005, **122**, 315.
90. A. Rasyid and A. Lelo, *Aliment. Pharmacol. Ther*, 1999, **13**, 245.

91. A. Rasyid, A. R. A. Rahman, K. Jaalam and A. Lelo, *Asia. Pac. J. Clin. Nutr.*, 2002, **11**, 314.
92. C. Niederau and E. Göpfert, *Med. Klin.*, 1999, **94**, 425.
93. D. Shoskes, C. Lapierre, M. Cruz-Corerra, N. Muruve, R. Rosario, B. Fromkin, M. Braun and J. Copley, *Transplantation*, 2005, **80**, 1556.
94. P. Khajehdehi, M. Pakfetrat, K. Javidnia, F. Azad, L. Malekmakan, M. H. Nasab and G. Dehghanzadeh, *Scand. J. Urol. Nep.*, 2011, **45**, 365.
95. T. Cai, S. Mazzoli, A. Bechi, P. Addonisio, N. Mondaini, R. C. Pagliai and R. Bartoletti, *Int. J. Antimicrob. Agents*, 2009, **33**, 549.
96. M. Srinivasan, *Indian J. Med. Sci.*, 1972, **26**, 269.
97. J. Wickenberg and S. L. Ingemansson, *Nutr. J.*, 2010, **9**, 43.
98. L. X. Na, Y. Li, H. Z. Pan, X. L. Zhou, D. J. Sun, M. Meng, X. X. Li and C. H. Sun, *Mol. Nutr. Food Res.*, 2013, **57**, 1569.
99. N. M. K. Selvi, M. Sridhar, R. Swaminathan and R. Sripradha, *Indian. J. Clin. Biochem.*, 2015, **30**, 180.
100. J. Heshmati, A. Moini, M. Sepidarkish, M. Morvaridzadeh, M. Salehi, A. Palmowski, M. F. Mojtahedi and F. Shidfar, *J. Phymed.*, 2020, **80**, 153395.
101. H. R. Rahimi, A. H. Mohammadpour, M. Dastani, M. R. Jaafari, K. Abnous, M. G. Mobarhan and R. K. Oskuee, *Avicenna J Phytomed.*, 2016, **6**, 567.
102. J. Biswas, D. Sinha, S. Mukherjee, S. Roy, M. Siddiqi and M. Roy, *Hum. Exp. Toxicol.*, 2010, **29**, 513.
103. R. Kuttan, P. Sudheeran and C. Josph, *Tumori*, 1987, **73**, 29.
104. K. Hastak, N. Lubri, S. Jakhi, C. More, A. John, S. Ghaisas and S. Bhide, *Cancer. Lett.*, 1997, **116**, 265.
105. C. Hsieh, *Anticancer. Res.*, 2001, **21**, 2895.
106. N. Chainani-Wu, S. Silverman Jr, A. Reingold, A. Bostrom, C. Mc Culloch, F. Lozada-Nur and J. Weintraub, *Phytomedicine*, 2007, **14**, 437.
107. M. H. Mirjalili, E. Moyano, M. Bonfill, R. M. Cusido and J. Palazon, *Molecules*, 2009, **14**, 2373.
108. J. Salve, S. Pate, K. Debnath and D. Langade, *Cureus*, 2009, **11**, 6466.
109. D. Choudhary, S. Bhattacharyya and S. Bose, *J. Diet. Suppl.*, 2017, **14**, 599.
110. R. Nasimi Doost Azgomi, H. Nazemiyeh, H. Sadeghi Bazargani, S. M. B. Fazljou, F. Nejatbakhsh, A. Moini Jazani, Y. Ahmadi AsrBadr and A. Zomorrodi, *Andrologia*, 2018, **50**, 13041.
111. S. Dongre, D. Langade and S. Bhattacharyya, *Biomed. Res. Int.*, 2015, **2015**, 284154.
112. V. R. Ambiye, D. Langade, S. Dongre, P. Aptikar, M. Kulkarni and A. Dongre, *Evidence-Based Complementary Altern. Med.*, 2013, **2013**, 571420.
113. A. Gupta, A. A. Mahdi, K. K. Shukla, M. K. Ahmad, N. Bansal, P. Sankhwar and S. N. Sankhwar, *J. Ethnopharmacol.*, 2013, **26**, 208.

114. P. Mamidi and A. B. Thakar, *Ayu*, 2011, **32**, 322.
115. A. A. Mahdi, K. K. Shukla, M. K. Ahmad, S. Rajender, S. N. Shankhwar, V. Singh and D. Dalela, *Evidence-Based Complementary Altern. Med.*, 2009, **29**, 576962.
116. P. Sengupta, A. Agarwal, M. Pogrebetskaya, S. Roychoudhury, D. Durairajanayagam and R. Henkel, *Reprod. Biomed. Online*, 2018, **36**, 311.
117. T. N. Zeigenfuses, A. W. Kedia, J. E. Sandrock, B. J. Raub, C. M. Kerksick and H. L. Lopez, *Nutrients*, 2018, **10**, 1807.
118. B. Choudhary, A. Shetty and D. G. Langade, *Ayu*, 2015, **36**, 63.
119. S. Wankhede, D. Langade, K. Joshi, S. R. Sinha and S. Bhattacharyya, *J. Int. Soc. Sports Nutr.*, 2015, **12**, 43.
120. S. Shenoy, U. Chaskar, J. S. Sandhu and M. M. Paadhi, *J. Ayurveda. Integr. Med.*, 2012, **3**, 209.
121. J. S. Sandhu, B. Shah, S. Shenoy, S. Chauhan, G. S. Lavekar and M. M. Padhi, *Int. J. Ayurveda Res.*, 2010, **1**, 144.
122. U. Pingali, R. Pilli and N. Fatima, *Pharmacogn. Res.*, 2014, **6**, 12.
123. D. Choudhary, S. Bhattacharyya and S. Bose, *J. Diet. Suppl.*, 2017, **14**, 599.
124. K. N. Chengappa, C. R. Bowie, P. J. Schlicht, D. Fleet, J. S. Brar and R. Jindal, *J. Clin. Psychiatry*, 2013, **74**, 1076.
125. G. S. Ramakanth, C. Uday Kumar, P. V. Kishan and P. Usharani, *J. Ayurveda Integr. Med.*, 2016, 7, 151.
126. B. M. Biswal, S. A. Sulaiman, H. C. Ismail, H. Zakaria and K. I. Musa, *Integr. Cancer Ther.*, 2013, **12**, 312.
127. A. L. Lopresti, S. J. Smith, H. Malvi and R. Kodgule, *Medicine*, 2019, **98**, e17186.
128. A. P. Agnihotri, S. D. Sontakke, V. R. Thawani, A. Saoji and V. S. Goswami, *Indian. J. Pharmacol.*, 2013, **45**, 417.
129. S. Nayak, S. Nayak, B. K. Panda and S. Das, *Ayushdhara*, 2015, **2**, 413.
130. S. P. Jahanbakhsh, A. A. Manteghi, S. A. Emami, S. Mahyari, B. Gholampour, A. H. Mohammadpour and A. Sahebkar, *Complementary Ther. Med.*, 2016, **27**, 25.
131. D. Langade, S. Kanchi, J. Salve, K. Debnath and D. Ambegaokar, *Cureus*, 2019, **11**, e5797.
132. S. B. Kelgane, J. Salve, P. Sampara and K. Debnath, *Cureus*, 2020, **12**, e7083.
133. B. M. Myers, J. L. Smith and D. Y. Graham, *Am. J. Gastroenterol*, 1987, **82**, 21.
134. H. H. Ahmed, A. A. Abd-Rabou, A. Z. Hassan and S. E. Kotob, *Asian. Pac. J. Cancer Prev.*, 2015, **16**, 7179.
135. P. K. Gupta, S. M. Samarakoon, H. M. Chandola and B. Ravishankar, *Ayu*, 2011, **32**, 478.
136. J. R. Streffer, M. Bitzer, M. Schabet, J. Dichgans and M. Weller, *Neurology*, 2001, **56**, 1219.

137. F. Sharifzadeh, M. Kashanian, J. Koohpayehzadeh, F. Rezaian, N. Sheikhansari and N. Eshraghi, *J. Matern. Fetal. Neonatal. Med.*, 2018, **31**, 2509.
138. L. B. Martins, A. M. D. S. Rodrigues, D. F. Rodrigues, L. C. Dos Santos, A. L. Teixeira and A. V. M. Ferreira, *Cephalalgia*, 2019, **39**, 68.
139. H. Mozaffari-Khosravi, Z. Naderi, A. Dehghan, A. Nadjarzadeh and H. Fallah Huseini, *J. Nutr. Gerontol. Geriatr*, 2016, **35**, 209.
140. N. Aryaeian, F. Shahram, M. Mahmoudi, H. Tavakoli, B. Yousefi, T. Arablou and S. Jafari Karegar, *Gene*, 2019, **698**, 179.
141. M. Makhdoomi Arzati, N. Mohammadzadeh Honarvar, A. Saedisomeolia, S. Anvari, M. Effatpanah, R. Makhdoomi Arzati, M. S. Yekaninejad, R. Hashemi and M. Djalali, *Int. J. Endocrinol. Metab.*, 2017, **15**, e57927.
142. F. Kashefi, M. Khajehei, M. Alavinia, E. Golmakani and J. Asili, *Phytother. Res.*, 2015, **29**, 114.
143. F. Kashefi, M. Khajehei, M. Tabatabaeichehr, M. Alavinia and J. Asili, *Pain Manag. Nurs.*, 2014, **15**, 826.
144. C. D. Black, M. P. Herring, D. J. Hurley and P. J. O'Connor, *J. Pain*, 2010, **11**, 894.
145. M. A. van Tilburg, O. S. Palsson, Y. Ringel and W. E. Whitehead, *Complementary Ther. Med.*, 2014, **22**, 17.
146. K. Nadkarni and A. Nadkarni, *Indian Materia Medica with Ayurvedic, Unani-Tibbi, Siddha, Allopathic, Homeopathic, Naturopathic & Home Remedies*, Popular Prakashan Private Ltd, Bombay, India, 1982, vol. 2.
147. J. Giri, B. Suganthi and G. Meera, *Indian. J. Nutr. Diet*, 1987, **24**, 193.
148. M. Ahmad, A. A. Faraazi and M. N. Aami, *Bull. Chem. Soc. Ethiop.*, 2013, **27**, 469.
149. S. Sarvaiya, Studies on hypertension with special reference to hypotensive effect of Ocimum sanctum, PhD thesis, Sardar Patel University, Anand, India, 1986.
150. M. Venu Prasad, *Antifatigue and Neuroprotective Properties of Selected Species of Ocimum L.*, University of Mysore, Mysore, India, 2014.
151. K. A. Habib-ur-Rehman, M. A. Yasin, N. Choudhary and M. I. Atta-ur-Rahman, *Nat. Prod. Res.*, 2007, **21**, 775.
152. L. D. Kapoor, *Handbook of Ayurvedic Medicinal Plants: Herbal Reference Library*, CRC Press, 2000, ISBN 9780849329296.
153. P. Usharani, N. Fatima and N. Muralidhar, *Diabetes. Metab. Syndr. Obes*, 2013, **6**, 275.
154. K. Walia, R. Boolchandani, S. Dhand and B. Antony, *Int. J. Basic. Appl. Med. Sci.*, 2015, **5**, 251.
155. M. S. Akhtar, A. Ramzan, A. Ali and M. Ahmad, *Int. J. Food Sci. Nutr.*, 2011, **62**, 609.
156. Y. K. Pandit, P. K. Garg, K. Agarwal, S. Balakrishna and P. S. Mehta, *Int. Res. J. Pharm.*, 2015, **6**, 199.
157. A. M. K. Sherwani, M. Zulkifle and R. M. Saleha, *J. Islam. Med. Assoc. North. Am*, 2012, **44**, 1.

158. N. Fatima, U. Pingali and R. Pilli, *Pharmacogn. Res*, 2014, **6**, 29.
159. R. Kumar, V. Saxena, M. B. Shamsi, S. Venkatesh and R. Dada, *Indian J. Urol.*, 2011, **27**, 357.
160. R. Colucci, F. Dragoni, R. Conti, L. Pisaneschi, L. Lazzeri and S. Moretti, *Dermatol. Ther.*, 2015, **28**, 17.
161. S. Grover, S. Tewari, R. K. Sharma, G. Singh, A. Yadav and S. C. Naula, *Phytother. Res.*, 2016, **30**, 956.
162. A. Velmurugan, M. M. Madhubala, S. Bhavani, K. S. Satheesh Kumar, S. S. Sathyanarayana and N. Gurucharan, *J. Conserv. Dent.*, 2013, **16**, 546.

CHAPTER 15

Ayurvedic and Traditional Systems of Medicine in Clinical Trials: An Overview

SHINTU JUDE,* KARTHIK VARMA A. C.,
SASIKUMAR KUTTAPPAN AND AUGUSTINE AMALRAJ

R&D Centre, Aurea Biolabs (P) Ltd, Kolenchery, Cochin-682 311, Kerala, India
*Emails: shintujude2@gmail.com; shintu.jude@plantlipids.com

15.1 Introduction

Since the very beginning mankind has faced many issues regarding health and well-being. As such, there have been attempts to find solutions from every possible means, including plants, animals, and other available natural substances. The resulting healthcare systems (including the knowledge, approaches, medicines, exercises, *etc.*), which have been effectively applied to diagnose, treat, or prevent illness and to maintain well-being are considered as traditional medical systems. They were developed in different parts of the world in ancient times, mostly related to cultural systems. Some of the major traditional medical systems are Ayurveda, Yunani, and Siddha from India, Koryo medicine from Korea, Traditional Chinese Medicine from China, Kampo medicine from Japan, traditional medicine of Sri Lanka, traditional Arabic and Islamic medicine in Arab countries and Traditional Medicine from Africa (especially from Egypt). All of these indigenous systems included the natural and locally available materials for the treatment protocols, and as such there are variations according to the regions.

Chemistry, Biological Activities and Therapeutic Applications of Medicinal Plants in Ayurveda
Edited by Augustine Amalraj, Sasikumar Kuttappan and Karthik Varma A. C.

Published by the Royal Society of Chemistry, www.rsc.org

Traditional medicines are developed by the efforts of many years. As per the recorded history, a Sumerian clay slab, referring to different plants and their preparation recipes is considered as the available earliest evidence for traditional medicine, which dates back 4000 years.[1] During the developing stage, the etiological factors behind diseases were unknown and therefore the systems followed trial and error methods. They completely relied on the efficacy of the systems rather than any other evidence.

As per the definition from the WHO, traditional medicine is a term, encompassing diverse health practices, approaches, knowledge, and beliefs incorporating plant, animal, and/or mineral-based medicines, spiritual therapies, manual techniques, and exercises applied singularly or in combination to maintain well-being, as well as to treat, diagnose, or prevent illness.[2] In an extension, these traditional medicines are termed complementary or alternative medicines: if they are used along with a western medicinal treatment, they are termed complementary, while the use of plant-derived medicines instead of Western medicine is known as alternative medicine.

Traditional medicine systems are now studied in a more scientific way, and the treatment protocols are subjected to experiments in order to obtain evidence. The knowledge and practices of traditional medicine in many countries are followed by many people, even today. It is true that many modern medicines have taken their model from natural components. Besides, natural therapeutics have been used as the models for pharmacologically active compounds with higher activity and less toxicity than the synthetic counterparts.[3] For example, the potential drugs **morphine** from opium plants (*Papaver somniferum*), **paclitaxel** from *Taxus brevifolia,* **Camptothecin** from *Camptotheca acuminata*, **Quinine** from *Cinchona officinalis*, **Vincristine** from *Catharanthus roseus,* and **Homoharringtonine** from *Cephalotaxus harringtonii etc.* are inevitable parts of the modern medicine culture, which were primarily derived from plants used for traditional medicines.[4] An important anti-hypertensive agent named **Reserpine**, proven to reduce the mortality rate is actually derived from an Ayurvedic herb '*Rauwolfia serpentine*' which is used to treat snakebites. The potential anti-malarial molecule **Arteminisine** is one of the most successful therapeutic products derived from herbs. It is the widely used conventional antimalarial drug of today, which took its path from the long history of usage of its parent plant, *Artemisia annua* (Quinhaosu) for treating fevers in Traditional Chinese Medicine (TCM). Fabricant and Farnsworth have reviewed the possibilities and different approaches of considering higher plants as leads for drug development. In their study, they identified that among a group of 122 compounds originally derived from plants and used widely as drugs, almost 80% have an ethno-medical use related to the current therapeutic assignments.[5] Some of the plants, which serve as the key players in traditional medicine systems, are given in Table 15.1.

Natural products seek attention as healthcare products and while considering the parameters such as safety profiles, side effects, toxicity levels,

Table 15.1 Some important herbs used in traditional medicine systems and their properties.

Plant	Common name	Parts used	To treat ailments	Used in the medicinal system	Major active moieties	Ref.
Artemisia herba-alba	Wormwood, desert Wormwood, Shih, Armoise blanche	Oil, leaves, barks, roots, aerial parts,	Arterial hypertension, diabetes, bronchitis, diarrhea, hypertension, stomach disorders, neuralgias	Traditional medicine system of Africa	Monoterpenes	6 and 7
Acacia senegal (*Senegalia senegal*)	Gum acacia, Gum arabic tree, Sudan gum, Hashab gum	Gum, bark	Bronchitis, diarrhea, dysentery, leprosy, typhoid fever, bleeding, upper respiratory tract infections, bacterial and fungal infections of the skin and mouth, cough, catarrh, gonorrhea	Ayurveda, traditional medicine system of Africa	Phenolic compounds, flavonoids, tannins	8 and 9
Acorus calamus	Cinnamon sedge, Flag root, Gladdon, Myrtle flag, Myrtle grass, Myrtle sedge, Sweet cane, Sweet myrtle, Sweet root, Sweet rush	Rhizome, stem, leaf, root, bark	Kidney and liver ailments, obesity, diabetes, heart disease, osteoarthritis, digestive disorders, stomach pain, bloating, fever, nausea, vomiting, fatigue, exhaustion, mouth ulcers, cuts, wounds, smallpox, cough, cold, toothache, snake bite, indigestion, epilepsy, hysteria, depression, hemorrhoids	Ayurveda, traditional Chinese medicine	Flavonoids, alkaloids, phenolic compounds, tannins, steroids, saponins, terpenoids, glycosides	10 and 11
Aloe ferox	Bitter aloe, Kaapse aalwyn, Umhlaba, Lekhala, Ikhala, Bergaalwyn, Bitteraalwyn	Leaf sap, leaf exudate, Stem, roots leaf	Laxative, conjunctivitis, eczema, arthritis, stress, hypertension, diabetes, sinusitis, ophthalmia, wounds, herpes, shingles, sore throat, venereal sores, sores, burns, ulcers, redwater, sheep scab	Traditional medicine system of Africa	Chromones, anthraquinones, anthrones, phenolic compounds, anthrone-C-glycosides	12–18

Table 15.1 (*Continued*)

Plant	Common name	Parts used	To treat ailments	Used in the medicinal system	Major active moieties	Ref.
Aloe vera	Kattar vazha,	Leaf, gel	Wounds, vertigo, colds, asthma, stomach ulcer, constipation, varicella zoster, herpes simplex type i and ii, influenza, pseudo rabies, aids, cardiovascular problems, inflammation, genital herpes, psoriasis, burns, skin irritations, headaches, immune-system deficiencies, coughs, arthritis,	Ayurveda	Amino acids, enzymes, vitamins, minerals, saponins, anthraquinones, lignin, and salicylic acid	19 and 20
Alpinia galanga	Greater galangal, Malaya Vacha	Rhizomes, tubers, seeds, roots	Inflammation, HIV, diabetes, ulcer, rheumatic pains, chest pain, burning of liver, kidney disease, disinfectants	Ayurveda	Flavonoids	21
Alpinia officinarum Hance	Lesser Galangal	Rhizomes, leaves, aerial parts, roots	Cold, pain, inflammation, stomach ache, cancer, microbial infection, Alzheimer's	Traditional Chinese Medicine	Flavonoids, diarylheptanoids, flavonols	22 and 23
Artemisia annua	Sweet wormwood	Leaves, aerial parts	Malaria, cancer, tuberculosis, lice, wounds, scabies, dysentery, acute convulsions, hemorrhoids, pain and swelling around tooth, pus in ear, rhinopolyp, eyesight problems, hemostasis, analgesic activities	Traditional Chinese medicine	Sesquiterpenoids, flavonoids, coumarins, triterpenoids, phenolics	24
Aspalathus linearis	Rooibos	Leaves,	HIV infections, glucose homeostasis, type 2 diabetes, age-related accumulation of lipid peroxidases in the brain	Traditional medicine system of Africa	Flavonoids	25
Asparagus racemosus	Satavar, Shatavari, Shatamull	Roots	Gastric ulcers, indigestion, reproductive and hormonal issues in women, lactation, colds	Ayurveda, Australian aboriginal	Saponins, mucilage, alkaloids, isoflavone	26 and 27

Azadirachta indica	Indian lilac, Neem, Aryaveppu, Veppu, Dogoyaro	Leaves, fruits, flowers, and stem bark	Cancer, wounds, diabetes, inflammation, liver problems, neuro-diseases, microbial infections, gingivitis, dental caries, nephrotoxicity, psoriasis	Ayurveda, Siddha, Unani, Homeopathy	Azadirachtin, flavonoids	28
Bacopa monnieri	Brahmi, Water hyssop	Leaf, whole plant	Cardiac diseases, cancer, hepatic disorders, anxiety, epilepsy neurodegenerative diseases, neurological and neuropsychiatric diseases	Ayurveda, Siddha	Saponins, alkaloids	29 and 30
Boophone disticha	Gifbol, Bushman poison bulb, Leshoma	Dried outer scales, bulbs	Boils, septic wounds, post circumcision wounds, headaches, eye conditions, weakness, headaches, chest pains, cancer, abdominal complaints, psychotic problems, gynecological conditions	Traditional medicine system of Africa	Alkaloids, phenolic compounds, flavonoids, acetovanillone	3, 31 and 32
Catharanthus roseus	Madagascar periwinkle, isisushlungu	Leaves	Diabetes, cancer	Traditional medicine system of Africa	Alkaloids	3 and 33
Cedrus deodara	Deodar cedar, Himalayan cedar, Deodar	All parts including inner wood	Fever, inflammation, pain, ulcer, apoptosis, spasmodic, hyperglycemia, infections, insomnia, disorder of mind, disease of skin and blood, rheumatism, piles, palsy, epilepsy, prolapsus recti, pulmonary and urinary disorder, insomnia, bruises, injuries to joint, boils, tuberculous glands, leprosy	Ayurveda	Taxifolin	34
Centaurea solstitialis	Yellow starthistle	Flowers	Peptic ulcers	Turkish folk medicine	Sesquiterpene lactones	35

Table 15.1 (*Continued*)

Plant	Common name	Parts used	To treat ailments	Used in the medicinal system	Major active moieties	Ref.
Centella asiatica	Gotu Kola, Kodavan, Indian pennywort, Asiatic pennywort	Leaves, roots, whole plant	Wound healing, burns, ulcers, leprosy, tuberculosis, lupus, skin diseases, eye diseases, fever, inflammation, asthma, hypertension, rheumatism, syphilis, epilepsy, diarrhea, mental illness	Ayurveda, Chinese traditional medicine, Siddha, Kampo, African traditional medicine	Pentacyclic triterpenoids	36
Cinnamomum camphora	Camphor tree	Wood, leaves	Heart conditions, colds, fevers, respiratory complaints, pneumonia, inflammatory conditions, infections, diarrhea, hysteria, topical irritations, fibrositis, neuralgia, skin diseases, joint and muscle pains	Traditional medicine system of Africa	Phenolic compounds, terpenes, camphor, linalool, safrole, cineole	37
Clerodenum serratum	Clerodendrum, Bharangi, cheruthekku	Roots, leaves	Pain, inflammation, rheumatism, respiratory disorders, fever and malarial fever, cancer, allergy, liver disorders	Ayurveda	Saponins, flavonoids, phenolics	38
Clitoria ternatea L.	Butterfly pea, conch flower, Aparajit, Kakkattan, shankapushpi	Roots, seeds, flowers, leaves	Body aches, animal stings, infections, urinogenital disorders, indigestion, constipation, fever, arthritis, eye ailments, ascetics, sore throat, skin diseases, chlorosis, liver and intestinal problems, hemicrania, rheumatism, swollen joints, ear-diseases	Ayurveda, Cuban traditional medicine system	Triterpenoids, flavonol glycosides, anthocyanins and steroids	39

Convolvulus pluricaulis	Sankhapuspi, Bilikanthisoppu, Aparajit, kakkurattai, Sanghupushpam	Leaf, flower, whole plant, roots	Convulsions, depression, anxiety, thyroid, ulcer, hepatic disorders	Ayurveda, Traditional Chinese medicine,	Alkaloids, flavonoids	40
Cyclopia genistoides	Honeybush	Leafy shoots, flowers	Weak digestion, urinary system, chronic catarrh, heartburn, nausea, pulmonary tuberculosis	Traditional medicine system of Africa	Xanthones, flavanones	15 and 41
Datura stramonium	Thornapple, jimsonweed, ummam	Seeds, flowers, leaves	Boils, gout, abscesses, rheumatism, asthma, wounds, motion sickness, parkinsonism, ulcers, inflammation, sciatica, toothache, fever	Traditional medicine system of Africa	Tropane alkaloids	42
Ephedra altissima	High-climbing joint fir	Stem, whole plant	Asthma, respiratory issues	Folk medicine of Iran	Alkaloids	43
Ginkgo biloba	Maidenhair tree, Ayurvedic ginseng,	Leaf, seeds,	Tinnitus, Alzheimer's, cardiovascular, diabetic nephropathy, cancer, neuro problems, lung and kidney disorders, bronchitis, stomach pain, asthma, cognitive dysfunction, tuberculosis, cognitive dysfunction	Ayurveda, traditional Chinese medicine	Flavonoids, flavonols, terpenoids, terpene trilactones, lignans, pro-anthocyanidins	44 and 45
Harpagophytum procumbens	devil's claw, sengaparile, grapple plant	Secondary tubers	Arthritis, rheumatism, digestive complaints, fevers, anorexia, skin complaints, allergies, analgesia, antiarrhythmic, blood diseases, antidiabetic, antiphlogistic, back pain, antipyretic, appetite stimulant, arteriosclerosis, bitter tonic, boils (topical), childbirth difficulties, choleretic, diuretic,	Traditional medicine system of Africa, traditional healthcare system of Europe	Glycosides, harpagoside	46

Table 15.1 (*Continued*)

Plant	Common name	Parts used	To treat ailments	Used in the medicinal system	Major active moieties	Ref.
			climacteric (change of life) problems, dysmenorrhea, dyspepsia, edema, fever, headache, fibromyalgia, fibrositis, gastrointestinal disorders, heartburn, indigestion, liver and gall bladder tonic, malaria, migraines, myalgia, neuralgia, nicotine poisoning, sedative, tendonitis, urinary tract infections			
Hedychium spicatum	Spiked ginger lily, Van haldi, Kapoorkachari, Shati	Rhizome, fruit, roots	Inflammation, pain, asthma, foul breath, vomiting, diarrhea, bronchitis, hiccough, fever, blood diseases, snake bite, cancer	Ayurveda, Traditional Chinese medicine, Unani, Tibetan Medicine system	Flavonoids, terpenes	47
Hypericum perforatum	St John's wort, tipton weed, klamath weed, goat weed, enola weed	Flowers, leaves, seeds	Diarrhea, gout, rheumatism, wounds, anxiety, depression, burns, cancer, inflammation-related disorders	Traditional medicine system of Africa	Flavonoids, naphthodianthrone	48
Inula recemosa	Puskara, Oriss root, Padmapatra	Roots, rhizome	Precordial chest pain, cough, stress, bronchoconstriction, eosinophilia, dyspnea	Ayurveda	Lactones	49
Ledebouria ovatifolia (*Bak.*) Jessop	Icubudwana, ntangana, flat leaved African hyacinth	Bulbs	Gastro-enteritis, influenza, backache, diarrhea, skin irritations, wounds, lumbago, heart conditions	Traditional medicine system of Africa	Bufadienolides	50

Momordica charantia Linn.	Bitter melon, balsam pear , paavakka,	Leaf, fruit, root, seeds, whole plant	Diabetes, cancer, hypertension, obesity, microbial infections, joint pain, burns, boils, rashes, chronic fever, gastric ulcers, inflammation, menorrhea, syphilis, rheumatism, measles, chicken pox, scabies, malaria, dysentery, high blood pressure	Traditional medicine system of Africa, Ayurveda, Turkish folk medicine	Charantin	51
Moringa oleifera	Drumstick tree, muringa	Leaves, bark, fruits, seeds, roots, flowers, immature pods	Cardiac problems, rheumatism, inflammations, articular pains, lower, back or kidney pain, constipation, sores, headaches, piles, fevers, sore throat, bronchitis, eye and ear infections, scurvy, catarrh, glandular swelling, tumors, ulcers, tooth cavity, intestinal complaints, dysentery, asthma, syphilis, muscle diseases, hysteria	Ayurveda, traditional medicine system of Africa, Siddha	Alkaloids, Phenolic compounds, flavonoids, amino acids	52
Mucuna pruriens L.	Seeds, roots,	Seeds, roots,	Inflammation, aging, rheumatoid arthritis, diabetes, neurodegenerative diseases, male virility	Ayurveda, traditional Chinese medicine	Levodopa, alkaloids	53
Oscimum sanctum	Tulsi, Krishna thulasi	Leaves, stem, root, flower, seeds, whole plant	Cough, cold, bronchitis, skin diseases, bronchial asthma, diarrhea, malaria, dysentery, arthritis, eye diseases, chronic fever, insect bite, fertility, cancer, diabetes, microbial infections, cardiac problems	Ayurveda	Eugenol	54
Pelargonium sidoides DC.	Umckaloabo, African geranium	Roots	Acute respiratory infections, acute rhino sinusitis, common cold, dysentery, diarrhea, sore throat, congestion	Traditional medicine system of Africa	Polyphenols, flavan-3-ols, proanthocyanidins, tannins	55 and 56

Table 15.1 *(Continued)*

Plant	Common name	Parts used	To treat ailments	Used in the medicinal system	Major active moieties	Ref.
Piper longum	Indian long pepper, Pipli	Roots, catkins, fruits	Chronic bronchitis, asthma, constipation, gonorrhea, paralysis of the tongue, diarrhea, cholera, chronic malaria, viral hepatitis, respiratory infections, stomach ache, bronchitis, diseases of the spleen, cough, tumors	Ayurveda	Alkaloids	57
Rauwolfia serpentina	Devil pepper, Serpentine wood, Indian snakeroot	Roots, leaves	Hypertension, anxiety, insomnia, schizophrenia	Ayurveda	Monoterpene indole alkaloids	58
Salacia kraussii	Ibhonsi, Ihelehele, Ubangalala, Umbhonsi , Umgunguluzane, Umnozane	Roots, stem bark	Bilharzia, dysentery, malarial infections, tuberculosis, diarrhea	Traditional medicine system of Africa	Quinone, isoiguesterol	59–61
Scilla natalensis	Wild squill, blue squill, blue hyacinth, blouberglelie, blouslangkop, inguduza	Bulbs, bulb scales	Wound-healing, sprains, fractures, boils, sores, chest pain, kidney troubles, gastro-intestinal ailments, eczema, internal tumors, menstrual pains, a purgative, a laxative	Traditional medicine system of Africa	Homoisoflavanones, cardiac glycosides, saponins, bufadienolides	12 and 62
Solanum surattense	Surattense nightshade, thai green eggplant, kantakari	Leaves, fruits (berries), seeds, roots	Bruised fingers, dyspepsia, earache, hemorrhoids	Oman traditional medicine	Glycoalkaloids	63
Sparganium stoloniferum	Sparganii Rhizoma	Rhizomes	Dysmenorrhea, amenorrhea, mass in the abdomen, abdominal distension, blood stasis, stomach diseases, uterine, hyperplasia of mammary glands, ovarian cysts,	Traditional Chinese Medicine	Phenylpropanoids, flavonoids, alkaloids, volatile oils, antraquinones, steroids	64

			endometriosis, infertility, bad circulation, postpartum abdominal pain, stomach ache, abortion, dyspepsia, gastric cancer			
Sutherlandia frutescens	Cancer bush, Turkey bush, Blaasbossie, Bitterbos, Insiswa, Phetola	Stems, leaves, flowers, roots, pods	Cancers, stomach complaints, colds, influenza, varicose veins, chicken pox, piles, liver problems, inflammation, kidney problems, backache, eye conditions, wounds, stress, heart failure, arthritis, diabetes, urinary tract infections	Traditional medicine system of Africa	Flavonols, free amino acids, triterpenes	65 and 66
Terminalia arjuna	Arjuna, Neermaruth	Bark, leaves, fruits, seeds	Heart failure, ischemic, cardiomyopathy, atherosclerosis, myocardium necrosis, blood diseases, anemia, venereal and viral diseases, fractures, ulcers	Ayurveda, Siddha	CoQ10, tannins, triterpenoid saponins, flavonoids	67
Tinospora cordifolia	Giloy, Amrita, Guduchi, Gulancha, Ambervel, Gulvel	Stem, root, whole plant	Fever, diabetes, anxiety, immunodeficiency, cancer, asthma, dysentery, leprosy, diarrhea, jaundice, skin infections, memory deficit, neurodegenerative diseases, neuro-inflammation, sleep disorders, psychological problems	Ayurveda, Chinese traditional medicine	Alkaloids, glycosides, diterpene lactones, steroids, phenolics, sesquiterpenoids	4
Vinca minor	Myrtle, small periwinkle, common periwinkle, creeping myrtle	Leaves, stems, flowers	Skin problems, hair problems, brain health, diarrhea, vaginal discharge, throat ailments, tonsillitis, chest pain, high blood pressure, sore throat, intestinal pain and swelling (inflammation), toothache, water retention, wound	Ayurveda, Traditional medicine system of Africa	Alkaloids (vincamine)	68

Table 15.1 (*Continued*)

Plant	Common name	Parts used	To treat ailments	Used in the medicinal system	Major active moieties	Ref.
Withania somnifera	Ashwagandha, Indian winter cherry, Punir, Asgand, Amukkuram	Leaf, flowers, fruits, seeds, roots, twigs	Piles, goiter, leukoderma, nervous breakdown, insomnia, constipation, stress, swellings, fever, tumors, cancer, post-traumatic stress disorder, heart problems, neuro diseases, epilepsy, carbuncles, psychological and behavioral disorders	Ayurveda	Alkaloids, steroidal lactones, flavonoids, steroids	69

etc., they have some dominance over synthetic medicines. But, while comparing with modern medicine, these formulations are not very standardized. The quality as well as effectiveness of ingredients are subject to many factors like the harvesting season and age, the geographic location, the agro-climatic conditions, the choice and stage of drying conditions, the extraction method, *etc.* Besides, different traditional medicine systems may utilize the same herb for different purposes. The thing is, traditional medicines lack proof, rather having the incidents of curing. Hence, the medicines of traditional systems need to be studied in a more scientific way and they have to be standardized for many parameters. Different studies have to be carried out according to the nature of the data required. Some of them are discussed below.

(1) The medicinal applications of plants and plant-derived products are attributed to the presence of a wide range of active compounds belonging to different classes such as flavonoids, saponins, phenolic compounds, terpenoids, steroids, alkaloids *etc.* Knowledge of these active ingredients behind the therapeutic properties of the plant make the processes a lot easier. These active components, mostly the secondary metabolites of plants are currently a vast topic under study. Bioassay-guided fractionation, docking studies, chemical characterization, *etc.*, may provide a complete picture of the plant. Besides, the physico-chemical profiling and metabolic fingerprinting contribute towards the assurance of quality evaluation. (2) The key research areas on plant secondary metabolites include the isolation, characterization, activity verification, safety profiling, toxicity studies, dosage verification, *etc.* These studies also furnish a strong platform for the development of synthetic drugs/drug components. (3) However, in the traditional systems, the herbal formulations are prepared from multiple ingredients and when tested individually, these ingredients exhibit much lower efficacy than the formulation.[70] So, it is not only the individual bio-actives that provide the therapeutic potential, but the synergistic and combinatorial efficacy of these components also play an important role. Hence, studies on the precise mechanisms of action, interactions with other compounds, and the structure–activity relationships of these formulations have to be conducted. (4) Different modes of pre-clinical studies provide better knowledge on the bioavailability, pharmacokinetics, toxicity, and metabolic parameters of the formulations. (5) Controlled clinical trials are considered as an epic method to determine the clinical efficacy, dose, and route of administration.

15.2 Different Traditional Medicine Systems

The traditional medicine systems in different countries follow different practices. In most of the countries they are acknowledged and approved by the governing systems. Let us consider some important traditional medicine systems, briefly.

15.2.1 Indian Traditional Medicine Systems

India has a long history of usage of indigenous healthcare systems. A ministry in its own right, the ministry of Ayush (Ayurveda, Yoga,

Naturopathy, Unani, Siddha, Sowa-Rigpa, and Homeopathy) is functioning in India, in order to handle the studies, research, and healthcare programs related to different indigenous alternative medicine systems in India. As the name indicates, the Ministry of Ayush takes care of seven indigenous as well as alternative healthcare systems: Ayurveda, Yoga, Naturopathy, Unani, Siddha, Sowa-Rigpa, and Homeopathy.

15.2.1.1 Ayurveda

Ayurveda is considered as the traditional therapeutic system rooted in the ancient Indian Vedas. The term means 'knowledge of life'. Most of the Ayurveda treatment protocols follow medications of natural origin and lifestyle disciplines. Various diseases in human beings are cured by Ayurvedic systems. According to Ayurveda, the human body is made up of *'panchamahabhuta'* (five classical elements of air, water, fire, earth, and ether) and the balance of the body with nature is everything and hence suppressing the natural urges will cause illness. Ayurveda incorporates medicine from minerals, herbal preparations, and metal substances along with lifestyle control for curing diseases. There is evidence available for even the surgical techniques practiced by Ayurveda systems in Ancient Ayurveda texts.[71] Ayurveda is divided into eight major clinical subdivisions – *Kayacikitsa* (internal medicine), *Kaumarabhrtya or balacikilsa* (pediatrics, obstetrics and gynecology), *Salya Tantra* (surgery), *Agada Tantra* (toxicology), *Salakya* (diseases of supraclavicular origin), *Bhutavidya* (psychiatry), *Rasayana Tantra* (rejuvenation and geriatrics), and *Vajikarana* (aphrodisiology and eugenics).[72]

15.2.1.2 Siddha

The Siddha system of medicine has evolved in the southern parts of India, and has been practiced since ancient times. The system possesses its own philosophy as well as concept and follows a holistic approach and lifestyle-oriented measures. The Siddha system is based on the theories of *Aimpootham* [five elements (*air, water, earth, fire, sky*), of which the universe and man are basically made up of] and *Mukkutram* [Three Forces/Faults (*Vali, Azhal, Aiyam*), which determine the physiological functions of the body]. The Siddha system follows *Envakai Thervukal* [eight methods of examination (*Nadi, Parisam, Naa, Niram, Mozhi, Vizhi, Malam, Moothiram*)] for the diagnosis and etiology. Oral medications, topical therapeutics, dietetics, *etc.* play major roles in Siddha treatment.[73]

15.2.1.3 Unani

The medicine system of Unani originated in ancient Greco-Arabic culture and developed as part of Arab civilization. Now, it is utilized much in India. According to Unani, human body health is maintained by the harmony of the seven basic physiological principles of elements, temperament, bodily humors, organs and systems, vital spirit, faculties or powers and functions.

Similarly, the four primary elements (earth, water, fire, and air) and four essential temperaments (hot, cold, moist, and dry) determine the health of a human body, and help to recognize the nature of disease. Basically, the Unani system tries to establish the balance between these factors to restore health. Unani consists of medicines derived from herbal, mineral, and animal origins.[74]

15.2.2 African Traditional Healthcare

Africa is known as the cradle of mankind and has huge diversity in its biological and cultural modes, which is also observed in its healing practices. In Africa, herbal medicine is part of the lifestyle as well as the existence of many people and it is estimated that almost 80% of the people depend on traditional medicine as their primary healthcare source.[75] The African traditional healthcare system is holistic and considers both the mind and body as its parts. In African traditional medicine more than 5000 plant species are used as medicine. Even though the extent of accessibility of traditional medicine is huge and makes it popular across Africa, the lack of documentation regarding efficacy, safety, and authentication of the traditional medicines of Africa make it harder to validate them and study them further.[75]

15.2.3 Traditional Chinese Medicine

Traditional Chinese Medicine (TCM) has a documented history of 5000 years.[76] TCM is based on *Yinyang* as well as *Wuxing* concepts and strictly follows *fang ji* (formulas) to prepare the prescriptions from herbs and other ingredients.[76] Normally, a TCM formula contains a group of ingredients which act synergistically. In any classic TCM formula the ingredients can be categorized into four: monarch, minister, assistant, and servant, according to their role of action. The data from clinical experiments make TCM more definite in terms of safety and efficacy. With the help of modern analytical techniques, both the procedures and products from TCM have been analyzed for their activities and are now becoming strong candidates with scientific evidence of their theoretical background, therapeutic principles, and associated technologies to stand alongside modern medicine. There are many studies ongoing to unleash the science behind their activities. Interestingly, TCM moves for new drug combinations as per the requirements.[77]

15.2.4 Kampo, the Traditional Medicine of Japan

Kampo is rooted in TCM, dating back to 500–600 AD. But now, Kampo is a distinct system with significant differences and is a major therapeutic platform in Japan. One of the most interesting factors with Kampo is that it can be well integrated with western medicine systems. For example, some Kampo medicines can be utilized along with chemotherapy or radiotherapy for treating cancer.[76] According to Kampo, every human being is considered as a complete and self-controlled whole in which body and mind impact mutually. Hence the treatments are given for the patient as a whole, not only for the illness. Besides, the diagnostic methods directly relate the symptoms to the therapy.[77]

15.2.5 Traditional Aboriginal Medicine from Australia

Aboriginal and Torres Strait Islander people are considered as the Indigenous people and original inhabitants of Australia. As per the Indigenous Australians view, ill health can be a result of a natural physical cause, a spirit causing harm, or sorcery. So, the healing process involves the physical, mental, emotional, and spiritual parts of the patient and may include traditions and ceremonies, along with the medicine. Mysticism and secrecy are kept with these practices and therefore there is not much written information available on such healing processes.[78]

15.3 Clinical Study in Traditional Medicine

As we have discussed already, traditional medicine systems are based on trial-and-error practices. So, the protocols would definitely involve 'clinical trials.' In traditional systems, knowledge on all the parameters of medicine preparation such as selection and identification of plant, proper selection of plant parts, methods of preparation, awareness on toxicity, regulations for dosage, and even the best time for harvesting to get good results are acquired by experience, involving what is now known as **clinical studies.** But in many cases however these were not recorded. As the modern system requires evidence, there are many research studies carried out all over the world on the therapeutic leads from the traditional systems. They include clinical studies to assess parameters such as safety, efficacy, toxicity, dosage, *etc.* Let us consider some of the systems and their current positions in evidence-based formulations.

15.3.1 Clinical Evidence for Ayurveda

Historically, medications were selected based majorly on the effectiveness of the ingredients, formulations, and the procedures followed. But in the modern world, it is really important to have evidence based on scientific research, in order to support the medicines and procedures followed in Ayurveda. Ayurveda involves the whole process of learning, research, and practice, by following *Pramanas* (Grandhas), which were derived from different approaches such as *pratyaksa* (direct perception), *Anumana* (logical inference), *Aptopadesa* (authentic documentary testimony), and *yukti* (experimental evidence). The primary concern of Ayurveda is the prevention of diseases and the complete well-being of health by following *dinacharya* (daily regimen), *ritucharya* (seasonal regimen), and *sadvritta* (behavioral and ethical considerations). As mentioned in the 'Ayurveda Grandhas' such as

Charaka Samhita, Susruta Samhita, Astanga Hridaya etc., Ayurveda follows different therapeutic modalities such as *panchakarma* (the bio-purification therapy), *ksharsutra* (*para* surgical measures), as well as *rasayana* (clinical medicine).

The evidence resulting from randomized controlled trials (RCTs) involving human subjects is considered as the highest level of its kind, for any medication. Furthermore, it is possible to conduct randomized, controlled studies by testing individualized classic Ayurvedic *versus* allopathic treatment by making the measurements acceptable to western standards and to Ayurvedic physicians. Moreover, many studies have derived specific, tailored features for a feasible comparison of parameters.[79]

The medications from Ayurveda can be of two types:

(1) The Ayurveda drugs or the classical preparations, which are prepared as per the recipes from the *Grandhas* (authoritative books), which provide a reference of the ingredients, preparation methods, dosage, indications, *etc.*
(2) Preparations made out of the ingredients mentioned in the *Grandhas*, but the combinations are different from the *Grandha* recipes. These are known as proprietary Ayurvedic medicines (PAM).

Hence, the evidence-based Ayurveda holds them both. Currently, a huge number of studies are going on to determine the efficacy of Ayurvedic drugs in the treatment of many ailments ranging from mild infections to complicated diseases. Some of them are discussed henceforth, just to deliver an idea, to what extend the studies are carried out and how they resulted.

Arthritis is a field of thriving research, where all the possible modes of medication have been studied due to the drastic necessity. In a study, three different treatments, Ayurveda, Allopathy, and a combination of both, were compared for their efficacy in rheumatoid arthritis (RA) patients. All three groups were comparable in every measured characteristic, without any statistically significant differences, but with a lesser number of reported adverse events for Ayurveda; this suggests a possible safe and effective treatment system for RA.[79] Similarly, treatment with Ashwagandha (*Withania somnifera*) powder and Sidh Makardhwaj was found to be effective in relieving the RA factor as well as all the assessment scores.[80] In osteoarthritis (OA; *Sandhigatavata* in Ayurveda) patients, management of the disease was achieved by integrating Ayurveda with allopathic systems. The use of Ashwagandha, Yogaraja Guggulu, and Narayan taila along with the dietary and lifestyle recommendations for eight weeks was useful for improving the symptoms and signs of OA.[81] Compared with multimodal conventional care, Ayurvedic treatment also provided more pronounced results, both for the primary and secondary outcomes.[82] Ayurvedic multimodal management of OA is a comparatively more effective concept in the field. In a study, the OA patients were given *Svedana, Snehana, Matrabasti, Mriduvirechana*, and *Jalukavacharana* along with oral medications such as *Ashwagandha Churna* and *Yogaraja Guggulu.* This protocol of multimodal OA management has beneficially impacted ($P < 0.001$) all the clinical features of *Sandhigatavata* and was recommended to be a better option to non-steroidal anti-inflammatory drugs (NSAIDs).[83]

Many clinical studies have been carried out for Ayurveda therapies, which are implemented as complementary to the conventional medicine system. In cancer patients, radiotherapy is accepted as one of the most effective treatments. One Ayurvedic formulation, given for gargling in the oral cavity of cancer patients who had undergone radiotherapy, was found to be effective in reducing the bacterial load in the oral cavity of cancer patients.[84] Similarly, in breast cancer patients undergoing chemotherapy, the administration of Ashwagandha root extracts furnished lower fatigue scores and improvement in the quality of life.[85] A recent review on the Ayurvedic formulations for breast cancer suggests that they are not only beneficial in terms of reducing the side effects of chemotherapeutic agents, but also improving their efficacy.[86] *Tinospora cordifolia* is such an important indigenous herb from the Ayurvedic realm, which was proven to inhibit breast cancer, cervical cancer, skin cancer, and brain cancers. It exhibited anti-proliferative, differentiation, as well as senescence inducing and anti-migratory activities even at lower concentrations, and has been proven to be safe for primary hippocampal and astrocytic cells.[4]

Lifestyle diseases are another important area where Ayurvedic interventions have been hugely studied. Many of the medicaments prescribed by Ayurveda can effectively manage type 2 diabetes mellitus (DM) and its intermediate stage – prediabetes by a reduction in blood sugar, decrease in insulin resistance, and improvement in beta cell activity. A double-blinded, placebo-controlled, randomized clinical trial has been conducted in order to study the potential of a polyherbal drug (that consists of *T. cordifolia, Gymnema sylvestre, Pterocarpus marsupium, Momordica charatia*, and *Zingiber officinale*) in the prevention of DM, while comparing it to a placebo. The intervention group exhibited a significant risk reduction, with a decrease in their blood glucose levels (both fasting and postprandial), fasting serum insulin, HbA_{1C} and HOMA-IR values.[87] However, another Ayurvedic formulation named *pancreas tonic* could only significantly improve the glucose levels of type 2 diabetic patients with HbA1c levels between 10.0% to 12.0%; in the case of 8% to 9.9%, there were not any significant changes, compared to the placebo.[88] *Ayush* 82 is recommended as a first line treatment for Type 2 Diabetics, as per a study published in 2017.[89]

Ayurveda serves effectively in cardiac disorders. A combination of *W. somnifera, Terminalia arjuna, Phyllanthus emblica*, and *Ocimum tenuiflorum* could improve the coronary artery disease, mediated through the reduction of systolic and diastolic blood pressure, serum cholesterol, and triglyceride levels as well as the increase of high-density lipoprotein (HDL) cholesterol.[90]

Asthma is a chronic inflammatory disease of the airways, and is one of the most common diseases in industrialized and developed countries. Ayurveda provides remedies for asthma (*Tamaka Swasa*) by oral medications, procedures such as *shodhana*, as well as *shamana* and lifestyle and diet control. As can be seen from the studies, the management of asthma in Ayurveda is at clinical as well as immunological levels. The significant whole system management of asthma by Ayurveda is found to be mediated through the reduction in circulating levels of IgE and eosinophil, associated with the decrease in levels of cytokines, Th_2 and Th_1, and changes of peak expiratory flow rate (PEFR).[91–93]

The importance and scope of Ayurveda in neuro/mental health is a widely studied area. Ayurveda gives great importance to brain health and more than 80 central nervous system (CNS) disorders are handled under the name '*Vatavyadhi*'. Besides, Ayurveda explains different brain conditions/diseases: Alzheimer's disease (AD) as *Smritibhramsha*, Autism as spectrum disorder *Unmada*, Parkinson's disease (PD) as *Kampvata*, and attention deficit/hyperactivity disorder (ADHD) as *Dhee, Dhriti, Smriti, Vibramsha*. The rasayana herbs specifically used for brain health are called medha rasayana herbs, and the important medha rasayana herbs are *T. cordifolia, W. somnifera, Gingko biloba, Oscimum sanctum, Nardostachys jatamansi, Convolvulus pluricaulis, Centella asiatica, Glycyrrhiza glabara, Clitoria ternatea L., Acorus calamus, Celastrus panniculatus, Bacopa monnieri etc.*[94,95] Significant anti-stress-like activities exhibited by *T. cordifolia* in male human subjects are found to be mediated through the activation of the parasympathetic nervous system.[96] The biochemical and behavioral improvements associated with the herb in mentally stressed patients are also attributed to the antioxidant and anti-inflammatory activities[46]. The efficacy and safety of Ashwagandha were studied in a trial for the management of stress-related disorders especially mood disorders. It was found that, as a rasayana, Ashwagandha improves the total quality of life without any adverse effects.[97] These results are in agreement with the results of another study, where a highly concentrated extract of full spectrum Ashwagandha roots effectively reduced all the stress assessment parameter scores compared to a placebo, within 60 days.[98] In line with these studies, the sleep quality improvement and sleep-inducing capacity of Ashwagandha root extract in insomnia patients, even at a dose of 300 mg extract twice daily is another important potential.[99] The effects of another important rasayana herb, *C. ternatea* L. on learning, memory, amnesia, biosynthesis, and content of acetylcholine, cholinergic activity, age-related memory disorders *etc.* were studied and found to induce significant permanent changes.[39] Hypertension can be relieved by the administration of *Rauwolfia serpentine*.[100,101] *B. monnieri* has been proved to alter the levels of acetylcholine as well as the monoamine neurotransmitter and improve attention, working memory, anxiety, depression, and cognitive processing.[29,30]

Mostly, Ayurveda treatments are a combination of medication and medical procedures. Research supports the fact that the combinations are more effective than individual interventions. For example, a combination of Integrated Yoga and *Ayurveda Rasayana* consisting of *brahmi ghrit* provided significantly better results than individual interventions for cognitive abilities such as learning, attention, and working memory in elderly persons with mild cognitive impairment (MCI).[102] Similarly, an integrated intervention of Ayurveda oral medicines, panchakarma, and yoga therapy was effective in reducing the pain and symptoms in migraine patients, while compared to NSAID's treatment.[103] As reported in a case study, Traumatic Optic Neuropathy (*Abhighatajanya Vataprakopaj Drishtinash*) could be successfully treated with a one-year treatment plan including internal medication, panchakarma, and local treatment. It had a significant improvement in the recovery of damaged optic nerve fibers and quality of vision.[104] Similarly, in patients with depression, ayurvedic therapy, including *snehana, svedana, shodhana, virecana,* and *samsarjanakarma* followed by oral medication of *medhyarasayana* and *Shirodhara* has a positive influence on parameters such as psychopathology, cardiac autonomic function, and endocrinal parameters.[105] Idiopathic Parkinson's Disease can be managed by the integrative applications of a natural source of levodopa (the seeds of *Mucuna pruriens* L.), *Panchakarma* regimens such as *nasya* as well as *basti* and intervention of rasayanas.[106]

A retrospective cohort study was conducted for liver cirrhosis with ascites from the hospital records of patients diagnosed through medical ultrasonography. Records of 55 cases with stage 3 decompensated liver cirrhosis were considered, for which interventions, including *nitya virechana* (daily therapeutic purgation), oral medicaments, diet, salt, and fluid restrictions were carried out. These procedures improved the parameters such as clinical profile, liver function, renal function, prothrombin time, and international normalized ratio (INR) in patients of ascites with decompensated cirrhosis. Also, the treatment regimen conducted in the clinical scenario involved a whole Ayurvedic approach.[107]

There are many clinical studies that have successfully been completed for Ayurveda in many different ailments. A single-arm exploratory open-label clinical trial was conducted on *Urdhwaga Amlapitta* (Dyspepsia) patients, by giving a different combination of Ayurveda drugs, according to ones *Prakriti* (psychosomatic constitution). The medication was found to be well-tolerated, successful, and safe.[108] In a similar way, a one-year intervention, including *Patolakaturohinyadi kashaya* (Herbal decoction), *Kaishor guggulu* (Tablet), *Mahatiktaka ghrita* (Capsule of medicated ghee), *Gandhaka rasayana* (Tablet), *Khadirarishta* (Herbal fermented liquid), *Winsoria oil* (Herbal coconut base oil), and *Pathyahara* (Strict dietary plan) could successfully relieve the complex pathophysiology of psoriasis.[109] Uncomplicated psoriasis can be treated in a better and quicker way by using a neem leaves (*Azadirachta indica*) extract, as shown in a double-blind drug trial.[28] In another study, *T. cordifolia* reduced hyperlipidemic and oxidative stress conditions in the urine and blood of a chronic alcohol consumer by increasing dopamine levels and cAMP activation as well as suppressing oxidative stress.[4] A recent open label, non-randomized, single center, single group pilot study has evaluated the potential of an Ayurvedic formulation in treating influenza and influenza like illness (ILI), along with its safety profiles in 34 patients. The formulation, AYUSH-64 composed of *Alstonia scholaris* R. Br., *Picrorhiza kurroa Royle* ex. Benth, *Swertia Chirata* Pexbex. Karst and *Caesalpinia crista* L. was found to be safe and the intervention of one week is efficacious for ILI.[110]

During the recent pandemic of Covid-19, there were many attempts made to derive new treatment protocols and medicines. Of course, from Ayurveda, a number of medications were proposed. One of them contained *Swasari Ras*, along with *T. cordifolia, W. somnifera*, and *O. sanctum*, with the administration of *Anu taila* (traditional nasal drop). The treatment group has expedited virological clearance, and reduced the inflammation markers, proving a 40% better recovery of the treatment group.[111] In another study, better symptomatic relief from COVID-19 was observed in patients treated with Ayurvedic medicines than those treated with both Ayurvedic and allopathic medicines.[112] Similarly, in another study, *Vyaghryadi Kashaya,* made out of *Kantakari* (*Solanum xanthocarpum* Schrad, & Wendl.), *Shunthi* (*Z. officinale* Rosc.), and *Guduchi* (*T. cordifolia* Thrunb.), along with *Pippali* (*Piper longum* L.) powder successfully improved the symptoms, vital parameters, and time for viral clearance, when compared with the combination of vitamin C tablets and paracetamol tablets, as well as the combination of fine powder of Shunthi and paste of *Rasona* (*Allium sativum* L.) *kalka.*[113]

In India, where Ayurveda is successfully administered as a major health care system under the government, an apex body known as the central council for research in Ayurvedic sciences (CCRAS) has been established and is dedicated to the research-related activities on the field. CCRAS has furnished specific systematic guidelines for ensuring the safety, efficacy, and quality control parameters for Ayurvedic formulations and procedures. Many randomized controlled double-blind clinical studies have been conducted and recorded evidences under CCRAS. Just to state some studies, *W. Somnifera* was effective for altering the levels of hemoglobin, hair melanin, red blood cell (RBC) count, serum cholesterol and erythrocyte sedimentation rate (ESR), *A. calamus* produced significant improvements in respiratory distress and chest pain, *Commiphora wightii* furnished a decrease in lipid fractions and caused regression of atherosclerosis and was found useful in treating ischemic heart diseases and *Boswellia serrata* is beneficial in rheumatoid arthritis.[114]

15.3.2 Clinical Evidence for Kampo

RCTs conducted for Kampo also compared the efficacies of both formulations and treatments. Rikkunshito, a combination of eight herbal medicines mentioned in the Japanese Pharmacopoeia (*Atractylodis Lanceae Rhizoma, Ginseng Radix, Pinelliae Tuber, Hoelen, Zizyphi Fructus, Aurantii Nobilis Pericarpium, Glycyrrhizae Radix* and *Zingiberis Rhizoma*) is a standardized Kampo herbal medicine, which is investigated for its therapeutic effects towards the gastrointestinal tract disorders and improving apetite.[115] In a double blind two arm study with 296 patients having dysmotility-like dyspepsia, the patients are orally administered either with the Kampo drug

TSUMURA Rikkunshito Extract granules (TJ-43), 2.5 g or with its 40-fold diluted low dose as a control. This double-blind study has validated the effectiveness of TJ-43 Rikkunshito and evaluated it as clinically useful.[116] Another randomized, multicenter, placebo-controlled study has assessed the effect of Rikkunshito in 247 patients suffering from functional dyspepsia. Administration of the drug for 8 weeks was really effective in reducing dyspepsia, as seen in earlier studies.[117] There have been many studies conducted so far, proving the effect of Rikkunshito for dyspepsia. But a systematic review conducted by Nobuaki Hoshino and crew in 2019 put forward the requirement of high-quality studies on Rikkunshito.[118]

15.3.3 Clinical Evidence for TCM

When considering the mutual impact of traditional medicine and modern science, TCM is the one with the most influence. Many research studies have been carried out on the efficacy of TCM products, in comparison with either a placebo or control (such as modern medicine). For example, a number of RTCs were conducted on the effects of TCM formulations on asthma. Yes, there were positive effects. But, in most of these studies, the control was a modern medicine, and the test dose was a combination of modern medicine and TCM formulation, and in almost all the studies, the treatment group provided the best results. Even though the results could not draw a specific distinction between these two, they pave the way for further research in the field.[119]

Furthermore, in TCM the procedures also hold equal importance to the medicines. From many of the RCTs, it is evidenced that TCM massage therapy is effective in treating congenital muscular torticollis (CMT) in infants and children.[120]

St. John's wort (*Hypericum perforatum*) is a popular TCM medication for treating depression. It was reported in a comparative review that the extent and frequency of the adverse effects obtained from the extracts of St John's wort are significantly lower than that for the conventional antidepressants, while they exhibit similar beneficial effects.[121] Interventions with formulations of St. John's wort were found to be more significant than a placebo and produced comparable results with standard antidepressants for short term treatment of depressive disorders.[48] Also, the herb evidenced its activity against different enveloped viruses in clinical trials.[122] Its anticancer, anti-inflammatory, neuroprotective, antioxidant, and wound healing effects were also proven clinically in humans.[48] Interestingly, the herb can amazingly treat opiate withdrawal symptoms in humans.[123]

Similarly, a review was published on the trials of Saw Palmetto (*Serenoa repens*) for lower urinary tract symptoms associated with benign prostatic hyperplasia (BPH). Saw palmetto, commonly known as *Ju Ye Zong* in Traditional Chinese Medicine is widely used for treating infertility, inflammation, and many other disorders. The review puts forward the conclusion that it exhibited significant improvement in urinary symptoms and flow, compared to finasteride while associated with lesser adverse treatment events.[124]

15.3.4 African Traditional Healthcare System

Compared to any other continent, Africa follows traditional healthcare systems more than the conventional medicines. So, extensive scientific studies have been carried out on the safety and efficacy of the herbs and procedures associated with the system. The substantial studies happening for *Sutherlandia frutescens* is an example of the recent interest in the indigenous plants of the African traditional healthcare system.[125–127] It is observed that, when taken with a proper diet, *S. frutescens* can improve the life quality of patients affected with human immunodeficiency virus (HIV), without any adverse effects.[127] Scientific studies have evidenced the analgesic, anti-inflammatory, and anti-rheumatic effects of Devil's claw (*Harpagophytum procumbens*) also.[128] The proprietary medicine EPs7630 is derived from the root extract of a herb with a long history of usage in African traditional medicine – *Pelargonium sidoides.* The safety and effectiveness of EPs7630 are evidenced in more than 20 clinical studies.[56]

15.3.5 Australian Aboriginal Traditional Medicine

A study published under the title "*If you don't believe it, it won't help you*" was about the usage of traditional medicine for treating cancer in Aboriginal people. The study has evidenced that bush medicine can play an important role in relieving the symptoms of chemotherapy or the stress associated with the situation. But the use depends on the access to the medicines, knowledge of the systems, concerns on the integrative medical systems *etc.*[78] Compared to the other traditional medicine systems, Aboriginal medicine has been less studied in a scientific manner.

15.4 Tailoring of Study Designs

There are, of course, paucities in the comprehensive evaluation of the complementary/alternative medicine systems, by the tool of clinical trials. Most of the traditional medicines involve complex interventions and diagnostic assessments, individualized therapy, *etc.* which make it difficult to conduct a clinical study, which is based on standardized interventions. However, there are specified intervention patterns and protocols developed for Ayurvedic studies; some of which are discussed below.

a. **N-of-1 RCTs** include single subjects, which undergoes different interventions separated by washout periods. Both the subject and the physician are blinded about the interventions and hence these kinds of trials are conducted only for the interventions which provide rapid onset of action, and stops the effects when the intervention is discontinued. This design allows the personalized assessment of interventions.[129]
b. In **Blackbox RCTs**, the patients are allocated randomly into different intervention groups (Ayurveda, allopathy, and/or integrated), based on biomedical diagnoses. In case of Ayurveda intervention, it is allowed to diagnose and treat individual cases in Ayurvedic paths, without any

restrictions on medicine/modality. The outcomes are measured and assessed based on the allopathic parameters which make it possible for comparisons and evaluations. This pattern provides a better design for evaluating the effectiveness of an Ayurvedic approach. However, it fails to explain the mode of action and efficiency of any particular single Ayurvedic intervention.[130–132]

c. In **stepped wedge patterns**, a sample is chosen and from this initial sample, the subjects/clusters move from the control towards the intervention in a phased manner. This study pattern is helpful in evaluating the efficacy of an intervention over time.[133]
d. **Adaptive trials** are found to resemble real time clinical practice, since they involve the modification/change of one or more factors of the intervention, during the course of the study.[134]
e. **In single subject research,** a single subject or a small group of subjects are administered with the treatment, where each subject serves as its own control.[130,135]

The availability of different study designs and aspects designed for real-time assessments of Ayurvedic interventions provide a wide array of options to select according to the objective of the study.

15.5 Conclusion

There has always been conflicts between modern medicine and traditional medicine systems. Modern medicine provides solutions for ailments and their symptoms, while traditional medicine systems believe in the overall well-being of the body. Moreover, conventional medicines are associated with side effects and reduced efficacy during a long course of use. Besides, they are reported to suffer from resistance and, in some cases, are highly expensive. But interestingly, modern research looks for newer remedies from the natural sources. Recent research in this field evidences the safety and efficacy, even while examining under clinical trials. The therapeutics derived from herbal sources are proved to have synergistic effects from a number of bioactive moieties. Scientific studies on due course pave the way towards a brighter, evidence-based practice of traditional medicine systems. Clinical trials are considered as the most reliable scientific evidence of a therapeutic and this chapter has outlined some of the clinical trials which have been conducted in this arena. If the concern is traditional medicine, the clinical trials are to be designed according to the aspects and mode of information needed.

References

1. B. B. Petrovska, *Pharmacogn. Rev.*, 2012, **6**, 1.
2. J. J. M. Miranda, Medicinal plants and their traditional uses in different locations, in *Phytomedicine*, ed. R. A. Bhat, K. R. Hakeem and M. A. Dervash, Academic Press, 2021, pp. 207–223.

3. A. M. Reid, C. B. Oosthuizen, B. D. Fibrich, D. Twilley, I. A. Lambrechts, M. N. de Canha, S. Rademan and N. Lall, Traditional medicine: the ancient roots of modern practice, in *Medicinal Plants for Holistic Health and Well-Being*, ed. N. Lall, Academic Press, 2018, pp. 1–11.
4. A. Sharma, P. Bajaj, A. Bhandari and G. Kaur, *Neurochem. Int.*, 2020, **141**, 104891.
5. D. S. Fabricant and N. R. Farnsworth, *Environ. Health Perspect.*, 2001, **109**, 69.
6. R. Segal, I. Feuerstein and A. Danin, *Biochem. Syst. Ecol.*, 1987, **15**, 411.
7. A. Ziyyat, A. Legssyer, H. Mekhfi, A. Dassouli, M. Serhrouchni and W. Benjelloun, *J. Ethnopharmacol.*, 1997, **58**, 45.
8. S. O. Okoro, A. H. Kawo and A. H. Arzai, *Bayero J. Pure Appl. Sci.*, 2011, **5**, 163.
9. R. Jain, P. Sharma, T. Bhagchandani and S. C. Jain, *Pharm. Res.*, 2012, **5**, 4934.
10. O. Nasir, *Saudi J. Biol. Sci.*, 2021, **28**, 2701.
11. M. S. Elshikh, E. Rani, D. A. Al Farraj, F. M. Al-Hemaid, M. R. A. Gawwad, T. J. Malar, L. Dyona and P. Vijayaraghavan, *Physiol. Mol. Plant Pathol.*, 2022, **117**, 101743.
12. J. M. Watt and M. G. Breyer-Brandwijk, *The Medicinal and Poisonous Plants of Southern and Eastern Africa*, E&S Livingston Ltd, Edinburgh, 1962.
13. D. S. Grierson and A. J. Afolayan, *J. Ethnopharmacol.*, 1999, **67**, 327.
14. A. Smith, *A Contribution to the South African Materia Medica*, Lovedale, South Africa, 1st edn, 1888.
15. B. E. Van Wyk and N. Gericke, *People's Plants: A Guide to Useful Plants of Southern Africa*, Briza Publications, Pretoria, South Africa, 2000.
16. C. Bishop, *Farmers Weekly,* 1997, p. 6.
17. R. B. Bhat and T. V. Jacobs, *J. Ethnopharmacol.*, 1995, **48**, 7.
18. N. Bizimana, *Traditional Veterinary Practice in Africa*, Deutsche Gesellschaft für Technische Zusammenarbeit GmbH, Eschborn, 1994.
19. J. Misir, F. H. Brishti and M. M. Hoque, *Am. J. Food Technol.*, 2014, **2**, 93.
20. A. A. Maan, A. Nazir, M. K. I. Khan, T. Ahmad, R. Zia, M. Murid and M. Abrar, *J. Herb. Med.*, 2018, **12**, 1.
21. R. K. Verma, M. Garima, S. Pradeep, K. K. Jha and R. L. Khosa, *Der Pharma Chem.*, 2011, **2**, 142.
22. A. Mukherjee, G. K. Chouhan, S. Singh, K. Chatterjee, A. Kumar, A. K. Gaurav, D. K. Jaiswal and J. P. Verma, Alpinia officinarum. in *Naturally Occurring Chemicals Against Alzheimer's Disease*, Academic Press, 2021, pp. 453–461.
23. A. S. Nair and A. Paliwal, Systems pharmacology and molecular docking strategies prioritize natural molecules as antiinflammatory agents. in *Inflammation and Natural Products*, Academic Press, 2021, pp. 283–319.
24. X. Feng, S. Cao, F. Qiu and B. Zhang, *Pharmacol. Ther.*, 2020, 107650.
25. E. Joubert and D. de Beer, *S. Afr. J. Bot.*, 2011, 77, 869.

26. P. Y. Hayes, A. H. Jahidin, R. Lehmann, K. Penman, W. Kitching and J. J. De Voss, *Phytochemistry*, 2008, **69**, 796.
27. V. K. Saxena and S. Chourasia, *Fitoterapia*, 2001, **72**, 307.
28. A. A. Mohammad, *Evidence-Based Complementary Altern. Med.*, 2016, 7382506.
29. J. M. Brimson, S. Brimson, M. I. Prasanth, P. Thitilertdecha, D. S. Malar and T. Tencomnao, *Sci. Rep.*, 2021, **11**, 1.
30. T. Peth-Nui, J. Wattanathorn, S. Muchimapura, T. Tong-Un, N. Piyavhatkul, P. Rangseekajee, K. Ingkaninan and S. Vittaya-Areekul, *Evidence-Based Complementary Altern. Med.*, 2012, 606424.
31. J. S. Neergaard, J. Andersen, M. E. Pedersen, G. I. Stafford, J. Van Staden and A. K. Jäger, *S. Afr. J. Bot.*, 2009, **75**, 371.
32. S. Tonisi, K. Okaiyeto, L. V. Mabinya and A. I. Okoh, *Saudi J. Biol. Sci.*, 2020, **27**, 3559.
33. M. Koel, M. Kuhtinskaja and M. Vaher, *Sep. Purif. Technol.*, 2020, 117438.
34. A. K. Chaudhary, S. Ahmad and A. Mazumder, *Pharmacogn. J.*, 2011, **3**, 12.
35. İ. Gürbüz and E. Yesilada, *J. Ethnopharmacol.*, 2007, **112**, 284.
36. B. Brinkhaus, M. Lindner, D. Schuppan and E. G. Hahn, *Phytomedicine*, 2000, 7, 427.
37. R. Singh and T. Jawaid, *Pharmacogn. J.*, 2012, **4**, 1.
38. J. J. Patel, S. R. Acharya and N. S. Acharya, *J. Ethnopharmacol.*, 2014, **154**, 268.
39. P. K. Mukherjee, V. Kumar, N. S. Kumar and M. Heinrich, *J. Ethnopharmacol.*, 2008, **120**, 291.
40. P. Agarwa, B. Sharma, A. Fatima and S. K. Jain, *Asian Pac. J. Trop. Biomed.*, 2014, **4**, 245.
41. J. Bowie, *S. Afr. Q. J*, 1830, 27.
42. G. Bhakta Prasad and S. Lalita, *J. Integr. Med.*, 2013, **11**, 73.
43. A. R. Tricker, C. D. Wacker and R. Preussmann, *Cancer Lett.*, 1987, **35**, 199.
44. L. Liu, Y. Wang, J. Zhang and S. Wang, *J. Pharm. Biomed. Anal.*, 2021, **193**, 113704.
45. I. D. Boateng and X. M. Yang, *Ind. Crops Prod.*, 2022, **176**, 114281.
46. S. Brien, G. T. Lewith and G. McGregor, *J. Altern. Complementary Med.*, 2006, **12**, 981.
47. S. Rawat, A. K. Jugran, I. D. Bhatt and R. S. Rawal, *J. Pharm. Pharmacol.*, 2018, **70**, 687.
48. K. M. Klemow, A. Bartlow, J. Crawford, N. Kocher, J. Shah and M. Ritsick, Medical Attributes of St. John's Wort *(Hypericum perforatum)*, in *Herbal Medicine: Biomolecular and Clinical Aspects*, 2nd edn, CRC Press/Taylor & Francis, Boca Raton (FL), 2011, p. 211.
49. M. Kalachaveedu, D. Raghavan, S. Telapolu, S. Kuruvilla and B. Kedike, *J. Ethnopharmacol.*, 2018, **210**, 408.
50. A. Hutchings, A. H. Haxton Scott, S. G. Lewis and A. Cunningham, *Zulu Medicinal Plants. An Inventory*, University of Natal Press, Pietermaritzburg, 1996.

51. M. Bortolotti, D. Mercatelli and L. Polito, *Front. Pharmacol.*, 2019, **10**, 486.
52. F. Anwar, S. Latif, M. Ashraf and A. H. Gilani, *Phytother. Res.*, 2007, **21**, 17.
53. S. N. Rai, H. Birla, W. Zahra, S. S. Singh and S. P. Singh, *J. Chem. Neuroanat.*, 2017, **85**, 27.
54. P. A. G. N. Prakash and N. Gupta, *Indian J. Physiol. Pharmacol.*, 2005, **49**, 125.
55. A. Timmer, J. Guenther, E. Motschall, G. Ruecker, G. Antes and W. V. Kern, *Cochrane Database Syst. Rev.*, 2013, 10.
56. D. Brown, *Nat. Med. J.*, 2009, **1**, 1.
57. S. Kumar, J. Kamboj and S. Sharma, *J. Acupunct. Meridian Stud.*, 2011, **4**, 134.
58. B. Mukerji, 2016, History of Rauwolfia serpentina and our early investigations, in Pharmacology of Oriental Plants: Proceedings of the First International Pharmacological Meeting, Stockholm, ed. K. K. Chen and B. Mukerji, Elsevier, 1961, p. 79.
59. S. O. Bandeira, F. Gaspar and F. P. Pagula, *Pharm. Biol.*, 2001, **39**, 70.
60. J. N. Figueiredo, B. Räz and U. Séquin, *J. Nat. Prod.*, 1998, **61**, 718.
61. H. E. Zimila, A. L. Matsinhe, E. Malayika, Á. I. Sulemane, V. N. C. Saete, S. C. Rugunate, P. J. Cumbane, I. Magaia and F. Munyemana, *Biocatal. Agric. Biotechnol.*, 2020, 101862.
62. A. Hutchings, *Bothalia*, 1989, **19**, 111.
63. M. Narayanan, T. Jayashree, S. Kandasamy, D. Natarajan, G. Liu, B. H. Elesawy, A. Elfasakhany and A. Pugazhendhi, *Process Biochem.*, 2021, **109**, 178.
64. J. Jia, X. Li, X. Ren, X. Liu, Y. Wang, Y. Dong, X. Wang, S. Sun, X. Xu, X. Li and R. Song, *J. Ethnopharmacol.*, 2020, 113571.
65. B. E. Van Wyk and C. Albrecht, *J. Ethnopharmacol.*, 2008, **119**, 620.
66. O. M. Aboyade, G. Styger, D. Gibson and G. Hughes, *J. Altern. Complementary Med.*, 2014, **20**, 71.
67. A. Amalraj and S. Gopi, *J. Tradit. Complementary Med.*, 2017, 7, 65–78.
68. M. Moudi, R. Go, C. Y. S. Yien and M. Nazre, *Int. J. Prev. Med.*, 2013, **4**, 1231.
69. N. Singh, M. Bhalla, P. de Jager and M. Gilca, *Afr. J. Tradit. Complementary Altern. Med.*, 2011, **8**, 208–213.
70. F. S. Li and J. K. Weng, *Nat. Plants*, 2017, **3**, 1.
71. Y. S. Jaiswal and L. L. Williams, *J. Tradit. Complementary Med.*, 2017, 7, 50.
72. S. Sen and R. Chakraborty, *J. Tradit. Complementary Med.*, 2017, 7, 234–244.
73. J. J. Thas, *Clin. Dermatol.*, 2008, **26**, 62.
74. The Editors of Encyclopaedia Britannica, Unani medicine, Encyclopedia Britannica, 2017, July 21, https://www.britannica.com/science/Unani-medicine.
75. M. F. Mahomoodally, *Evidence-Based Complementary Altern. Med.*, 2013, **2013**, 617459.

76. H. Yuan, Q. Ma, L. Ye and G. Piao, *Molecules*, 2016, **21**, 559.
77. A. Ishibashi, H. Kosoto and S. Ohno, General introduction to Kampo, in *Introduction to Kampo, Japanese Traditional Medicine, The Japan Society for Oriental Medicine*, Elsevier, Tokyo, Japan, 2005, pp. 2–13.
78. S. J. Oliver, *J. Ethnobiol. Ethnomed.*, 2013, **9**, 1.
79. D. E. Furst, M. M. Venkatraman, M. McGann, P. R. Manohar, C. Booth-LaForce, R. Sarin, P. G. Sekar, K. G. Raveendran, A. Mahapatra, J. Gopinath and P. R. Kumar, *J. Clin. Rheumatol.*, 2011, **17**, 185.
80. G. Kumar, A. Srivastava, S. K. Sharma, T. D. Rao and Y. K. Gupta, *Indian J. Med. Res.*, 2015, **141**, 100.
81. S. Bhat, V. Gupta, N. Srikanth, M. M. Padhi, R. Rana, R. Singhal and K. S. Dhiman, *J. Res. Ayurvedic Sci.*, 2017, **1**, 40.
82. C. S. Kessler, K. S. Dhiman, A. Kumar, T. Ostermann, S. Gupta, A. Morandi, M. Mittwede, E. Stapelfeldt, M. Spoo, K. Icke and A. Michalsen, *Osteoarthr. Cartil.*, 2018, **26**, 620.
83. M. R. Sharma, C. S. Mehta, D. J. Shukla, K. B. Patel, M. V. Patel and S. N. Gupta, *Ayu*, 2013, **34**, 49.
84. S. Chavan, S. Bhuvad, B. Kumbhlakar, J. Auti, T. Walunj, S. Pathak, R. Tanpure, S. Gujar, J. Shinde, A. Kulkarni and V. Gupta, *J. Herb. Med.*, 2021, **30**, 100510.
85. B. M. Biswal, S. A. Sulaiman, H. C. Ismail, H. Zakaria and K. I. Musa, *Integr. Cancer Ther.*, 2013, **12**, 312.
86. M. K. Dash, N. Joshi, D. N. S. Gautam, R. Jayakumar and Y. B. Tripathi, *J. Herb. Med.*, 2021, **29**, 100490.
87. A. Nakanekar, K. Kohli and P. Tatke, *J. Ayurveda Integr. Med.*, 2019, **10**, 284.
88. S. H. Hsia, M. Bazargan and M. B. Davidson, *Metabolism*, 2004, **53**, 1166.
89. A. K. Panda, K. K. Ratha and M. M. Rao, *J. Tradit. Med. Clin. Natur.*, 2017, **6**, 2.
90. S. Dwivedi, D. Gupta and K. K. Sharma, *J. Med. Aromat. Plant Sci.*, 2000, **22**, 616.
91. A. K. Panda and S. R. Doddanagali, *J. Ayurveda Integr. Med.*, 2011, **2**, 85.
92. S. J. Jayandan, C. M. Kundagol and J. Chacko, *J. Ayurveda Integr. Med.*, 2018, **9**, S5.
93. K. S. Joshi, T. M. Nesari, A. P. Dedge, V. R. Dhumal, S. A. Shengule, M. S. Gadgil, S. Salvi and M. V. S. Valiathan, *J. Ethnopharmacol.*, 2017, **197**, 110.
94. R. H. Singh, K. Narsimhamurthy and G. Singh, *Biogerontology*, 2008, **9**, 369.
95. R. Kulkarni, K. J. Girish and A. Kumar, *Phcog. Rev.*, 2012, **6**, 147.
96. B. A. Salve, R. K. Tripathi, A. U. Petare, A. A. Raut and N. N. Rege, *Ayu*, 2015, **36**, 265.
97. S. Shilpi and Y. K. Sharma, *Int. J. Ayurveda Res.*, 2016, **4**, 707–712.

98. K. Chandrasekhar, J. Kapoor and S. Anishetty, *Indian J. Psychol. Med.*, 2012, **34**, 255.
99. D. Langade, S. Kanchi, J. Salve, K. Debnath and D. Ambegaokar, *Cureus*, 2019, **11**, E5797.
100. R. S. Green and D. Davolos, *Am. J. Med.*, 1956, **20**, 760.
101. R. J. Vakil, *Br. Heart J.*, 1949, **11**, 350.
102. S. Chobe, S. K. Patra, M. Chobe and K. Metri, *J. Ayurveda Integr. Med.*, 2020, 100373.
103. V. M. Sharma, N. K. Manjunath, H. R. Nagendra and C. Ertsey, *Complementary Ther. Clin. Pract.*, 2018, **32**, 85.
104. P. M. Bhat, *J. Ayurveda Integr. Med.*, 2021, 100494.
105. R. K. Kishore, H. A. Abhishekh, K. Udupa, J. Thirthalli, G. S. Lavekar, B. N. Gangadhar, T. R. Raju and T. N. Sathyaprabha, *Asian J. Psychiatr*, 2014, **12**, 100.
106. N. Pathak-Gandhi and A. D. Vaidya, *J. Ethnopharmacol.*, 2017, **197**, 46.
107. B. R. Tubaki, S. C. Gawas and H. Negi, *J. Ayurveda Integr. Med.*, 2022, 100508.
108. K. Meenakshi, N. Vinteshwari, J. Minaxi and S. Vartika, *J. Ayurveda Integr. Med.*, 2021, **12**, 87.
109. G. C. Nille and A. K. Chaudhary, *J. Ayurveda Integr. Med.*, 2021, **12**, 172.
110. M. S. Gundeti, L. W. Bhurke, P. S. Mundada, S. Murudkar, A. Surve, R. Sharma, S. Mata, R. Rana, R. Singhal, N. Vyas, S. Khanduri, B. S. Sharma, N. Srikanth and K. S. Dhiman, *J. Ayurveda Integr. Med.*, 2022, **13**, 100325.
111. G. Devpura, B. S. Tomar, D. Nathiya, A. Sharma, D. Bhandari, S. Haldar, A. Balkrishna and A. Varshney, *Phytomedicine*, 2021, **84**, 153494.
112. A. Balkrishna, A. B. Bhatt, P. Singh, S. Haldar and A. Varshney, *J. Herb. Med.*, 2021, 100472.
113. A. Rais, D. S. Negi, A. Yadav, H. Arya, R. Verma, R. Galib, A. Ahmad, M. K. Yadav and P. N. Ahirwar, *J. Ayurveda Integr. Med.*, 2021, **13**, 100393.
114. National Conference on Diseases & Drugs: Emerging Trends and Challenges At: Department of Zoology, Zakir Hussain Delhi College, University of Delhi, Volume, 67–68.
115. H. Takeda, S. Muto, K. Nakagawa, S. Ohnishi and M. Asaka, *Curr. Pharm. Des.*, 2012, **18**, 4827.
116. S. Harasawa, A. Miyoshi, T. Miwa, O. Masamune, Y. Matsuo and H. Mori, *J. Clin. Exp. Med.*, 1998, **187**, 207.
117. H. Suzuki, J. Matsuzaki, Y. Fukushima, F. Suzaki, K. Kasugai, T. Nishizawa, Y. Naito, T. Hayakawa, T. Kamiya, T. Andoh and H. Yoshida, *Neurogastroenterol. Motil.*, 2014, **26**, 950.
118. N. Hoshino, D. Nishizaki, K. Hida, K. Obama and Y. Sakai, *Complementary Ther. Med.*, 2019, **42**, 255.
119. H. Li, H. Chen, L. Morgan, W. Li and B. G. Oliver, *Complementary Ther. Clin. Pract.*, 2021, **44**, 101433.
120. S. C. Chen, Y. S. Ho, L. Kwai-Ping Suen, J. Yu, W. Tang, J. F. Jiang, X. Y. Qu and W. F. Yeung, *Complementary Ther. Clin. Pract.*, 2020, **39**, 101112.

121. K. Linde, C. D. Mulrow, M. Berner and M. Egger, *Cochrane Database Syst. Rev.*, 2005, **18**, CD000448.
122. C. Holden, *Science*, 1991, **254**, 522.
123. A. Feily and N. Abbasi, *Phytother. Res.*, 2009, **23**, 1549.
124. J. Tacklind, R. MacDonald, I. Rutks, J. U. Stanke and T. J. Wilt, *Cochrane Database Syst. Rev.*, 2009, **2**, CD001423.
125. E. Mills, B. C. Foster, R. van Heeswijk, E. Phillips, K. Wilson, B. Leonard, K. Kosuge and I. Kanfer, *AIDS*, 2005, **19**, 95.
126. Q. Johnson, J. Syce, H. Nell, K. Rudeen and W. R. Folk, *PLoS Clin. Trials*, 2007, **2**, 16.
127. C. Jenkins, *Mercury*, 2005, **12**, 6.
128. N. Mncwangi, W. Chen, I. Vermaak, A. M. Viljoen and N. Gericke, *J. Ethnopharmacol.*, 2012, **143**, 755.
129. C. L. Backman and S. R. Harris, *Am. J. Phys. Med. Rehabil.*, 1999, **78**, 170.
130. E. Ernst and D. E. Furst, *Focus Altern. Complementary Ther.*, 2011, **16**, 49.
131. C. M. Witt, A. Michalsen, S. Roll, A. Morandi, S. Gupta, M. Rosenberg, L. Kronpaß, E. Stapelfeldt, S. Hissar, M. Müller and C. Kessler, *Trials*, 2013, **14**, 149.
132. P. A. Gautama, *J. Ayurveda Integr. Med.*, 2021, **12**, 556.
133. K. Hemming, T. P. Haines, P. J. Chilton, A. J. Girling and R. J. Lilford, *BMJ*, 2015, 350.
134. R. Mahajan and K. Gupta, *Indian J. Pharmacol.*, 2010, **42**, 207.
135. A. E. Kazdin, *Single-case research designs, methods for clinical and applied settings*, Oxford University Press, New York, 1982.

Subject Index